Advances in
BOTANICAL RESEARCH

VOLUME 4

Advances in

BOTANICAL RESEARCH

Edited by

R. D. PRESTON

Department of Physics,
The University,
Leeds, England.

and

H. W. WOOLHOUSE

Department of Plant Sciences,
The University, Leeds, England

VOLUME 4

1977

ACADEMIC PRESS

London   New York   San Francisco

A Subsidiary of Harcourt Brace Jovanovich, Publishers

ACADEMIC PRESS INC. (LONDON) LTD.
24/28 Oval Road,
London NW1 7DX

U.S. Edition published by
ACADEMIC PRESS INC.
111 Fifth Avenue
New York, New York 10003

Library of Congress Catalog Card Number: 62-21144
ISBN: 0-12-005904-5

Printed in Great Britain by
Whitstable Litho Ltd., Whitstable, Kent

CONTRIBUTERS TO VOLUME 4

J. A. CALLOW, *Department of Plant Sciences, University of Leeds, England* (p. 1).

I. R. COWAN, *Department of Enviromental Biology, Research School of Biological Sciences, Australian National University, Canberra, Australia* (p. 117).

P. JOHN, *Botany School, University of Oxford, South Parks Road, Oxford, England* (p. 51).

J. D. LOVIS, *Department of Plant Sciences, University of Leeds, England* (p. 229).

F. R. WHATLEY, *Botany School, University of Oxford, South Parks Road, Oxford, England* (p. 51).

PREFACE

It is seven years since the last volume in this series appeared; circumstances beyond the control of the editors have contributed to this long gap but it is our hope that subsequent volumes in the series will appear at more frequent intervals.

John and Whatley, to whom our apologies are particularly due with respect to delays in the publication of their paper, present an interesting contribution concerning the possible ancestry of mitochondria. Whatever ones views on this problem there can be no doubting the great stimulus to comparative biochemical studies which has derived from the revival of the hypothesis of a prokaryotic origin for mitochondria and plastids.

J. A. Callow discusses host-parasite interactions with particular reference to genetic aspects of resistance and the location of defence mechanisms. There appear to be exciting reasons for a much more thorough characterisation of cell surfaces as a basic requisite for progress in this field for, as Callow suggests, some aspects of pathogen interactions involve recognition phenomena dependant on cell surface interactions.

J. D. Lovis discusses the vexed question of the evolution of the ferns including more detailed analysis of particular taxa. His article has much to offer for both the general reader and the specialist.

I. R. Cowan's article presents a re-examination of leaf water relations in relation to stomatal movements. The substantial mathematical content of Cowans paper should not deter the reader; in many instances the equations may be "taken as read" in that they serve to formalise important points made in the text.

The Editors are deeply indebted to Miss J. M. Denison for secretarial assistance in the preparation of this volume.

R. D. Preston
H. W. Woolhouse

Leeds 1977

CONTENTS

Stomatal Behaviour and Environment

I. R. COWAN

Evolutionary Patterns and Processes in Ferns
J. D. LOVIS

Recognition, Resistance and the Role of Plant Lectins in Host-Parasite Interactions

J. A. CALLOW

Department of Plant Sciences, University of Leeds, England

I. INTRODUCTION

Recent years have witnessed something of a revolution in the study of the properties of the cell surface. In the case of animal cells this has had considerable impact in the areas of cell biology, developmental biology and immunology, since it is now apparent that the animal cell surface plays a major role in those various developmental and physiological phenomena that involve some aspect of cellular or molecular recognition. These include morphogenesis and differentiation, tissue incompatibility, gamete recognition, lymphocyte-homing, toxin- and hormone-binding, and virus infection. The study of plant cell surfaces on the other hand has lagged considerably behind its animal counterpart. It is now apparent, however, that several associative reactions involving plant cells are also mediated by specific complementary molecular mechanisms located at the cell surface. Rather special examples of recognition involving plant cells, of vital concern to mankind, are those interactions between plant cells and microorganisms which result in parasitism and disease, and in which there is a greater or lesser degree of host-parasite specificity. It is the particular intention of this review to consider how varietal specificity in such host-parasite interactions, involving mutual recognition of host and parasite, may be mediated by surface-localized complementary molecular mechanisms. To this end it is necessary to evaluate in some detail current concepts of the structure of surface membranes, the role of the cell surface in other biological recognition phenomena, and to review in broad terms our current understanding of the nature of host-parasite specificity. It is not the intention of this article to consider in detail the actual mechanisms by which the host plant may restrict access or growth of the potential pathogen, but rather to study the triggering events that induce or elicit such mechanisms, although in certain cases, such as the varietal-specific toxins, it will become clear that the act of recognition itself is an intimate part of the resistant or susceptible response.

II. SPECIFICITY IN HOST-PATHOGEN INTERACTIONS

Two broad types of interaction may occur between higher plant cells and fungal pathogens. Those parasites that quickly kill their hosts, deriving energy from dead cells are termed *necrotrophs*, whereas *biotrophic* parasites do not quickly kill their hosts but rather, derive their energy from living host cells (Thrower, 1966). As a nutritional group, necrotrophic fungi may be facultatively or obligately symbiotic in nature (Lewis, 1973), but are generally easily cultured and appear to derive much of their potential for disease from the secretion of toxins and degradative enzymes (Wood, 1967). Most biotrophs, however, are obligate, i.e. they cannot complete their life-cycle in the absence of the host plant, several of them such as the powdery and downy mildews cannot be cultured in defined media at any stage of their life-cycle, and there are several features of the host-parasite interaction which imply a certain "delicacy" in the relationship with the host plant (Brian, 1966; Lewis, 1973).

Plants, like all other living organisms, have evolved immune or resistance mechanisms of various types by which they are able to counteract the advance of foreign organisms, the result being that a given parasite can usually only infect a distinct range of host plants. A frequent observation of reviewers in this area (Kuć, 1966, 1972; Albersheim et al., 1969; Matta, 1971; Day, 1974) is that plant disease, i.e. susceptibility, is a relatively rare phenomenon in nature, despite the fact that the environment of the plant contains many potential pathogens. Thus there is a greater or lesser degree of specificity in host-parasite interactions, ranging from diseases caused by unspecialized, necrotrophic pathogens such as *Botrytis cinerea* that are able to attack a relatively wide range of host plants, to diseases caused by highly specialized pathogens, usually biotrophs that are restricted to a few closely related hosts. For example, *Phytophthora infestans* will infect susceptible potato cultivars, or close relations of the potato such as the tomato, but will not infect corn or wheat. This type of specificity is even more extreme in the case of those pathogens that exist as physiologic races, such as rusts, smuts and powdery mildews (but including some necrotrophic parasites also), that are restricted to a few host genotypes. The ultimate expression of the latter form of specificity is described by the so-called gene-for-gene relationship (Flor, 1956, 1971; Person, 1959), where resistance or susceptibility may be controlled by one gene pair (i.e. the host resistance gene and its complementary gene for avirulence in the parasite, see section IIC). In terms of host resistance, it is conventional to recognize "non-host resistance", where plants are quite clearly outside the range of hosts a fungus is able to infect, and "host-resistance", the result of genetic modification to the host which renders it resistant to pathogens which would otherwise infect it (Day, 1974). Thus there are cultivars of wheat that are either resistant or susceptible to wheat stem rust (*Puccinia graminis tritici*). A resistant cultivar may differ from its susceptible relative by essentially one gene, i.e. they are near-isogenic but for the resistance gene in question. Those wheat cultivars that are resistant to a given race of *P.g. tritici* are termed "resistant hosts", whereas all varieties of corn, for example, which are resistant to all races of wheat stem rust are termed "non-hosts" (Day, 1974) or "uncongenial hosts" (Matta, 1971).

One important question that may be asked here concerns the comparative nature of non-host and host resistance. For example, is the mechanism of resistance exhibited by corn to wheat stem rust in any way similar to the mechanism of resistance exhibited by a resistant cultivar of wheat challenged by an avirulent race of wheat stem rust? Alternatively, does the mechanism of resistance exhibited by a resistant cultivar of wheat to wheat stem rust have anything in common with the mechanism of resistance exhibited by wheat to the maize smut fungus? Such questions are pertinent to the theme of this review and are currently being approached in several laboratories through biochemical and structural investigations of non-host and varietal-specific resistance, and through cross-inoculation experiments.

A. MECHANISMS OF RESISTANCE

Disease resistance can be divided into two primary components: protection and defence (Hooker and Saxena, 1971; Ingham, 1972). Static protection mechanisms are based on pre-formed barriers. These may be structural barriers provided by a thick cuticle or suberization of the epidermis, or chemical barriers due to the presence of various constitutive fungitoxic or fungistatic compounds such as alkaloids, phenols and glycosides (Rohringer and Samborski, 1967). Included here are the inhibitors of fungal polygalactouronases recently discovered by workers in Albersheim's group (Albersheim and Anderson, 1971; Anderson and Albersheim, 1972). These inhibitors are proteinaceous, similar in certain properties to plant lectins (see section IV), and appear to be associated with plant cell walls. Since the inhibitors are possessed in equal quantity by both resistant and susceptible cultivars of bean, and are equally effective against the enzymes produced by different races of *Colletotrichum lindemuthianum*, it is not believed that these proteins play any major role in the determination of varietal specificity (Albersheim and Anderson-Prouty, 1975). Whilst there is little doubt that static protection mechanisms play an important part in the general defence mechanisms by which host plants resist the majority of pathogens in the environment, it is unlikely that mere differences in chemical or structural composition of different host varieties can explain resistance to one physiological race of the pathogen, and susceptibility to another. There is, however, a rather special example of a form of disease resistance which does not appear to require induced biochemical activity on the part of the host, but which nevertheless involves a very high degree of specificity, i.e. the resistance exhibited by certain plants to phytotoxins secreted by some parasites (Wood, 1973; Day, 1974). In such cases the pathogen exists as a single race and secretes a toxin which is effective in damaging only those host plant varieties that are infected by the fungus. Although relatively few diseases have been shown to involve host-specific toxins as primary determinants of specificity (see section VC*1*) this may prove to be a more widespread phenomenon. It is considered that the host plant variety which is killed by the toxin must contain toxin receptors, whilst these are absent or non-functional in a resistant cultivar. Until recently little direct evidence was available to support this concept, but in the case of *Helminthosporium sacchari* infections of sugar cane the toxin receptor has been isolated and shown to be located within the host plasmalemma (Strobel and Hess, 1974; and see section VC*1*). In the case of a resistant host, the receptor protein is modified in such a way as to prevent toxin-binding.

    In contrast to the static protection mechanisms discussed above, defence mechanisms are active, dynamic phenomena involving cell-cell contact between host and pathogen, resulting in the inhibition of further fungal development. In many cases host cells are penetrated before such mechanisms are induced, in others contact between host and fungal cell walls appears to be sufficient. Defence mechanisms may involve both structural and chemical barriers. Thus Rohringer and Samborski (1967) showed that the synthesis of aromatic carbon skeletons was increased in wheat seedlings infected by wheat stem rust. Stahmann *et al.* (1966) have shown that ethylene is produced in diseased tissues and may be involved in the stimulation of peroxidase, phenylalanine ammonia lyase and polyphenoloxidase activities. Such changes would permit an increased synthesis of lignin-like materials which could form mechanical barriers around penetrating hyphae, as demonstrated for resistant cucumbers infected with *Cladosporium cucumerinum* (Hijwegen, 1963). A further effect of ethylene in plant tissues is to increase, up to 50-fold, the activity of chitinase and endo-1,3-glucanase (Abeles *et al.*, 1970). Chitin and 1,3-glucans are common constituents of many fungal cell walls, and Pegg and Vessey (1973) have shown that tomato chitinase will degrade the cell walls and thus cause lysis of *Verticillium albo-atrum* mycelium. In response to fungal infection, therefore, plants may synthesize fungal cell-wall degrading enzymes. It is not clear whether this constitutes a general defence mechanism against a wide variety of pathogens or whether there is a high degree of specificity involved. Albersheim and Valent (1974) have recently shown that *Colletrotrichum lindemuthianum* secretes a proteinaceous inhibitor of the endo-β-1-3-glucanase secreted by french bean leaves. Again the degree of specificity is not known but it is possible that the secretion of various types of degradative enzymes and inhibitors are all part of a delicate interplay between host and parasite.

Two further defence mechanisms that are attracting considerable interest are the so-called "hypersensitive response" and the induced synthesis of antifungal and antibacterial metabolites known as "phytoalexins".

B. HYPERSENSITIVITY AND PHYTOALEXINS

Plant cells challenged by many incompatible or avirulent races of biotrophic or necrotrophic fungal pathogens, frequently undergo a response known as hypersensitivity, in which fungal penetration of the first host cell is rapidly accompanied by host cell disorganization, browning and necrosis (Kuć, 1966; Hooker and Saxena, 1971; Wood, 1973; Day, 1974; Maclean *et al.*, 1974). Death of the host cell may then lead to the cessation of further fungal growth either due to a depletion in nutrient supply in the case of biotrophic pathogens which require living host cells, or to an accumulation of fungitoxic compounds known as phytoalexins (Müller and Borger, 1940; Rahe *et al.*, 1969; Bailey and Deverall, 1971; Tomiyama, 1971; Kuć, 1972; Skipp and Deverall, 1972; Day, 1974). Some workers have, however, claimed that the hypersensitive response is not the cause, but the consequence of an earlier death or inhibition of the fungus (Kiraly *et al.*, 1972), and the exact relative timing of the biochemical and structural changes, must still be considered somewhat controversial.

The earliest microscopically detectable changes associated with the hyper-sensitive response, such as granulation of the cytoplasm, changes in the appearance of membranes and onset of browning and necrosis, can occur within a matter of minutes or a few hours following cell penetration (Tomiyama, 1971, 1973; Kitizawa *et al.*, 1973; Tomiyama, 1973; Maclean *et al.*, 1974; Skipp and Samborski, 1974), and the physiological changes that underly these morpho-logical and anatomical changes will occur even sooner. In addition, the hypersensitive response is usually extremely localized, normally being limited to the penetrated cell (e.g. Maclean *et al.*, 1974). The rapidity and localization of the hypersensitive response suggests strongly that the primary event in determining the compatibility of the host and parasite is some form of recognition localized at the interface between host and pathogen. As a result of this recognition event, a series of metabolic changes are "triggered" in the host cell leading to the hypersensitive response. Kuć (1972) points out that top priority for research in this area should be the determination of the nature of the recognition processes involved, and this problem constitutes the major portion of this review.

The above interpretation of the course of events involved in the hyper-sensitive response suggests that specific resistance is a highly localized active, energy-requiring, induced process. More recent studies have suggested that this interpretation may not be entirely acceptable. Mercer *et al.* (1974) have shown that "healthy" cells surrounding necrotic hypersensitive cells contained more cytoplasm and nuclei which were more lobed than in control cells. A number of examples demonstrate that phytoalexins cannot be entirely localized within hypersensitively responding cells since pre-inoculation of host tissue with an avirulent fungal race, and the subsequent induction of hypersensitivity, protects the tissue against subsequent colonization by a virulent fungal race (Müller, 1950; Matta, 1971). Mansfield *et al.* (1974) examined epidermal strips of broad bean infected by *Botrytis cinerea*. Infected cells became necrotic, brown and absorbed u.v. light. Live cells adjacent to the necrotic cells showed a blue/green fluorescence, however, and a microspectrographic comparison of the emission spectrum of these cells with that of the authentic broad bean phytoalexin, wyerone acid, suggested that these adjacent, living cells were accumulating this phytoalexin. Cells surrounding hypersensitively responding cells thus appear to be able to synthesize phytoalexins, which is one possible explanation of the protective effect of prior inoculation with an avirulent race discussed above. It cannot, however, be concluded from the studies of Mansfield *et al.* (1974) that the hypersensitive cells do not synthesize phytoalexins since the characteristic fluorescence of wyerone acid in necrotic brown cells could have been substantially quenched.

How do these observations affect the concept of host-parasite recognition? It is possible that there are several ways by which the synthesis of phytoalexins may be induced. Phytoalexins can be synthesized in response to a wide variety of physical (wounding) treatments, or through the application of a range of inorganic or organic compounds (both of natural or synthetic origins) (Kuć, 1972). The initial recognition event between host and parasite might well involve specific compounds of parasite origin (termed "elicitors"; Keen *et al.*, 1972; Albersheim and Anderson-Prouty, 1975) which through their interactions with appropriate (resistant) host cells elicit the hypersensitive response and phyto-alexin accumulation. The products of cell necrosis, resulting perhaps from the release of lysosome-bound hydrolytic enzymes (Pitt and Coombes, 1969; Wilson, 1973), may then diffuse to the adjacent cells and act as non-specific "irritants" (Day, 1974) of the host genome leading to the loss of gene control and phytoalexin synthesis. However, whilst the connection between the hyper-sensitive response and phytoalexin synthesis is clearly in need of some clarification, the actual means by which the host restricts the growth of the fungus is irrelevant to the concept of a discrete recognition event occurring at the host-parasite interface which acts as a "trigger" for the subsequent metabolic changes.

C. THE GENETICS OF VARIETAL SPECIFICITY

The hypersensitive reaction is most characteristic of varietal specific resistance under oligogenic control, and frequently a relationship known as the "gene-for-gene" relationship exists between host and parasite (Hooker and Saxena, 1971; Day, 1974). In such a relationship genes for resistance or susceptibility in the host are associated with genes for avirulence or virulence in the parasite, as gene pairs. Each gene in either partner can only be identified through its counterpart in the other member of the relationship (Flor, 1956, 1971). This relationship is most easily described by the so-called "quadratic check" (Ellingboe, 1972; Day, 1974; Person and Mayo, 1974; Fig. 1).

The combination $R : A$ has been termed a "stop signal" by Person and Mayo (1974) since it triggers incompatibility (generally through a hypersensitive response), and in those cases where several gene loci are known to condition resistance the genetic evidence indicates that a single "stop signal" generated by any one of the $R : A$ interactions is sufficient to determine a resistant interaction.

There are two important considerations to be taken into account when attempting to devise biochemical models for the gene-for-gene concept. The first of these is that the genes for resistance in the host and avirulence in the parasite are both dominant, and normally dominant genes are expressed and are usually associated with selective advantage. Secondly, Fig. 1 shows that both resistance (R) and avirulence (A) genes can be associated with either plus (+) or minus (−) reactions. Thus expression of R and A is "conditional" on the presence of the other for a minus (−) reaction to be produced (Person and Mayo, 1974). Such genetic considerations lead directly then to the concept of induced resistance as dependent upon the association of products of the dominant but conditional R

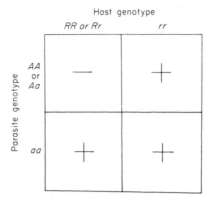

Fig. 1. The Quadratic check in gene-for-gene interactions. R and r represent host genes segregating for resistance (dominant) and susceptibility (recessive). A and a represent genes of the parasite segregating for avirulence (dominant) and virulence (recessive). Any gene combination containing both R and A will lead to a minus ($-$) or incompatible response. Plus ($+$) combinations indicate host susceptibility, parasite virulence, i.e. compatible reactions. (Ellingboe, 1972; Day, 1974; Person and Mayo, 1974.)

and A genes to form a hybrid molecule which constitutes the "stop signal". Day (1974) has developed a model of genetic regulation in host-parasite systems, through which such a "stop-signal" could lead to the biochemical expression of resistance, and this type of model has been further elaborated in Fig. 6. The major part of this review is concerned with the identification and localization of the "stop signals" resulting from recognition in varietal-specific pathogenesis.

D. NON-HOST RESISTANCE AND INDUCED SUSCEPTIBILITY

The nature of non-host (uncongenial host) resistance compared with oligogenic, varietal-specific host resistance was questioned earlier in this section. As discussed above, in many cases varietal specificity mediated by gene-for-gene systems is associated with the hypersensitive response, and genetic analysis provides a basis on which biochemical models may be constructed. Are we to suppose, then, that this form of resistance results only from a highly specific recognition response which permits the host plant to discriminate between different races of the same fungus, or can the same type of response result from the interaction between a potential pathogen and a non-host? If the latter, then this could clearly affect the development of biochemical models that purport to explain the basis of specificity. It was suggested earlier that much of the general resistance of plants to the thousands of potential pathogens to which they are exposed in nature may be due to the inability of the pathogen to overcome non-specific mechanical and chemical barriers. To answer the questions posed above, however, requires comparative studies on the nature of host versus non-host resistance. Unfortunately such studies, employing modern techniques, are relatively rare.

Staub *et al.* (1974) examined the development of barley powdery mildew (*Erysiphe graminis*) and cucumber powdery mildew (*E. cichoracearum*) on their respective host and non-host plants. On cucumber leaves the total number of germinating conidia of *E. graminis* and the rates of germination were reduced by 50% compared with the "congenial" situation, i.e. *E. graminis* on barley. Nevertheless, those spores which germinated did form appressoria and penetrated the cucumber epidermal cells. The invaded host epidermal cells responded with a form of hypersensitive response whereby haustorial development was arrested and the host cells showed different staining properties. In the case of cucumber powdery mildew conidia inoculated onto barley, conidial germination was unaffected but germ tubes were unable to dissolve epicuticular waxes and so did not penetrate the epidermal cells.

An extensive study of both varietal and non-host resistance to cowpea rust infection was made recently by Heath (1974). In neither case was uredospore germination affected, but in the case of non-hosts directional germ tube growth towards stomata was less pronounced, which may be the result of differences in leaf surface topography. In most of the non-host plants examined, germ tubes penetrated the stomatal pore and differentiated into substomatal vesicles and infection hyphae. Growth beyond this in non-host plants was then restricted so that few infection hyphae actually formed haustoria. Where haustoria were differentiated, death of both haustorium and host cell followed very rapidly. Absence of haustorial development could have been due to several possible mechanisms. In some cases, absence of haustoria was correlated with the accumulation of osmiophilic material on and within the adjacent non-host cell wall. In other cases the infection hyphae appeared to lose contact with the non-host cell wall, or fungal death occurred before haustoria could be initiated. In most cases of varietal resistance to cowpea rust examined, penetration of host cells and haustorial formation was achieved, but this was usually followed by rapid host cell and haustorial death. On the basis of these results, Heath (1974) suggested that in both compatible and incompatible cowpea rust-host plant associations, there are a number of "switching points" at which resistance or susceptibility may be controlled.

Leath and Rowell (1970) studied the interaction between stem rust fungus *Puccinia graminis tritici* and the non-host plant *Zea mays*. In mature corn leaf tissues the fungus was inhibited at the infection hypha stage, before haustorial formation. The presence of some phytoalexin-like compound was inferred, however, from cross-inoculation experiments with the compatible fungus *P. sorghi*. Prior inoculation of corn tissue with *P. g. tritici* rendered the corn tissue resistant to *P. sorghi*, implicating the induction of some type of fungitoxic material by *P. g. tritici*.

The affinity of a range of different non-hosts to *Phytophthora infestans* was examined by Müller (1950). Non-solanaceous hosts such as *Lactuca, Dahlia, Phaseolus* and *Brassica* reacted to penetration by means of a rapid hypersensitive response. In other cases the host cells appeared not to be penetrated, and in such cases resistance was probably due to protection rather than active defence mechanisms. Inoculation of potato tuber tissues with oligogenic resistance to *Phytophthora infestans* leads to hypersensitivity accompanied by accumulation of terpenoid phytoalexins, rishitin and phytuberin (Tomiyama *et al.*, 1968; Varns, Kuć and Williams, 1971). Varns, Currier and Kuć (1971) showed that physical and chemical damage to tubers resulted in necrosis without phytoalexin accumulation and that sonicates of virulent and avirulent races of *P. infestans* induced necrosis and phytoalexin accumulation when applied to both resistant and susceptible potato cultivars. Non-pathogens of potato, *Helminthosporium carbonum* and *Ceratocystis fimbriata*, both induced phytoalexin synthesis in the absence of hypersensitivity, further proof that these two aspects of resistance need not necessarily be connected.

Klarman and Gerdeman (1963) and Klarman (1968) showed that extracts from Harosoy, a variety of soyabean susceptible to *Phytophthora megasperma* var. *sojae*, did not contain any phytoalexin activity, whilst the near-isogenic variety Harosoy 63 did accumulate phytoalexins in response to infection. Two non-pathogens, *P. megasperma* and *P. cactorum*, also induced phytoalexin accumulation in both Harosoy and Harosoy 63. It must be concluded from studies with potato and soyabean that the potential for phytoalexin synthesis is contained within resistant and susceptible hosts, that both avirulent pathogens and non-pathogens can release this potential, but that the virulent pathogen can suppress or divert the response. This, then, is the concept of induced susceptibility, and is at variance with the concept of induced resistance proposed earlier for varietal-specific, gene-for-gene resistance. These matters will be returned to later, but an apparent solution to the problem is to propose that varietal-specific resistance is rather exceptional and that the widespread resistance of plants to the thousands of potential micro-organisms in the environment is more economically provided by a number of general resistance mechanisms which are non-specific, and that the successful pathogen is one which is able to avoid or divert these resistant responses. If so, then it is quite possible that gene-for-gene systems of varietal specificity evolved (or were devised by the plant breeder), from more general systems of defence in such a way that induced resistance systems are superimposed upon these more general mechanisms. There will be more extensive discussion of this problem in section V, but any explanation must take into account observations such as those made by Varns and Kuć (1971), who showed that potato tuber tissue inoculated 12 h previously with a virulent race of *P. infestans* did not accumulate rishitin and phytuberin when challenged with an incompatible, avirulent race, suggesting that after 12 h the host-parasite system is effectively behaving as "self" and no longer has the ability to either recognize or respond to an avirulent parasite race.

In this section the broad spectrum of host-parasite specificity has been examined. Gene-for-gene systems are of great value in the design and testing of biochemical hypotheses which seek to explain the nature of varietal specificity. These are the systems that biochemists will find most amenable to investigation, but it must be recognized that such systems represent only a most extreme form of the whole spectrum of host-parasite specificity and must eventually therefore be considered in the broader context.

As discussed above, a plausible biochemical interpretation of gene-for-gene systems of varietal specificity is that precise, surface-localized complementary molecular mechanisms are involved in a discrete act of recognition, which serves as a trigger for the induction of mechanisms leading to resistance or susceptibility. Such an act of recognition would not of course be unique to host-parasite interactions but rather would be one example of a whole range of biological phenomena which involve cell-cell or cell-molecule recognition. At this point it is germane therefore to consider the biochemistry and biology of these diverse phenomena in some detail, since the rationale and methodology utilized should have direct value in the future study of host-parasite interactions.

III. MOLECULAR ASPECTS OF RECOGNITION IN BACTERIA, ANIMALS AND PLANTS

A. THE PLASMA MEMBRANE

The important molecules that form the major structural, or matrical materials of plasma membranes or other biological membranes are amphipathic phospholipids and proteins (integral membrane proteins). The phospholipids are arranged in the form of a bimolecular leaflet as proposed originally by Gorter and Grendel (1925) and Danielli and Davson (1935), with the integral membrane proteins and glycoproteins intercalated into the lipid (Bretscher, 1973; Nicolson, 1974). On the cytoplasmic face of the plasma membrane, peripheral proteins, some of which may be enzymes, are found. The carbohydrate constituents of plasma membranes (up to 10% of the total membrane) are exclusively localized on the outer face in the form of glycoproteins and glycolipids (Bretscher, 1973). The term "glycocalyx" was devised by Bennett (1963) to describe this carbohydrate coating. The model plasma membrane depicted in Fig. 2 is consistent with the biochemical (Bretscher, 1973) and histochemical evidence (Bennett, 1963; Leblond and Bennett, 1974), and with a whole series of recognition phenomena to be described, which have as their common basis some form of interaction with these carbohydrate-containing receptors. Inevitably the vast majority of information which has led to such models has been derived from studies with animal cells. Faith in the concept of biological universality leads one to suppose that plant cell surface membranes will not be too dissimilar to those of animals, but definitive proof of this rests upon our skill in isolating plasma membranes or studying protoplast surfaces.

Transmission of signals from one side of the membrane to the other is provided for by the hydrophobic polypeptide portions of some glycoproteins spanning the lipid bilayer and extending through into the cytoplasm (Bretscher, 1971).

B. SURFACE ANTIGENS AND THE IMMUNE RESPONSE

It is now well established that the molecules containing the A, B and O antigens on the human erythrocyte membrane are glycoproteins or glycolipids (Koscielak, 1963; Watkins, 1966; Zahler and Wiebel, 1968). The genetic basis for these antigenic differences lies in the specificity of glycosyltransferase enzymes which transfer *N*-acetylgalactosamine, galactose or fucose to specific acceptors of the glycoprotein or glycolipid. If the terminal non-reducing sugars of the oligosaccharide antigen are *N*-acetylgalactosamine the blood group is A, if galactose the blood group is B, if both are lacking then the oligosaccharide ends in fucose and the group is O (Winzler, 1970).

These antigenic determinants may also be recognized by plant carbohydrate-binding proteins known as lectins or phytohaemagglutinins (see section IV). Some of these lectins are able specifically to discriminate between various blood groups. The major glycoprotein ("glycophorin") bearing blood group antigens and lectin receptors has been isolated from human erythrocyte plasma membranes. Glycophorin has a molecular weight of approximately 50 000, contains 60% carbohydrate by weight, has a central "domain" of hydrophobic amino acid groups and *N*- and *C*-terminal domains of hydrophilic amino acids. This glycoprotein is therefore able to traverse the lipid bilayer with the hydrophilic residues predominantly associated with the inner and outer faces, as depicted in Fig. 2 (Nicolson, 1974).

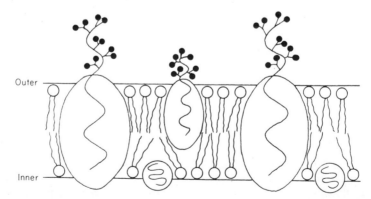

Fig. 2. Model of an idealized plasma membrane. The model illustrates the distribution of phospholipid, protein and glycoprotein components. Note the symmetrical lipid bilayer, the location of glycoprotein sugar residues on the outer membrane face, and the complete span of some glycoprotein molecules across the plane of the membrane so that the exterior of the cell is in direct molecular contact with the cytoplasm. (Modified after Nicolson, 1974.)

There are two important points here, as far as biological recognition phenomena are concerned. Firstly, protein-carbohydrate interactions are extremely specific, antibodies can recognize the difference between terminal *N*-acetylgalactosamine or galactose, or between either of these and fucose. Secondly, in the lectins, plants contain molecules with this type of discriminatory ability.

Defence against foreign molecules or invaders in vertebrates rests in the highly specific immune system by which antigens are recognized by lymphocytes, triggering antibody synthesis and the neutralization of the antigen. The theoretical basis of immunologic specificity is the "clonal selection theory" of Burnet (1959). Burnet suggested that individual immunocompetent cells possess unique surface receptors and that only a limited panel of antigens capable of interacting with that particular receptor would activate that cell. The vast potentiality of the organism was explained through the existence of large numbers of clones of individual cells each bearing different receptor molecules. The immune response involved the selective activation, proliferation and differentiation of a limited set of specific clones. In broad terms this theory appears to be substantially correct but recent research has revealed large ramifications in this basic theory. Lymphocytes can be divided into two classes, T (thymus-dependent) and B (bone marrow-derived), and these two cell types appear to co-operate in antibody production. Initial recognition of the antigen (possibly bound to macrophage surfaces) is mediated by an antigen-specific glycoprotein T factor localized on the T cell surface, controlled (in mice) by one of the two functionally distinct gene types in the I region of the histocompatibility complex (Munro and Taussig, 1975). Humoral release of the T factor antigen complex may then take place. This complex may then be recognized by acceptor sites on the surface of B cells, under the control of the second gene type. This second recognition event then triggers a whole series of events leading to immunoglobulin biosynthesis by the B cells (Fig. 3). Immunoresponsiveness thus involves several discrete acts of recognition at the cell surface, and in certain cases the surface interaction can trigger metabolic changes. Other aspects of surface antigenicity will be considered later.

C. INTERACTIONS BETWEEN CELLS AND VIRUSES

Animal cells can be agglutinated by viruses which bind to surface receptors on two different cells forming a bridge; the specific binding of viruses to such surface receptors probably constitutes the first step in the infection process. The surface receptors of animal cells appear to be the carbohydrate moieties of glycoproteins or glycolipids, and a variety of such receptors have now been isolated and purified (Winzler, 1970). Myxoviruses, for example, have surface glycoprotein "spikes" which bind specifically to red blood cell receptors. Since neuraminidase destroys binding it is concluded that neuraminic acid residues are critically involved in the recognition process (Gottschalk, 1966).

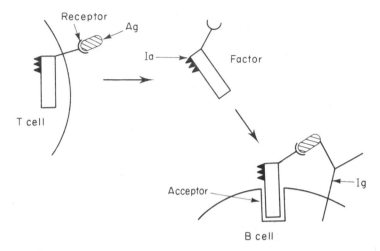

Fig. 3. Model for cell interaction in the antibody response. This tentative scheme shows the possible relationship between T cell receptor, specific T cell factor and B cell acceptor in cell co-operation and the mechanism for B cell triggering. "Ia" refers to the antigenic determinants carried by the T cell factor detected by antisera raised against the I region. The *Ir* genes in the major histocompatibility complex code for the T cell receptor, the T cell factor and the B cell acceptor. (Reproduced with permission, Munro and Taussig, 1975.)

A most spectacular example of carbohydrate-binding involved in specific virus attachment and recognition is that of lysogenic phage conversion of Salmonellae by epsilon phages ϵ^{15} and ϵ^{34}, recently reviewed by Barksdale and Arden (1974). The outermost wall layers of Gram-negative bacteria contain large amounts of a lipopolysaccharide which bears the somatic (O) antigens on which the Kauffmann-White serotypic method of classification is based (Kauffmann, 1972). The lipopolysaccharide consists of a polysaccharide core linked to structural lipid, and the O antigens are repeating unit polysaccharide chains linked glycosidically to the polysaccharide core (Ludertiz *et al.*, 1966).

The O antigens also serve as receptors for the attachment of bacteriophages. When *Salmonella anatum* [3,10] (i.e. bearing O antigens 3 and 10) is infected by ϵ^{15}, the synthesis of receptors by which ϵ^{15} gains entry into the cell is repressed, the O antigens change from 3,10 to 3,15 and the cell now becomes susceptible to ϵ^{34} (Uetake *et al.*, 1958; Uchida *et al.*, 1963). Factor 3 is a unit of D-mannosyl-L-rhamnose occurring as an integral part of the 6-β-D-mannosyl-1,4-α-L-rhamnosyl-1-1, 3-*o*-acetyl-α-D-galactosyl $\rightarrow$ repeating unit of the wall lipopolysaccharide. Factor 10 is the *o*-acetyl group of the *o*-acetyl-α-D-galactosyl fragment. The O-antigen side-chain unit made by *S. anatum* [3,15]. ϵ^{15} however, is 6-β-D-mannosyl-1, 4-α-L-rhamnosyl-1, 3-β-D-galactosyl $\rightarrow$. It has been shown that during conversion by the ϵ^{15} the host cell transacetylase responsible for antigen-10 synthesis is lost, an inhibitor to the α-polymerase of the cell is produced, and a β-linked polymerase substituted (Uchida *et al.*, 1963; Robbins *et al.*, 1965, 1967; Losick and Robbins, 1967). Thus the subtle difference involved in the distinction between antigen 10 and antigen 15, and the ability of ϵ^{15} to bind to the O-antigen receptors involves deacetylation and a change in the anomeric configuration of galactose. Furthermore, when *S. anatum* [3,15] · ϵ^{15} is superinfected with the previously avirulent phage ϵ^{34}, to yield double-lysogenic *S. anatum* [3,15,34] · ϵ^{15} · ϵ^{34}, synthesis of ϵ^{34} receptor sites is also repressed. The O-antigenic side chain in this case is modified by glucosylation of the β-galactose residues so that factors 3, 15 (mannose-rhamnose-β-galactose) can no longer serve as ϵ^{34} receptors (Wright and Barzilai, 1971).

D. RECEPTION OF HORMONES, TOXINS AND OTHER EFFECTORS

The polypeptide hormone insulin facilitates transmembrane sugar transport in adipose and muscle tissues. Studies using insulin covalently linked to agarose beads, which cannot enter cells, have shown that cells retain their insulin-directed transport activities, the conclusion being that insulin action involves a discrete plasma membrane receptor (Cuatracasas, 1969). The interaction probably involves carbohydrate-containing receptors since the plant lectins concanavalin A and wheat germ agglutinin can perturb the insulin binding and may themselves exert insulin-like properties on fat cells (Cuatracasas and Tell, 1973). In a similar way, the glycoprotein antiviral agent, interferon, appears to exert its effects by interaction with plasma membranes since agarose-interferon complexes retain full antiviral activity (Ankel *et al.*, 1973; Besancon and Ankel, 1974). It is not clear whether interferon acts by blocking viral receptor sites or by some other means.

It has been known for some time that various plant seeds such as those of *Abrus* and *Ricinus* contain high molecular weight protein toxins. Abrin and ricin both consist of two polypeptide chains of slightly different molecular weights, joined by disulphide bridges (Olsnes and Pihl, 1973a, b; Olsnes, Refsnes and Pihl, 1974; Olsnes, Saltvedt and Pihl, 1974). The A chains are termed "effectomers" since they enter the cell and inhibit protein synthesis. The B chains are termed "haptomers" since they actually bind to galactose-containing receptors on the cell surface through a single binding site. Abrin and ricin can therefore be considered as monovalent plant lectins.

E. CELL-CELL RECOGNITION IN PLANTS AND ANIMALS

As our understanding of the cell surface has advanced, it has been realized that certain components of cell membranes are involved in a range of cell-cell contact and recognition phenomena, e.g. fertilization, differentiation, phagocytosis, adhesion, cell communication and metastasis. In such cases, control is manifest initially through the specific association of complementary molecules at the cell surface; indeed the real function of virus, toxin, antibody and hormone receptors may be related to cell-cell recognition. Following the binding of an external ligand on one cell to the other cell, usually involving carbohydrate moieties, signals are generated within the membrane, transmitted to cytoplasm and nucleus, resulting in altered macromolecular synthesis and the appropriate cellular response. The association of cell-surface interactions with morphogensis in animals has been well documented elsewhere (Weiss, 1969; Winzler, 1970; Guidotti, 1972; Moscona, 1974; Nicolson, 1974). In this section one or two of these examples will be briefly considered before examining examples from the plant literature in more detail.

1. Phagocytosis by Acanthamoeba

The growth of the soil sarcodinid *Acanthamoeba castellanii* in axenic culture requires particulate matter which is ingested by phagocytosis. It has been shown that *A. castellanii* preferentially takes up horse red blood cells over other types in monaxenic culture (Brown *et al.*, 1975) and that this uptake is preferentially inhibited by mannose, suggesting that binding of particles proceeds through mannose-sensitive binding sites. The ability of *A. castellanii* to take up more naturally occurring micro-organisms in the soil environment also correlates well with the occurrence of mannose in the cell walls of these organisms. It is suggested that in the soil environment phagocytosis would be greatly facilitated by the preliminary binding of food organisms, such as mannan-containing fungi.

2. Sponge cell aggregation

Separated cells of certain sponges, e.g. *Microciona parthena*, are known to reassociate in a species-specific manner (Moscona, 1963). An aggregation-promoting factor has been isolated from *M. parthena* (Henkart *et al.*, 1973). The

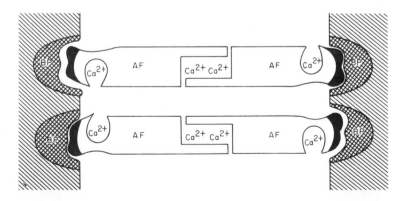

Fig. 4. Model for cell-cell interactions in sponge tissue construction. Two glycoprotein aggregating-factor (AF) complexes are illustrated. The black termini at each pole carry the carbohydrates which are recognized by the baseplate (BP) factors anchored on adjoining cell surfaces. The model also depicts the essential role played by calcium ions in the formation of the AF complexes, and as a co-factor for aggregating activity. (Reproduced with permission, Weinbaum and Burger, 1973.)

factor is an acidic proteoglycan complex of several million daltons molecular weight, contains 47% amino acids, 49% sugars, the major residues being glucuronic acid, and in the electron microscope a fibrous complex is revealed with the fibres arranged in a "sunburst" configuration. The activity *in vitro* of this factor reflects the same species-specificity as the cellular aggregation reactions. Its activity can be inhibited by glucuronic acid implying an important role for this sugar in the binding reaction, and it has been suggested (Turner and Burger, 1973) that the protein is a structural carrier of the carbohydrate groups providing the specificity. Weinbaum and Burger (1973) have also shown that a second macromolecule other than the agglutinating factor is involved in species-specific aggregation, which they term "baseplate". Figure 4 shows how the two factors may be involved in tissue construction.

3. *Embryonic cell aggregation*

Certain embryonic tissues such as chick neural retina dissociate on treatment with chelating agents or trypsin. Under suitable conditions such cells reaggregate and tissue patterns become defined (Moscona, 1974) within the aggregate. Such systems have been widely used as models of the mechanisms by which cells regulate their sorting out and mutual recognition during tissue formation.

Studies on the fundamental cell-linking or recognition mechanisms based on cell surface interactions have been extensively carried out by Moscona and co-workers (see Moscona, 1974 for a review). Reaggregation of cells requires *de novo* protein synthesis and coincides with the formation of cell surface glycoproteins. A hypothesis (the cell-ligand hypothesis) was developed whereby cell recognition and surface adhesion proceeded through the interaction of specific components of the cell surface. These cell-linking components were assumed to be protein-carbohydrate complexes since periodate oxidation inhibited reaggregation, and were referred to by the general term "cell-ligands" (Moscona, 1968). It was postulated that different ligand patterns would develop on cells destined to form different tissues. Also, during development the patterns would continually be altered or "tuned". It has now been demonstrated that cells from different tissues do in fact have different surface antigens (Goldschneider and Moscona, 1972), and that during the development of a single tissue type surface carbohydrate-containing lectin receptors do become modified (Kleinschuster and Moscona, 1972; Nicolson, 1974). Attempts have been made to isolate from the surfaces of cells, components with cell-ligand activity. It was found that aggregation factors released into the culture medium would cause tissue specific reaggregation of cells (Hausman and Moscona, 1973). Fully differentiated cells did not respond to the aggregation factor obtained from younger cells, suggesting that as cells complete their development and become specialized they are no longer capable of reforming the specific surface components required for morphogenesis (Moscona, 1962). Pure cell-ligand preparations from embryonic chick neural retina cells are glycoproteins of molecular weight approximately 50 000, containing mannose, galactose, glucosamine and sialic acids as major sugar constituents (Hausman and Moscona, 1973; McClay and Moscona, 1974). The nature of the receptors for this molecule has yet to be determined.

A more detailed proposal, which is consistent with the hypotheses discussed above, is that the carbohydrate-ligand binding components are glycosyl-transferase enzymes localized at the cell surface (Roseman, 1970, 1974; Roth *et al.*, 1971a, b). The glycosyl transferases may themselves be involved in the synthesis of the carbohydrate portions of the ligands. The diversity of complex carbohydrate ligands and the specificity exhibited by glycosyltransferases would provide the requisite amount of variation necessary to achieve differences in adhesion specificity between cells. In addition, this proposal has the advantage in that it permits the dissociation of cells which must occur in development since completion of the enzyme-substrate reaction with the appropriate sugar nucleotide would lead to cell-cell dissociation (Fig. 5).

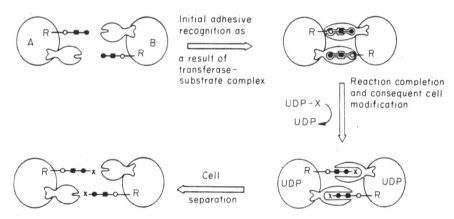

Fig. 5. Modification of intercellular adhesion by glycosyl-transferase reaction. The first step represents adhesion as a result of the binding of complex carbohydrate chain (–○–□–●) to the corresponding glycosyl transferase. In the second step internally generated sugar nucleotide (UDP-X) provides X to the enzyme on the internal face of the membrane. The transferase utilizes X to complete the reaction, and in the third step (cell separation) the product of the enzymatic reaction dissociates from the enzyme. (Reproduced with permission, Roseman 1974.)

There is as yet little evidence which unequivocally links surface glycosyl transferase activity to cell adhesion although results using model systems consisting of Sephadex beads linked to putative substrates, provide a promising approach to the problem (Roseman, 1974). Deppert *et al.* (1974) recommend the exercise of caution in the assignment of glycosyl transferase activities to cell surfaces, and Garfield *et al.* (1974) were unable to detect any galactosyl transferase activity in purified retina-specific aggregating factor.

4. Fertilization and incompatibility

(a) Animal fertilization.

Sea urchin eggs are covered with an extracellular jelly coat and a vitelline membrane through which the spermatozoan must pass before contacting the plasmalemma. The role of the jelly coat in fertilization is somewhat obscure, but it has been shown that sperm will bind to the vitelline membrane through a complementary macromolecular mechanism. Aketa (1967) and Aketa *et al.* (1968) were able to isolate a protein from the vitelline membrane of eggs which was capable of binding sperm of the same species. Homologous, but not heterologous, antiserum prepared against sperm-binding protein prevented fertilization when added to eggs (Aketa and Onitake, 1969), and sperm pre-treated with homologous sperm-binding protein were unable to fertilize eggs (Aketa, 1973). Sperm-egg binding is thus mediated by a highly specific complementary mechanism. Lectins have been used to investigate egg surfaces both in sea urchins and mammals (see Nicolson, 1974 for a review). In sea urchin the effects are complex but some lectins will inhibit fertilization, implicating carbohydrate-binding in the recognition process.

In mammalian eggs, the thick, transparent coat surrounding the egg, the zona pellucida, contains lectin-sensitive sperm-binding sites. Some lectins inhibit fertilization by cross-linking oligosaccharides in the zona pellucida, preventing sperm-borne lysins from depolymerizing the zona pellucida and gaining access to the egg membrane (Oikawa *et al.*, 1974), whilst other lectins, notably wheat germ agglutinin, appear actually to inhibit sperm-binding in the first place.

(b) Gamete fusion in algae.

In isogamous, dioecious species of *Chlamydomonas*, the specific contact components, the mating type substances, are glycoproteins localized at the tips of gamete flagella. On interaction between plus (+) and minus (−) gametes an instantaneous species-specific sexual agglutination occurs (Wiese, 1974). The agglutination of gametes of a single mating type can also be induced by the isolated mating factors obtained from culture filtrates of the opposite mating type. In *C. eugametos* and *C. moewusii* syngens I and II, the contact capacity of the minus (−) gametes is sensitive to proteolytic enzymes whereas the plus (+) gametes are resistant to proteolytic enzymes but sensitive to α-mannosidase and the plant lectin concanavalin A which binds to mannose residues (Wiese and Shoemaker, 1970; Wiese and Hayward, 1972; Wiese, 1974). It is suggested that in these particular species, sex-cell contact is mediated by a proteinaceous agglutinin on the (−)-gamete flagella, interacting with a mannose-containing carbohydrate or glycoprotein on the (+)-gamete flagella. Simple sugar residues such as α-mannose or α-methyl-mannoside did not inhibit the agglutination reaction, possibly indicating that the binding protein requires a more complex mannose-containing receptor.

The isoagglutinins isolated by fractionation of aged gamete culture supernatants have been shown to be particles of molecular weight approximately 10^8

(Wiese, 1974). In the most extensively studied species, *Chlamydomonas eugametos* and *C. moewusii* syngen II, the particles are sulphated glycoproteins differing quantitatively and qualitatively in their amino acid and sugar composition between both species and sexes. McLean *et al.* (1974) have shown that the *C. moewusii* particles are in reality receptor-bearing membrane vesicles which appear to bleb off the flagellar membranes, complete with flagellar hairs. The hairs do not carry any contact capacity. McLean and Bosmann (1975) have produced some evidence to support the hypothesis developed by Roseman (section IIIE*3*) that galactosyl transferase ectoenzyme systems are involved in cell-cell recognition. Glycosyl transferase enzyme systems were detected on the surfaces of both vegetative cells and gametes of *C. moewusii*. Gametes, however, had higher activities, and when plus (+) and minus (−) gametes were mixed there was an enhancement of activity, possibly indicating that gametes of one type provide the appropriate substrates for the surface enzymes of the opposite mating type gametes. No such enhancement occurred with vegetative cells, possibly suggesting that these enzyme systems were involved in the actual synthesis of carbohydrate-containing surface components and were therefore replete with acceptors. Isolated flagellar vesicles also contained a high glycosyl transferase activity which was again enhanced on mixing with vesicles of the opposite mating type.

(c) Sexual Compatibility in Yeasts

Cell fusion between haploid cells of *Hansenula wingei* is initiated by a strong sexual agglutination between the cells of opposite mating type (Brock, 1959; Crandall and Brock, 1968). Agglutination of mating types 5 and 21 is brought about by the complementary interaction of two glycoproteins present on the cell surfaces of the haploid mating strains. The mating factor from strain 5 is a multivalent mannan-protein, heterogeneous in molecular weight $(1.5 \times 10^4 - 10^8)$, and with a variable carbohydrate content (50–96%) (Brock, 1965; Crandall and Brock, 1968; Taylor and Orton, 1968; Crandall *et al.*, 1974). The mating factor from strain 21 cells is a univalent homogeneous mannan-protein (molecular weight approximately 40 000; Crandall *et al.*, 1974). *In vitro* a range of different complexes are formed between 21-factor and 5-factor substantiating the concept that these two molecules are complementary (Crandall *et al.*, 1974). The exact role of the mannan portions of these molecules in the binding process is not clear.

(d) Pollen-style incompatibility in plants

In both interspecific, and sporophytically controlled intraspecific incompatibility systems, the site of rejection of unacceptable pollen occurs at the stigma surface (for a review on self-incompatibility see Heslop-Harrison, 1975). Angiosperm pollen grains carry within their walls extracellular proteins and glycoproteins. These compounds are released onto the surface of the stigma, and whilst some of them may be enzymic, possibly being concerned with germination and penetration of the stigma, others are involved in the recognition or rejection responses occurring at the stigma surface. Thus extracellular protein and glycoprotein fractions from incompatible pollen grains are believed to bind to the stigma (Knox and Heslop-Harrison, 1971), where they elicit a rejection reaction characterized by the deposition of the β-1,3-glucan polymer, callose (Heslop-Harrison *et al.*, 1973). The role of these protein-containing wall fractions in recognition and rejection has been most convincingly demonstrated by Knox *et al.* (1973) in their studies on interspecific hybridization in *Populus*. Pollen of *P. alba* will not germinate on *P. deltoides* stigma surfaces. If, however, γ-irradiated (dead) *P. alba* pollen is mixed with live *P. deltoides* pollen, then the rejection reactions are apparently circumvented and cross-pollination will occur. The dead *P. alba* pollen can be replaced by a protein-containing leachate derived from *P. alba* pollen.

Studies on the stigmatic surfaces of plants have shown that the stigmatic papillae bear an external protein coat or pellicle overlying the cuticle (Matsson *et al.*, 1974). It has been suggested that this pellicle may play a role in the hydration of pollen grains as they alight on the stigma surface and that the pellicle may also contain macromolecules which are complementary to the pollen grain proteins or glycoproteins involved in the recognition response. It has been shown that these proteins and glycoproteins will bind to the pellicle in *Silene*. Preliminary evidence suggests that the complementary macromolecular mechanism mediating the recognition response may involve carbohydrate-binding. Watson *et al.*, (1974) examined the binding of the lectin concanavalin A (con A) to various pollen wall glycoproteins and polysaccharides. In double-diffusion agar studies, con A bound to and precipitated wall-held materials from a variety of grass pollens. Some degree of differentiation was noted between different groups of grasses, festucoid species being especially inactive. The reactions between con A and *Zea mays* pollen wall materials and those of its close relatives were distinguishable from the reactions with pollen wall materials of other grasses. From this limited study it would thus appear that there are differences in carbohydrate composition between secreted pollen wall poly-saccharides and glycoproteins. It has yet to be shown, however, that the molecules exhibiting lectin-detectable differences are involved in the recognition processes controlling incompatibility. It has also been suggested that the specific receptors on the stigma surface may in fact be lectins or lectin-like proteins.

5. Cellular adhesion in slime-mould morphogenesis

During periods of starvation, free-living amoeboid slime-mould cells aggregate to form a multicellular structure which eventually differentiates to form a fruiting body (Bonner, 1967). Although free-living cells are brought into close proximity by chemotaxis, the formation of stable cell—cell contacts proceeds through a surface-localized carbohydrate-binding cell-ligand. Carbohydrate-binding pro-teins have been isolated from the surfaces of cohesive cells of *Dictyostelium discoideum* and *Polysphondylium pallidum* (Rosen *et al.*, 1973, 1974; Simpson *et al.*, 1974). These proteins are localized on the surface of the slime moulds and will agglutinate erythrocytes; they are not secreted however. The evidence suggesting that these proteins are involved in adhesion is: (a) they are surface-localized and only present on the surfaces of cohesive cells; (b) addition of purified protein stimulates adhesion; (c) sugars which react with the binding sites of the proteins also block cell adhesion; (d) the proteins from *Dictyostelium* and *Polysphondylium* differ in their sugar-binding characteristics which might provide the specific molecular basis of intercellular affinities. "Discoidin", the *Dictyostelium* aggregation factor, has a molecular weight of approximately 100 000, is tetrameric, does not contain sugar residues, and its binding properties are most effectively inhibited by *N*-acetylglucosamine. The binding-protein from *Polysphondylium* has approximately the same weight and subunit structure; it is more effectively inhibited by lactose and galactose.

IV. PLANT LECTINS

A. INTRODUCTION

In an earlier section of this review, it was proposed that varietal specificity in gene-for-gene host-parasite interactions involves molecular recognition at or near the host-parasite interface. To carry the requisite amount of specificity and variability it is quite likely that the elements involved are macromolecules. Many (if not all) of the examples chosen in section III to illustrate recognition phenomena were based upon carbohydrate binding by proteins and glycoproteins. Now, in fact, much more is known of the existence of carbohydrate-binding proteins in plants than the few examples illustrated above would tend to suggest. Plant carbohydrate-binding proteins, known as lectins, are widely distributed, but despite their extensive study by medical scientists and biochemists they have been largely neglected by plant biochemists, and until recently there was little definitive proof for any role of lectins in plant physiological processes. It has now come to be realized, however, that plant lectins may be involved in various recognition phenomena although this may not be their only function (Callow, 1975). It is pertinent at this point therefore to consider briefly some aspects of lectin biology and biochemistry.

Many aspects of lectin biochemistry and interaction with animal cells have been reviewed elsewhere (Boyd, 1963, 1970; Sharon and Lis, 1972; Lis and Sharon, 1973; Nicolson, 1974). The taxonomic distribution of lectins has been reviewed by Toms and Western (1971), and Callow (1975) has considered other aspects of lectin biology and the potential functions of lectins.

Lectins are proteins, or more often glycoproteins, that have the ability to interact with different types of animal cells to produce various effects (Table I). In all of these examples of lectin action it can be shown that the biological effect of the lectin is due to the binding of the lectin to carbohydrate-containing receptors localized at the cell surface. In the majority of cases the lectin

TABLE I

Effects of Lectins on Animal Cells

1. Agglutination of erythrocytes and other cells
2. Preferential agglutination of tumour cells (Burger and Goldberg, 1967; Inbar and Sachs, 1969)
3. Induction of mitosis (Nowell, 1960)
4. Insulin-like effects on fat cells (Cuatracasas and Tell, 1973)
5. Induce platelet release reaction (Tollefson *et al.*, 1974)
6. Induce insulin-release from pancreatic cells (Lockhart-Ewart *et al.*, 1973)
7. Inhibit phagocytosis by granulocytes (Berlin, 1972)
8. Inhibit fertilisation of ovum by sperm (Nicolson, 1974)
9. Inhibit protein synthesis (Olsness *et al.*, 1974)

(After Kornfeld *et al.* 1974)

molecule does not enter the cell, hence the initial lectin binding constitutes a "trigger" releasing a cascade of intracellular changes in metabolism that lead to the particular response in question. The one exception to this concerns the inhibition of protein synthesis by the toxic, non-agglutinative lectins ricin and abrin, referred to in section IIID, where one part of the molecule binds to the cell surface whilst the other part actually enters the cell and inhibits protein synthesis (Olsnes and Pihl, 1973 a, b; Olsnes, Refsnes and Pihl, 1974; Olsnes, Saltvedt and Pihl, 1974). Presumably the variety of physiological effects exerted by different lectins on different cells are due to the availability of suitable receptors and the valency of the lectin.

Although the effects of lectins on animal cells are interesting and important model systems, one must suppose that such effects are fortuitous, that plant lectin receptors on animal cell surfaces exist for the purpose of hormone binding, cell adhesion and differentiation, and that the real functions of lectins in plants are probably related to their abilities to bind carbohydrate-containing molecules.

B. DETECTION AND ASSAY

Lectins are generally extracted from plant tissues (commonly seeds) by homogenizing the tissue in physiological saline. The lectin concentration is then normally estimated in terms of agglutinating units by reacting serial dilutions of the sample with red blood cells and noting the end-point dilution. This procedure has been criticized elsewhere (Callow, 1975) for the following reasons. Firstly, it assumes that all lectins are soluble, cytoplasmic proteins. Clearly some lectins are, but what of those which are cell-wall or membrane bound? (Kauss and Glaser, 1974; Bowles and Kauss, 1975.) Secondly, the assay is rather gross and insensitive. Microlocalizations of lectins at high effective concentrations but low absolute concentrations would possibly not be detected. Thirdly, the red blood cells may not contain the appropriate carbohydrate-containing receptors. The structure of lectin receptors may be determined by the types of sugar present, their configuration or their modification by methylation or acetylation. Where lectins bind to more complex receptors than specified by terminal sugar residues, the structure of the receptor may be also determined by the presence of neighbouring sugar residues and the protein backbone to which the receptors are attached. It is known, for example, that some lectins may preferentially interact with one type of animal cell rather than another (Nicolson, 1974). Finally, some lectins, such as abrin and ricin, are monovalent and therefore non-agglutinative (unless they undergo polymerization as in the case of wheat germ agglutinin (Allen *et al.*, 1973)).

At this time it is difficult to define the term "lectin" accurately and in a biologically meaningful way. Purists would reserve the term for haemagglutinins since this was the context in which the term was originally defined by Boyd (1963). In this author's opinion, however, it would be more valuable, from the point of view of elucidating the functions of these molecules in plants, to remove the artificiality of the haemagglutination criterion and simply to define lectins as plant carbohydrate-binding proteins of potentially diverse function. The problem here is that enzymes of carbohydrate metabolism, such as glycosyl transferases, amylases, lysozyme, all bind carbohydrates to their active sites prior to catalysis of the appropriate reaction. Are these also to be termed lectins? It is interesting in this context to speculate on the evolution of lectins or carbohydrate-binding proteins. Lectins may have evolved from enzymes of carbohydrate metabolism through the loss of their enzymic activity. To this author's knowledge, the only attempt to demonstrate any enzymic activity on the part of lectins is that of Goldstein *et al.* (1974), who were unable to detect any hexokinase, phosphorylase, or phosphatase activity on the part of concanavalin A. Horejsi and Kocourck (1974) recently showed that monovalent, i.e. non-agglutinative, lysozyme can be converted to a multivalent haemagglutinin by glutaraldehyde treatment. No information was provided on the retention of enzymic activity on the part of the converted lysozyme complexes. It is this author's opinion that lectins will eventually be found to have diverse functions in the plant, that some of them may be enzymes, others may play a structural role binding to cell ligands etc., and that at the present time a broad definition of lectins simply as proteins or glycoproteins with the ability to bind carbohydrates, will suffice.

C. DISTRIBUTION

Any information provided on the distribution of lectins in the plant kingdom and within different parts of the same plant is strictly qualified by the comments on the suitability of the assay general employed, as discussed above. Thus Toms and Western (1971) stated that 54% of higher plant familes tested contain lectin-positive species and they provide detailed information of the distribution of lectins in the Leguminosae. This information is largely based on assays of seed extracts for haemagglutination, and it may be that so-called lectin-negative individuals contain lectins at lower concentration, or which are wall- or membrane-bound, or which have sugar-binding specificities that are not matched by the receptors present on the red blood cell surface. Also lectin inhibitors may be present (Howard *et al*., 1972). Some of these difficulties might be overcome by the use of immunological techniques for detecting the lectin protein *per se* although the question of sensitivity at low lectin concentrations is still pertinent. Inadequate technique has led to some unfortunate, and premature, conclusions. Brucher *et al*. (1969) studied the genetics of lectin distribution in wild bean (*Phaseolus aborigineus*) populations. Most bean populations were lectin positive, but some were polymorphic containing positive and negative individuals. The authors suggested that no significant selection pressure exists for lectins and that therefore no vital role exists for lectins in plants. These conclusions can be criticized both on technical and theoretical grounds. Since the authors did not demonstrate that loss of lectin activity was correlated with loss of gene product (protein), it could be that lectin-negative individuals (with respect to haemagglutination) still possess the lectin gene in somewhat different form due to mutation. Whilst this mutation might well affect the ability of the lectin to bind to and agglutinate red blood cells, its real function in the plant could be unimpaired. Immunological detection procedures would have assisted here. Also, the discovery of a polymorphism does not mean that the character in question has no significance. It does mean that it is not obligatory to have one allele rather than another, the alternative allele being maintained within a balanced polymorphic population through heterotic superiority of the heterozygote, or a genuine effective neutrality of this particular allelic difference.

From the point of view of the biochemical plant pathologist, the distribution of seed lectins is of less interest (although see section IVD), than their distribution in vegetative parts of the plant. Some of the older studies on lectin distribution (reviewed in Callow, 1975) show that lectins are not restricted to seeds alone, but may occur as soluble factors in leaves, roots, tubers, fruits and bark. In terms of the types of mechanism being proposed for varietal-specific resistance in this article, it would be more meaningful, however, if it could be demonstrated that lectins were localized at the interface between host and pathogen, i.e. in the plant walls or plasma membranes. Some progress in this direction has recently been made. In section V an example is discussed where the species-specific recognition of *Rhizobium* bacteria appears to be mediated by root surface-localized lectins (Hamblin and Kent, 1973; Bohlool and Schmidt, 1974). Albersheim and Anderson (1971) and Anderson and Albersheim (1972) have demonstrated the existence of lectin-like proteins in the cell walls of various plants which have the ability to inhibit fungal polygalactouronases, although it is believed that these molecules may be involved in more general defence mechanisms (Albersheim and Anderson-Prouty, 1975). Kauss and Glaser (1974) have shown that mung bean hypocotyl cell walls contain molecules with haemagglutination activity, and Bowles and Kauss (1975) demonstrated that

lectins with different carbohydrate-binding specificities were present in mitochondrial, Golgi, endoplasmic reticulum and plasma membrane fractions of
mung bean hypocotyls.

D. CARBOHYDRATE-BINDING SPECIFICITIES OF LECTINS

If the lectins or lectin-like proteins play a role in varietal-specific pathogenesis,
then one important requirement is that the binding of lectins to carbohydrate-
containing molecules be both sufficiently variable and specific to permit
discrimination against different pathogen races. The carbohydrate-binding
specificity of lectins has been traditionally tested by the "hapten inhibition
technique" by which different sugars are tested for inhibitory activity against
the lectin. A check of the sugar antagonists of some extensively studies lectins,
provided by Sharon and Lis (1972) and Nicolson (1974), shows that 35 different
lectins are inhibited by only 10 simple sugars, i.e. different lectins may be
inhibited by the same sugar, for example potato lectin and wheat germ
agglutinin are both inhibited by chitobiose. However, Lis and Sharon (1973) and
Nicolson (1974) point out that the interaction of lectins with sugars is
considerably weaker than their interactions with cellular receptors and that the
binding specificities of lectins cannot be adequately described by simple
inhibition studies. Cell surface oligosaccharide lectin receptors have been isolated
and shown to be complex structures with specific linkages, sidechains, and
peptide attachments. Neighbouring oligosaccharide chains may also interact by
hydrogen bonding. In addition, lectin binding is not restricted to terminal sugars
but can involve internal binding sites.

Kornfeld and Kornfeld (1974) have extensively studied the structural features
of glycopeptides which determine the binding of different lectins. Trypsin
treatment of human erythrocytes releases a complex glycopeptide containing
two types of oligosaccharide side-chain. These side-chains may then be released
by alkali treatment or pronase digestion to yield two glycopeptide fragments,
GP I and GP II. The structure of these two glycopeptides is shown in Fig. 6. GP I
binds tightly to lectin E of *Phaseolus vulgaris* (E-PHA) and to the lectins of *Lens*
and *Robinia*. GP II, however, will only bind to the lectin from the common
mushroom *Agaricus bisporus*. Studies on the binding of GP I and II were carried
out in conjunction with sequential degradation techniques. It was found that in
the case of GP I binding to E-PHA and *Robinia* lectins, the galactose residues of
PG I were necessary for the binding but the inner mannose residues were also
important. In the case of lentil lectin, the mannose residues were again
important in GP I binding but it was the *N*-acetylglucosamine residues in the
outer chains that additionally determined binding rather than the galactose
residues. Binding of mushroom lectin to GP II is determined by the galactose-*N*-
acetylgalactosamine-Ser sequence of the receptor. The sialic acid residues
actually hinder lectin binding in this case. Both in the binding of GP I and GP II
to the various lectins, the application of simple sugars such as mannose,

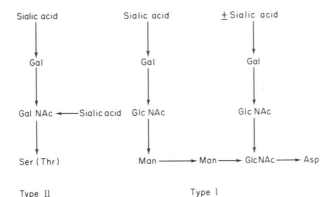

Type II   Type I

Fig. 6. The structures of the *N*- and *O*-glycosidically linked chains that form part of the two glycopeptides GP I (type I chains) and GP II (type II chains) which are released from the trypsin-released glycopeptide of human erythrocytes. (After Kornfeld and Kornfeld, 1974.)

galactose, *N*-acetylgalactosamine, or even appropriate disaccharides such as lactosamine (Gal-GlcNAc) did not inhibit binding to the glycopeptides. The binding of lectins to natural receptors is thus a complex affair not limited to simply the presence or absence of a particular sugar residue. In addition, the protein backbone appears to play an important role in the binding of glycopeptide receptors. The binding of the intact, solubilized membrane glycoprotein that contains the trypsin-released glycopeptide discussed above, was compared with the binding of the trypsin-released fragment itself. The intact molecule had a much greater ability to bind lectins. This was probably due to the fact that the intact molecule, with many oligosaccharide chains, would permit multivalent lectin binding to occur rather than the simple monovalent binding that occurs with small glycopeptides.

Studies such as that of Kornfeld and Kornfeld (1974) are relatively rare, however, and at the point in time it is difficult to answer the question of whether lectins are sufficiently variable and specific to account for varietal specific pathogenesis. More appropriate experiments in the context of this article would be to examine lectin-binding specificities when presented with fungal cell-wall glycopeptides for example.

E. FUNCTIONS OF LECTINS

In section III a range of recognition phenomena was discussed. In most of these cases it was clearly demonstrated that the interacting molecules were carbohydrate-containing receptors or ligands and carbohydrate-binding proteins or glycoproteins. Such binding proteins appear to be widely distributed in nature therefore, indeed some of them may be common types of enzyme such as glycosyl transferases. It is this author's view that there is no biologically legitimate distinction between these heterogeneous proteins involved in recognition that can bind to carbohydrate-containing molecules and what one might term the "traditional" lectins, i.e. those carbohydrate-binding molecules that have been discovered by gross extraction techniques and assays involving animal cells. It is suggested that the lectins are a part of the whole biological spectrum of enzymes and non-enzymic proteins with the basic ability to bind to carbohydrate-containing molecules and that, for some reason, some of these molecules may become localized in certain tissues of the plant, particularly the seed, and that they may have the appropriate specificity to enable them, quite fortuitously, to interact with animal cells. Now, in the context of this article, the most relevant "lectins" are almost certainly those carbohydrate-binding proteins which are localized in cell walls and membranes and which are not therefore easily demonstrated by conventional lectin assay techniques. Reference has already been made to the existence of such proteins in cell walls and membrane fractions.

The following major proposals for lectin function in plants have been made:
  (a) Plant antibodies (Punin, 1952; Saint-Paul, 1961)
  (b) Carbohydrate transporters (Ensgraber, 1958)
  (c) Control of cell division in embryogenesis (Howard *et al.*, 1972)
  (d) Control of extension growth (Kauss and Glaser, 1974)
  (e) Storage proteins
  (f) Recognition in various cell-cell interactions including those discussed in sections III and V

Concerning the suggestion that lectins are the equivalent of animal antibodies, being produced in response to antigenic stimulation by micro-organisms, experiments carried out to test this hypothesis have proved to be inconclusive (Saint-Paul, 1961). At the present time there is no evidence to show that plants are capable of manufacturing antibodies in a manner analogous to the immune system of animals. This suggestion does, however, have a bearing on proposals for a static role of lectins in general defence mechanisms against micro-organisms. Reference has already been made in section II to the existence in plant cell walls, of proteins which have certain properties of lectins, and which inhibit fungal polygalacturonase enzymes (Albersheim and Anderson, 1971; Anderson and Albersheim, 1972). A more direct interaction has been proposed by Jones (1964), who sought to explain the characteristic localization of lectins in seeds in high concentration. It is known that seeds of the Leguminosae are relatively impermeable to water, and that in nature germination is probably accelerated by micro-organisms that can degrade the impervious testa permitting imbibition. It was suggested that the lectins in the cotyledons of the seeds may serve to prevent further degradation of the seed once the testa has been broken down, and Mialonier *et al.* (1973) have shown that in *Phaseolus vulgaris* seeds lectins are present in cotyledons but not in testas. Semipurified lectin preparations were made from seeds of *Vicia cracca* and these were shown to inhibit the growth of a soil bacterium that was cultured on agar containing *Vicia* seed coats. This particular bacterium was able to accelerate the germination of *Vicia* seeds. No inhibition of growth was obtained with soil bacteria of less specific growth requirements. Further work is clearly required on this interesting proposal to see whether other legume seed lectins are able to inhibit micro-organisms involved in testa decay, the specificity exhibited in this type of interaction, and the nature of lectin receptors on the bacterial surface. Direct effects of certain lectins on fungal hyphal extension have also been demonstrated. Mirelman *et al.* (1975) and J. A. Callow (unpublished) have demonstrated that wheat germ agglutinin and potato lectin, both of which bind to polymers of N-acetylglucosamine, will inhibit hyphal extension and spore germination in *Trichoderma viride* and *Botrytis cinerea*. It is believed that these lectins may act by cross-linking chitin microfibrils at the fungal apex. The significance of these effects *in vivo* is uncertain, however, since not all fungal pathogens contain chitin in their cell walls, and not all lectins bind to polymers of N-acetylglucosamine. It is possible that other lectins with different sugar specificities could bind to fungal walls containing mannans for example, but the degree of specificity possible in such interactions would be rather low.

It has been proposed that lectins serve as carbohydrate fixers and may serve to transport carbohydrates or carbohydrate-containing materials in a manner analogous to the "transporter" function of animal globulins (Krupe, 1956; Ensgraber, 1958). The fact that the developing seed, a nutrient "sink" in plants, is a major source of lectins might provide circumstantial evidence in favour of this proposal but no experiments appear to have been designed to test this hypothesis.

The apparent restriction of some mitogenic lectins to the seed in certain plants has led to the suggestion that these lectins may be involved in differentiation and development processes in the seed embryo, possibly by controlling rates of cell division. There is as yet no evidence to indicate that lectins which are mitogenic against animal cells also exhibit this activity against plant cells. Single cell cultures from plants may be of use here. Further circumstantial evidence which suggests a potential role for lectins in cell division comes from the studies of Hamblin and Kent (1973) on *Rhizobium*-binding to bean roots (see section VB2). In addition to their studies on *Rhizobium* binding, they also showed that nodules were a rich source of lectins. In red kidney bean the phytohaemagglutinin (PHA) complex contains several components, at least one of which is mitogenic (Lis and Sharon, 1973). Growth of nitrogen-fixing legume nodules proceeds through cell division and cell enlargement and it may be that lectins are involved in the control of these growth processes.

Kauss and Glaser (1974) have recently suggested that lectins may be involved in the control of cell extension growth, by "glueing" wall polysaccharide components together. The evidence for this suggestion is rather slight, however.

Small differences in lectin content between cell walls of growing and non-growing mung beans were obtained.

In many seeds lectins are localized at high concentration (1—3% of the total protein, Callow, 1975). This suggests that seed lectins may have a storage function but obviously relegates the specific carbohydrate-binding capacity exhibited by all lectins to some type of fortuitous activity, and does not account for the observed distribution of lectins in higher plant tissues other than seeds, and in non-seed bearing plants.

In conclusion, although several suggestions have been put forward for the role of lectins in plants, at the present time there is no strong evidence for any of these except those which relate to recognition phenomena. Possibly the most useful way to look at lectins is that they are representative of plant carbohydrate-binding proteins in general, being both diverse in nature and function.

V. CELL SURFACE RECOGNITION AND VARIETAL SPECIFICITY

A. INTRODUCTION

In section II of this article the point was made that specificity in host-parasite interactions can be considered at several levels. It was observed that in plants susceptibility is the exception rather than the rule, and that therefore plants must possess many general protection and defence mechanisms which are effective against a large number of potential pathogens. Some pathogens are non-specific in their host range whilst others exist as a number of races each with characteristic host specificity. Thus we have the extreme form of specificity where cultivars of a single host species differ in their susceptibility to different races of the same pathogen. In many cases such highly specific interactions have been described in terms of the gene-for-gene concept (Flor, 1956, 1971; Day 1974; Person and Mayo, 1974). In the case of such interactions, expression of resistance in the host (R) and avirulence in the parasite (A) are under the control of inter-dependent genes, and it has been proposed that the products of the R and A genes must interact to form a "hybrid" molecule at or near the host-parasite interface. The resulting "stop signal" (Person and Mayo, 1974) generated by such a molecule could then lead to the biochemical expression of resistance through the hypersensitive response and the accumulation of phytoalexins for example. The initial interaction between host and parasite, resulting in the formation of a hybrid molecule, is thus quite analogous to other instances of recognition described in section III. It is also quite likely that the interacting principles are macromolecules since large molecules would be required in order to carry both the necessary degree of specificity and variability. What are these molecules, where are they localized, and how does their interaction trigger resistance responses?

B. CELL-SURFACE INTERACTIONS AND VARIETAL SPECIFICITY IN BACTERIAL
INFECTIONS

1. Surface receptors and crown gall induction
A direct correlation has been observed between the pathogenicity of different
strains of *Agrobacterium tumefaciens* and their resistance to the potent
antibacterial agents called bacteriocins. Conversion of a non-pathogenic strain to
pathogenicity is accompanied by a change from bacteriocin resistance to
bacteriocin susceptibility and vice versa (Kerr and Htay, 1974; Roberts and Kerr,
1974). Bacteriocins are known to exert their toxicity by binding to receptors
localized on the surface of susceptible cells. Little is known of the nature of
these receptors except that they are proteinaceous, contain a small amount of
carbohydrate, and that carbohydrate moieties are involved in the binding (Sabet
and Schnaitman, 1973; Reeves, 1972). Thus pathogenicity of a micro-organism
is correlated with a specific, surface-localized molecular structure or configuration. Roberts and Kerr (1974) suggest that in order to achieve crown gall
induction and DNA transfer, host cell-bacterial cell contact is necessary and that
this is mediated by a complementary association of molecules on plant and
bacterial surfaces. The bacteriocin receptors appear to form an integral part of
this recognition mechanism.

2. The role of lectins in the nodulation of legumes by Rhizobium
It has been proposed that lectins on the root surface may be involved in the
specific binding of *Rhizobium* bacteria to legume root surfaces. Hamblin and
Kent (1973) incubated roots of *Phaseolus vulgaris* in erythrocytes and observed
massive binding of cells to certain areas of the root, especially those most
suitable for nodulation such as the root hairs. It was also shown that *Rhizobium*
would bind the lectin from *P. vulgaris*. Bohlool and Schmidt (1974) examined
the specificity of this lectin binding by adding fluorescein isothiocyanate-
labelled soyabean lectin to suspensions of various *Rhizobium* strains. The lectin
bound to all but three of the 25 strains of the soyabean-nodulating *R. japonicum*
but did not bind to strains unable to nodulate soyabeans. The identity of the
carbohydrate-containing lectin receptors on the *Rhizobium* surface is not yet
clear. T. S. Brethauer and J. D. Paxton (personal communication) have
demonstrated that the impure soyabean lectin will react in agar with material
from culture filtrates of *R. japonicum*, precipitating it, but will not react with
culture filtrates of non-nodulating *R. leguminosarum*. P. Albersheim (personal
communication) has suggested that the lectin receptors may be the bacterial
O-antigens.

C. VARIETAL SPECIFICITY IN FUNGAL DISEASES
There are two well-characterized forms of varietal specificity in fungal diseases.
In one of these, specificity can be described in terms of the interaction between
the host plant and fungal toxins. Only those plants that possess receptors for the

toxin are parasitised. The other form of specificity is described by the gene-for-gene concept already discussed.

1. Host-specific toxins

It is now possible to explain host-parasite specificity in several diseases caused by necrotrophic pathogens in terms of host-specific toxins secreted by the fungus (Scheffer and Yoder, 1972). The best-known example of such a toxin is HV-toxin produced by *Helminthosporium victoriae*, the causal organism of Victoria blight of oats. HV-toxin is a low molecular weight peptide fragment linked to a sesquiterpenoid base. At low concentrations (9 ng/ml) the toxin inhibits seedling growth of susceptible host varieties, whereas non-hosts and resistant host varieties are only partially affected by concentrations as high as 3.6 mg/ml (Kuo *et al.*, 1970). HV-toxin also alters membrane permeability, and a body of evidence suggests that the toxin acts primarily at the host plasmalemma (Luke and Gracen, 1972). Specificity in resistance has been shown not to be related to any form of differential inactivation of the toxin by resistant hosts (Scheffer and Yoder, 1972), and the most likely explanation is that toxin receptors are localized in or near the host plasmalemma and that these are absent, or inactive, in toxin binding in resistant hosts (Luke and Gracen, 1972).

Other host-specific toxins may act in similar ways, but the most convincing demonstration of the molecular differences involved in this type of specificity has been obtained for the eyespot disease of sugar cane caused by *Helmintho-sporium sacchari* (Steiner and Strobel, 1971; Strobel and Steiner, 1972; Strobel, 1973 a, b, 1974; Strobel and Hess, 1974; Strobel and Hapner, 1975). Here it is claimed that specificity involves the selective binding of the toxin "helmintho-sporoside" (2-hydroxycyclopropylgalactopyranoside) to membranes of suscep-tible hosts. A binding protein of molecular weight 48 000 has been isolated from membranes of susceptible hosts and shown to consist of four subunits with at least two toxin-binding sites. The same protein has also been found in membranes of resistant hosts but in a form which does not bind to the toxin. Activity can, however, be restored by treating membranes from resistant plants with the detergent Triton X-100. Several lines of evidence indicate that the binding protein is localized in the plasma membrane (Strobel and Hess, 1974; Strobel and Hapner, 1975): (1) prior treatment of susceptible plant tissues with antiserum to the binding protein protects the tissue from the toxin; (2) pyridoxylation of proteins *in vivo*, a process essentially limited to external membrane proteins, followed by their reduction, resulted in the isolation of a protein with the same elution volume and mobility as authentic pyridoxylated-reduced binding protein; (3) sugar cane protoplasts from susceptible tissue were agglutinated by antiserum to the binding protein; (4) protoplasts incubated in ^{3}H-labelled helminthosporoside did not accumulate label in the protoplasm. Strobel and Hapner (1975) have shown that the binding protein is the primary site governing susceptibility to toxin since protoplasts of resistant sugar cane clones were able to bind the toxin following treatment with purified binding protein.

The toxin-binding protein also binds to other α galactosides such as raffinose and melibiose (Strobel, 1974), and it has been suggested that since there is presumably a strong selective disadvantage to the possession of the toxin-binding protein the real physiological function of the binding protein is more likely to be related to a α-galactoside transport between cells. A major difficulty with this proposal, however, is that resistant clones of sugar cane apparently thrive in the absence of this α-galactoside-binding protein. The mechanism by which toxin binding results in the physiological expression of disease symptoms has been examined. Strobel (1974) has shown that the toxin-binding activates by 30% the membrane-associated K^+ and Mg^{2+}-activated ATPase, and membrane glycosyl transferases by an unspecified amount. As a result of toxin binding, therefore, it is quite possible that ion flux regulation by affected cells would be impaired.

2. Varietal Specificity in Gene-for-Gene Systems

A brief reiteration of some of the genetic principles governing varietal specificity in gene-for-gene systems is appropriate here since they provide a framework within which biochemical hypotheses may be designed and tested. In a gene-for-gene interaction only the association between the products of dominant host resistance genes (R) and dominant parasite avirulence genes (A) results in resistance. Each gene is therefore "conditional" on the presence of the other for resistance to be expressed (Person and Mayo, 1974). Since the R and A genes are usually dominant, they are probably expressed, and it has been suggested that the specific interaction between R and A gene products results in the formation of a "hybrid" molecule which acts as a "stop signal" (Person and Mayo, 1974), triggering or "eliciting" a whole series of metabolic changes in the host culminating in resistance through the hypersensitive response and phytoalexin accumulation. The crux of varietal-specific resistance in gene-for-gene systems thus lies in the mutual recognition and interaction of products of host R genes and pathogen A genes at or near the host-parasite interface. Similar conclusions have been reached recently by Albersheim and Anderson-Prouty (1975).

What are these gene products? Most of the various examples of recognition described in this article are based on the highly specific interactions that occur between carbohydrate-containing molecules and carbohydrate-binding proteins. The presence of carbohydrate-binding proteins in plants has been considered. By direct analogy with other examples of biological recognition, it is suggested that the products of host resistance genes are surface-localized lectins or lectin-like molecules which serve as receptors, to recognize specific parasite signals which are most likely cell-wall localized, or secreted carbohydrate-containing molecules. It is known that fungal cell walls contain various polysaccharides, heteropolysaccharides, glycoproteins and peptidopolysaccharides in addition to the structual polysaccharides (Gander, 1974). The biosynthesis of such complex polysaccharides, or the carbohydrate portions of protein-polysaccharide complexes, proceeds through the stepwise addition of monosaccharides from their receptive sugar nucleotides, to the ends of oligosaccharide acceptors, each step being catalysed by a glycosyltransferase enzyme. The battery of glycosyltransferases required for the synthesis of each complete molecule forms a multiglycosyltransferase enzyme complex (Roseman, 1974). These glycosyltransferases show a high degree of specificity for the sugar nucleotide and the terminal and penultimate monosaccharide residues of the acceptor molecule (Bartholomew *et al.*, 1973). Albersheim and Anderson-Prouty (1975) have suggested that certain of these polysaccharides or protein-polysaccharide complexes are the parasite signals recognized by the products of the host resistance genes, forming a hybrid molecule. Such signals may have a basic common structure within a given species. These are then modified by glycosyl transferase or other enzymes coded for by the host avirulence genes to provide the requisite structural variability from race to race. It is also suggested that these parasite signals constitue "elicitors" of the host hypersensitive response and phytoalexin accumulation. Purified elicitors from *Colletotrichum lindemuthianum* and *Phytophthora megasperma* var. *sojae* appear to be heterogeneous polysaccharides, largely glucan in nature, which are secreted by the pathogens into the culture medium, and which when applied to the host plants induce the hypersensitive response and the synthesis of the appropriate phytoalexins. As yet, there is no evidence concerning the specificity of these glucan elicitors. Other evidence also indicates that fungal pathogen cell walls have variable carbohydrate-containing components. Serological techniques for the identification of fungi are based upon the antigenicity of these cell wall carbohydrates. J. G. Manners (personal communication) has for example demonstrated that there are sufficient differences in the surface antigens of *Erysiphe graminis tritici* and *E. graminis hordei* to permit their discrimination by the fluorescent antibody technique. Ballou and Raschke (1974) have demonstrated species specificity in the cell wall mannan glycoproteins of yeast. These mannan proteins constitute the principal immunogens of the yeast cell wall. The mannan portion consists of an α-1 $\rightarrow$ 6-linked backbone with variable α-1 $\rightarrow$ 2 and α-1 $\rightarrow$ 3 side-chains. Mutant yeasts synthesize different mannosyl transferases, and mannosyl phosphate transferases which alter the side-chain structures of the mannan. Kleinschuster and Baker (1974) have used various sugar-specific plant lectins to demonstrate that conidia of different species of *Fusarium* contain different types of carbohydrate-containing lectin receptors in their walls, although it would have been more appropriate if lectin receptors on vegetative hyphae had been examined. Although such studies are few, it would appear that fungal cell walls do exhibit considerable variability. It must be remembered, however, that the actual wall "elicitors" or carbohydrate-containing molecules recognized by the host might constitute only a small proportion of the total cell wall carbohydrate-containing material and that studies on gross cell wall fractions could prove misleading. More progress is likely to be made through the study of specific carbohydrate-containing molecules that are able to elicit the hypersensitive response, or phytoalexin accumulation, as described by Albersheim and Anderson-Prouty (1975).

How does this initial event of recognition trigger a resistant response? In section II it was shown that a very common resistant response in gene-for-gene interactions is the hypersensitive response, usually accompanied by phytoalexin biosynthesis. The whole gamut of recognition phenomena described above testify to the fact that widespread changes in cell metabolism are the consequence of initial surface interactions. Studies on the architecture of the plasma membrane (section IIIA) have shown that some receptor glycoprotein molecules traverse the lipid bilayer, i.e. there is direct molecular contact between the environment and the cytoplasm through these receptors. Roseman (1974) suggested that membrane receptors are associated in their cytoplasmic faces with "membrane messengers" which could be RNA molecules involved in gene regulation, effector proteins, or enzymes such as adenyl cyclase which generate cyclic nucleotide messengers. On the binding of the appropriate ligand by the membrane receptor or sensor, each receptor would release a different "membrane messenger" which could diffuse to the nucleus and there exert effects on gene regulation.

Some of the ideas expressed above have been incorporated (Fig. 7) into a modification of a recent model for gene-for-gene interactions developed by Day (1974). In this model certain concepts in eukaryotic gene regulation described by Britten and Davidson (1969) have been taken into account. These aim to ascribe the large amount of genetic redundancy that exists in eukaryotes to regulatory functions. The model also owes much to other induction hypotheses based on gene regulation (e.g. Hadwiger and Schwochau, 1969). Four distinct types of gene may be involved:

1. Sensor genes (R) bind, either directly or indirectly, through a protein intermediate or sensor, to those external agents such as hormones or effectors, which induce a specific pattern of metabolic activity.
2. Integrator genes (I) link directly to sensor genes and specify the synthesis of activator RNA molecules.
3. The activator RNAs (or their protein products) bind to receptor genes (R) activating them to turn to promote the transcription of linked
4. producer genes (P) which are the structural genes on which messenger RNA templates for new enzyme synthesis are formed.

The binding of an external agent to the sensor gene, either directly or indirectly through the protein product of the sensor gene and "membrane messengers", thus results in the sequential activation of regulatory genes governing the producer genes. A series of such gene batteries might be involved in the induction of several enzymes in a new metabolic pathway and their

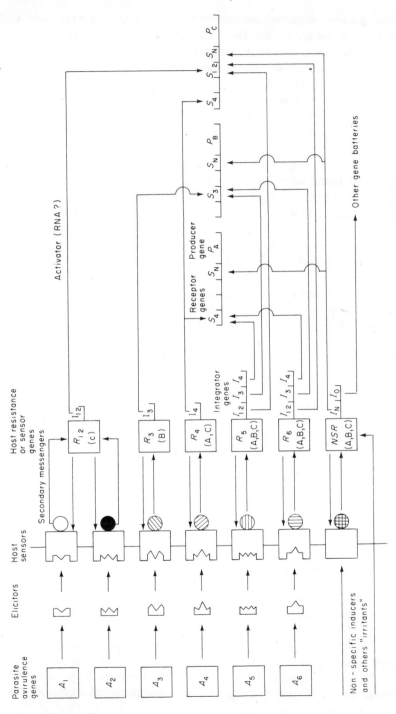

Fig. 7. Proposed model for recognition and gene control in those instances of varietal specificity governed by gene-for-gene interactions. (Based on Day, 1974.)

co-ordinated activity might be achieved in a variety of ways. Through redundancy of receptor genes, different producer genes might be associated with the same receptor genes. For example, in Fig. 7, both P_A and P_C can be activated through S_4. Alternatively, simultaneous synthesis of activator RNA molecules from a set of linked integrator genes results in co-ordinated synthesis of a number of producer genes not sharing a common receptor gene (e.g. in Fig. 7, $P_{A,B,C}$, are all activated through the simultaneous activation of a set of integrator genes common to a single sensor gene (R_5 or R_6)).

In the model presented in Fig. 7, three producer genes are shown, A, B, C. Each of these notional genes might code for enzymes involved in phytoalexin synthesis for example, or enzymes involved in some, as yet, unknown aspect of the hypersensitive response. Each producer gene is under the control of linked receptor genes (S), which are in turn activated by the products of integrator genes (I). The notional R genes for resistance are shown here as coding for membrane-localized carbohydrate-binding proteins or sensors. Each R gene thus specifies a different sensor, and according to the gene-for-gene hypothesis the particular sensor for any individual R gene should only be capable of binding or recognizing the carbohydrate-containing product specified by the corresponding parasite avirulence gene. Six specific R genes are depicted. R_1 and R_2 are allelic, R_{3-6} are separate genes. R_{1-2}, R_3 and R_4 have different integrator genes controlling different receptor genes. R_5 and R_6 are shown as being closely linked and controlling the same battery of integrator genes (although this is not necessarily obligatory for closely linked genes). In any appropriate $R : A$ resistant, or incompatible interaction, the end production of particular avirulence gene activity, i.e. the resistance "elicitor", binds to the appropriate sensor protein at the host-parasite interface, inducing the release of a specific membrane messenger that interacts with the sensor gene in question. Sequential activation of the appropriate regulatory genes thus results in activation of the producer gene or genes, resulting in the physiological expression of resistance.

In addition, the gene batteries can be regulated by non-specific sensor genes (NSR) which, through protein sensors, are capable of interacting with a variety of non-specific inducers of resistant reactions such as elicitors from non-pathogens or other "irritants" (Day, 1974). The non-specific sensor genes are thus shown as being responsible for some aspects of general disease resistance. Where general resistant responses differ markedly from those exhibited in examples of varietal specificity it could be that the non-specific sensors regulate quite different gene batteries.

The model also takes into account the following important considerations:
1. Sato et al. (1968) showed that potato cultivars containing R_2, R_3, R_4, major resistance genes all exhibited a common form of resistance characterised by accumulation of the phytoalexin rishitin, when inoculated with a common avirulent race of Phytophthora infestans. This is accounted for in the model by linking the notional R genes R_5 and R_6 to

the same set of integrator genes. Hence, irrespective of whether the host-parasite gene pairing involves R_5-A_5 or R_6-A_6, integrator genes $I_A I_B$ and I_C are activated resulting in the transcription of all three producer genes, P_A, P_B, and P_C, and presumably the identical physiological response.

2. In the case of multiple allelism, as in flax (Flor, 1956), each allele would specify a slightly different sensor protein, but the sensor or resistance genes could be linked to the same integrator gene thus activating the same producer genes and resulting in identical physiological responses, affecting the same stage in the development of the parasite, as predicted by Ellingboe (1972).

3. Ellingboe (1972) also showed that different gene pairs for powdery mildew resistance block development of the parasite at different times or stages of development. Some alleles, however, may regulate fungal development at more than one stage or time. In the model both of these situations are taken into account. So, hypothetical resistance genes with notionally distinct effects on parasite development such as R_{1-2} and R_3 are depicted as controlling the transcription of two different producer genes, P_B and P_C. On the other hand, one notional R gene (R_4) is shown as controlling two different producer genes (P_A and P_C), which would result in the synthesis of two different enzymes that might exert effects on fungal development at two different levels or times.

4. In flax rust, five gene loci are involved in the determination of resistance, each one controlling a separate gene-for-gene relationship. Person (1959) has shown that a single "stop signal" generated at any one of the five different loci takes precedence over all other interactions that do not read "stop". In the model this situation is most easily accounted for by the co-ordinated synthesis of the various enzymes required for the full expression of resistance, i.e. the gene products of $P_{A,B,C}$, and is achieved in the case of hypothetical resistance genes R_5 and R_6 by redundancy at the level of the integrator genes so that activation of either R_5 or R_6 sensor genes results in an identical physiological effect.

There is, however, one major problem associated with this view of gene-for-gene varietal specificity. Parasite avirulence genes are generally dominant and, in classical genetics, dominant genes are usually expressed in the form of functional gene products whereas recessive genes are often associated with an absent or defective gene product. It seems reasonable to suggest that the primary role of avirulence genes is to support the success of the parasite as a viable organism (although not necessarily in the direct support of parasitism in the case of necrotrophs) since the avirulence genes would presumably have been lost otherwise. Indeed, many parasites tend to regain those avirulence genes that do not correspond to resistance genes in the particular host population at the time (Flor, 1956, 1971; Van der Plank, 1968; Watson, 1970; Albersheim and

Anderson-Prouty, 1975). In those cases where fitness and virulence are not negatively associated (Watson, 1970) it may be that the virulence character is genuinely neutral. However, in the interaction of R with A, avirulence genes clearly function against parasite success since the host uses the products of avirulence gene action to recognize and resist the avirulent parasite. This apparent contradiction may be resolved by considering how gene-for-gene systems have evolved (Day, 1974; Person and Mayo, 1974; Albersheim and Anderson-Prouty, 1975).

Oligogenic, gene-for-gene systems of varietal specificity have probably evolved from systems of more general resistance through the attentions of the plant breeder (all gene-for-gene systems involve domestic crops although Day (1974) suggests that they may also be found in nature). It is likely that some aspects of parasite success in the initial interaction between an avirulent parasite (A) and a susceptible host (r) rely on the diversion or suppression of host-resistant responses. Some of these may involve forms of hypersensitive host cell death and phytoalexin synthesis (see section IID) and may be under the control of non-specific sensor genes that are triggered by non-specific products, "elicitors", "inducers", or "irritants" of the parasite (Fig. 7; and Day, 1974). In this initial interaction, then, the avirulence gene is associated with virulence and parasite success. If, as suggested by Albersheim and Anderson-Prouty (1975), the product of the parasite avirulence genes are glycosyl transferases or other enzymes capable of modifying oligo- or polysaccharide chains of fungal cell wall polymers, then these enzymes could be responsible for the insertion or addition of specific sugar residues into specific configurations of a large number of cell wall polysaccharides, or polysaccharide-protein complexes similar to the mannan proteins described in section VC2, or wall-bound or secreted glycoprotein enzymes.

In the case of the former suggestion, Albersheim and Anderson-Prouty (1975) have suggested that wall polysaccharides may serve as "elicitors" of the hypersensitive response and phytoalexin synthesis in the interaction with resistant hosts (i.e. in $R : A$ interactions). As yet, however, little evidence is available on the specificity of these elicitors and one must ask what essential role these wall polysaccharides serve in the $r : A$ interaction when A is not being conditioned for avirulence by the R gene. It could be that such elicitors serve to divert or suppress the normal host-resistant response through some narcotic or toxic activity, or that they play an important part in determining cell wall properties involved in hyphal extension for example. Suppression of resistant responses in this latter case would be due to other molecules secreted by the pathogen. If the product of avirulence gene action is the synthesis of an important part of the glycan moiety of a glycoprotein exoenzyme such as invertase, centrally involved in parasite nutrition (Long et al., 1975), then again it must be supposed that the suppression of host resistance responses is caused by some other pathogen molecule. In the latter case, what is the significance of

Albersheim's cell wall polysaccharide elicitor? Until more is known of the specificity of these elicitors it is difficult to answer this question. If such molecules prove to be non-specific then it may be that their role is essentially structural or connected with the regulation of cell wall growth, or they may be involved in other aspects of recognition such as in hyphal anastomosis. Their ability to induce phytoalexin synthesis could then represent a fortuitous irritant action on the non-specific host receptors controlling synthesis.

In this initial $r : A$ interaction, then, the products of the avirulence genes are serving to further parasite success within a system of induced susceptibility. The host plant is under pressure to evolve resistance rapidly by adapting its general defence mechanisms to respond to specific parasite signals. This might be most readily achieved by mutation of a non-specific sensor gene that already controls resistant responses to a form (R) which is now able to both control resistant responses such as phytoalexin synthesis and recognize specific signals produced by the parasite. These specific signals could quite fortuitously be those surface-localized or secreted molecules that are modified by avirulence gene action, and the recognition would be most likely to proceed through surface-localized protein intermediates, the products of host sensor genes. Alternatively the R genes could be introduced into the host by the plant breeder.

Thus a system of general resistance and induced susceptibility has evolved into a system of specific induced resistance. In both cases the actual mechanism of inhibition of fungal development is the same but the means by which it is controlled have changed. Where the biochemical expression of non-specific resistance differs from varietal-specific resistance it might be supposed that, in the conversion of a non-specific gene sensor to a varietal-specific sensor, certain of the producer gene's controlling functions may have been lost.

In this model, the R gene is able to control the regulatory and producer genes involved in a resistant response, it is able to produce a specific carbohydrate-binding protein, and it is able to react with the messenger released when this protein binds to the products of the parasite's avirulence gene. In this situation the avirulence gene is "conditioned" through the host R gene and is associated with lack of parasite success or avirulence. If the parasite is to survive it is now under selection pressure to lose the product of the avirulence gene that stimulates host defences. Mutation of dominant A to recessive a (avirulence to virulence) would result in the inability of the pathogen to synthesize the appropriate enzyme involved in the modification or synthesis of that part of the pathogen elicitor recognized by the host R gene sensors. The new product is no longer recognized by the host sensors and hence does not trigger the host R genes into operation. At the same time the replacement of a system of general defence by a specific system of induced resistance may mean that the host plant is no longer so resistant to other pathogens in the environment, which may explain why r genes conferring host susceptibility to a particular race of pathogen are found in host populations at all, since they should not apparently be favoured by selection.

While virulence has now been regained by the parasite it has been achieved at the expense of reduced parasite vigour and fitness. Where the loss of vigor is substantial or total then the corresponding host resistance genes have been termed "strong" genes (Van der Plank, 1968). In practice, it would appear, however, that most host R genes result in a loss of vigour which is less than total (Day, 1974), and host genes with small or no effects on parasite vigour are termed "weak" genes. Where parasite vigour has been reduced it may be regained either by back mutation to avirulence (A), which is obviously only effective when the host populations which are available lack the corresponding R gene, or by development of another avirulence gene. As described by Albersheim and Anderson-Prouty (1975) the product of a second avirulence gene would presumably replace the first non-functional enzyme resulting in a modified carbohydrate-containing surface ("elicitor") not recognized by host R gene sensors, and renewal of pathogen vigour. The pathogen has thus differentiated into a second race. Albersheim and Anderson-Prouty (1975) proceed to describe how this type of co-evolution of host and parasite could then result in the formulation of gene-for-gene systems (Fig. 7), with multiallelic host R genes clustered close to each other through tandem gene duplication, and parasite avirulence genes tending to be non-allelic and dispersed through the genome.

D. SUMMARY AND CONCLUSIONS

1. Genetic analysis of many instances of varietal specificity in host plant-fungal parasite interactions reveals that the resistance and virulence characteristics are under oligogenic control and that the genes for resistance-susceptibility and virulence-avirulence are interdependent (i.e. the "gene-for-gene" interaction).

2. In most of these examples the genes for resistance and avirulence are dominant, and only the association of $R : A$ leads to resistance.

3. Resistance in cases of varietal specificity is generally an induced, active, energy-requiring process, characterized by the so-called hypersensitive reaction, frequently associated with the synthesis of fungitoxic phytoalexins.

4. The hypersensitive reaction is initiated within the first few minutes of the interaction, i.e. when host and parasite first come into intimate contact. Varietal-specific resistance can therefore be viewed as a form of surface-surface interaction. In a large range of other cell-cell contact phenomena, the initial events leading to the appropriate cellular reactions are mediated by associations between surface-localized complementary macromolecules, involving some form of carbohydrate binding.

5. By direct analogy, therefore, it is suggested that the interdependent or "conditional" nature of R and A genes is a direct result of the formation of a "hybrid" molecule at the host-parasite interface, between the products of the appropriate R and A genes. As in other cell-cell associations, this "hybrid" molecule may then constitute a "stop signal" or trigger to a cascade of changes in metabolism culminating in resistance.

6. Higher plants contain proteins or glycoproteins known as "lectins" with the ability to bind in a highly specific manner, to carbohydrate-containing receptors and these have been shown to be contained within cell wall and membrane fractions of plant cells. Fungal cell walls contain a variety of carbohydrate-containing macromolecules as glycoproteins, and polysaccharides. It has been demonstrated that in certain instances these cell wall, carbohydrate-containing macromolecules, when applied to host plants, induce or "elicit" the hypersensitive response and phytoalexin accumulation. As yet, the specificity of these elicitors is unknown.

7. It is suggested, therefore, that the products of host resistance genes are surface-localized, lectin-like, carbohydrate-binding proteins or "sensors". The specific interaction of these sensors with the end-products of avirulence gene action, i.e. the carbohydrate-containing elicitors, then constitutes the $R : A$ interaction or "stop signal". The initial products of the parasite avirulence genes may be enzymes such as glycosyl transferases which serve to modify in a race-specific manner, the essentially common basic structure of fungal elicitors, through the introduction of specific sugar residues into specific configuration.

8. The above suggestions have been incorporated into a model of gene regulation which aims to describe certain important aspects of gene-for-gene systems of varietal specificity.

9. The mutually contradictory roles of the parasite avirulence genes may be resolved by considering the means by which gene-for-gene systems may have evolved. It is suggested that they arose from systems of more general resistance (perhaps based on systems of induced susceptibility) through the need of the host to recognize the parasite. The most effective recognition (leading to resistance) would occur if the parasite signals recognized by the host (products of potential avirulence genes) were molecules that are essential to parasite vigour. When not conditioned by host R genes the elicitors may serve as enzymes, or may be involved in cell-wall metabolism.

The biochemical identification of the products of host resistance genes and parasite avirulence genes, and the ways in which they interact at the host-parasite interface, is crucial to our understanding of host-parasite specificity. In the past, attention has been concentrated on the means by which host plants inhibit the growth of potential parasites, and while our understanding of these mechanisms is far from complete, our understanding of the interactions which trigger such responses is even less so. In this review, an attempt has been made to provide a background for an attractive current hypothesis which seeks to explain varietal specificity in gene-for-gene interactions as a surface-surface interaction biochemically similar to many other biological recognition phenomena, i.e. based on the interaction of complementary macromolecules, involving some aspect of carbohydrate binding. How can these suggestions be tested experimentally? Since one is in effect attempting to identify the products of certain host and parasite genes, the use of isogenic host and parasite lines, with the exception of the virulence-avirulence and susceptibility-resistance traits, may prove to be essential in order to be able to pick out specific effects from background variations due to other genes not concerned in pathogenicity.

The hypothesis proposes that a given avirulent fungal race will differ from its virulent counterpart in that a surface-localised carbohydrate-containing macromolecule will have been modified, perhaps by the addition of a specific sugar in a certain configuration, in such a way that it is now recognized as a parasite signal by the products of host resistance genes, or sensors. This may be revealed in a variety of ways, by comparative chemical analysis of sugar compositions, by structural analysis of oligosaccharides resulting from acetolysis or enzymic digestion (Yen and Ballou, 1974), or by immunological means. These procedures could be employed on whole hyphal walls, or on fractions derived from them by various extraction techniques. In the case of biotrophic parasites possessing haustoria, such as rusts and powdery mildews, recently developed techniques for isolating haustoria (Dekhuizen and van der Scheer, 1969) would have to be employed since it is the haustorium which comes into intimate contact with the host protoplast.

An alternative approach, employed by Albersheim and Anderson-Prouty (1975), would be to prepare fungal wall macromolecular fractions and to test these for activity in the elicitation of the hypersensitive response and phytoalexin synthesis. Results so far appear to indicate that both *Phytophthora megasperma* var. *sojae* and *Colletotrichum lindemuthianum* contain highly active, branched glucan elicitors, but no evidence is yet available on the specificity of these elicitors. Having identified specific elicitors, the site of avirulence gene action could again be determined by comparative structural analysis of the elicitors and their corresponding non-active equivalents in virulent races.

An interesting approach to the study of host-parasite surface effects might be to examine host response in systems *in vitro* consisting of washed fungal wall fragments and living host protoplasts. If, as suggested, host surface-localized carbohydrate-binding proteins bind to fungal wall elicitors, the nature of the carbohydrate-binding could be studied in such systems by techniques such as the sugar inhibition technique described in section IVC or by the use of plant lectins. The products of host resistance genes may in fact be lectin-like proteins. Clearly we need to know much more of the microlocalization of lectins in cellular organelles and membranes. If the products of host resistance genes are plasma membrane-localized binding proteins, then one potential approach would be to examine whether plasma membrane vesicles or fragments, or indeed solubilized membrane proteins, will interact with the binding of homologous antiserum prepared to fungal walls.

The above suggestions represent only a small portion of the various experimental approaches to the hypothesis outlined in this article. The development of hypotheses and models is inevitably a hazardous process, many of the ideas presented above will never be substantiated. If, however, this article serves to initiate new experimental approaches to the problem of understanding plant disease then some useful purpose will have been served.

REFERENCES

Abeles, F. B., Bosshart, R. P., Forrence, L. E. and Habig, W. H. (1970). *Pl. Physiol.* **47**, 129-134.

Aketa, K. (1967). *Embryologia* **9**, 238-243.

Aketa, K. (1973). *Exp. Cell Res.* **80**, 439-441.

Aketa, K. and Onitake, K. (1969). *Exp. Cell Res.* **56**, 84-86.

Aketa, K., Tsuki, T. and Onitake, K. (1968). *Exp. Cell Res.* **50**, 676-679.

Albersheim, P. and Anderson, A. J. (1971). *Proc. Natn. Acad. Sci. U.S.A.* **68**, 1815-1819.

Albersheim, P. and Anderson-Prouty, A. J. (1975). *A. Rev. Pl. Physiol.* **26**, 31-52.

Albersheim, P. and Valent, B. S. (1974). *Pl. Physiol.* **53**, 684-687.

Albersheim, P., Jones, T. M. and English, P. P. (1969). *A. Rev. Phytopathol.* **7**, 171-194.

Allen, A. K., Neuberger, A. and Sharon, N. (1973). *Biochem. J.*, **131**, 155-162.

Anderson, A. J. and Albersheim, P. (1972). *Physiol. Pl. Pathol.*, **2**, 339-346.

Ankel, H., Chang, C., Garriot, B., Chevalier, M. J. and Robert, M. (1973). *Proc. Natn. Acad. Sci. U.S.A.*, **70**, 2360-2363.

Bailey, J. A. and Deverall, B. J. (1971). *Physiol. Pl. Pathol.* **1**, 435-449.

Ballou, C. E. and Raschke, W. C. (1974). *Science, N.Y.* **184**, 127-134.

Barksdale, L. and Arden, S. B. (1974). *A. Rev. Microbiol.* **28**, 265-299.

Bartholomew, B. A., Jourdain, G. W. and Roseman, S. (1973). *J. biol. Chem.* **248**, 5751-5762.

Bennet, H. S. (1963). *J. Histochem. Cytochem.* **11**, 14-23.

Berlin, R. D. (1972). *Nature, Lond.* **235**, 44-45.

Besancon, F. and Ankel, H. (1974). *Nature, Lond.* **250**, 784-786.

Bohlool, B. B. and Schmidt, E. L. (1974). *Science, N.Y.* **185**, 269-271.

Bonner, J. T. (1967). "The Cellular Slime Moulds", 2nd Edn. Princeton University Press, Princeton, New Jersey.

Bowles, D. J. and Kauss, H. (1975). *Pl. Sci. Lett.* **4**, 411-418.

Boyd, W. C. (1963). *Vox Sang.* **8**, 1-32.

Boyd, W. C. (1970). *Ann. N.Y. Acad. Sci.* **169**, 168-196.

Bretscher, M. S. (1971). *Nature, Lond.* **231**, 229-232.

Bretscher, M. S. (1973). *Science, N.Y.* **181**, 622-629.

Brian, P. W. (1966). *Trans. Br. mycol. Soc.* **49**, 3-9.

Brian, P. W. (1973). *Proc. R. Soc. B.* **168**, 101-118.

Britten, R. J. and Davidson, E. H. (1969). *Science, N.Y.* **165**, 349-357.

Brock, T. D. (1959). *J. Bact.* **90**, 1019-1025.

Brock, T. D. (1965). *Proc. Natn. Acad. Sci. U.S.A.* **54**, 1104-1112.

Brown, R. C., Bass, H. and Coombes, J. P. (1975). *Nature, Lond.* **254**, 434-435.

Brucher, D., Wecksler, M., Levy, A., Palozzo, A. and Jaffé, W. G. (1969). *Phytochemistry* **8**, 1739-1743.

Burger, M. M. and Goldberg, A. R. (1967). *Proc. Natn. Acad. Sci. U.S.A.* **57**, 359-366.

Burnet, F. M. (1959). "The Clonal Selection Theory of Acquired Immunity". Cambridge University Press, London and New York.

Callow, J. A. (1975). *Curr. Adv. Pl. Sci.* **7**, 181-193.

Crandall, M. A. and Brock, T. D. (1968). *Bact. Rev.* **32**, 139-163.

Crandall, M. A., Lawrence, L. M. and Saunders, R. M. (1974). *Proc. Natn. Acad. Sci. U.S.A.* **71**, 26-29.

Cuatracasas, P. (1969). *Proc. Natn. Acad. Sci. U.S.A.* **63**, 450-457.

Cuatracasas, P. and Tell, A. P. E. (1973). *Proc. Natn. Acad. Sci. U.S.A.* **70**, 485.

Danielli, J. F. and Davson, H. (1935). *J. Cell Physiol.* **5**, 495-508.

Day, P. R. (1974). "Genetics of Host-Parasite Interaction", W. H. Freeman and Co., San Francisco.

Dekhuizen, H. M. and van der Scheer, C. (1969). *Neth. J. Pl. Pathol.* **75**, 169-177.

Deppert, W., Werchau, H. and Walter, G. (1974). *Proc. Natn. Acad. Sci. U.S.A.* **71**, 3068-3072.

Ellingboe, A. H. (1972). *Phytopathology* **62**, 401-406.

Ensgrabcr, A. (1958). *Ber. dt. bot. Ges.* **71**, 349-361.

Flor, H. H. (1956). *Adv. Genet.* **8**, 29-54.

Flor, H. H. (1971). *A. Rev. Phytopath.* **9**, 275-296.

Gander, J. E. (1974). *A. Rev. Microbiol.* **28**, 103-119.

Garfield, S., Hausman, R. E. and Moscona, A. A. (1974). *Cell Differentiation* **3**, 215-219.

Goldschneider, I. and Moscona, A. A. (1972). *Cell Biol.* **53**, 435-449.

Goldstein, I. J., Reichert, C. M. and Misaki, A. (1974). *Ann. N.Y. Acad. Sci.* **234**, 283-295.

Gorter, E. and Grendel, F. (1925). *J. exp. Med.* **41**, 439-443.

Gottschalk, A. (1966). *In* "Glycoproteins" (Ed. A. Gottschalk), pp. 543-547. Elsevier Publishing Co., Amsterdam.

Guidotti, G. (1972). *A. Rev. Biochem.* **41**, 731-752.

Hadwiger, L. A. and Schwochau, M. E. (1969). *Phytopathology,* **59**, 223-227.

Hamblin, J. and Kent, S. P. (1973). *Nature, Lond.* **245**, 28-30.

Hausman, R. E. and Moscona, A. A. (1973). *Proc. Natn. Acad. Sci. U.S.A.* **70**, 3111-3114.

Heath, M. C. (1974). *Physiol. Pl. Pathol.* **4**, 403-414.

Henkart, S., Humphreys, S. and Humphreys, T. (1973). *Biochemistry, N.Y.* **12**, 3045-3050.

Heslop-Harrison, J. (1975). *A. Rev. Pl. Physiol.* **26**, 403-426.

Heslop-Harrison, J., Heslop-Harrison, Y., Knox, R. B. and Howlett, B. (1973). *Ann. Bot.* **37**, 403-412.

Hijwegen, T. (1963). *Neth. J. Pl. Pathol.* **69**, 314-319.

Hooker, A. L. and Saxena, K. M. S. (1971). *A. Rev. Genet.* **5**, 407-424.

Horejsi, V. and Kocourek, J. (1974). *Experientia* **30**, 1348-1349.

Howard, I. K., Sage, H. J. and Horton, C. B. (1972). *Archs Biochem. Biophys.* **149**, 323-326.

Inbar, M. and Sachs, L. (1969). *Proc. Natn. Acad. Sci. U.S.A.* **63**, 1418-1425.

Ingham, J. L. (1972). *Bot. Rev.* **38**, 343-424.

Jones, D. A. (1964). *Heredity, Lond.* **19**, 459-469.

Kauffman, F. (1972). "Sociological Diagnosis of *Salmonella* Species. Kauffman-White Schema". Williams and Wilkins Co., Baltimore.

Kauss, H. and Glaser, C. (1974). *FEBS Let.* **45**, 304.

Keen, N. T. (1971). *Physiol. Pl. Pathol.* **1**, 265-275.

Keen, N. T., Partridge, J. E. and Zaki, A. I. (1972). *Phytopathology.* **62**, 768.

Kerr, A. and Htay, K. (1974). *Physiol. Pl. Pathol.* **4**, 37-44.

Kiraly, Z., Barna, B. and Ersek, T. (1972). *Nature, Lond.* **239**, 456-458.

Kitizawa, K., Inagaki, H. and Tomiyama, K. (1973). *Phytopath. Z.* **76**, 80-86.

Klarman, W. L. (1968). *Neth. J. Pl. Pathol.* **74** (Suppl. 1), 171-175.

Klarman, W. L. and Gerdemann, J. (1963). *Phytopathology* **53**, 1317-1320.

Kleinschuster, S. J. and Baker, R. (1974). *Phytopathology* **64**, 394-399.

Kleinschuster, S. J. and Moscona, A. A. (1972). *Exp. Cell Res.* **70**, 397-410.

Knox, R. B. and Heslop-Harrison, J. (1971). *J. Cell Sci.* **9**, 238-251.

Knox, R. B., Willing, R. R. and Ashford, A. G. (1973). *Nature, Lond.* **237**, 381-438.

Kornfeld, S., Adair, W. L., Gotlieb, C. and Kornfeld, R. (1974). *In* "Biology and Chemistry of Eukaryotic Cell Surfaces" (Eds. E. Y. C. and E. E. Smith), pp. 291-316. Academic Press, London and New York.

Kornfeld, R. and Kornfeld, S. (1974). *Ann. N.Y. Acad. Sci.* **234**, 276-281.

Koscielak, J. (1963). *Biochim. biophys. Acta* **78**, 313-328.

Krupe, M. (1956). "Blutgruppen spezifische pflanzenliche Eiweiskorper (Phyto-agglutinine)". Enke, Stuttgart.

Kuć, J. (1966). *A. Rev. Microbiol.* **20**, 337-370.

Kuć, J. (1972). *A. Rev. Phytopathol.* **10**, 207-232.

Kuo, M., Yoder, O. C. and Scheffer, R. P. (1970). *Phytopathology* **60**, 365-368.

Leath, K. T. and Rowell, J. B. (1970). *Phytopathology* **60**, 1097-1100.

Leblond, C. P. and Bennett, G. (1974). *In*: "The Cell Surface in Development" (Ed. A. A. Moscona), pp. 29-50, John Wiley and Sons, New York and London.

Lewis, D. H. (1973). *Biol. Rev.* **48**, 260-278.

Lis, H. and Sharon, N. (1973). *A. Rev. Biochem.* **42**, 541-574.

Lockhart-Ewart, R. B., Kornfeld, S. and Kipnis, D. M. (1973). *Clin. Res.* **21**, 622- .

Long, D. E., Fung, A. K., McGee, E. E., Cooke, R. C. and Lewis, D. M. (1975). *New Phytol.* **74**, 173-182.

Losick, R. and Robbins, P. W. (1967). *J. molec. Biol.* **30**, 445-455.

Luderitz, O., Staub, A. M. and Westphal, O. (1966). *Bact. Rev.* **30**, 192-255.

Luke, H. H. and Gracen, V. E. (1972). *In*: "Microbial Toxins" (Eds S. S. Kadir, A. Ciegler and S. J. Ajil), pp. 139-168. Academic Press, New York and London.

Maclean, D., Sargent, J. A., Tommerup, I. C. and Ingram, D. S. (1974). *Nature, Lond.* **249**, 186-187.

McLean, R. J. and Bosmann, H. B. (1975). *Proc. Natn. Acad. Sci. U.S.A.* **72**, 310-313.

McLean, R. J., Laurendi, C. J. and Braun, R. M. (1974). *Proc. Natn. Acad. Sci. U.S.A.* **71**, 2610-2613.

McClay, D. R. and Moscona, A. A. (1974). Cited in Moscona, A. A. (1974). *In*: "The cell surface in Developments" (Ed. A. A. Moscona), pp. 67-100. John Wiley and Sons, New York and London.

Mansfield, J. W., Hargreaves, J. A. and Boyle, F. C. (1974). *Nature, Lond.* **252**, 316-317.

Matsson, O., Knox, R. B., Heslop-Harrison, J. and Heslop-Harrison, Y. (1974). *Nature, Lond.* **247**, 298-299.

Matta, A. (1971). *A. Rev. Phytopathol.* **9**, 387-410.

Mercer, P. C., Wood, R. K. S. and Greenwood, A. D. (1974). *Physiol. Pl. Pathol.* **4**, 291-306.

Mialonier, A., Privat, J. P., Monsigny, M., Kahlem, G. and Durand, R. (1973). *Physiol. Veg.* **11**, 519-537.
Mirelman, D., Galun, E., Sharon, N. and Lotan, R. (1975). *Nature, Lond.* **256**, 414-415.
Moscona, A. A. (1962). *Int. Rev. exp. Pathol.* **1**, 371-529.
Moscona, A. A. (1963). *Proc. Natn. Acad. Sci. U.S.A.,* **49**, 724-747.
Moscona, A. A. (1968). *Devel.*
Moscona, A. A. (1974). *In:* "The Cell Surface in Development" (Ed. A. A. Moscona), pp. 67-100. John Wiley and Sons, New York and London.
Müller, K. O. (1950). *Nature, Lond.* **166**, 392-395.
Müller, K. O. and Borger, H. (1940). *Arb. biol. BundAnst. Land-u. Forstw. Berlin* **23**, 189-231.
Munro, A. J. and Taussig, M. J. (1975). *Nature, Lond.* **256**, 103-106.
Nicolson, G. L. (1974). *Int. Rev. Cytol.* **39**, 89-190.
Nowell, P. C. (1960). *Cancer Res.* **20**, 462-466.
Oikawa, T., Nicolson, G. L. and Yanagimachi, R. (1974). *Exp. Cell Res.* **83**, 239-246.
Olsnes, S. and Pihl, A. (1973a). *Eur. J. Biochem.* **12**, 3121-3126.
Olsnes, S. and Pihl, A. (1973b). *Eur. J. Biochem.* **35**, 179-185.
Olsnes, S., Refsnes, K. and Pihl, A. (1974). *Nature, Lond.* **249**, 627-631.
Olsnes, S., Saltvedt, E. and Pihl, A. (1974). *J. biol. Chem.* **249**, 803-810.
Pegg, G. F. and Vessey, J. C. (1973). *Physiol. Pl. Pathol.* **3**, 207-222.
Person, C. O. and Mayo, G. M. E. (1974). *Can. J. Bot.* **52**, 1339-1347.
Person, C. O. (1959). *Can. J. Bot.* **37**, 1101-1130.
Pitt, D. and Coombes, C. (1969). *J. gen. Microbiol.* **56**, 321-329.
Punin, W. (1952). *Z. Naturf.* **7b**, 48-50.
Rahe, J. E., Kuć, J., Chien-Mei, Chuang, and Williams E. B. (1969). *Neth. J. Pl. Pathol.* **75**, 58-71.
Reeves, P. (1972). "The Bacteriocins". Springer-Verlag, Berlin, Heidelberg and New York.
Robbins, P. W., Bray, D., Dankert, M. and Wright, A. (1967). *Science, N.Y.* **158**, 1536-1542.
Robbins, P. W., Keller, J. M., Wright, A. and Bernstein, R. L. (1965). *J. biol. Chem.* **240**, 384-390.
Roberts, W. P. and Kerr, A. (1974). *Physiol. Pl. Pathol.* **4**, 81-91.
Rohringer, R. and Samborski, D. J. (1967). *A. Rev. Pl. Pathol.* **5**, 77-86.
Roseman, S. (1970). *Chem. Phys. Lipids* **5**, 270-297.
Roseman, S. (1974). *In* "The Cell Surface in Development" (Ed. A. A. Moscona), pp. 255-271. John Wiley and Sons, New York and London.
Rosen, S. D., Kafka, J. A., Simpson, D. L. and Barondes, S. H. (1973). *Proc. Natn. Acad. Sci. U.S.A.* **70**, 2554-2557.
Rosen, S. D., Simpson, D. L., Rose, J. E. and Barondes, S. H. (1974). *Nature, Lond.* **252**, 128-150.
Roth, S., McGuire, E. J. and Roseman, S. (1971a). *J. Cell Biol.* **51**, 525-535.
Roth, S., McGuire, E. J. and Roseman, S. (1971b). *J. Cell Biol.* **51**, 536-547.
Sabet, S. F. and Schnaitman, C. A. (1973). *J. biol. Chem.* **248**, 1797-1806.
Saint-Paul, M. (1961). *Transfusion* **4**, 3-37.
Sato, N. and Tomiyama, K. (1969). *Ann. Phytopathol. Soc. Japan* **35**, 202-207.
Sato, N., Tomiyama, K., Katsui, N. and Masmune, T. (1968). *Ann. Phytopathol. Soc. Japan* **34**, 140-147.
Sato, N., Kitazawa, K. and Tomiyama, K. (1971). *Physiol. Pl. Pathol.* **1**, 289-295.

Scheffer, R. P. and Yoder, O. C. (1972). In: "Phytotoxins in Plants Diseases", (Eds R. K. S. Wood, A. Ballio and A. Graniti), pp. 251-272. Academic Press, London and New York.

Sharon, N. and Lis, H. (1972). *Science, N.Y.* **177**, 949-959.

Simpson, D. L., Rosen, S. D. and Barondes S. H. (1974). *Biochemistry, N.Y.* **13**, 3487-3493.

Skipp, R. A. and Deverall, B. J. (1972). *Physiol. Pl. Pathol.* **2**, 357-374.

Skipp, R. A. and Samborski, D. J. (1974). *Can. J. Bot.* **52**, 1107-1115.

Stahmann, M. A., Clare, B. and Woodbury, W. (1966). *Pl. Physiol.* **41**, 1505-1512.

Staub, T., Dahmen, H. and Schwinn, F. J. (1974). *Phytopathology* **64**, 364-372.

Steiner, A. W. and Strobel, G. A. (1971). *J. biol. Chem.* **246**, 4350-4357.

Strobel, G. A. (1973a). *J. biol. Chem.* **248**, 1321-1328.

Strobel, G. A. (1973b). *Proc. Natn. Acad. Sci. U.S.A.* **70**, 1693-1696.

Strobel, G. A. (1974). *Proc. Natn. Acad. Sci. U.S.A.* **71**, 4232-4236.

Strobel, G. A. and Hapner, K. D. (1975). *Biophys. biochem. Res. Comm.* **63**, 1151-1156.

Strobel, G. A. and Hess, W. M. (1974). *Proc. Natn. Acad. Sci. U.S.A.* **71**, 1413-1417.

Strobel, G. A. and Steiner, G. W. (1972). *Physiol. Pl. Pathol.* **2**, 129-132.

Taylor, N. W. and Orton, W. L. (1968). *Archs Biochem. Biophys.* **126**, 912-921.

Thrower, L. B. (1966). *Phytopathol. Z.* **56**, 258-259.

Tollefson, D., Feagler, J. R. and Majerus, P. W. (1974). cited in Kornfeld, S. *et al.* In: "Biology and Chemistry of Enkaryotic Cell Surfaces", (Eds. E. Y. C. Lee and E. E. Smith), pp. 291-316. Academic Press, London and New York.

Tomiyama, K. (1971). *In* "Morphological and Biochemical Events in Plant-Parasite Interaction" (Eds. S. Akai and E. Ouchi), pp. 387-399. Phytopathological Society of Japan, Tokyo.

Tomiyama, K. (1973). *Ann. Phytopath. Soc. Japan* **39**, 73-78.

Tomiyama, K., Ishizaka, N., Sato, N., Masamune, T. and Katsui, N. (1968). *In* "Biochemical Regulation in Diseased Plants or Injury" (Eds. T. Hirai, Z. Hidaka and I. Uritani), pp. 287-292. Phytopathological Society, Japan, Tokyo.

Toms, G. C. and Western, A. (1971). In "Chemotaxonomy of the Leguminosae" (Eds. J. B. Harborne, D. Boulter and B. L. Turner), pp. 367-462. Academic Press, London and New York.

Turner, R. S. and Burger, M. M. (1973). *Nature, Lond.* **244**, 509-510.

Uchida, T., Robbins, P. W. and Luria, S. E. (1963). *Biochemistry, N.Y.* **2**, 663-668.

Uetaka, H., Luria, S. E. and Burrows, J. W. (1958). *Virology* **5**, 68-91.

Van der Plank, J. E. (1968). "Disease Resistance in Plants". Academic Press, New York and London.

Varns, J. and Kuć, J. (1971). *Phytopathology* **61**, 178-181.

Varns, J., Currier, W. W. and Kuć, J. (1971). *Phytopathology* **61**, 968-971.

Varns, J., Kuć, J. and Williams, E. (1971). *Phytopathology* **61**, 174-177.

Watkins, W. M. (1966). *Science, N.Y.* **152**, 172-181.

Watson, I. A. (1970). *A. Rev. Phytopathol.* **8**, 209-230.

Watson, L., Knox, R. B. and Creaser, E. H. (1974). *Nature, Lond.* **249**, 574-576.

Weinbaum, G. and Burger, M. M. (1973). *Nature, Lond.* **249**, 510-512.

Weiss, L. (1969). *Int. Rev. Cytol.* **26**, 68-121.

Wiese, L. (1974). *Ann. N.Y. Acad. Sci.* **234**, 383-395.
Wiese, L. and Hayward, P. C. (1972). *Am. J. Bot.* **59**, 530-536.
Wiese, L. and Shoemaker, D. (1970). *Biol. Bull.* **138**, 88-95.
Wilson, C. L. (1973). *A. Rev. Phytopath.* **11**, 247-272.
Winzler, R. J. (1970). *Int. Rev. Cytol.* **29**, 77-125.
Wood, R. K. S. (1967). "Physiological Plant Pathology," Blackwell, Scient. Publ., Oxford.
Wood, R. K. S. (1973). *In* "Fungal Pathogenicity and the Plant's Response" (Eds. R. J. W. Byrde and C. T. Cutting). Academic Press, London and New York.
Wright, A. and Barzilai, N. (1971). *J. Bact.* **105**, 937-939.
Yen, P. H. and Ballou, C. E. (1974). *Biochemistry, N.Y.* **13**, 2420-2427.
Zahler, P. and Wiebel, E. (1968). *In*: "Protides of Biological Fluids", (Ed. H. Peeters), Vol. 16, p. 71, Pergamon Press, Oxford.

Paracoccus denitrificans Davis
(*Micrococcus denitrificans* Beijerinck)
as a Mitochondrion

P. JOHN AND F. R. WHATLEY

Botany School, University of Oxford, South Parks Road, Oxford, England

I. INTRODUCTION

There is much evidence in the literature of a structural and functional homology between bacteria and mitochondria (Nass, 1969). These similarities have provided support for the newly reintroduced endosymbiotic hypothesis of the evolutionary origin of mitochondria (Margulis, 1970), which now receives widespread acceptance (Cohen, 1970; Raven, 1970; Stanier, 1970; Flavell, 1972; Taylor, 1974; but see also Raff and Mahler, 1972).

In outline the hypothesis proposes that the mitochondrion evolved from a prokaryote similar to a present-day aerobic bacterium, which was taken up endocytotically (Stanier, 1970) by a plastid-free protoeukaryote. The prokaryote maintained itself within the host cytoplasm, obtaining a ready supply of oxidizable substrates from its host. In time, there was a progressive loss of autonomy and the prokaryote became adapted to a symbiotic existence; its proliferation became synchronized with the cell division of its host; its cell wall was lost; many of its biosynthetic capabilities were lost or taken over by the host; and its limiting membrane, the prokaryotic plasma membrane, acquired the appropriate carriers necessary for its metabolism to be integrated with the metabolism of its host.

This theory implies that the inner mitochondrial membrane, and its invaginations, are homologous with the plasma membrane of a present-day aerobic bacterium and that the outer mitochondrial membrane is derived from the protoeukaryote, its original function being to contain and isolate physiologically the endosymbiotic prokaryote.

An evolutionary origin of mitochondria such as that outlined above is supported by the essentially prokaryotic rather than eukaryotic nature of the nucleic acid metabolism and protein synthesis mechanism by which mitochondria maintain their semiautonomous existence in the cytoplasm. Bacteria and mitochondria resemble each other, and differ from the extramitochondrial components of the eukaryotic cell, in a number of ways. Both bacterial and mitochondrial DNA is devoid of histone, whereas nuclear DNA is characteristically closely associated with histone. Both bacterial and mitochondrial protein synthesis resemble each other, and differ from eukaryotic helix (see Nass, 1969; Stanier, 1970). Bacterial protein synthesis and mitochondrial protein synthesis resemble each other, and differ from eukaryotic cytoribosomal protein synthesis, in being initiated by formylmethionyl tRNA, in being carried out on 70s ribosomes, and in being insensitive to cycloheximide but sensitive to chloramphenicol (Ashwell and Work, 1970; Schnepf and Brown, 1971).

Many examples of bacteria existing, supposedly symbiotically, within the cytoplasm of eukaryotes are to be found today, appropriately in the least differentiated of eukaryotes, the amoebae (see reviews by Chapman-Andresen, 1971; Daniels, 1973). In most species of amoebae, e.g. in *Amoeba proteus*, the bacteria exist alongside fully developed mitochondria. However, the giant amoeba *Pelomyxa palustris* presents us with a most interesting situation, which appears to be of significance to the postulated evolutionary origin of mitochondria. This large, herbivorous, multinucleate amoeba lacks mitochondria but always contains within its cytoplasm large numbers of well-defined bacteria (Figs 1 and 2), which are seen as membrane-bound rods or cocci about 3 μm long and 0.3 μm in diameter encircling the nuclei and glycogen bodies (Daniels *et al.*, 1966; Andresen *et al.*, 1968; Bovee and Jahn, 1973). *Pelomyxa palustris*

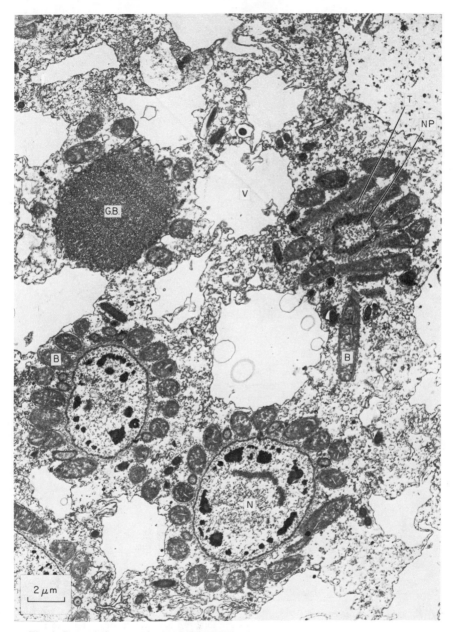

Fig. 1. Part of the cytoplasm of the multinucleate amoeba *Pelomyxa palustris*, showing bacteria surrounding nuclei and glycogen body. A glancing section of one nucleus shows a network of tubuli linking the nucleus with the vacuoles enveloping the bacteria. Fixed in glutaraldehyde stained with OsO$_4$ and post-stained with lead citrate. B, bacterium; GB, glycogen body, N, nucleus; NP, nuclear pore; T, tubuli; V, vacuole. (Courtesy of Dr J. M. Whatley.)

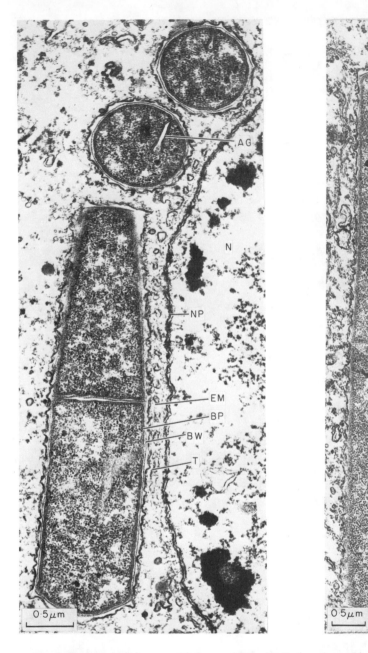

Fig. 2. Bacteria in *Pelomyxa*. (a) Cross- and longitudinal sections of bacteria adjacent to a nucleus. The longitudinal section shows a bacterium dividing. The cross-sections show the presence of axial grooves, a characteristic feature of the large bacteria of *Pelomyxa*. (b) A longitudinal section of a dividing bacterium. The characteristic axial groove is present in both daughter cells. Fixation and staining as described under Fig. 1. AG, axial groove; BP, bacterial plasma membrane; BW, bacterial cell wall; EM, enveloping membrane; N, nucleus; NP, nuclear pore; T, tubuli. (Courtesy of Dr J. M. Whatley.)

also lacks Golgi membranes, endoplasmic reticulum and any "9 + 2" organelle (centriole, basal body, centrosome or flagellum). Thus it appears to be one of the most primitive of present-day amoebae (Bovee and Jahn, 1973). *Pelomyxa palustris* cannot tolerate long periods of anaerobiosis, oxygen being necessary for its continued existence (Andresen *et al.*, 1968). Lactic acid and ATP are produced glycolytically, and presumably the amoeba supports itself anaerobically by a fermentative metabolism (Leiner *et al.*, 1968). Under aerobic conditions there is a low rate of oxygen uptake which is sensitive to carbon monoxide, cyanide, azide and to 2-*n*-heptyl-4-hydroxyquinoline-*N*-oxide but is insensitive to antimycin A. This inhibitor sensitivity, combined with fractionation studies, indicates that the bacteria are responsible for the observed respiration (Leiner *et al.*, 1968). At present, the degree of interdependence of the bacterium and its host is unknown, and we can only speculate on the stage in the postulated evolution of the mitochondrion which is represented by this present-day arrangement. The possession by the bacterium of a cell wall and the containment of the bacterium within a membrane vesicle suggest that the bacterium is physiologically external to the amoeba; the metabolic exchange between bacterium and amoeba is far less than that which occurs between a mitochondrion and its eukaryotic "host". If it be assumed that the bacterium does not supply the amoeba with ATP (the principal export of the mitochondrion) then a possible explanation of the requirement of the amoeba for oxygen lies in the ability of the bacterium, by virtue of its respiratory chain, to oxidize completely fermentation products such as lactic acid, which may otherwise accumulate to toxic levels in such a relatively large and undifferentiated cell. It is possible that herein lies an explanation of the selective advantage which was gained originally by the protoeukaryote when it first maintained within its cytoplasm the aerobic prokaryotic ancestor of the mitochondrion. It was only with the later change in the permeability of the enveloping eukaryotic vesicle, to become the freely permeable outer mitochondrial membrane, and the acquisition of the appropriate carriers in the prokaryotic plasma membrane, to become the inner mitochondrial membrane, that the incipient mitochondrion would have been able to supply its host with ATP and generally integrate its metabolism more fully with that of the host.

There are already a number of excellent accounts of the similarities between mitochondria and bacteria in general (Nass, 1969; Cohen, 1970; Margulis, 1970; Raven 1970; Stanier, 1970; Schnepf and Brown, 1971; Flavell, 1972). These accounts stress the important similarities exhibited by the nucleic acid metabolism and mechanisms of protein synthesis of mitochondria and bacteria already mentioned. However, the main function of the mitochondrion in the eukaryotic cell is the synthesis of ATP by oxidative phosphorylation. In the present essay we propose to compare the respiration and energy transduction processes in a mitochondrion and in a particular bacterium, *Paracoccus denitrificans* (*Micrococcus denitrificans*). This bacterium, which appears in all previous biochemical literature as *Micrococcus denitrificans*, has been renamed by Davis *et al*., (1969) because it stains as a Gram-negative organism (unlike typical members of the genus *Micrococcus*) and cannot be assigned to another genus of Gram-negative aerobic cocci (for example *Neisseria*) on account of its nutrition, physiology and DNA base composition. We have therefore adopted the name *Paracoccus* in this article.

It has already been noted (Imai *et al*., 1967; Scholes and Smith, 1968a, b) that the respiratory chain of *Paracoccus denitrificans* more closely resembles the respiratory chain of mammalian mitochondria than does the respiratory chain of other bacteria studied so far. This generalized similarity has been confirmed and extended by recent, more detailed studies of the amino acid composition (Scholes *et al*., 1971) and the X-ray crystal structure (Timkovich and Dickerson, 1972, 1973) of a purified cytochrome *c* isolated from *P. denitrificans*; and by studies of the energy transduction processes both in intact cells of *P. denitrificans* (Scholes and Mitchell, 1970a, b) and in membrane preparations derived from the plasma membrane (John and Whatley, 1970; John and Hamilton, 1970, 1971; John, 1973).

In this essay it will be demonstrated that there is no *single* feature of *Paracoccus denitrificans* which sets it apart from other bacteria in its resemblance to a mitochondrion, in the way that the ability, unique among prokaryotes, of the cyanophytes to use water as an electron donor in photosynthesis places them in a special position with respect to the evolutionary origin of the chloroplast. But rather, it will be demonstrated that when the widest range of characters permitted by our present knowledge is examined, we find that *P. denitrificans* occupies a unique position among bacteria in that it collects in a single bacterium those features of the mitochondrion which are otherwise distributed randomly among a wide variety of aerobic bacteria. We have summarized elsewhere (John and Whatley, 1975) the comparison of *P. denitrificans* with a mitochondrion which is presented in detail in this essay, and have also briefly discussed the evolutionary significance of the similarity revealed by this comparison.

Bacterial and mitochondrial oxidative phosphorylation and related phenomena will be discussed within the theoretical and experimental framework of the chemiosmotic hypothesis (Mitchell, 1966). We appreciate that the mechanism of oxidative phosphorylation and of active transport in bacteria remain the subject of considerable debate by the leading schools of study in these fields (Slater, 1971; Chance, 1972; Kaback, 1974). However, the chemiosmotic hypothesis has the following features which make it attractive as a currently acceptable working hypothesis.

1. Its four main postulates have been observed experimentally to be features of membranes capable of supporting oxidative phosphorylation (Mitchell and Moyle, 1967a).

2. It provides for a simple energetic integration of oxidative phosphorylation and active transport (Mitchell, 1970; Harold, 1972, 1973; Hamilton, 1975).

3. It has received experimental support from the widest possible range of naturally occurring respiratory and photosynthetic systems, and also from model and synthetic membrane systems capable of energy transduction (Jagendorf, 1967; Greville, 1969; Skulachev, 1971, 1972; Harold, 1972, 1973; Racker, 1972; Hinkle, 1973; Oesterhelt and Stoeckenius, 1973).

II. *PARACOCCUS DENITRIFICANS*

Paracoccus denitrificans is an aerobic, non-spore forming immotile coccus or short rod of about 1 μm diameter, commonly isolated from the soil (Verhoeven, 1957; Vogt, 1965; Kocur *et al.*, 1968) (Fig. 3). Although *P. denitrificans* stains Gram-negatively, actively growing cells are as sensitive to osmotic lysis after lysozyme treatment as are Gram-positive cells (Scholes and Smith, 1968a). However, treatment with lysozyme causes no difference in the appearance of the cell wall when thin sections are examined under the electron microscope (Scholes and Smith, 1968a). Electron micrographs of thin sections (Fig. 3) reveal a layered structure to the cell wall (Kocur *et al.*, 1968; Scholes and Smith, 1968a) which is more characteristic of Gram-negative bacteria such as *Escherichia coli* than of Gram-positive bacteria such as *Micrococcus lysodeikticus.*

Paracoccus denitrificans is unable to ferment but it can adapt to grow under anaerobic conditions if it is provided with nitrate. The nitrate is used as a terminal electron acceptor alternative to oxygen and is reduced to gaseous nitrogen via nitrite and nitrous oxide, both of which can also be used as terminal electron acceptors to support anaerobic growth (see Kluyver, 1956). When nitrate is used instead of oxygen the process is known as "nitrate respiration" and the conversion of nitrate-N to gaseous N_2 is known as "denitrification".

Cells grown anaerobically with nitrate as added terminal electron acceptor are able to use oxygen immediately it becomes available since the oxidase is a constitutive feature of the cell (Kluyver and Verhoeven, 1954b). When cells which have adapted to use nitrate are provided with both nitrate and oxygen they use oxygen in preference to nitrate; only when oxygen is not available do they carry out nitrate respiration (Pichinoty and d'Ornano, 1961). A clear demonstration of this preference for oxygen is provided in Fig. 4, where it can be seen that the rate of oxygen uptake is not affected by the presence of nitrate, whereas nitrate respiration proceeds only in the absence of oxygen. The nature of the control mechanism by which reducing equivalents are directed to oxygen rather than to nitrate is unknown, but it is pertinent to note that this control is inoperative when the cells are broken, since it has been observed (Lam and

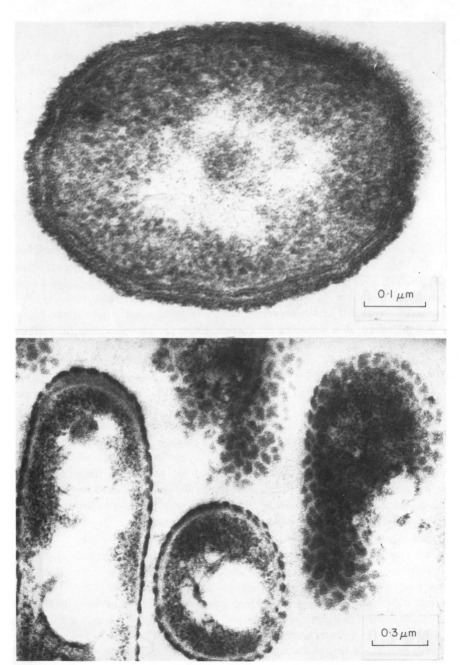

Fig. 3. Electron micrographs of thin sections of *Paracoccus denitrificans*. Cells grown to the logarithmic phase with succinate as substrate and nitrate as terminal electron acceptor were fixed either in (a) OsO_4 or in (b) glutaraldehyde. In both cases the sections were stained with OsO_4 and post stained with uranyl acetate and lead citrate. In (a) the plasma membrane is seen as a distinct layer underlying the cell wall. In (b) the plasma membrane is less distinct and the cell wall takes on a patterned appearance (cf. Scholes and Smith, 1968a). (Courtesy of Dr J. M. Whatley.)

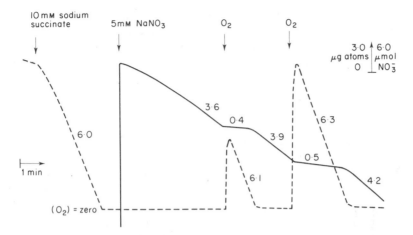

Fig. 4. Oxygen and nitrate as alternative terminal electron acceptors for respiration in *Paracoccus denitrificans*. Cells grown with succinate as substrate and nitrate as terminal electron acceptor were washed with water and added to a closed reaction vessel containing 10 mM potassium phosphate (pH 8) and 2 μl of catalase (Sigma C-100). The total volume was 30 ml and the temperature was 30°C. The reaction vessel was equipped with a nitrate-specific electrode (Orion Research Inc., Cambridge, Mass., U.S.A.) and a Clark-type oxygen electrode. Oxygen was reintroduced into the vessel by adding small aliquots (5 and 10 μl) of hydrogen peroxide solution (6%). The numbers alongside the traces refer to the rates of oxygen uptake (broken line) expressed as μg atoms/min, and to the rates of nitrate disappearance (solid line) expressed as μmol/min, respectively.

Nicholas, 1969a) that nitrate respiration catalysed by cell-free extracts of *Paracoccus denitrificans* can occur in the presence of oxygen. Cells grown with adequate aeration in the presence or absence of nitrate are able to use nitrate as a terminal electron acceptor only after a period of adaptation under anaerobic conditions, the synthesis of the nitrate reductase being repressed by oxygen (Pichinoty, 1965).

The adaptibility of *Paracoccus denitrificans* with respect to the terminal electron acceptor used is paralleled by its ability to use a wide variety of carbon and energy sources. Davis *et al.* (1969) found that, of a total of 143 different organic compounds tested, 64 could be utilized as the sole source of carbon. These carbon sources range from the C_1 compound formate (Harms, 1969) to the C_{12} compound sucrose (Bovell, 1957; Davis *et al.*, 1969).

In *Paracoccus denitrificans* glucose is broken down, not by glycolysis, but rather via the Entner-Doudoroff or pentose-phosphate pathways or by a combination of these two latter pathways (Forget, 1968). Two key enzymes of the glycolytic sequence, phosphofructokinase and aldolase, are absent from the organism, or present at only low levels (Forget, 1968; Slabas and Whatley, 1974). However the key enzymes of the Entner-Doudoroff pathway, 6-phosphogluconate   dehydratase   and   2-keto-3-deoxy-6-phosphogluconate aldolase and the key enzymes of the pentose phosphate pathway, glucose-6-phosphate dehydrogenase, 6-phosphogluconate dehydrogenase and transketolase have been detected. Cells grown on glucose show increased levels of these enzymes compared to cells grown with either succinate or acetate as substrate (Forget, 1968). When glucose-6-phosphate labelled at specific carbons is supplied to intact cells, the subsequent distribution of label is compatible with the operation of the Entner-Doudoroff pathway and incompatible with the operation of glycolysis (Slabas and Whatley, 1974).

Pyruvate formed by the breakdown of glucose, or succinate supplied in the growth medium, are oxidized to carbon dioxide via the tricarboxylic acid cycle, which has been shown to operate in *Paracoccus denitrificans* (Forget and Pichinoty, 1965). When acetate is the sole carbon and energy source it is metabolized by the glyoxylate cycle (Kornberg *et al.*, 1960). Addition of the "preferred" substrate, succinate, to cells utilizing acetate results in a depression of the synthesis of isocitratase, a key enzyme of the glyoxylate cycle. Succinate also inhibits the activity of isocitratase (Kornberg *et al.*, 1960). When *P. denitrificans* is grown on glycolate as sole carbon and energy source a metabolic sequence operates, the β-hydroxyaspartate pathway, which is apparently unique to this bacterium (Kornberg and Morris, 1965). Glycolate is oxidized to glyoxylate which then combines with glycine, formed by the reductive amination of some of the glyoxylate, giving rise to β-hydroxyaspartate which, upon deamination to oxaloacetate, is available for entry into the tricarboxylic acid cycle. This pathway effects the direct condensation of two $2C_2$ compounds to a $4C_4$ dicarboxylic acid. *Paracoccus denitrificans* can also utilize formate as the sole source of carbon and energy. Harms (1969) has provided evidence that the carbon requirement of cells grown with formate is met by the fixation of carbon dioxide rather than by the direct incorporation of formate. In the absence of an organic source of carbon *P. denitrificans* can grow autotrophically with hydrogen as reductant and either oxygen or nitrate as the terminal electron acceptor (Kluyver and Verhoeven, 1954a), carbon dioxide being fixed by the operation of the reductive pentose cycle (Kornberg *et al.*, 1960).

The main intracellular reserve material in *Paracoccus denitrificans* is poly-β-hydroxybutyric acid (Vogt, 1965; Scholes and Smith, 1968a). When growth is limited by the availability of nitrogen or phosphorus this polymer can constitute 60% of the cell dry weight (Vogt, 1965). Presumably poly-β-hydroxybutyrate acts as a reserve of fixed carbon and of energy; the latter being released by the oxidative breakdown of the polymer. Phosphate is stored as polyphosphate (Kaltwasser *et al.*, 1962).

III. A MITOCHONDRION

Mitochondria are found in the cells of all aerobic eukaryotes (Lehninger, 1964). They are bounded by two membranes: an inner membrane which is periodically invaginated to form cristae, and an enveloping outer membrane (Fig. 5). No specific function has been ascribed to the cristae and they appear to be simply devices for increasing the surface area of the inner membrane. The mitochondria of different organisms vary in size and morphology, as also do the mitochondria of different tissues and organs of higher plants and animals. Because of the relative ease of isolation and the high yields of mitochondria which are obtained, mammalian liver has proved to be the most useful source of mitochondria for metabolic studies. In thin sections these mitochondria appear to be about 3 μm long and 1 μm in diameter, and the cristae have the appearance of simple stacked membranes intruding into the internal aqueous phase, the matrix. In tissues where the respiration rate is high, such as mammalian heart tissue and insect flight muscle, the cristae are correspondingly more extensive and can often be seen almost completely to fill the matrix.

Undoubtedly the principal function of the mitochondrion is to carry out cell respiration and coupled oxidative phosphorylation. Cell fractionation studies (see Roodyn, 1967) have revealed that the mitochondrion is the only cell component which contains the entire respiratory chain and the complete complement of enzymes responsible for the operation of the tricarboxylic acid cycle. The respiratory chain consists of NADH and succinate dehydrogenases,

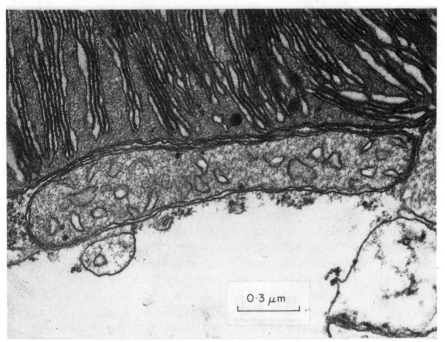

Fig. 5. Electron micrograph of a thin section of a mitochondrion from a microphyll of *Selaginella apus*. The sections were fixed in glutaraldehyde and stained as in Fig. 3b. (Courtesy of Dr J. M. Whatley.)

non-haem iron, flavoprotein, quinone and cytochromes of the *b*, *c* and *a* type. It is located on the inner mitochondrial membrane, where the transhydrogenase and ATPase are also to be found. A membrane-bound oxidation-reduction system, NADH-cytochrome *c* reductase, bearing a superficial resemblance to an abbreviated respiratory chain is present on the outer mitochondrial membrane and on the endoplasmic reticulum (see Ernster and Kuylenstierna, 1970). It differs from the analogous portion of the respiratory chain in being insensitive to the respiratory inhibitors, rotenone, amytal and antimycin A, in lacking an associated ATP synthesis, and in lacking a terminal cytochrome oxidase (see Ernster and Kuylenstierna, 1970).

Although the complete complement of tricarboxylic acid cycle enzymes is found only in the mitochondrial matrix, some of the individual enzymes involved, such as fumarase, isocitrate dehydrogenase and aconitase, are also to be found in the "soluble" fraction of disrupted cells (Roodyn, 1967), and it appears that it is the specifically mitochondrial localization of the succinate and α-ketoglutarate dehydrogenases which limit the operation of the tricarboxylic acid cycle to the mitochondrion (Roodyn, 1967).

The range of compounds which can be utilized as sources of energy and carbon is far more restricted for a mitochondrion than for a bacterium. Thus, while *Paracoccus denitrificans* is capable of utilizing at least 64 different organic compounds as sole sources of energy and carbon (Davis *et al.*, 1969), a mitochondrion is capable of utilizing only those compounds which are either fed directly into the tricarboxylic acid cycle (e.g. succinate, malate) or are readily converted to intermediates of the tricarboxylic acid cycle (e.g. glutamate and pyruvate, which are converted in the mitochondrial matrix by single reactions to α-ketoglutarate and acetyl CoA, respectively).

The particular respiratory "substrate" of the mitochondrion depends upon the nutritional status of the cell in which it is found. In those cells where carbohydrate is an energy source, the mitochondrion is furnished with pyruvate from the glycolytic breakdown of hexose in the cytosol, the mitochondrion itself being incapable of metabolizing sugars. Upon oxidative decarboxylation the pyruvate is converted to acetyl CoA which enters the tricarboxylic acid cycle, the operation of which results in the reduction of NAD^+ to NADH and the reduction of succinic dehydrogenase flavoprotein. The transfer of reducing equivalents from these reduced coenzymes to oxygen is mediated by the respiratory chain in a stepwise manner with the concomitant formation of ATP from ADP, by the process of oxidative phosphorylation (see Lehninger, 1964). In those cells where lipid is metabolized as an energy source, the mitochondria take up long-chain fatty acids and degrade them by a series of β-oxidation reactions to acetyl CoA units which are then fed into the tricarboxylic acid cycle. As well as the enzymes of the tricarboxylic acid cycle and β oxidation pathway, the mitochondrial matrix contains enzymes involved in the metabolism of amino acids and nucleic acids (Roodyn, 1967; Borst, 1972).

The function of the outer mitochondrial membrane is not clear (see Ernster and Kuylenstierna, 1970). Enzymatically it resembles the endoplasmic reticulum. In rat liver mitochondria the most conspicuous activities associated with the outer mitochondrial membrane are monoamine oxidase and NADH-cytochrome c reductase. It is freely permeable to small molecules such as nucleotides, sugars and salts, although it constitutes a barrier to large molecules, such as albumin, with a molecular weight exceeding 10 000. By contrast the inner mitochondrial membrane is permeable only to small uncharged molecules and to those metabolites for which a carrier system is present. Thus it is the inner mitochondrial membrane which constitutes the physiological barrier separating the mitochondrial interior from the cytosol.

Carriers are present in the inner mitochondrial membrane, which allow for the uptake of fatty acids, phosphate and pyruvate (Meijer and Van Dam, 1974). There also exists a carrier, the adenine nucleotide carrier, which allows an equimolar exchange of ADP for ATP across the inner mitochondrial membrane (see Klingenberg, 1970). Translocation of tricarboxylic acid cycle intermediates is mediated by carriers which allow dicarboxylate-inorganic phosphate, dicarboxylate-dicarboxylate, and dicarboxylate-tricarboxylate exchanges across the inner mitochondrial membrane (Chappell and Haarhoff, 1967).

Mitochondria isolated from different organisms, and from different tissues and organs of higher plants and animals, resemble one another in most features but differ in some others. Thus, we may note that while all functional mitochondria so far examined possess internally orientated NADH and succinate dehydrogenases, mitochondria from higher plants and lower organisms also possess an externally orientated NADH dehydrogenase which is lacking in mitochondria from higher animals (see Hughes *et al.*, 1970). It is with those mitochondrial features which are of widespread distribution that the following comparison between *Paracoccus denitrificans* and a mitochondrion is mainly concerned.

IV. CHEMICAL COMPOSITION OF THE MEMBRANES

Analyses of the plasma membrane of *Paracoccus denitrificans* reveal that about 95% of the dry weight of the plasma membrane can be accounted for by protein and lipid (Scholes and Smith, 1968b). The protein/lipid ratio of this membrane is 1 : 0.5. For guinea-pig liver mitochondria the protein/lipid ratio is 1 : 0.3 for the inner, and 1 : 0.8 for the outer membrane (Parsons *et al.*, 1967). Thus the plasma membrane of *P. denitrificans* contains relatively less protein compared to lipid than does the inner mitochondrial membrane.

The composition of the phospholipid fraction of *Paracoccus denitrificans* and of mammalian membranes is compared in Table I, which also shows the composition of the phospholipid fraction of the Gram-positive aerobe *Micrococcus lysodeikticus*, and of the Gram-negative facultative aerobe

TABLE I

The major phospholipids of bacterial and mammalian membranes.
Phospholipid contents are expressed as percentages of total phospholipid.

Phospholipid	Bacteria			Guinea-pig Liver[d]		
				Mitochondrial membranes		
	Escherichia coli[a]	*Micrococcus lysodeikticus*[b]	*Paracoccus denitrificans*[c]	Inner	Outer	Microsomes
Phosphatidyl glycerol	12	46	52	2	3	1
Phosphatidyl choline	–	–	31	45	55	63
Cardiolipin	13	39	3	22	3	–
Phosphatidyl ethanolamine	75	–	6	28	25	18
Phosphatidyl inositol	–	13	–	4	14	13
Phosphatidyl serine	–	–	–	–	–	5

[a] From Cronan (1968); data obtained with aerobically grown cells.
[b] From Whiteside et al. (1971).
[c] From Wilkinson et al. (1972).
[d] From Parsons et al. (1967).

Escherichia coli. The most significant similarity between the phospholipid composition of *P. denitrificans* and of the inner mitochondrial membrane appears to be that each type of membrane contains phosphatidyl choline as a major phospholipid component. Phosphatidyl choline is also present in the outer mitochondrial membrane and in the microsomes, but is rarely found in bacteria (see Wilkinson *et al.*, 1972). Phosphatidyl inositol is absent from *P. denitrificans* and, similarly, it is present as only a minor component of the phospholipid fraction of the inner mitochondrial membrane, although present in significantly larger amounts in the outer mitochondrial membrane and in the microsomes.

The most significant difference between the phospholipids of *Paracoccus denitrificans* and of the inner mitochondrial membrane appears to be in the distribution of cardiolipin. Among the mammalian membranes this phospholipid is effectively restricted to the inner mitochondrial membrane, where it accounts for almost 30% of the total phospholipid. In *P. denitrificans* cardiolipin constitutes only about 3% of the total phospholipid, although it is present as a major phospholipid component in *Escherichia coli* and in *Micrococcus lysodeikticus.* Phosphatidyl glycerol, the predominant phospholipid in *P. denitrificans*, is less common in mitochondrial membranes, although it is a common constituent of bacterial membranes.

Little is known of the phospholipid composition of the mitochondria of lower animals, protozoa or plants, although mitochondria isolated from potato and cauliflower have been shown to resemble mammalian mitochondria in the types and relative amounts of phospholipid present (Ben Abdelkader and Mazliak, 1970).

The fatty acid composition of the membranes of *Paracoccus denitrificans* and of mammalian mitochondrial and microsomal membranes is compared in Table II, which also shows the fatty acid composition of the Gram-positive aerobe *Micrococcus lysodeikticus* and the Gram-negative facultative aerobe *Escherichia coli.* From this comparison it is apparent that *P. denitrificans* and mammalian mitochondria are similar in possessing only straight-chain, saturated and unsaturated fatty acids, while *M. lysodeikticus* and *E. coli* possess, in addition, branched-chain or cyclopropane fatty acids. In *P. denitrificans* oleic acid (18 : 1) is the sole unsaturated fatty acid present (Girard, 1971; Wilkinson *et al.*, 1972), while in mammalian mitochondria oleic acid occurs together with a variety of longer chain mono- and polyunsaturated fatty acids (Getz *et al.*, 1962). In both *P. denitrificans* (Girard, 1971; Wilkinson *et al.*, 1972) and in mammalian mitochondria (Getz *et al.*, 1962) the saturated fatty acids are stearic (18 : 0) and palmitic (16 : 0) acids.

In mammalian liver the fatty acid composition of the inner mitochondrial membrane is not significantly different from that of the outer mitochondrial membrane (Stoffel and Schiefer, 1968) or from that of the microsomes of the same tissue (Getz *et al.*, 1962). However, variation does occur in the fatty acid composition of liver mitochondria of different animal species, and this variation

TABLE II

The major fatty acids of bacterial and mammalian membranes.
Fatty acid contents expressed as percentages of total fatty acid

| | Bacteria | | | Rat Liver[c] | |
Fatty acid	*Escherichia coli*[a]	*Micrococcus lysodeikticus*[b]	*Paracoccus denitrificans*[b]	Mitochondria	Microsomes
Straight chain, saturated, 14:0–18:0	39	7	21	34	41
Straight chain, unsaturated, 16:1–20:6	48	2	78	63	55
Branched chain, 15:0Br–17:0Br	—	87	—	—	—
Cyclopropane, 17:0V	10	—	—	—	—

[a] From Knivett and Cullen (1965); data obtained with aerobically grown cells.
[b] From Girard (1971).
[c] From Getz *et al.* (1962).

has been correlated with the diet and temperature at which the organism normally functions (see Chapman and Leslie, 1970). Variation in the fatty acid composition of plant mitochondria has been correlated with the susceptibility of the particular species to injury by freezing (Lyons *et al.*, 1964). Bacterial membranes are also able to alter their fatty acid composition and these variations can be brought about by changes in the growth phase of the bacterial culture, or by altering the growth temperature (see, for example, Cronan and Vagelos, 1972).

V. RESPIRATORY CHAIN

The essential function of the respiratory chain is to conduct reducing equivalents from relatively reducing electron donors to a relatively oxidizing terminal electron acceptor in such a way as to allow some of the free energy released to be conserved in forming the terminal phosphate bond of ATP.

In mitochondria the respiratory chain is an integral part of the inner membrane. The bacterial respiratory chain is similarly an integral part of the plasma membrane, or consists of infoldings derived from the plasma membrane such as are found particularly in autotrophic and lithotrophic bacteria (see Gel'man *et al.*, 1967). However, the infoldings of the bacterial membrane characterized as mesomes appear to be devoid of significant respiratory activity (see Reusch and Burger, 1973). Similarly the lipoprotein fraction of the cell wall of Gram-negative bacteria appears to be devoid of significant respiratory activity (White *et al.*, 1972).

Although the respiratory chains of mitochondria isolated from different tissues of the same organism or from different species are not identical, there is a far greater variety of respiratory chains among bacteria of different species or among the same bacterial species grown under different cultural conditions. This variation in the bacterial respiratory chain is manifested as a variation both in the types of respiratory components present and in their relative amounts. In the respiratory chains of bacteria, but not in those of mitochondria, it is possible to distinguish on the one hand components which are present under all conditions of growth ("constitutive" components) and, on the other hand, components which are synthesized only under certain conditions of growth ("adaptive" components). For example, in *Paracoccus denitrificans* cytochrome aa_3 occurs both in cells grown with oxygen as the terminal acceptor and in cells grown anaerobically with nitrate as the terminal electron acceptor (Scholes and Smith, 1968b). The cytochrome aa_3 has no known function in the absence of oxygen. While cytochrome aa_3 is a constitutive component of the respiratory chain of *P. denitrificans*, cytochrome *cd* is an adaptive component (Fig. 6). This latter two-haem cytochrome has both nitrite reductase and cytochrome oxidase activities, and is present in cells grown anaerobically with nitrate as the added terminal electron acceptor but is absent from cells grown aerobically in the

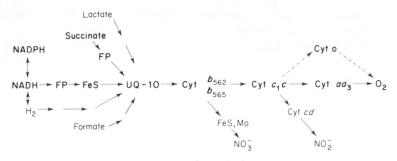

Fig. 6. The respiratory chain of a mitochondrion and of *Paracoccus denitrificans*. Components of the mitochondrial respiratory chain, and of the constitutive respiratory chain of *P. denitrificans* are in heavy print; the additional adaptive components of the respiratory chain of *P. denitrificans* are in lighter print. Cytochrome *o* occurs in the mitochondria of only a few organisms, and its significance in the constitutive respiratory chain of *P. denitrificans* is not known (see text). (After Asano *et al.* (1967b), Scholes and Smith (1968b), Lam and Nicholas (1969a, b), John and Whatley (1970), Knobloch *et al.* (1971), Forget and Dervartanian (1972), Wikström (1973).)

presence or absence of nitrate (Newton, 1969). The function of cytochrome *cd* is to reduce further the nitrite produced by nitrate reduction. Since nitrate reduction does not occur under aerobic conditions (Fig. 4), and the cells always possess an effective cytochrome oxidase in cytochrome aa_3, cytochrome *cd* would be redundant under aerobic conditions.

A. TRANSHYDROGENASE

The enzyme nicotinamide nucleotide transhydrogenase (NADPH-NAD$^+$ oxido-reductase, EC 1.6.1.1) catalyses the reversible transfer of hydrogen (formally, a pair of reducing equivalents) between the reduced forms of NAD$^+$ and NADP$^+$:

$$NADPH + NAD^+ \longleftrightarrow NADP^+ + NADH$$

In mitochondria the transhydrogenase is located on the inner mitochondrial membrane. Its activity is associated with an electrogenic translocation of protons across the inner mitochondrial membrane via the transhydrogenase: protons are taken up by the mitochondrion with the reduction of NADP$^+$ and NADH, and protons are expelled from the mitochondrion with the oxidation of NADPH by NAD$^+$ (Mitchell and Moyle, 1965; Moyle and Mitchell, 1973a). A transhydrogenase-catalysed oxidation of NADPH by NAD$^+$ has been observed to be coupled to ATP synthesis in beef heart submitochondrial particles (Van de Stadt *et al.*, 1971), and ATP hydrolysis or respiration can drive a transhydrogenase-catalysed reduction of NADP$^+$ by NADH (see Lee and Ernster, 1966). The stoichiometry of the mitochondrial transhydrogenase has been determined to be 1 mole of ATP hydrolysed per mole of NADP$^+$ reduced by NADH (Lee and Ernster, 1966); moreover 2 equivalents of H$^+$ are translocated per mole of

NADPH oxidized by NAD^+ (Moyle and Mitchell, 1973a). The equilibrium constant of the transhydrogenase reaction is about 1, the midpoint redox potential of the $NADPH/NADP^+$ couple being only about 4 mV more negative than that of the $NADH/NAD^+$ couple (Lee and Ernster, 1966). However, when no external proton motive force is applied there is an asymmetry to the catalytic capability of the transhydrogenase, so that the reduction of NAD^+ by NADPH occurs at a faster rate than the reduction of $NADP^+$ by NADH. This property has been ascribed to a conformational change in the enzyme determined by the relative concentrations of the oxidized and reduced nicotinamide nucleotides (Rydström *et al.*, 1971). When an external proton motive force is applied, either by ATP hydrolysis or by respiration, the properties of the transhydrogenase are altered so that the rate of $NADP^+$ reduction by NADH is increased (Rydström *et al.*, 1971).

A membrane-bound nicotinamide transhydrogenase activity has been detected in *Paracoccus denitrificans* (Asano *et al.*, 1967a), *Nitrosomonas europaea* (Aleem, 1966a), *Thiobacillus novellus* (Aleem, 1966b), *Mycobacterium phlei* and *Escherichia coli* (Murthy and Brodie, 1964).

The transhydrogenase of *Paracoccus denitrificans* was studied by Asano *et al.*, (1967a), using membrane particles derived from the plasma membrane of cells grown aerobically with succinate as substrate. The transhydrogenase of these particles and the transhydrogenase of beef heart submitochondrial particles are similar; both catalyse the reduction of NAD^+ by NADPH in the absence of an energy source, and the reduction of $NADP^+$ by NADH driven by the hydrolysis of ATP or by succinate respiration. The molar ratio of ATP hydrolysed per $NADP^+$ reduced was found to be about 0.6 which, considering the conditions employed, is comparable with the ratio of 1 obtained with mitochondria. The transhydrogenase of *P. denitrificans* was found to further resemble the mitochondrial and *Escherichia coli* enzyme (Murthy and Brodie, 1964) in being inhibited by thyroxine and triiodothyronine.

The transhydrogenase of *Paracoccus denitrificans* also appears to resemble the transhydrogenase of a mitochondrion in the stoichiometry of proton translocation. Scholes and Mitchell (1970b) found that intact cells of *P. denitrificans* gave an H^+/O ratio of 8 when oxidizing endogenous substrates. This ratio is higher than the H^+/O ratio obtained with mitochondria oxidizing NADH-linked substrates, which give an H^+/O ratio of 6 (Mitchell and Moyle, 1967a, b). To explain this difference, Scholes and Mitchell (1970b) suggested that NADPH accumulates during the anaerobic incubation which precedes the addition of oxygen required for the H^+/O determinations; and furthermore it was suggested that the transhydrogenase is coupled to the translocation of 2 proton equivalents per mole of NADPH oxidized. Thus the inferred stoichiometry of the transhydrogenase of *P. denitrificans* is similar to the observed stoichiometry of the mitochondrial transhydrogenase (Moyle and Mitchell, 1973a).

The role of the mitochondrial and bacterial nicotinamide nucleotide transhydrogenases *in vivo* is unknown. A role for the transhydrogenase of *E. coli* in supplying NADPH for the biosynthesis of amino acids has been suggested by the work of Bragg *et al.* (1972), who demonstrated that cells supplied with a range of amino acids in the growth medium had a lower transhydrogenase activity than similar cells grown in the absence of the amino acids; the implication is that synthesis of the transhydrogenase was repressed when the end-products of the NADPH-dependent pathways of amino acid synthesis were supplied.

B. DEHYDROGENASES

It is convenient to divide the dehydrogenases of mitochondria and bacteria into two classes: (a) soluble, NAD^+-linked dehydrogenases, which are mainly located in the mitochondrial matrix and bacterial cytosol, and function to channel reducing equivalents from a wide variety of substrates into NADH; and (b) membrane-bound, flavoprotein dehydrogenases linked directly to the respiratory chain. The latter are bound to the inner mitochondrial membrane or bacterial plasma membrane, and function in the transfer of reducing equivalents from a limited number of substrates, commonly NADH or succinate, to the rest of the respiratory chain. In mammalian mitochondria, soluble NAD^+-linked de-hydrogenase involved in the oxidation of tricarboxylic acid cycle intermediates, fatty acids and glutamate are present (see Ernster and Kuylenstierna, 1970). Little is known of the soluble dehydrogenase of *Paracoccus denitrificans*, but it can be inferred from the wide range of oxidizable substrates which can support growth that this bacterium possesses the ability to synthesize a wider variety of dehydrogenases than are found in mitochondria.

The most important of the membrane-bound, flavoprotein dehydrogenases are NADH dehydrogenase and succinate dehydrogenase, providing as they do the principal ports of entry of reducing equivalents into the respiratory chains of mitochondria and aerobic bacteria. These dehydrogenases are found in all types of mitochondria (see Singer, 1963) and, as in most aerobic bacteria, they are present as constitutive features of the respiratory chain of *Paracoccus denitrificans*. These two flavoprotein dehydrogenases have been found on plasma membrane preparations derived from cells grown with succinate, glucose, D,L-lactate, formate or hydrogen as energy sources (Imai *et al.*, 1967; Scholes and Smith 1968b; John and Whatley, 1970; Knobloch *et al.*, 1971; P. John, unpublished). Of limited distribution among mitochondria are the flavoprotein dehydrogenases choline dehydrogenase of mammalian liver mitochondria, α-glycerophosphate dehydrogenase of brain and of skeletal muscle tissues of higher animals and in insect flight muscle mitochondria, and the L-lactate and D-lactate dehydrogenases of aerobically grown yeast (see Singer, 1963; Frisell and Cronin, 1971). D- and L-lactate dehydrogenases linked to the respiratory chain of *P. denitrificans* have been detected (Pascal *et al.*, 1965; Imai *et al.*, 1967), but the presence of the other mitochondrial flavoprotein dehydrogenases, α-glycerophosphate and choline dehydrogenases, have not been determined.

Succinate-dependent respiration in membrane preparations of *Paracoccus denitrificans*, and in mitochondria, is specifically inhibited by the systemic fungicide carboxin (Tucker and Lillich, 1974). The effect of carboxin on other bacteria is not known.

The NADH dehydrogenase-ubiquinone region of the respiratory chain of *Paracoccus denitrificans* has been found to resemble closely the same span of the mitochondrial respiratory chain (Asano *et al.*, 1967b; Imai *et al.*, 1968a, b). Both respiratory chains include flavin, two sulphydryl groups, non-haem iron, and a rotenone-sensitive site. When membrane particles of *P. denitrificans* are reduced with NADH, e.s.r. spectra show a "$g = 1.94$" signal similar to that originating from the non-haem iron component of the mitochondrial respiratory chain (Imai *et al.*, 1968b). NADH-dependent respiration in *P. denitrificans*, as in a mitochondrion, is specifically inhibited by rotenone (Imai *et al.*, 1967; Scholes and Smith, 1968b), although membrane preparations of *P. denitrificans* are less sensitive than comparable mitochondrial preparations. Thus to obtain complete inhibition, 200 nmol rotenone are required per mg of *P. denitrificans* membrane protein (calculated from data of Scholes and Smith, 1968b; S. J. Ferguson and P. John, unpublished) while 0.2 nmol rotenone are required per mg of mitochondrial membrane protein (Ernster *et al.*, 1963; Burgos and Redfearn, 1965). *Paracoccus denitrificans*, like *Mycobacterium flavum*, is unusual among bacteria in being sensitive to rotenone (Erickson, 1971); most bacteria, e.g. *M. phlei* (Asano and Brodie, 1964), are quite insensitive to this inhibitor.

The NADH dehydrogenase of animal mitochondria and the succinate de-hydrogenase of all mitochondria are located on the inner mitochondrial membrane so that they are accessible to their respective substrates only from the mitochondrial matrix. In the mitochondria of higher plants (Coleman and Palmer, 1972), algae, fungi and protozoa (see Hughes *et al.*, 1970) there is an additional NADH dehydrogenase orientated towards the outer face of the inner mitochondrial membrane which permits the direct oxidation of exogenous NADH. The NADH and succinate dehydrogenases of *Paracoccus denitrificans* appear to be orientated towards the inner face of the plasma membrane (Scholes and Smith, 1968b; John and Hamilton, 1971), and there appears to be no evidence for the presence in bacteria of an externally facing NADH de-hydrogenase. Since the intermediary metabolism of a bacterium is confined by the plasma membrane to the cytosol, it would be difficult to ascribe a function to an outwardly facing NADH dehydrogenase.

In common with many other bacteria (see Smith, 1968) *Paracoccus denitri-ficans* can possess respiratory chain-linked dehydrogenases not present in mitochondria, for example, formate dehydrogenase (P. John, unpublished) and hydrogenase (Fewson and Nicholas, 1961; Knobloch *et al.*, 1971). These adaptive dehydrogenases are synthesized when the cells are presented with the appropriate substrate, and in the absence of alternative sources of reducing potential; thus glucose represses the induction of hydrogenase (Fewson and Nicholas, 1961; see also Schlegel and Eberhardt, 1972).

C. QUINONES

Ubiquinone is an essential redox component of the mitochondrial respiratory chain (see Ernster *et al.*, 1969). Its location in the chain appears to be on the substrate side of the *b* cytochromes where it forms a common pool collecting reducing equivalents from the various dehydrogenases (Klingenberg and Kröger, 1967). Chemically, ubiquinones have a 2,3-dimethoxy-5-methylbenzoquinone nucleus bearing a polyisoprenoid side-chain of varying length. Naturally occurring ubiquinones have from 6 (UQ-6) to 10 (UQ-10) isoprenoid units in the side chain (Crane, 1965). Analyses of representatives of the major phyla of eukaryotes reveal the almost universal distribution of UQ-10 as the sole quinone, the exceptions being the Protozoa, Fungi and Algae, which contain instead the lower homologues UQ-7, UQ-8 and UQ-9 (Crane, 1965).

Paracoccus denitrificans resembles the majority of mitochondria and differs from most bacteria in containing UQ-10 as its sole quinone (Imai *et al.*, 1967; Scholes and Smith, 1968b). Although the quinones of a wide variety of bacteria have been characterized (see Crane, 1965), relatively few bacteria are known to resemble *P. denitrificans* and most mitochondria in containing UQ-10. These bacteria include *Pseudomonas denitrificans* (see Crane, 1965), *Acetobacter xylinum* (Benziman and Goldhamer, 1967), *Agrobacterium tumifaciens* (Whistance *et al.*, 1969), *Achromobacter cycloclastes* (Watanuki *et al.*, 1972), *Rhodopseudomonas spheroides* and *Rhodospirillum rubrum* (Carr and Exell, 1965). By contrast, other Gram-negative bacteria contain lower homologues, usually UQ-8, as, for example, does *Azotobacter vinelandii*; while Gram-positive bacteria contain no ubiquinone, but have instead naphthoquinones, e.g. *Bacillus megaterium* (see Crane, 1965; Gel'man *et al.*, 1967). The UQ-10 content of *P. denitrificans* was determined by Scholes and Smith (1968b) to be about 0.175 μmol/g dry weight of cells, and the UQ-10 content of the isolated membranes of aerobically grown cells of *P. denitrificans* measured by Imai *et al.*, (1968a) was 2.50 μmol/mg membrane protein, a value which is similar to that obtained with fragments of beef heart mitochondria when expressed on a mg membrane-protein basis (Table III). The cytochrome *a* and *b* contents of the *P. denitrificans* membranes used by Imai *et al.* (1967, 1968a) and of comparable preparations of beef heart mitochondria are also given in Table III. In both types of membrane the molar ratios of UQ-10 to total cytochrome *b* are similar.

D. CYTOCHROME aa_3

The cytochromes, by virtue of the characteristic absorption bands of the reduced form in the visible region of the spectrum, are the most easily identified components of the respiratory chain. In membrane preparations these absorption bands are most conveniently seen in reduced *minus* oxidized difference spectra, obtained by measuring the difference in absorption between an anaerobic sample of the membrane suspension reduced with substrate (or with dithionite) and a similar but aerobic sample containing no reductant. If the

TABLE III

Ubiquinone and cytochrome contents of mitochondrial membranes and the plasma membrane of Paracoccus denitrificans

	UQ-10	Cytochromes $\dfrac{aa_3 \quad b\ \text{type}}{\text{(nmol/mg protein)}}$		$\dfrac{\text{UQ-10}}{\text{Cyt } b}$	References
		aa_3	b type		
Keilin-Hartree beef heart preparation	5	0.57	0.59	8.5 ⎫	From Klingenberg (1968)
Rat liver mitochondria	2.4	0.3	0.30	8.0 ⎭	
Paracoccus denitrificans membrane particles	2.5	0.035	0.47	5.3	Imai *et al.* (1967, 1968a)

temperature is reduced to that of liquid N_2 ($77°K$), the absorption peaks are intensified and cytochromes with closely adjacent absorption peaks can be more easily resolved. Unfortunately this procedure results in an approximately 3 nm shift in the absorption peaks towards the blue end of the spectrum, the heights of the absorption peaks are not reproducible, and more than one peak may be observed due to a splitting of the room temperature absorption peak of a single cytochrome component. An increased spectroscopic resolution of low temperature spectra has been sought by plotting fourth-order finite difference spectra (Shipp, 1972a, b). This technique has been claimed to be useful in the recognition of multiple c- and b-type cytochromes in enteric bacteria such as *Escherichia coli* (Shipp, 1972a). However, the presence of cytochrome species detected only by fourth-order difference spectra must be verified by independent techniques because of the possibility of generating spurious derivative bands by higher derivative analysis (Butler and Hopkins, 1970). For an excellent account of spectrophotometric methods which have been used to determine bacterial cytochromes the reader is referred to White and Sinclair (1971).

Cytochrome aa_3 has been found in all mitochondria so far examined, and in almost all these mitochondria it is the only cytochrome capable of reacting directly with oxygen. The presence of cytochrome aa_3 may be detected in membranes by the characteristic absorption peaks at about 605 nm and 444 nm in reduced *minus* oxidized difference spectra. Mitochondrial cytochrome aa_3 has been isolated and found to be a large single protein bearing two haem groups. Haem a_3 reacts directly with oxygen (as well as with carbon monoxide and cyanide) and haem a reacts directly with reduced cytochrome c. There has been no direct physicochemical characterization of the haem groups isolated from the protein moiety, and the above functional definitions are all we have to distinguish the two haem groups (Lemberg and Barrett, 1973).

The presence of cytochrome aa_3 in *Paracoccus denitrificans* is indicated by reduced *minus* oxidized difference spectra of membrane preparations (Imai *et al.*, 1967; Scholes and Smith, 1968b; John and Whatley, 1970; Sapshead and Wimpenny, 1972) and intact cells (Lam and Nicholas, 1969a) which show absorbance peaks at 605—610 nm in the α region. The 444 nm peak, characteristic of cytochrome aa_3 in the γ region of the spectrum, which appears as a shoulder in the very large absorbance peak in this region due to b- and c-type cytochromes, has been observed at 443—446 nm in membranes isolated from aerobically grown cells of *P. denitrificans* (Imai *et al.*, 1967; Scholes and Smith, 1968b).

While cytochrome aa_3 can easily be detected spectrophotometrically in membranes isolated from aerobically grown cells of *Paracoccus denitrificans* its presence in membranes derived from cells grown with nitrate is sometimes difficult to show (John and Whatley, 1970; Sapshead and Wimpenny, 1972). Cytochrome aa_3 was detected by Scholes and Smith (1968b) in membranes from cells grown with nitrate although the amount present per mg membrane protein was one-third that of membranes from aerobically grown cells. The cytochrome aa_3 content of *P. denitrificans* grown either aerobically or anaerobically with nitrate and expressed on a membrane protein basis or as a molar ratio with *b*- and *c*-type cytochromes is low compared with mitochondria (Table III). Undoubtedly it is this low concentration which has made detection difficult in the membranes of cells grown with nitrate.

Cytochrome aa_3 also occurs in a number of other aerobic bacteria, e.g. in *Mycobacterium phlei* (Asano and Brodie, 1964), in *Micrococcus (Sarcina) lutea* (Erickson and Parker, 1969), in *Bacillus megaterium* (Kröger and Dadák, 1969) and in *Rhodopseudomonas spheroides* (Dutton and Wilson, 1974); but cytochrome aa_3 is absent from the obligate aerobes *Azotobacter vinelandii* (Jones and Redfearn, 1966) and *Haemophilus parainfluenzae* (White and Smith, 1962)—where cytochromes a_1, a_2 and *o* function as oxidases—and from the facultative aerobes *Escherichia coli*, *Staphylococcus aureus* and *Rhodospirillum rubrum*—where cytochrome *o* functions as an oxidase (see Kamen and Horio, 1970).

E. CYTOCHROME *o*

Cytochrome *o* is identified by the absorption spectrum of its carbon monoxide derivative which shows peaks at about 418, 535 and 570 nm (see Kamen and Horio, 1970; Lemberg and Barrett, 1973).

Cytochrome *o* is rarely observed in mitochondria although it is a common constituent of bacteria, including *Paracoccus denitrificans* (see Kamen and Horio, 1970). In the facultative aerobe *Staphylococcus aureus*, where cytochrome *o* is the principal oxidase, the prosthetic group has been characterized chemically as a protohaem (like cytochrome *b*), and its physiological role as an oxidase has been shown by the action spectrum of the photoreversal of the carbon monoxide inhibition of respiration (Taber and Morrison, 1964).

Much less is known of the properties and role of carbon monoxide-binding cytochromes whose spectra are similar to the cytochrome *o* of *Staphylococcus aureus*. It has been suggested (see Smith, 1968) that at least some of these cytochromes are derived from a previously non-reactive cytochrome *b* which has been dislocated from its position *in vivo* and thus rendered carbon monoxide reacting, since it has been observed that cytochrome *b* can become oxygen- (and carbon monoxide-)reacting upon isolation (see Lemberg and Barrett, 1973).

Mitochondrial cytochrome *o* appears to be confined to parasitic protozoa, e.g. *Trypanosoma mega* (Ray and Cross, 1972) and *Crithidia fasciculata* (Hill and White, 1968), and to several species of yeast (Mok *et al.*, 1969).

The presence of cytochrome *o* in *Paracoccus denitrificans* is concluded from the observation of an absorption peak at about 415 nm in CO-reduced *minus* reduced difference spectra obtained with intact cells (Sato, 1956) and with cell-free preparations (Porra and Lascelles, 1965; Imai *et al.*, 1967; Scholes and Smith, 1968b; Knobloch *et al.*, 1971; Sapshead and Wimpenny, 1972).

The available evidence is insufficient to decide on the status and physiological significance of the cytochrome o of *Paracoccus denitrificans*. A physiological role is suggested both by its detection in intact cells (Sato, 1956) and by its reducibility by NADH, succinate and hydrogen in membrane preparations (Scholes and Smith, 1968b; Knobloch *et al.*, 1971). However, an origin of this cytochrome o from a labile b-type cytochrome is indicated by an increased absorbance after treatment with the detergent Triton X-100, by the shift in the absorption peak from 423 to 418 nm upon treatment with Triton X-100 (Porra and Lascelles, 1965), and also by the variability of the absorbance peak (contrast Imai *et al.* (1967) at 421 nm with Scholes and Smith (1968b) at 418 nm).

Bacteria which resemble *Paracoccus denitrificans* in containing cytochrome o in addition to an a-type cytochrome include *Bacillus megaterium, Mycobacterium phlei*, and *Azotobacter vinelandii* (see Lemberg and Barrett, 1973). However, in some bacteria, notably *Acetobacter suboxydans* (Daniel, 1970) cytochrome o is the sole oxidase.

F. CYTOCHROME b

Cytochromes of the b type have protohaem as prosthetic group, absorption maxima in the α region of the absorption spectrum at 556-565 nm, and are universally distributed among mitochondria and aerobic bacteria (see Lemberg and Barrett, 1973).

In mammalian mitochondria at least two b cytochromes can be distinguished: cytochrome b_{562}, which can be readily reduced by succinate, and cytochrome b_{556}, which is reducible by succinate only in the presence of antimycin A but readily reduced by dithionite alone. It is not clear at present whether a third peak at 558 nm represents an additional b-type cytochrome or whether this peak is due to a split α peak of the 566 component (Wikström, 1973). By using an anaerobic potentiometric titration of the mitochondrial respiratory components, in which a variety of dyes of appropriate midpoint potentials are employed to mediate electron transfer between the membrane-bound cytochromes and a platinum electrode, it has been shown by Dutton *et al.* (1970) that the b_{562} component has a midpoint potential of +40 mV and the b_{566} component has a lower midpoint potential of −30 mV. With a variety of plant mitochondria reduced *minus* oxidized difference spectra show an absorbance maximum at about 556 nm which is due to an additional b-type cytochrome with a midpoint potential of about 75 mV (Lance and Bonner, 1968; Dutton and Storey, 1971; Passam *et al.*, 1973).

Like mitochondria, membrane preparations of *Paracoccus denitrificans* contain at least two b-type cytochromes: a cytochrome b_{560} which is reduced by NADH or by succinate (Imai *et al.*, 1967; Scholes and Smith, 1968b; Lam and Nicholas, 1969a; John and Whatley, 1970), and a cytochrome b_{566} which is easily reduced by dithionite but not readily reduced by NADH or succinate (Scholes and Smith, 1968b; Gray *et al.*, 1973). At 77°K fourth-order finite difference spectra, obtained by comparing the absorbance of a dithionite-reduced suspension of *P. denitrificans* cells with a similar cell suspension oxidized with ferricyanide or with oxygen, show absorbance peaks at 559 and 564 nm (Shipp, 1972b) which are probably due to cytochromes b_{560} and b_{566}. Thus *P. denitrificans* appears to possess the two essential cytochrome *b* components of the mitochondrial respiratory chain. This pattern appears to be common among aerobic bacteria (Shipp, 1972b; Cohen *et al.*, 1973; and see Lemberg and Barrett, 1973). However, with the exception of the *b*-type cytochromes of *Escherichia coli* (Hendler *et al.*, 1975) and of *Rhodopseudomonas sphaeroides* (Dutton and Wilson, 1974), the midpoint potentials of the bacterial *b*-type cytochromes are unknown, and it remains to be seen whether the *b*-type cytochromes function in the same manner in the mitochondrial and bacterial respiratory chains.

In those bacteria like *Paracoccus denitrificans* which have cytochrome *o* there is a contribution by the cytochrome *o* to the absorption in the 560 nm region of reduced *minus* oxidized difference spectra (Cheah, 1969; Daniel, 1970). This complication does not arise in most mitochondria, which lack cytochrome *o*.

When *Paracoccus denitrificans* is grown anaerobically with nitrate as the terminal electron acceptor the amount of the two *b*-type cytochromes per mg membrane protein increases about threefold when compared with membranes from aerobically grown cells (Scholes and Smith, 1968b). Presumably this increase is in response to the new requirement for nitrate reduction, since nitrate reductase interacts directly with the *b*-type cytochrome present (Vernon, 1956; Lam and Nicholas, 1969a; John and Whatley, 1970).

Conversely, the amount of cytochrome aa_3 per mg membrane protein decreases when cells are grown anaerobically with nitrate to about one-third of the amount in membranes of aerobically grown cells (Scholes and Smith, 1968b). Thus in *Paracoccus denitrificans* the transition from oxygen to nitrate as a terminal electron acceptor is accompanied by an approximately ninefold increase in the ratio of *b*-type cytochrome to cytochrome aa_3. By contrast, mitochondria from a variety of animal sources show little difference in the ratio of *b*-type cytochrome to cytochrome aa_3 (see Klingenberg, 1968).

Mitochondrial electron transport between cytochrome *b* and cytochrome c_1 is inhibited by low concentrations of the antibiotic antimycin A (see Chance and Williams, 1956). *Paracoccus denitrificans* (Imai *et al.*, 1967, 1968a) and *Mycobacterium flavum* (Erickson, 1971) are unusual among bacteria in their sensitivity to this inhibitor (Gel'man *et al.*, 1967). *Mycobacterium phlei* (Asano and Brodie, 1964) and *Micrococcus lysodeikticus* (Ishikawa and Lehninger, 1962) are quite insensitive, while *Azotobacter vinelandii* (Jones and Redfearn, 1966) is inhibited only at higher concentrations than those necessary to obtain complete inhibition with membrane preparations of *P. denitrificans* (Scholes and Smith, 1968b) and with comparable submitochondrial particles (Brandon *et al.*, 1972).

G. CYTOCHROME c

Cytochromes of the c type are characterized by an α-band absorption maximum in the reduced form at 550-555 nm. Animal mitochondria, and probably also plant mitochondria (Lance and Bonner, 1968), possess two distinct c-type cytochromes: one, called cytochrome c, has a smaller molecular weight (about 12 000), is easily solubilized by washing mitochondria in salt solutions, and has absorption maxima in the reduced form at 550, 521 and 415 nm. The other c-type cytochrome is called cytochrome c_1 and is larger (mol. wt about 37 000), is less easily solubilized, and it has absorption maxima in the reduced form at 554, 524 and 418 nm (see Lemberg and Barrett, 1973).

Paracoccus denitrificans possesses two c-type cytochromes with similar properties to cytochromes c and c_1 of mitochondria: separate peaks at 550 and 553 nm have been observed in reduced *minus* oxidized difference spectra of plasma membrane preparations of *P. denitrificans* (Scholes and Smith, 1968b), and fourth-order finite difference spectra obtained at 77°K with intact cells of *P. denitrificans* show distinct peaks at 550 and 554 nm (Shipp, 1972b). The presence of a c-type cytochrome resembling mitochondrial cytochrome c_1 is also suggested by the absorption spectrum of a "cytochrome oxidase" preparation of *P. denitrificans* described by Vernon and White (1957) which showed absorption maxima in the reduced form at 553, 522 and 421 nm. No further studies have been made of this c_1-like cytochrome, but, at least spectrally, it resembles the mitochondrial cytochrome c_1.

As in *Paracoccus denitrificans*, reduced *minus* oxidized difference spectra in a number of other bacteria which possess cytochromes aa_3 and o as oxidase show two absorption maxima separated by several nanometres in the 550-555 nm region, indicative of the presence of two c-type cytochromes which correspond spectrally to the mitochondrial cytochromes c and c_1, e.g. in *Mycobacterium phlei*, *Bacillus subtilis* and *Micrococcus lutea* (Lemberg and Barrett, 1973). By contrast, in some facultatively aerobic bacteria like *Escherichia coli* (Gray *et al.*, 1963), *Staphylococcus aureus* (Taber and Morrison, 1964) and *Klebsiella* (*Aerobacter*) *aerogenes* (Knook *et al.*, 1973) reduced *minus* oxidized difference spectra of preparations derived from aerobically grown cells provide no evidence of any c-type cytochromes. However, with the increased resolution of low temperature fourth-order finite difference spectra the presence of at least one c-type cytochrome can be recognized in these bacteria (Shipp, 1972a, b).

Compared to other proteins the soluble c-type cytochrome of mitochondria (and of bacteria) is easily isolated and purified, and is relatively stable (Margoliash and Shejter, 1966). These properties have enabled extensive physical, physiological and structural studies to be made of both mitochondrial cytochrome c and the corresponding bacterial c-type cytochromes (see Lemberg and Barrett, 1973), including cytochrome c_{550} of *Paracoccus denitrificans* (Kamen and Vernon, 1955; Vernon, 1956; Smith *et al.*, 1966; Scholes *et al.*, 1971; Timkovich and Dickerson, 1973). Taken together, these studies reveal that the cytochrome c_{550} of *P. denitrificans* resembles the mitochondrial cytochrome *c* more closely than do other comparable *c*-type cytochromes of bacterial origin.

The cytochrome c_{550} of *Paracoccus denitrificans* resembles mitochondrial cytochrome *c* in that both cytochromes have an absorption maximum in the α band at 550 nm in the reduced form, and both cytochromes have a midpoint potential at about +250 mV (Table IV). An α-band absorption maximum at 550 nm is also observed with reduced *c*-type cytochromes of a variety of bacteria, although absorption maxima in the α band at 551 and 552 nm are commonly observed with other aerobic bacteria (Table IV). A midpoint potential at about +250 mV is unusual in bacterial *c*-type cytochromes. Although the cytochrome c_{550} of *Bacillus megaterium* has a midpoint potential at +250 mV, the *c*-type cytochrome of *B. subtilis* has a lower midpoint potential at +210 mV, *c*-type cytochromes isolated from sulphate-reducing species of *Desulphovibrio* and from nitrite-reducing *Escherichia coli* have much lower midpoint potentials at −200 to −220 mV (see Gel'man *et al.*, 1967), and the *c*-type cytochromes of *Pseudomonas denitrificans, Pseudomonas aeruginosa, Rhodospirillum rubrum* (light or dark grown) and *Spirillum itersonii* have higher midpoint potentials at +290 to +320 mV (Table IV).

The cytochrome c_{550} of *Paracoccus denitrificans* resembles many other soluble cytochromes isolated from bacteria (Table IV) in being an acidic protein (Kamen and Vernon, 1955) with an isoelectric point at about pH 4 (Timkovich and Dickerson, 1973). By contrast, most mitochondrial cytochromes *c* are basic proteins, e.g. horse heart and yeast cytochromes *c* have an isoelectric point at about pH 10 (Margoliash and Shejter, 1966). However, mitochondrial cytochromes *c* with an acid or neutral isoelectric point have been reported from the fungi *Aspergillus oryzae* and *Ustilago sphaerogena* (Yamanaka *et al.*, 1963). As expected from its acidic nature, amino acid analysis of the cytochrome c_{550} of *P. denitrificans* reveals a preponderance of dicarboxylic over diamino amino acids, the ratio of dicarboxylic to diamino amino acids being twice the ratio in horse heart cytochrome *c* (Scholes *et al.*, 1971).

The spectrophotometric and potentiometric similarities between mitochondrial cytochrome *c* and the cytochrome c_{550} of *Paracoccus denitrificans* are extended by the interchangeability of the two cytochromes in their reactivity with mitochondrial cytochrome oxidase and with the cytochrome oxidase of *P. denitrificans*.

Mitochondrial cytochromes *c* isolated from a wide variety of eukaryotic phyla readily act as reductants for the purified cytochrome oxidase of beef heart mitochondria (cytochrome aa_3) (Yamanaka and Okunuki, 1964; Yamanaka, 1973). Of the bacterial *c*-type cytochromes that have been tested for their reactivity with purified mitochondrial cytochrome oxidase, the *c*-type cytochromes of *Pseudomonas aeruginosa, Pseudomonas saccharophila, Pseudomonas stutzeri, Nitrosomonas europaea*, and *Azotobacter vinelandii* show no significant

TABLE IV

Properties of c-type cytochromes isolated from a mitochondrion and from bacteria. The data for the reactivity with purified beef heart cytochrome oxidase is taken from Yamanaka and Okunuki (1964) and Yamanaka (1973),.

Cytochrome	Absorption maxima of reduced form			Isoelectric point (pH)	Molecular weight	Midpoint potential at pH 7 (mV)	Reactivity with purified beef heart cytochrome oxidase	References
	α	β	γ					
Mitochondrial								
Horse heart cytochrome c	550	520	416	10.6	12 400	+260	+++	Margoliash and Shejter (1966)
Bacterial								
Paracoccus denitrificans c_{550}	550	522	416	4	14 300	+250	+	Kamen and Vernon (1955) Scholes *et al.* (1971)
Bacillus megaterium c_{550}	550	520	415	?	?	+250	?	Vernon and Mangum (1960)
B. subtilis c_{550}	550	520	414	8.7	13 000	+210	+/−	Miki and Okunuki (1969a, b)
Spirillum itersonii c_{550}	550	522	416	9.9	10 400	+300	?	Clark-Walker and Lascelles (1970)
Rhodospirillum rubrum c_2	550	521	415	6.0	13 000	+310	+	Kamen and Horio (1970)
Azotobacter vinelandii c_4	551	522	414	4.4	11 200	+300	−	Swank and Burris (1969)
Pseudomonas aeruginosa c_{551}	551	521	416	4.7	8 100	+290	−	Horio *et al.* (1960)
P. denitrificans c_{552}	552	525	418	?	?	+320	?	Kamen and Vernon (1955)

Symbols: +++, +, reactive; −, not reactive; ?, not tested.

activity, while the c-type cytochromes of *Paracoccus denitrificans, Thiobacillus novellus* and *Rhodospirillum rubrum* do show significant activity, although the reduced bacterial cytochromes are oxidized at lower rates than the reduced mitochondrial cytochromes (Yamanaka and Okunuki, 1964; Yamanaka, 1973). But with a crude cytochrome oxidase preparation of pig heart, Kamen and Vernon (1955) reported that reduced cytochrome c_{550} of *P. denitrificans* was oxidized at almost the same rate as reduced horse heart cytochrome c, while reduced cytochrome c_{552} of *Pseudomonas denitrificans* and reduced cyto- chrome c_2 of *Rhodospirillum rubrum* gave no detectable rates of oxidation under the same conditions. However, Smith *et al.* (1966) observed that the turnover rate of the cytochrome oxidase in Keilin-Hartree particles prepared from beef heart was considerably lower when reduced cytochrome c_{550} of *P. denitrificans* was the reductant than when beef heart cytochrome c was used.

In their reaction with the cytochrome oxidase (cytochromes aa_3, o) of membrane particles of *Paracoccus denitrificans*, mitochondrial cytochrome c and the cytochrome c_{550} of *P. denitrificans* appear to be completely interchange- able, since similar rates of oxidation are observed with either cytochrome (Smith *et al.*, 1966; Lam and Nicholas, 1969b). In this respect *P. denitrificans* resembles *Azotobacter vinelandii* (Swank and Burris, 1969), *Bacillus subtilis* (Miki and Okunuki, 1969b) and heterotrophically grown *Rhodospirillum rubrum* (Taniguchi and Kamen, 1965), since membrane particles of these bacteria also oxidize mitochondrial cytochrome c as effectively as they oxidize their own soluble c-type cytochromes.

When *Paracoccus denitrificans* is grown anaerobically with nitrate or nitrite as terminal electron acceptor, a cytochrome with two haem groups (cytochrome cd) is present in addition to the constitutive cytochrome oxidase (Lam and Nicholas, 1969a, b; Newton, 1969). This cytochrome cd has both nitrite reductase and cytochrome oxidase activities. *In vivo* it probably acts as a nitrite reductase, accepting electrons from c-type cytochrome(s). Purified cytochrome cd of *P. denitrificans* oxidizes mitochondrial cytochrome c and cytochrome c_{550} of *P. denitrificans* with equal effectiveness (Lam and Nicholas, 1969b), whereas a similar cytochrome cd purified from *Pseudomonas aeruginosa* uses a cytochrome c_{551} from *Ps. aeruginosa* but not mitochondrial cytochrome c (Yamanaka and Okunuki, 1964). The cytochrome c_{550} of *P. denitrificans* resembles mitochondrial cytochrome c and differs from the soluble c-type cytochromes of many other bacteria in being ineffective with the cytochrome cd of *Ps. aeruginosa* (Yamanaka, 1973).

Thus it is clear that mitochondrial cytochrome c and cytochrome c_{550} of *Paracoccus denitrificans* are largely interchangeable in their cross-reactivity with both mitochondrial cytochrome oxidase (cytochrome aa_3) and with the con- stitutive (cytochrome aa_3, o) and adaptive (cytochrome cd) cytochrome oxidases of *P. denitrificans.*

Cytochromes c from a wide range of eukaryotes, encompassing yeasts, protozoa, plants and animals, show a marked homology in their amino acid sequences (see Dayhoff, 1972), and analyses of the X-ray crystal structure of fish and horse heart cytochrome c show the presence of a characteristic folding of the polypeptide chain, known as the "cytochrome fold" (see Timkovich and Dickerson, 1973). Comparable information is available for only a few bacterial c-type cytochromes, but, fortunately for our present purposes, the amino acid composition (Scholes et $al.$, 1971) and the X-ray crystal structure (Timkovich and Dickerson, 1973) of cytochrome c_{550} from $P.$ $denitrificans$ are known, although the determination of the amino acid sequence of this cytochrome is yet to be completed (E. Margoliash, personal communication). The other bacterial c-type cytochromes for which comparable information is available are cytochrome c_2 of $Rhodospirillum$ $rubrum$, where the amino acid sequence (Dus et $al.$, 1968) and X-ray crystal structure (see Timkovich and Dickerson, 1973) are known, and cytochrome c_{551} of $Pseudomonas$ $aeruginosa$, (formerly $Ps.$ $fluorescens$), where the amino acid sequence is known (Ambler, 1963) and a three-dimensional structure has been proposed (Dickerson, 1971).

The cytochrome c_{550} of $Paracoccus$ $denitrificans$ has a chain length of about 135 amino acid residues (Scholes et $al.$, 1971), which makes it longer than mitochondrial cytochromes c, which range in length from 103 (fish) to 112 (plant) amino acid residues (see Dayhoff, 1972), and longer than both the cytochrome c_{551} of $Pseudomonas$ $denitrificans$, which has 82 (Ambler, 1963), and the cytochrome c_2 of $Rhodospirillum$ $rubrum$, which has 112 amino acid residues (Dus et $al.$, 1968).

X-ray structure analysis reveals that the polypeptide chain of the cytochrome c_{550} of $Paracoccus$ $denitrificans$ is folded in a similar way to the polypeptide chain of both the mitochondrial cytochrome c and the cytochrome c_2 of $Rhodospirillum$ $rubrum$ (Timkovich and Dickerson, 1973). The latter cytochrome has an amino acid sequence which shows broad homology with that of mitochondrial cytochrome c, differing essentially in the deletion of four single residues, and in the insertion of two amino acid sequences, one of nine and the other of four residues, to replace two single residues in the mitochondrial cytochrome c. It is inferred from X-ray crystal structure analysis of the cytochrome c_{550} of $P.$ $denitrificans$ (Timkovich and Dickerson, 1973) that this cytochrome resembles cytochrome c_2 of $R.$ $rubrum$ in possessing these insertions. Furthermore, the 25 amino acid residues present in the cytochrome c_{550} of $P.$ $denitrificans$ in excess of those in both mitochondrial cytochrome c and cytochrome c_2 of $R.$ $rubrum$ are distributed on the surface of cytochrome c_{550} in three loops of about five residues each and in one loop of about ten residues. Two of these additional five-residue loops are found at each end of the chain. The cytochrome c_{550} of $P.$ $denitrificans$ and mitochondrial cytochrome c resemble one another, and differ from cytochrome c_2 of $R.$ $rubrum$, in the arrangement of aromatic groups in the right-hand side channel. The homology apparent between mitochondrial cytochrome c and both cytochrome c_{550} of $P.$

denitrificans and cytochrome c_2 of *R. rubrum* is greater than the homology apparent between mitochondrial cytochrome *c* and cytochrome c_{551} of *Pseudomonas aeruginosa*, which differs from mitochondrial cytochrome *c* in large segments of its amino acid sequence (see Needleman and Blair, 1969; Dickerson, 1971). Cytochrome c_{551} of *Ps. aeruginosa* does, however, share with mitochondrial cytochrome *c* and cytochrome c_2 the typical haem-binding pentapeptide sequence, the presence of which in cytochrome c_{550} of *P. denitrificans* seems probable from structural analysis (Timkovich and Dickerson, 1973).

It has been concluded from the extensive sequence analyses, the X-ray structure analyses and the studies of cross-reactivity of mitochondrial cytochromes *c* and their oxidases that "mitochondrial cytochromes *c*, no matter what their source, have the same molecular structure and polypeptide chain folding, and are evolutionary descendents of a common ancestor" (Timkovich and Dickerson, 1973). It is among the bacteria that this ancestral cytochrome *c* has been sought. From the limited knowledge which is currently available of the primary structure and tertiary folding of bacterial *c*-type cytochromes, it is already clear that cytochrome c_{550} of *Paracoccus denitrificans* and cytochrome c_2 of *Rhodospirillum rubrum* are more closely related to the ancestor of mitochondrial cytochrome *c* than is cytochrome c_{551} of *Pseudomonas aeruginosa*. Timkovich and Dickerson (1973) have argued, on the basis of their structural analyses, that the most plausible evolutionary relationship for cytochrome c_{550} of *P. denitrificans*, mitochondrial cytochrome *c*, and cytochrome c_2 of *R. rubrum* would be one in which a cytochrome c_2-like ancestor gave rise to mitochondrial cytochrome *c* by deletions and to cytochrome c_{550} of *P. denitrificans* by insertions. Whatever the homology revealed by the determination of the complete amino acid sequence of cytochrome c_{550} of *P. denitrificans*, it is already clear that future extensions of the phylogenetic tree of cytochrome *c* back from mitochondrial to bacterial types will position cytochrome c_{550} of *P. denitrificans* close to the transition stage when bacteria became mitochondria.

VI. OXIDATIVE PHOSPHORYLATION

A. STOICHIOMETRY OF RESPIRATION AND OXIDATIVE PHOSPHORYLATION

The traditional procedure employed for determining the efficiency of energy conservation during oxidative phosphorylation is to measure the amount of P_i esterified during the consumption of a measured amount of oxygen. The stoichiometry of oxidative phosphorylation so determined is usually expressed as a P/O ratio, which is the ratio of P_i esterified (or ATP produced), expressed in moles, to oxygen consumed expressed in g atoms. In isolated mitochondria supplied with ADP and an oxidizable substrate the P/O ratio may be determined directly by simultaneously measuring the changes in the levels of oxygen and P_i. In bacteria the obstacles to a direct determination of the P/O ration are similar to those which would be encountered if one were to attempt to determine P/O ratios not, as is customary, with isolated mitochondria, but rather with a suspension of cells which contained mitochondria. The enzyme complex, the ATPase, responsible for the actual phosphorylation reaction is orientated in the inner mitochondrial membrane so that its active centre is accessible to ADP, ATP and P_i only from the mitochondrial matrix (see Racker, 1970). Similarly in the bacterial cell the ATPase of the plasma membrane is orientated so that it is accessible to its reactants only from the cytoplasmic side of the membrane (see Harold, 1972). During oxidative phosphorylation the accumulation of ATP in the medium, and the uptake of ADP into the mitochondrion, is made possible by the presence of an adenine nucleotide carrier in the inner mitochondrial membrane (see Klingenberg, 1970). This carrier allows the reversible, equimolar exchange of ADP and ATP across the otherwise impermeable membrane. It has an obvious role in cell metabolism by linking the site of ATP synthesis on the inner surface of the inner mitochondrial membrane with the main sites of ATP utilization in the cytoplasm. No analogous carrier system has been detected in the bacterial cell, and the lack of an obvious physiological role for such a carrier makes its discovery unlikely. Thus the lack of an adenine nucleotide carrier physiologically isolates the ATPase of the bacterial cell from ADP added to the suspending medium, and continued oxidative phosphorylation is dependent on the continuing turnover of ATP and ADP in the bacterial cell. This turnover of the endogenous adenine nucleotide pool is due to the activity of cytoplasmic enzymes (Harrison and Maitra, 1969).

Thus there are two main obstacles to the direct measurement of oxidative phosphorylation in intact bacteria: the absence of an adenine nucleotide carrier, and the presence of interfering cytoplasmic reactions. Both of these obstacles are removed by breaking the plasma membrane and then separating the particulate, membrane fraction from the soluble, cytoplasmic fraction. However, it is clear from much recent research (and it is one of the main postulates of the chemiosmotic theory) that in order to support oxidative phosphorylation a membrane must exist as a closed vesicle (see Henderson, 1971; Harold, 1972). Such a requirement is implicit in the large amount of data which is available on the uncoupling ability of ionphorous antibiotics (Henderson, 1971). Fortunately most membranes, including mitochondrial and bacterial membranes, have a natural ability for spontaneously reforming into closed vesicles after breakage. Thus, upon rupture of the bacterial cell, the fragments of the plasma membrane released reform themselves into closed vesicles. On the face of it, such vesicles could have either an inside-out or a right-side-out configuration. Right-side-out vesicles have the side of the membrane which originally faced towards the interior retaining this orientation in the vesicle. Inside-out vesicles have resealed into vesicles in which the original outside of the membrane faces the interior of the vesicle. Only the inside-out vesicles would be expected to show continued oxidative phosphorylation since only they would possess an ATPase orientated towards the medium and thus accessible to added ADP. On this basis it is clearly the inside-out vesicles which are responsible for the oxidative phosphorylation observed in "membrane preparations" of bacteria disrupted by any of the usual techniques such as sonication, French Press treatment or by osmotic lysis. In right-side-out vesicles the ATPase would be inaccessible to added ATP, and in membranes which had not resealed to form vesicles with the necessary ionic tightness (i.e. without a physiological sidedness) the ATPase would be perfectly accessible but no transmembrane ionic gradient could be maintained and hence no ATP could be synthesized. Both in their mode of formation and in their sidedness these phosphorylating, inside-out bacterial particles resemble sub-mitochondrial particles prepared by the sonication of mitochondria (see Racker, 1970). P/O ratios observed with bacterial particles should therefore be compared with those observed with submitochondrial particles rather than with the P/O ratios of intact mitochondria.

Phosphorylating bacterial particles from a variety of bacterial species, and produced by a variety of disruptive techniques, can support P/O ratios of the same order as those observed with submitochondrial particles (Table V). In this respect phosphorylating particles prepared from *Paracoccus denitrificans* are no different from particles from a number of other bacteria. However, in contrast to the situation in submitochondrial particles the P/O ratios of *intact mito-chondria* frequently closely approach the values of 3, 2 and 1 which are expected for the respective substrates NADH, succinate and reduced cytochrome *c* in a completely coupled system (see Lehninger, 1964). Therefore the preparation of submitochondrial particles must involve damage to the oxidative phosphorylation system sufficient to reduce the efficiency with which the free energy released by the oxido-reduction reactions of the respiratory chain is conserved as the terminal pyrophosphate bond of ATP. Thus the P/O ratios of phosphorylating baterial particles can be relied upon only as lower limits to the P/O ratios which may be possible in intact bacteria.

As outlined above, *direct* determinations of the type used with mitochondria are not feasible for the estimation of P/O ratios in intact bacterial cells. Consequently three rather different approaches have been adopted to estimate the P/O ratios of intact bacterial cells: (i) molar growth yield studies (Stout-hamer, 1969), (ii) short-term determinations of P_i esterification and NADH oxidation (Hempfling, 1970a), and (iii) determinations of H^+/O ratios. The first two of these approaches have yet to be applied to *P. denitrificans* but H^+/O ratios have been measured in both *P. denitrificans* and mitochondria.

Determinations of molar growth yields give a P/O ratio of 1.7 for *Aerobacter aerogenes* and for *Proteus mirabilis* (Stouthamer and Bettenhaussen, 1973), 3.4 for *Bacillus subtilis*, and 3.1 for *Escherichia coli* (Meyer and Jones, 1973). P/O ratios of about 3 have also been obtained for *E. coli* (Hempfling, 1970a) and *Azotobacter vinelandii* (Baak and Postma, 1971) by simultaneously measuring changes in the levels of NADH, P_i and adenine nucleotides during short-term

TABLE V

P/O ratios of phosphorylating submitochondrial and bacterial particles

	Preparation technique	P/O ratios		References
		NADH	Succinate	
Submitochondrial particles				
Beef heart	Sonication	1.7-2.7	1.0-1.8	Beyer (1967), Vallin (1968) Christiansen *et al.* (1969)
Yeast	Sonication	0.38	0.55	Schatz (1967)
Jerusalem artichoke	Sonication	0.46	0.55	Passam and Palmer (1971)
Bacterial particles				
	Sonication	1.0	0.40	Imai *et al.* (1967)
Paracoccus denitrificans	Osmotic lysis	1.5	0.48	John and Whatley (1970)
	French Press	1.4	0.41	Knobloch *et al.* (1971)
Pseudomonas saccharophila	French Press	0.92	0.41	Ishaque *et al.* (1973)
Azotobacter vinelandii	French Press	0.98	—	Ackrell and Jones (1971)
Mycobacterium phlei	Sonication	1.2	0.75	Brodie (1959)
Escherichia coli	Sonication	0.3	0.5	Kashket and Brodie (1963)
Nitrobacter agilis	Sonication	1.8	—	Aleem (1968)

exposure of cells to aerobiosis (Hempfling, 1970a). However, the P/O ratio of 3 obtained for *E. coli* by both of these approaches is not in agreement with the P/O ration inferred for *E. coli* by measurements of H^+/O ratios, where a value of 2 is indicated (Lawford and Haddock, 1973). Because of the many uncertainties inherent both in molar growth yield studies (Forrest and Walker, 1971; Stouthamer and Bettenhaüssen, 1973) and in measurements of oxidative phosphorylation over short time periods (Van der Beek and Stouthamer, 1973), an accurate estimation of P/O ratios in intact bacterial cells by these techniques is not yet possible. However, the P/O ratios which have been suggested are significantly higher than those obtained with subcellular preparations of bacteria (Table V), and more comparable with the P/O ratios obtained with mitochondria.

According to the chemiosmotic theory (Mitchell, 1966) the respiratory chains of mitochondria and bacteria are so arranged within the energy-transducing membrane that the flow of reducing equivalents along the respiratory chain results in an outward, electrogenic translocation of protons from the mitochondrial matrix or bacterial cell. The translocation of two protons results from the flow of each pair of reducing equivalents across a single coupling site of the respiratory chain, so that the entire span of the mitochondrial respiratory chain from NADH to oxygen is associated with the translocation of six protons. The stoichiometry of the ATPase is postulated to be $2H^+/ATP$ (in either the synthetic or hydrolytic direction), in agreement with a P/O ratio of 3 for the aerobic oxidation of NADH.

Experimental confirmation of the postulated stoichiometry of respiration-driven proton translocation has been obtained for the mitochondrial system by the technique developed by Mitchell and Moyle (1967b), in which an anaerobic suspension of mitochondria supplied with the appropriate reductant is provided with a small, known quantity of oxygen and the resultant pH changes in the lightly buffered suspending medium are recorded by a sensitive pH meter. According to Mitchell and Moyle (1967b) the two main conditions which must be satisfied to allow an accurate measurement of the limiting H^+/O ratios are (a) that the mitochondria respond to the addition of oxygen by a short burst of respiratory activity and then quickly return to an anaerobic state, and (b) that the back flow of protons across the membrane is minimized. The first condition is satisfied by supplying a sufficiently small quantity of oxygen; the high activity and high affinity of the mitochondrial cytochrome oxidase ensures that all of the oxygen supplied is utilized rapidly. The second condition is satisfied by providing an appropriate ion which will move electrophoretically across the membrane and thus electrically neutralize the membrane potential which is set up by the initial proton translocation and which, if maintained, would both retard further proton translocation and accelerate the return of protons across the membrane. Potassium ions, in the presence of valinomycin, and Ca^{2+} have been employed as charge-neutralizing counterions for the mitochondrial proton translocation (Mitchell and Moyle, 1967b). In bacteria, K^+ in the presence of valinomycin has been employed as a counterion, and SCN^- as a co-transported ion (Scholes and Mitchell, 1970b; Lawford and Haddock, 1973; Meyer and Jones, 1973). Calcium is not effective in bacteria since they lack the Ca^{2+} transport system present in the mitochondrial membrane.

For mitochondria, it has been clearly established that, as postulated, the flow of each pair of reducing equivalents across a single coupling site results in the translocation of two protons across the inner mitochondrial membrane (Mitchell and Moyle, 1967a, b; Hinkle and Horstman, 1971; Lawford and Garland, 1972; Downie and Garland, 1973). A H^+/O ratio of 6 is observed with NAD^+-linked substrates and a H^+/O ratio of 4 with succinate oxidation (Mitchell and Moyle, 1967a, b). Furthermore, it has been observed that the hydrolysis of each molecule of ATP at the mitochondrial ATPase results in the translocation of two protons across the inner mitochondrial membrane (Moyle and Mitchell, 1973b; Thayer and Hinkle, 1973).

Thus the H^+/O ratios of mitochondria, determined under the conditions specified by Mitchell and Moyle (1967), faithfully reflect the P/O ratios obtained, and hence H^+/O ratios obtained with bacteria under the specified conditions may be relied upon to provide us with a measure of bacterial P/O ratios.

Paracoccus denitrificans was the first bacterium for which an accurate determination of H^+/O ratios was made. After a careful investigation of the conditions necessary for obtaining stoichiometric proton translocation, Scholes and Mitchell (1970b) determined the H^+/O ratio to be 8.0 ± 0.1, with SCN^- as the charge neutralizing anion and with endogenous substrate(s) as the reductant. They proposed that this H^+/O ration was attributable to the operation of a transhydrogenase and a respiratory chain operating in series, both of which have the same proton translocation stoichiometry shown by the mitochondrial transhydrogenase and respiratory chain, with the translocation of 2 protons per pair of reducing equivalents traversing the transhydrogenase and of six protons per pair of reducing equivalents traversing the respiratory chain from NADH to oxygen (see Mitchell, 1972). In the cells used in these experiments it is assumed that NADPH accumulates anaerobically and that, when an oxygen pulse is introduced, reducing equivalents are drawn from this NADPH via the trans-hydrogenase to NADH, and thence via the respiratory chain to oxygen. Thus the effective respiratory reductant in the cells of *P. denitrificans* used in the experiments of Scholes and Mitchell (1970b) is NADPH.

It has been reported recently (Mitchell, 1972; Moyle and Mitchell, 1973a) that H^+/O ratios close to 8 can also be observed in rat liver mitochondria after treatment with *N*-ethylmaleimide. The *N*-ethylmaleimide probably inactivates succinate dehydrogenase and NAD^+-linked enzymes without affecting the mitochondrial NADP-linked isocitrate dehydrogenase, NADH oxidase or transhydrogenase (Mitchell, 1972). Thus NADPH accumulates under anaerobic conditions and acts as the main respiratory substrate (Mitchell, 1972; Moyle and Mitchell, 1973a). The *N*-ethylmaleimide treatment effectively alters the availability of reductants for the mitochondrial respiratory chain, and "allows" the mitochondrion to give similar H^+/O ratios to those observed in cells of *Paracoccus denitrificans.*

Although determinations of the proton translocating stoichiometry of the respiratory chain of *Paracoccus denitrificans* using different reductants have yet to be made, it is already clear that the respiratory chain of whole cells of *P. denitrificans* from NADH to oxygen shows the same H^+/O ratio as mammalian mitochondria, and by inference they may be expected to have the same P/O ratio.

When the results of similar determinations of H^+/O ratios made with other bacterial species are examined (Table VI) it is found that *Hydrogenomonas eutropha* H-16 (Beatrice and Chappell, 1974; Drozd and Jones, 1974), and possibly *Bacillus subtilis* (Meyer and Jones, 1973), share with *Paracoccus denitrificans* the possession of the mitochondrial stoichiometry; while

TABLE VI
H^+/O ratios of mitochondria and bacterial cells

	Substrate	H^+/O Ratio	References
Mitochondria			
Rat liver	β-Hydroxybutyrate (NADH)	6.0	Mitchell and Moyle (1967a, b)
	Succinate	4.0	
	Reduced TMPD	1.6	
Yeast (*Candida utilis*)	Pyruvate + malate	5.8	Downie and Garland (1973)
Beef heart	NADH	6.0	Hinkle and
submitochondrial	Succinate	4.0	Horstman (1971)
particles	Reduced PMS	1.8	
Bacteria			
Paracoccus denitrificans	Endogenous (NADPH?)	8.0	Scholes and Mitchell (1970b)
Hydrogenomonas eutropha	Endogenous (NADPH?)	7.8	Beatrice and Chappell (1974)
Escherichia coli	Malate	3.9	Lawford and
	Succinate	2.4	Haddock (1973)
Bacillus megaterium	Endogenous (NADH?)	4.1	Downs and Jones (1974)

Escherichia coli (Lawford and Haddock, 1973), *Klebsiella pneumoniae* (Brice *et al.*, 1974) and *B. megaterium* (Downs and Jones, 1974) exhibit a H^+/O ratio of 4. The H^+/O ratio of 6 for *E. coli* reported by Meyer and Jones (1973) has been retracted (Brice *et al.*, 1974) and they now agree with a previous finding (Lawford and Haddock, 1973) that *E. coli* exhibits a H^+/O ratio of 4. *Acetobacter* T71 and *Kurthia zopfii* exhibit even lower H^+/O ratios of 1.70 and 1.94 respectively (Meyer and Jones, 1973).

The span of the respiratory chain of *Paracoccus denitrificans* from NADH to oxygen (i.e. the constitutive portion) and the mitochondrial respiratory chain consist of essentially similar components (Fig. 6). The similarity of the H^+/O ratios exhibited by this segment of the respiratory chain in both *P. denitrificans* and in a mitochondrion suggests that these components are arranged in a similar way in the plasma membrane of *P. denitrificans* and in the inner mitochondrial membrane, since it is the particular orientation of the alternating electron and hydrogen carriers of the respiratory chain which appears to be responsible for proton translocation (Mitchell, 1966; Racker, 1972; Hinkle, 1973). However, despite this inferred similarity, oxidative phosphorylation coupled to electron transport in the cytochrome *c*-oxygen segment of the respiratory chain of submitochondrial particles and of phosphorylating particles from *P. denitrificans* shows two significant differences.

Firstly, in submitochondrial particles (and in mitochondria) the oxidation of reduced tetramethyl-p-phenylenediamine (TMPD) is coupled to ATP synthesis (Vallin, 1968), whereas in phosphorylating particles from *Paracoccus denitrificans* the oxidation of reduced TMPD consistently fails to show coupled ATP synthesis (Imai *et al.*, 1967; John and Whatley, 1970; Knobloch *et al.*, 1971), despite a similarity in the P/O ratios obtained with NADH oxidation by the two types of particle (Table V). This absence of significant phosphorylation is observed with a wide range of TMPD concentrations (0.25-50 μM) (P. John, unpublished), with coupled particles prepared by three different isolation procedures, and in coupled particles prepared from cells grown under a variety of conditions (autotrophically, heterotrophically, aerobically and anaerobically) (see Imai *et al.*, 1967; John and Whatley, 1970; Knobloch *et al.*, 1971). In particles prepared from *Azotobacter vinelandii* (Ackrell and Jones, 1971), from *Mycobacterium phlei* (Orme *et al.*, 1969) and from *Nitrobacter agilis* (Aleem, 1968) oxidation of reduced TMPD is coupled to significant phosphorylation, as in submitochondrial particles, whereas in otherwise coupled particles from *Pseudomonas saccharophila* (Ishaque *et al.*, 1973), as in particles from *P. denitrificans*, oxidation of reduced TMPD is not coupled to significant phosphorylation.

The second difference between oxidative phosphorylation coupled to electron transport in the cytochrome c-oxygen span of the respiratory chain of submitochondrial particles and of phosphorylating particles from *Paracoccus denitrificans* is that only in the latter (Vernon and White, 1957; Racker, 1970; Knobloch *et al.*, 1971) is phosphorylation coupled to the oxidation of reduced mammalian cytochrome c.

The only known difference between the terminal portions of the respiratory chains of *Paracoccus denitrificans* and of a mitochondrion is the presence of cytochrome o in *P. denitrificans* and its absence in mitochondria of most eukaryotes (Fig. 6).

It has been suggested by Knobloch *et al.* (1971) that the cytochrome o present in *Paracoccus denitrificans* acts as an alternative oxidase to cytochrome aa_3, reducing equivalents going by way of cytochrome o bypass the terminal phosphorylating site. Assuming the cytochrome o to have a lower cyanide and azide sensitivity than cytochrome aa_3, this hypothesis explains the marked decrease in P/O ratios observed by Knobloch *et al.* (1971) on addition of low concentrations of cyanide and azide. It is, as yet, not known precisely how the presence of cytochrome o in *P. denitrificans* could be responsible for the absence of phosphorylation coupled to the oxidation of reduced TMPD, and for the presence of phosphorylation coupled to the oxidation of reduced cytochrome c. What is clear from the H^+/O ratios measured in intact cells is that the terminal phosphorylation site is present in *P. denitrificans*, as in a mitochondrion, but that *P. denitrificans* contains, in addition, a cytochrome o which may account for some peculiar features of the oxidative phosphorylation observed with cell-free preparations.

B. RESPIRATORY CONTROL

In mitochondria supplied with sufficient respiratory substrate, the rate of respiration is dependent upon the availability of ADP and P_i. When the concentration of ADP or P_i is essentially zero then only a slow rate of respiration is observed; the maximum rate, limited by the rate at which reducing equivalents can flow down the respiratory chain, is observed in the presence of ADP and P_i, i.e. when the phosphorylation reaction is allowed to proceed. The control exerted by the phosphorylation reaction on the respiratory rate is known as respiratory control (see Lehninger, 1964). This control is routinely observed by continuously monitoring mitochondrial respiration in an oxygen electrode. When the mitochondria are supplied with excess substrate and P_i the relatively slow rate observed in the absence of ADP is known as the state 4 rate, the increased rate on the addition of ADP is known as the state 3 rate, and a return to the state 4 rate is observed when all the added ADP has been phosphorylated (Chance and Williams, 1955). The degree of control exerted by ADP is given by the respiratory control ratio, which is obtained by dividing the state 3 rate by the state 4 rate observed after the added ADP has been phosphorylated (Chance and Baltscheffsky, 1958). Tightly coupled mitochondria have high respiratory control ratios, which approach infinity when the state 4 rate approaches 0, although a respiratory control ratio of the order of 10, such as is routinely obtained with "well coupled" mitochondria, is sufficient to indicate "good" control (Chappell and Hansford, 1972). Loosely coupled mitochondria have lower respiratory control ratios and a ratio of 1 indicates no respiratory control at all.

Since the pioneering studies of Chance and Williams (1956) the ADP-induced stimulation of mitochondrial respiration has been widely employed as a convenient means of measuring mitochondrial P/O ratios, which are calculated from the total amount of oxygen consumed during the phosphorylation of the known amount of ADP added. ADP is used in preference to P_i in such measurements of respiratory control since mitochondria have a much higher affinity for ADP than for P_i. When virtually all the ADP added has been phosphorylated there is therefore a corresponding sharp decrease in the respiratory rate.

The physiological importance of ADP in controlling cellular respiration by its role in respiratory control has been shown by Chance and others in muscle, yeast and ascites tumour tissue (Chance and Williams, 1956).

A contemporary, chemiosmotic interpretation of oxidative phosphorylation would be represented in the simplest possible form thus:

$$\text{Respiratory electron transport} \longrightarrow \Delta H^+ \longrightarrow \text{ATP}$$

where ΔH^+ stands for the proton motive force set up across the inner mitochondrial membrane (Mitchell, 1966, 1972). From this representation it follows that the *essence* of respiratory control is the back pressure exerted by the proton motive force on the oxidation-reduction reactions of respiration. Thus the effect of uncoupling agents, e.g. 2,4-dinitrophenol, carbonyl cyanide m-chlorophenyl-hydrazone (CCCP) and valinomycin (in the presence of K^+), in stimulating respiration in the absence of ATP synthesis can be explained by the operation of the same mechanism as in the stimulation of respiration by the onset of the phosphorylation reaction, viz a reduction in the back pressure of the proton motive force on the proton-translocating respiratory chain (Mitchell, 1966). The validity of this interpretation of respiratory control has been experimentally reinforced by the studies of Hinkle (1973) and of Racker and co-workers (see Racker, 1972) on uncoupler-induced stimulations of electron transport in simple, defined artifical electron transport systems and resynthesized mito-chondrial respiratory systems.

Among bacteria it is only with *Paracoccus denitrificans* that we have direct, unequivocal evidence of a mitochondrial type of respiratory control. This evidence was obtained by John and Hamilton (1970) with phosphorylating particles prepared from the plasma membrane of cells of *P. denitrificans* which had been broken by osmotic lysis after treatment with lysozyme (John and Whatley, 1970). In the presence of P_i, the NADH-dependent rate of respiration by these particles is increased on the addition of a small amount of ADP, and then returns to a slow rate when all the added ADP has been converted to ATP. This slow rate can be restimulated by a further addition of ADP (Fig. 7). The ADP/O ratio, which could be calculated as for mitochondria (Chance and Williams, 1955), is found (John and Hamilton, 1970) to be similar to the P/O ratios previously determined for these particles by a direct estimation of esterified phosphate (John and Whatley, 1970). The rate of respiration observed in the absence of ADP can also be increased by the addition of uncoupling agents, e.g. CCCP (John and Hamilton, 1970, 1971) and bis-(hexafluoroacetonyl) acetone ("1799") (Fig. 7), while the rate of respiration observed in the presence of ADP can be decreased to the rate observed in the absence of ADP by the addition of venturicidin (John and Whatley, 1975). This antibiotic (an analogue of the well-known inhibitor of mitochondrial ATPase, oligomycin) is also effective in mitochondria (Walter *et al.*, 1967).

An increased rate of respiration of the particles from *Paracoccus denitrificans* is also observed upon addition of combinations of ionophorous antibiotics and cations, for example, by trinactin *plus* NH_4^+ (Fig. 7) and by valinomycin *plus* nigericin *plus* K^+ (John and Hamilton, 1971). In their sensitivity to uncoupling by these antibiotics, in the presence of the appropriate cations, the particles from *P. denitrificans* resemble submitochondrial particles, and differ from mitochondria (John and Hamilton, 1971). Thus, while mitochondria can be uncoupled by valinomycin in the presence of K^+, submitochondrial particles and the particles from *P. denitrificans* are uncoupled by valinomycin in the presence of K^+, only upon the further addition of nigericin (John and Hamilton, 1971). This difference has been attributed (Mitchell, 1968; Montal *et al.*, 1970) to the

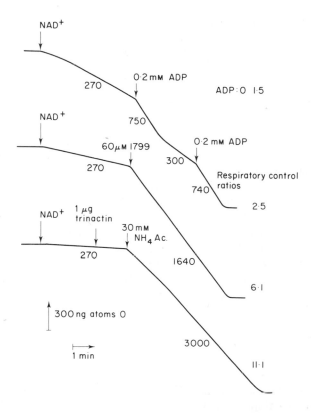

Fig. 7. Respiratory control in membrane particles from *Paracoccus denitrificans*. Membrane particles were prepared by a modification of a previously published procedure (John and Hamilton, 1970; John and Whatley, 1970) from cells grown with succinate as substrate and nitrate as terminal electron acceptor. The reaction mixture contained in a total volume of 3 ml : 5 mM tris phosphate (pH 7.3), 5 mM magnesium acetate, 30 μl ethanol, and 0.1 mg alcohol dehydrogenase (Sigma, A-7011). Additions of 0.6mM NAD$^+$ and of the other reagents were made as indicated. Oxygen uptake was measured in a Clark-type oxygen electrode maintained at 30°C. The ADP/O ratio was calculated as described for mitochondria by Chance and Williams (1956). The respiratory control ratios were obtained, as for mitochondria (Chance and Baltscheffsky, 1958; Chappell and Hansford, 1972) either by dividing the respiratory rate observed after ADP addition by the rate observed after all the ADP has been esterified (top trace), or by dividing the rate observed after addition of uncoupler by the rate observed before addition of uncoupler (middle and bottom traces). The numbers alongside the traces refer to the rates of oxygen uptake in ng atoms/min per mg protein.

difference in the direction of respiration-driven proton translocation in mitochondria on the one hand, and in submitochondrial particles and the particles from *P. denitrificans* on the other hand. These results support the arguments we have made earlier for comparing the phosphorylating particles from *P. denitrificans* with submitochondrial particles rather than with intact mitochondria.

A curious feature of the respiratory control observed with the particles from *Paracoccus denitrificans* is that while addition of ADP, in the presence of P_i (or arsenate) causes a two- to threefold stimulation of respiration, addition of uncouplers such as CCCP or 1799, or ionophorous antibiotics in combination with cations, give up to a tenfold stimulation of respiration (Fig. 7) (John and Hamilton, 1971; John, 1973). In general, intact mitochondria do not show such a large disparity between the ADP and uncoupler-induced respiration rates, although the respiratory rate of phosphorylating submitochondrial particles is rarely affected by ADP (Vallin, 1968; Van de Stadt *et al.*, 1973), and yet can be stimulated significantly by uncouplers (Lee *et al.*, 1969). Thus it is possible that the respiration of intact cells of *P. denitrificans* is controlled by the availability of ADP to the same extent as the respiration of intact mitochondria, but we have no direct evidence that this is so.

In the vast majority of phosphorylating bacterial preparations (see Gel'man *et al.*, 1967; Smith, 1968) the respiratory rate is not significantly affected by the presence or absence of an accompanying phosphorylation reaction. However, in phosphorylating particles of *Nitrobacter winogradskyi* NADH-dependent respiration is stimulated twofold by the addition of ADP, with a return to a slow rate when phosphorylation of the added ADP is complete (Cobley and Chappell, 1974). The extraordinary feature of these particles is that when nitrite is the respiratory reductant a "reverse respiratory control" can be observed: addition of ADP causes a decrease in the rate of oxygen uptake, which returns to a faster rate when phosphorylation of the added ADP is complete. Consistently with this ADP effect, CCCP causes a stimulation of NADH oxidation and an inhibition of nitrite oxidation. A mitochondrial type of respiratory control has also been observed in extracts of *Pseudomonas denitrificans* oxidizing succinate (Ohnishi, 1963), although it was observed in only a few of the preparations studied. Indications of respiratory control in phosphorylating particles of other bacteria have come from the observation of increased rates of respiration on the inclusion of P_i or of P_i-acceptor with preparations from *Escherichia coli* (Hersey and Ajl, 1950), *Micrococcus lysodeikticus* (Ishikawa and Lehninger, 1962), *Mycobacterium phlei* (Revsin and Brodie, 1967) *Alcaligenes faecalis* (Scocca and Pinchot, 1968), and *A. vinelandii* (Eilermann *et al.*, 1970; Jones, Erickson and Ackrell, 1971). However, the degree of respiratory stimulation is lower than that observed with particles from *Paracoccus denitrificans*, and a return to a slow rate on phosphorylation of all the added ADP (an essential feature of a "control" process) is usually not observed.

The maximum rates of NADH-dependent respiration by particles from *Paracoccus denitrificans* are two- to threefold greater than those observed with submitochondrial particles. The degree of respiratory stimulation produced by the addition of uncoupling agents is also two- to threefold greater with the particles from *P. denitrificans* than with submitochondrial particles. In order to determine whether there was any relationship between the degree of uncoupler stimulation and the maximum respiratory rates in these two types of particle, the respiratory control ratios of representative particle preparations of *P. denitrificans* and of representative submitochondrial particle preparations were plotted against the uncoupler-induced respiratory rates observed with the same preparations. From the resultant graph (Fig. 8) it can be seen that the points obtained with the particles of *P. denitrificans* oxidizing NADH fall roughly on an extrapolation of the points obtained with the submitochondrial particles oxidizing either NADH or succinate. However, the points obtained with the same preparations of the particles of *P. denitrificans* oxidizing succinate fall outside this area (Fig. 8). In this respect succinate oxidation by particles of *P. denitrificans* resembles NADH oxidation by phosphorylating particles of *Azotobacter vinelandii* (Jones, Erickson and Ackrell, 1971; Jones, Ackrell and Erickson, 1971) (Eilermann *et al.*, 1970, 1971) (see Fig. 8). Particles of *A.*

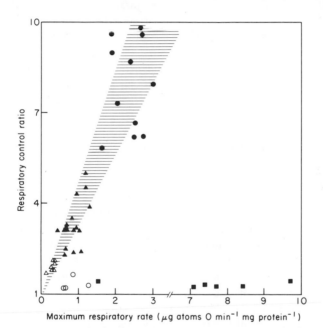

Fig. 8. Relationship between the respiratory control ratios and the maximum respiratory rates observed with submitochondrial particles and phosphorylating particles from *Paracoccus denitrificans* and from *Azotobacter vinelandii*. Data for the particles of *P. denitrificans* was obtained as described for Fig. 7, using a combination of NH_4^+ and either gramicidin or trinactin to obtain the maximum respiratory rates; for submitochondrial particles from Lee *et al.* (1969) and Montal *et al.* (1970), the maximum respiratory rates being taken as those obtained after addition of the uncoupler carbonylcyanide *p*-trifluoromethoxy phenylhydrazone; and for particles of *A. vinelandii* from Eilermann *et al.* (1970) and Jones, Ackrell and Erickson (1971), the maximum respiratory rates being taken as those obtained after addition of ADP (see text p. 94). Key to symbols: submitochondrial particles oxidizing NADH (▲) or succinate (△); *P. denitrificans* particles oxidizing NADH (●) or succinate (○); *A. vinelandii* particles oxidizing NADH (■).

vinelandii have much higher rates of respiration, but much lower respiratory control ratios compared to particles of *P. denitrificans* and submitochondrial particles. The particles of *A. vinelandii* also differ from particles of *P. denitrificans* and submitochondrial particles in that the maximum respiratory rate is obtained with the addition of either ADP or of uncoupler (Jones, Ackrell and Erickson, 1971).

The variation between the points obtained with a particular type of preparation (Fig. 8) is presumably due to uncontrolled variation in the preparation procedure, and, for the bacterial particles, in the culture conditions. The possible relationship between respiratory rates and the degree of respiratory control observed has already been discussed by Jones *et al.* (Jones, Erickson and Ackrell, 1971; Jones, Ackrell and Erickson, 1971) with reference to the phosphorylating particles of *A. vinelandii.* It is sufficient for our present purposes to point out that mitochondrial respiratory control appears to resemble more closely the respiratory control observed with particles of *P. denitrificans* oxidizing NADH than the respiratory control observed with particles of *A. vinelandii* oxidizing NADH.

In intact bacteria the most direct evidence of respiratory control has come from the work of Oishi *et al.* (1970) and Oishi and Aida (1970). These authors observed that addition of small quantities of P_i to cells of *Brevibacterium ammoniogenes* which had been grown under conditions of phosphate limitation resulted in an increased rate of respiration; this later returned to a low rate which could be restimulated by a further addition of P_i. When the resulting oxygen electrode trace (e.g. Fig. 7 of Oishi *et al.*, 1970) is analysed in the same way as a trace obtained with respiring mitochondria to which ADP had been added (Chance and Williams, 1956), then for glucose oxidation a P/O ratio of 0.3 and a respiratory control ratio of 4 are obtained. The ability of P_i to stimulate the respiration of P_i-deficient cells was also observed with *Escherichia coli, Aerobacter aerogenes*, and *Serratia marcescens*, but not with *Azotobacter vinelandii, Bacillus subtilis* and *B. megaterium* Oishi *et al.*, 1970).

Other evidence which implies respiratory control in intact bacteria has come from the stimulation of the respiration of bacterial cells observed on the addition of uncouplers. In *Escherichia coli* stimulation of glucose-dependent oxygen uptake occurs only in those cells which are able to carry out oxidative phosphorylation (Hempfling, 1970b). In the autotroph *Hydrogenomonas* H-16 (probably *H. eutropha*, Davis *et al.*, 1969) the rate of hydrogen and oxygen consumption is decreased in the absence of carbon dioxide (when presumably the ATP/ADP ratio is high), and this slow rate of respiration is increased by the addition of CCCP to the level observed in the presence of carbon dioxide; CCCP has no significant effect on the rate of respiration in the presence of carbon dioxide (Hippe, 1967). These results are consistent with the operation of respiratory control in *E. coli* and *Hydrogenomonas* H-16, but it is only in *Paracoccus denitrificans* that an alternative explanation, that the effect of the uncoupler is to allow the internal pH of the bacterial cell to change to a pH more suitable for the activity of the rate-limiting, internally orientated dehydrogenase, has been experimentally excluded (Scholes and Mitchell, 1970a).

In none of the bacteria examined, including *Paracoccus denitrificans*, does the uncoupler-induced stimulation of respiration in whole cells exceed about twofold. Yet it is known that the respiration of membrane particles of *P. denitrificans* observed in the absence of ADP can be stimulated by tenfold on the addition of an uncoupler (Figs 7 and 8). The most probable explanation for this discrepancy is that the respiration of intact cells before the addition of uncoupler is already released from respiratory control. This respiration could be linked to ion transport (see Harold, 1972) or to continuous ATP synthesis linked to ATP-utilizing metabolic reactions. Both ion transport and ATP synthesis could reduce the effective backpressure of the proton motive force on the respiratory chain. Thus the small degree of uncoupler-induced respiratory stimulation in intact bacterial cells cannot be used as evidence for poor respiratory control in bacteria.

Investigations of the relationship between the generation and utilization of energy in bacteria have consistently failed to provide evidence for control systems analogous to respiratory control in mitochondria (see reviews by Gunsalus and Shuster, 1961; Senez, 1962; Forrest and Walker, 1971). If the respiratory rate of bacteria is controlled by the availability of ADP, as in mitochondria, then it should be decreased when there is a decreased requirement of ATP for biosynthesis, since the ATP level would then be high and the ADP level correspondingly low. However, bacteria do not reduce their rate of metabolism of energy-yielding substrates in response to a decreased requirement of energy for biosynthesis (see Senez, 1962). This applies equally to bacteria which depend on oxidative phosphorylation and to those which depend on fermentation for ATP synthesis, e.g. *Desulphovibrio desulphuricans* (Senez, 1962) and *Streptococcus faecalis* (Rosenberger and Elsden, 1960) respectively. Since there is no evidence that in fermentative bacteria glycolysis can proceed without a concomitant synthesis of ATP from ADP, it is concluded that when the demand for ATP to be used in biosynthesis is decreased, ATP is now used with a regeneration of ADP by more cryptic processes (Gunsalus and Shuster, 1961). If these processes occurred in aerobic bacteria they would account for the failure to observe a decreased respiratory rate when biosynthesis is decreased. Mechanisms which have been proposed to account for this non-biosynthetic ATP utilization are: (i) the formation of polymers, such as poly-β-hydroxybutyrate and polyphosphate, and (ii) energy-dissipating reactions coupled to ATP hydrolysis (Gunsalus and Shuster, 1961).

Thus, while there is evidence that some type of respiratory control is of general occurrence in bacteria, an unequivocal demonstration of a classical mitochondrial type of respiratory control has been made only with phosphorylating particles from *Paracoccus denitrificans*.

C. ADENOSINE TRIPHOSPHATASE

The ATPase complex located in the inner mitochondrial membrane and in the bacterial plasma membrane is responsible for the catalysis of the terminal step in oxidative phosphorylation: the condensation of ADP and P_i to form ATP. The current concept of the ATPase complex has arisen largely from the extensive studies of Racker and his colleagues (see reviews of Racker, 1970, 1972). The two major components of the ATPase complex are the so-called F_1 and CF_0 components. The F_1 components can be seen in negatively stained preparations viewed under the electron microscope as a row of stalked spheres lining the inner face of the mitochondrial membrane. F_1 components isolated from the membrane ("solubilized") can catalyse ATP hydrolysis and when reintroduced to F_1-depleted submitochondrial particles they catalyse ATP synthesis. The CF_0 component consists of hydrophobic proteins which when added to an artifical membrane vesicle act as a "proton channel", allowing protons to move across the otherwise proton-impermeable membrane (Racker, 1972). *In situ* the CF_0 component constitutes that region of the inner mitochondrial membrane to which the F_1 component is bound. ATP hydrolysis but apparently not ATP synthesis catalysed by the ATPase of beef heart mitochondria is inhibited by a trypsin-sensitive protein of small molecular weight which is found tightly bound to the F_1 (Pullman and Monroy, 1963). This naturally occurring ATPase inhibitor confers on the ATPase complex an apparent kinetic asymmetry in its catalytic capabilities. The interaction of this inhibitor with the F_1 component of the ATPase is loosened during coupled substrate oxidation and when the ATP/ADP ratio is low (Van de Stadt *et al.*, 1973). Although the mechanism by which the ATPase inhibitor operates is unknown, it is clear that it functions in beef heart mitochondria as a "directional regulator of respiratory chain-linked energy transfer, controlling the backflow of energy from ATP to the mito-chondrial electron- and ion-transport systems" (Asami *et al.*, 1970). The ATPase inhibitor protein has been isolated from beef heart mitochondria, but it has not been detected in the ATPases of yeast and rat liver mitochondria (Senior, 1973).

Less detailed knowledge is available of bacterial ATPases but the information obtained suggests that the mitochondrial and bacterial ATPases are, in essence, similar both structurally and functionally (see Harold, 1972). Thus the ATPase of both *Paracoccus denitrificans* and beef heart mitochondria is inactivated by the chemical modifying agent 7-chloro-4-nitrobenzo-2-oxa-1,3-diazole (Ferguson *et al.*, 1974). In both types of ATPase the rates of inactivation are similar, there is a similar pH dependence, and in both cases the inhibition is relieved by the addition of dithiothreitol. We do not attach great significance to this observation as indicating a special similarity between the ATPases of *P. denitrificans* and a mitochondrion, since it is probable that future studies with a wider variety of organisms will reveal that this chemical modifying agent is "a general inhibitor of ATPases involved in oxidative phosphorylation" (Ferguson *et al.*, 1974).

In mitochondria the ATPase complex is orientated on the inner membrane with the F_1 component facing the mitochondrial matrix and thus directly accessible to ATP, ADP and P_i from the matrix side of the membrane (Racker, 1970). In *Paracoccus denitrificans* it has been inferred (John and Hamilton, 1970, 1971) from the ready accessibility of the ATPase in inside-out membrane vesicles that the ATPase of this bacterium, like that of other bacteria (see Harold, 1972), is similarly orientated towards the cytosolic (inside) face of the plasma membrane of *P. denitrificans*.

The rate of ATP hydrolysis of phosphorylating membranes of *Paracoccus denitrificans* is only one-tenth of the rate of phosphorylation observed with NADH as substrate (Imai *et al.*, 1967; Ferguson *et al.*, 1974), and ATP hydrolysis is not stimulated by the addition of uncouplers (Imai *et al.*, 1967). Thus the ATPase of *P. denitrificans* exhibits a kinetic asymmetry such that it synthesizes ATP much more rapidly than it hydrolyses ATP, even though ATP hydrolysis, assayed in the presence of an uncoupler, is thermodynamically highly favourable. A unidirectionality of catalysis by the mitochondrial ATPase of beef heart mitochondria has been ascribed to the presence of a trypsin-sensitive ATPase inhibitor protein. In bacterial preparations which exhibit a similar unidirectionality it is probable that an analogous protein is responsible. Although no such protein has yet been isolated from bacteria, the ATPase activity of membranes from *Mycobacterium phlei* (Bogin *et al.*, 1970), from *Azotobacter vinelandii* (Eilermann *et al.*, 1971), and from *Micrococcus lysodeikticus* (Ishikawa, 1970) is greatly stimulated by treatment with trypsin. These bacteria, like *P. denitrificans*, are wholly dependent on a respiratory metabolism and cannot live by a fermentative metabolism. Hence *in vivo* the ATPase operates in the direction of ATP synthesis rather than in the direction of ATP hydrolysis. On the other hand, in a bacterium such as *Streptococcus faecalis* which is devoid of a respiratory chain (Harold, 1972) the reverse applies: ATP synthesis occurs not at the membrane, but is coupled to fermentative metabolism in the cytosol, and ATP hydrolysis at the membrane generates a transmembrane pH gradient and membrane potential to drive the accumulative uptake of metabolites (Harold, 1972). As expected from this rationale the ATPase activity of *S. faecalis* is not stimulated by trypsin (see Harold, 1972) and presumably lacks the ATPase inhibitor of aerobic bacteria and of beef heart mitochondria. It may also be noted that the ATPase activity of the facultative aerobe *Escherichia coli* grown aerobically is stimulated to a limited extent by treatment with urea or trypsin.

Thus it appears that the ATPase of *Paracoccus denitrificans* resembles that of beef heart mitochondria in being a strongly unidirectional enzyme, preferentially catalysing ATP synthesis rather than ATP hydrolysis. It would be of interest to determine whether a protein analogous to that isolated from the ATPase of beef heart mitochondria (Pullman and Monroy, 1963) is responsible for the unidirectionality of the ATPase of *P. denitrificans*.

VII. MEMBRANE TRANSPORT

While detailed and extensive studies have been made of the transport of metabolites across the inner mitochondrial membrane (Chappell, 1968; Meijer and Van Dam, 1974), the transport properties of the plasma membrane of *Paracoccus denitrificans* remain largely unknown except for those involved in the uptake of the amino acids glycine, alanine, glutamine and asparagine (White *et al.*, 1974), the uptake of sulphate and P_i (Burnell *et al.*, 1975), and in the chelation and uptake of iron (Tait, 1975). However, there is no reason to suppose that *P. denitrificans* differs significantly from better-known bacteria, such as the obligate aerobe *Azotobacter vinelandii* and the facultative aerobe *Escherichia coli*, in the types of transport systems it uses and in their general properties.

A comparison of transport systems in mitochondria and in bacteria (Table VII) shows that, while the substances transported by mitochondria and bacteria are often similar, the mode of operation of the mitochondrial carriers is often different from the mode of operation of the bacterial carriers.

The only carrier which appears to operate by the same mechanism in mitochondria and in bacteria is the P_i/H^+ symporter (equivalent to a P_i/OH^- exchange carrier). In both rat liver mitochondria and in the facultative aerobe *Staphylococcus aureus* this carrier has an alkaline pH optimum, reacts with arsenate as well as phosphate, and is inhibited by mercurials and other sulphydryl group reagents (see Mitchell, 1970). A similar mechanism of P_i uptake appears to operate in *Paracoccus denitrificans* (Burnell *et al.*, 1975).

In mitochondria from a variety of animal sources dicarboxylic and tricarboxylic acid intermediates of the citric acid cycle traverse the inner mitochondrial membrane via a series of specific exchange mechanisms (Chappell and Haarhoff, 1967). From studies made of the uptake of dicarboxylates and tricarboxylates into cells of *Azotobacter vinelandii* (Postma and Van Dam, 1971; Visser and Postma, 1973), of *Pseudomonas* spp. (Lawford and Williams, 1971), and of *Bacillus subtilis* (Ghei and Kay, 1972) it seems unlikely that similar exchange reactions occur in bacteria, and a mechanism involving the simultaneous uptake of protons seems more likely. Similarly, while sulphate enters the mitochondrion in exchange for carboxylates (Crompton *et al.*, 1974), sulphate uptake into *Paracoccus denitrificans* appears to operate by a SO_4^{2-}/H^+ symport mechanism (Burnell *et al.*, 1975).

The mechanism of amino acid uptake into *Paracoccus denitrificans* (White *et al.*, 1974) is not yet known, but in *Staphylococcus aureus* it appears that the mitochondrial type amino acid exchange carriers are not present (Hamilton, 1975). Glutamate uptake, for example, occurs by a mechanism of glutamate/H^+ symport.

So far, the differences between the transport systems of bacteria and mitochondria can be attributed to differences in the mechanism involved.

TABLE VII

Comparison of mitochondrial and bacterial transport systems

Permeant	Carrier systems	
	Mitochondria	*Bacteria*
Anions		
P_i	Present	Present
	P_i/H^+ symport, e.g. rat liver; P_i/carboxylate exchange, e.g. rat liver[a]	P_i/H^+ symport, e.g. *Paracoccus denitrificans*,[b] *Staphylococcus aureus*[c]
Carboxylates, e.g. malate, citrate	Present	Present
	Dicarboxylate/P_i, dicarboxylate/tricarboxylate exchanges, e.g. rat liver[a]	Dicarboxylate/H^+ symport(?) e.g. *Azotobacter vinelandii*[d,e]
SO_4^{2-}	Present	Present
	SO_4^{2-}/carboxylate exchange, e.g. rat liver[f]	SO_4^{2-}/H^+ symport, e.g. *P. denitrificans*[b]
Adenine nucleotides (ATP, ADP)	Present	Absent
	ATP/ADP exchange, e.g. rat liver[g]	
Cations		
Ca^{2+}	Present	Present
	Active uptake in animal, less active in plant[h]	Active efflux, e.g. *Escherichia coli*[i] active uptake, e.g. *Bacillus megaterium*[j]
K^+	Present	Present
	Active uptake in Plant[k], less active in animal[l,m]	Active uptake, e.g. *E. coli*[e]
Na^+	Present	Present
	Low activity by Na^+/H^+ exchange, e.g. rat liver[m]	Active Na^+ efflux by Na^+/H^+ exchange, e.g. *E. coli*[n]
Fe^{3+}	Present	Present
	e.g. rat liver[o]	e.g. *P. denitrificans*[p]
Amino acids	Present	Present
	e.g. glutamate/aspartate exchange in rat liver[a]	e.g. *P. denitrificans*[q] glutamate/H^+ symport, e.g. *S. aureas*[e]
Sugars		
e.g. glucose, lactose	Absent	Present
		e.g. *E. coli*[e]

References:
[a] Meijer and Van Dam (1974). [b] Burnell *et al.* (1975). [c] Mitchell (1970). [d] Postma and Van Dam (1971). [e] Hamilton (1975). [f] Crompton *et al.* (1974). [g] Klingenberg (1970). [h] Lehninger *et al.* (1967). [i] Rosen and McClees (1974). [j] Golub and Bronner (1974). [k] Kirk and Hanson (1973). [l] Hansford and Lehninger (1972). [m] Mitchell and Moyle (1969). [n] West and Mitchell (1974). [o] Romslo and Flatmark (1973). [p] Tait (1975). [q] White *et al.* (1974).

However, bacteria and mitochondria are quite different in their transport capabilities in two respects. Firstly, bacteria, but not mitochondria, are capable of the concentrative uptake of sugars (Kaback, 1974). Secondly, mitochondria, but not bacteria, possess an adenine nucleotide carrier. This carrier allows the exchange of ATP for ADP across the inner mitochondrial membrane. The significance of this latter difference between mitochondria and bacteria has already been discussed in relation to the measurement of oxidative phosphorylation in the two systems. Its possible significance to the evolution of the mitochondrion is discussed later in this article.

VIII. SUMMARY AND CONCLUSIONS

In their nucleic acid metabolism and in their mode of protein synthesis mitochondria resemble bacteria and differ from the extramitochondrial components of the eukaryotic cell in a number of ways. These similarities and differences, which were outlined in the Introduction of this article, provide strong evidence in support of a prokaryotic origin for the mitochondrion. However, this evidence gives no indication of the type of prokaryote which was the most likely evolutionary ancestor of the mitochondrion, since all prokaryotes appear to resemble the mitochondrion to a similar extent with respect to their nucleic acid metabolism and mode of protein synthesis. On the other hand, when the energy metabolism of a mitochondrion is considered, it seems likely that the free-living ancestor of the mitochondrion was an aerobic prokaryote similar to the aerobic bacteria found today, rather than an anaerobic prokaryote similar to present-day anaerobic bacteria such as *Clostridium* spp. or *Streptococcus faecalis*, which lack a respiratory chain. This hypothesis is supported by the presence in the respiratory system of present-day aerobic bacteria of those features which are characteristic of present-day mitochondria, e.g. NADH, and succinate dehydrogenases, and cytochromes of *a, b* and *c* types. There are almost no features of the mitochondrial respiratory system which, having been sought, have not been found in at least some aerobic bacteria. However, these mitochondrial features are scattered among the different types of aerobic bacteria, apparently at random. The purpose of the present article is to demonstrate that *Paracoccus denitrificans* occupies a unique position among present-day aerobic bacteria in that it alone effectively collects in a single bacterium those features of the mitochondrial respiratory system which are otherwise distributed among a variety of aerobic bacteria.

Those mitochondrial features which *Paracoccus denitrificans* has but which are of limited distribution among other aerobic bacteria include: (a) phosphatidyl choline as a major component of the membrane phospholipid fraction, (b) straight-chain saturated and unsaturated fatty acids accounting for essentially all the membrane fatty acids, (c) ubiquinone-10 as the functional quinone of the respiratory chain, (d) two *b*-type and two *c*-type cytochromes as easily distinguishable components of the respiratory chain of aerobically grown cells, (e) a sensitivity to low concentrations of rotenone and antimycin A, and (f) cytochrome aa_3 as the terminal cytochrome.

In addition mitochondrial cytochrome c resembles more closely cytochrome c_{550} isolated from *Paracoccus denitrificans* than the c-type cytochromes which have been isolated from other aerobic bacteria, both in its structural and physical properties. The cytochrome c_{550} of *P. denitrificans* is unusual among bacterial c-type cytochromes in that it is to a large extent interchangeable with mitochondrial cytochrome c in its reactivity with mitochondrial cytochrome oxidase and with the two cytochrome oxidases of *P. denitrificans*: the constitutive cytochrome aa_3 and the inducible cytochrome cd.

Paracoccus denitrificans also resembles a mitochondrion, and differs from some other aerobic bacteria, in the stoichiometry of its oxidative phosphorylation as determined by measuring H^+/O ratios. Finally, particles prepared from *P. denitrificans* show a mitochondrial type of respiratory control, which is rarely observed in other bacterial preparations.

For our present purposes it is obviously of great significance that *Paracoccus denitrificans* possesses all these features of the mitochondrion, which have a limited distribution among other aerobic bacteria. However, *P. denitrificans* also possesses many features of the mitochondrial respiratory system which have a widespread (or possibly universal) distribution among aerobic bacteria: for example, the presence of nicotinamide nucleotide transhydrogenase, NADH and succinate dehydrogenases and a tricarboxylic acid cycle.

We should like to stress that while none of these mitochondrial features is unique to *Paracoccus denitrificans* no other bacterium possesses as many mitochondrial features as does *P. denitrificans*. For example, while *Mycobacterium phlei*, *Micrococcus lutea*, *Bacillus megaterium*, *Rhodopseudomonas spheroides* and *P. denitrificans* all resemble a mitochondrion in that they contain cytochrome aa_3, of these bacteria only *P. denitrificans* and *Rps. spheroides* contain the same type of quinone as a mitochondrion (UQ-10); and conversely of the relatively few bacteria which are known to contain UQ-10 only *P. denitrificans* and *Rps. spheroides* are known to contain cytochrome aa_3. For example, *Acetobacter xylinum* (Daniel, 1970) and aerobically grown *Rhodospirillum rubrum* (Kamen and Horio, 1970) contain cytochrome o as the sole functional oxidase.

The stoichiometry of oxidative phosphorylation in aerobically grown cells of *Rhodopseudomonas spheroides* has yet to be determined, and evidence of a mitochondrial type of respiratory control in *Rps. spheroides* is also lacking. However, the marked resemblance between the respiratory chain of *Rps. spheroides* which has been revealed by recent spectrophotometric and potentiometric studies (Dutton and Wilson, 1974) leads us to suggest that future studies of *Rps. spheroides* will reveal that oxidative phosphorylation in this bacterium, like that in *Paracoccus denitrificans*, has a similar stoichiometry to mitochondrial oxidative phosphorylation, and that *Rps. spheroides* will show a mitochondrial type of respiratory control. It is probable that future research will also reveal that *P. denitrificans* is not unique in the degree to which it resembles a mitochondrion, and that it will then be viewed as a representative of a small group of bacteria (presumably including *Rps. spheroides*) all of which will have an obvious affinity with the mitochondrion.

In addition to the mitochondrial features found in *Paracoccus denitrificans* and described above, it may be noted that there appears to be no significant feature of the mitochondrial respiratory chain which has been shown to be absent from *P. denitrificans* but present in another aerobic bacterium.

Most of the differences which exist between the respiratory chains of *Paracoccus denitrificans* and of a mitochondrion can be accounted for by the presence in *P. denitrificans* of adaptive components which are absent from mitochondria. These adaptive components may be viewed as additional components "plugged into" the constitutive portion of the respiratory chain, which is essentially similar to the mitochondrial respiratory chain (Fig. 6). This constitutive portion consists of nicotinamide nucleotide transhydrogenase, NADH and succinate dehydrogenases, flavoprotein, iron sulphur proteins, UQ-10, cytochromes of the *b* and *c* type, and cytochrome aa_3. The adaptive components of the respiratory chain of *P. denitrificans* are hydrogenase, formate and lactate dehydrogenases, nitrate reductase, and nitrite reductase (cytochrome *cd*). The synthesis of these adaptive components in response to the presence of their respective substrates in the environment of *P. denitrificans* obviously enables this bacterium to utilize a wider range both of electron acceptors (oxygen, nitrate and nitrite) and of electron donors (hydrogen, formate, reduced carbon compounds) than a mitochondrion, which is restricted to oxygen as a terminal electron acceptor and to a narrow range of reduced carbon compounds as a source of reducing power. In *P. denitrificans* the synthesis of these adaptive components is readily suppressed in the presence of alternative electron donors or acceptors which are "more acceptable"; e.g. hydrogenase is present only when cells are grown in the presence of hydrogen and in the absence of organic compounds such as glucose (Fewson and Nicholas, 1961), and nitrate reductase is absent from cells grown aerobically in the presence of nitrate (Pichinoty, 1965). This preference for oxygen over nitrate is also observed in *P. denitrificans* cells, which, although capable of using either oxygen or nitrate as terminal electron acceptor, use nitrate only when oxygen is not available (Fig. 4).

In general, eukaryotic cells are nutritionally less adaptable than bacteria, and in eukaryotic cells much of the carbon metabolism of the cell is performed in the cytosol, notably the glycolytic breakdown of sugars. The cytosol provides the mitochondrion with a more stable environment than that likely to be encountered by a free-living bacterium, such as *Paracoccus denitrificans*, whose natural environment is the soil (Verhoeven, 1957).

An evolutionary transition from *Paracoccus denitrificans* to a mitochondrion as envisaged by the endosymbiotic theory (Margulis, 1970) would involve a loss of the genetic information required for the synthesis of the adaptive components of the respiratory chain, and a retention of the constitutive components which would then form, with relatively little modification, the mitochondrial respiratory chain.

This transition does not require the adoption of new features by the respiratory chain of *Paracoccus denitrificans*, neither does it involve a significant degree of modification of those features already present in the respiratory chain of *P. denitrificans*. However, when the respective transport systems of bacteria and of mitochondria are compared (Table VII) it becomes apparent that there are mitochondrial features not found in any bacteria, as well as bacterial features not found in mitochondria. Furthermore, some carriers operate in a different way in bacteria and in mitochondria.

As with those features of the respiratory system which are present in *Paracoccus denitrificans* but are absent from mitochondria, so also the transport systems which are uniquely bacterial can be related to the potentially more varied and unstable environment of bacteria compared with that of mitochondria. Those bacterial transport systems, which would be lost in the transition to a mitochondrion, include carriers responsible for sugar transport, the periplasmic-binding proteins and the extracellular iron chelators (Hamilton, 1975; Tait, 1975).

Both the plasma membrane of *Paracoccus denitrificans* (Burnell *et al.*, 1975) and the inner mitochondrial membrane contain a sulphydryl-sensitive, phosphate carrier which mediates the uptake of P_i coupled to the simultaneous uptake of protons by an electroneutral process equivalent to the proton-symport of Mitchell (1970). Thus this carrier, already present in the plasma membrane of *P. denitrificans*, would require no modification to operate in the mitochondrion.

On the other hand carboxylates, sulphate and acidic amino acids are taken up by mitochondria via exchange systems and by bacteria, probably via proton symport systems (Table VII). The exchange carriers of the inner mitochondrial membrane clearly function to integrate the operation of the tricarboxylic acid cycle and amino acid metabolism in the mitochondrial matrix with extramitochondrial metabolism (Chappell, 1968; Meijer and Van Dam, 1974), whereas the carboxylate, sulphate and amino acid carriers of the bacterial membrane function simply in the accumulative uptake of these substances from the bacterial environment. Thus in an evolutionary transition from an aerobic bacterium resembling *P. denitrificans* to a mitochondrion modification would be necessary to the mechanism of these carriers from a proton symport to a heterologous exchange.

The principal export of the mitochondrion is ATP produced by oxidative phosphorylation. This ATP is synthesized at the ATPase located in the inner mitochondrial membrane so that it is directly accessible to ATP and to ADP only from the matrix (inner) side of the inner mitochondrial membrane. The efflux of ATP from the mitochondrial matrix to the sites of ATP utilization in the rest of the cell, and the transfer of ADP in the opposite direction, is made possible by an adenine nucleotide carrier which catalyses the equimolar exchange of ATP for ADP across the inner mitochondrial membrane. An analogous adenine nucleotide carrier has not been found in the plasma membrane of bacteria, including that of *Paracoccus denitrificans*, and the lack of an obvious function for such a carrier in a free-living organism makes its discovery unlikely. No obvious candidate exists among known bacterial carriers for an "ancestral" adenine nucleotide carrier. The adenine nucleotide carrier is the only entirely new component necessary in the evolutionary transition from the plasma membrane of *P. denitrificans* to the inner mitochondrial membrane. The acquisition of the adenine nucleotide carrier would have marked an important stage in the postulated evolution of the mitochondrion from a free-living prokaryotic ancestor, since the initial absence of this carrier prevents the symbiotic prokaryote (or protomitochondrion) from making the ATP produced by its oxidative phosphorylation available for the eukaryotic "host". If the plasma membrane of the protomitochondrion were to have become unspecifically permeable to adenine nucleotides it would necessarily have become permeable to other, smaller ions (such as protons) and thus it would have been incapable of supporting oxidative phosphorylation. If it be assumed that the protomitochondrion was tolerated by its host for a sufficient length of time for a stable, symbiotic relationship to have developed before the acquisition of the adenine nucleotide carrier by the protomitochondrion, then it is necessary to consider what advantage this protomitochondrion conferred on its host for the symbiotic relationship to have survived the selective pressures to which it was inevitably exposed. The contribution which the protomitochondrion made to its host prior to the acquisition of the adenine nucleotide carrier would probably have been substantially smaller than after it could supply ATP. A suggestion as to the nature of this contribution has been made in the Introduction of the present work, i.e. that the protomitochondrion, by virtue of its tricarboxylic acid cycle and its respiratory chain was able to oxidize completely fermentation products which would otherwise have accumulated to potentially noxious levels in a relatively large and undifferentiated proto-eukaryote. This speculative suggestion is prompted by observations which have been made on the possible role of the endosymbiotic, aerobic bacteria which are always present in the cytoplasm of the primitive, giant amoeba *Pelomyxa palustris* (Chapman-Andresen, 1971; Daniels, 1973). Lactic acid accumulates in the amoeba, presumably as a result of fermentative reactions, and, although this amoeba lacks mitochondria, oxygen is necessary for its continued existence.

The most conspicuous morphological features of the mitochondrion are the invaginations of the inner membrane, termed cristae (Fig. 5). By contrast the plasma membrane of *Paracoccus denitrificans* does not project into the interior of the cell, but remains applied to the internal surface of the cell wall (Fig. 3). Presumably the development of cristae would be associated with the progressive specialization of the aerobic bacterium resembling *P. denitrificans* essentially into a generator of ATP by oxidative phosphorylation. As this bacterium assumed its new role as a mitochondrion, the plasma membrane, as the site of oxidative phosphorylation, would increase in importance relative to the bacterial cytoplasm, which, as the site of redundant metabolic activities, would decrease in importance. Reflecting these changes, we could expect an expansion in the surface area of the plasma membrane, and a contraction in the volume of the cytoplasm, thus leading to the development of cristae.

The endosymbiotic hypothesis of the origin of the mitochondrion implies an evolutionary origin of the inner mitochondrial membrane from the plasma membrane of the ancestral prokaryote. The outer mitochondrial membrane may well have developed from the enveloping membrane produced by the eukaryotic host, since the present-day outer mitochondrial membrane closely resembles in its chemical composition and in its enzyme activities the microsomal membranes of the eukaryotic cell (Tables I and II). The original function of this enveloping membrane was probably to isolate the "invading" prokaryote. By contrast, the present-day outer mitochondrial membrane is freely permeable, allowing all but molecules of a molecular weight exceeding 10 000 to gain free access to the inner mitochondrial membrane from the cytosol. In *Pelomyxa palustris* the endosymbiotic bacteria have a distinct cell wall and each of the many bacteria present in the amoeba is bounded by a membrane, which is presumably produced by *Pelomyxa palustris* to contain the bacteria; the synthesis of this membrane keeps pace with the growth and multiplication of the bacteria.

We have demonstrated in this article that *Paracoccus denitrificans* resembles a mitochondrion more closely than does any other known bacterium. This resemblance is based not on the presence in *P. denitrificans* of characteristically mitochondrial features which are not found elsewhere among aerobic bacteria, but on the concentration in *P. denitrificans* of characteristically mitochondrial features which are otherwise distributed at random among other aerobic bacteria. The relatively close resemblance between *P. denitrificans* and a mitochondrion, compared to that which exists between other aerobic bacteria and a mitochondrion, obviously places *P. denitrificans* in a special position among aerobic bacteria. Our conclusion is that *P. denitrificans* is the closest existing free-living relative of the prokaryote which, on entering the protoeukaryote, evolved into the present-day mitochondrion. This necessarily speculative conclusion will be either supported or weakened when future determinations of the amino acid sequences of the *c*-type cytochromes of a sufficiently wide variety of bacteria have enabled the phylogenetic tree of mitochondrial cytochrome *c* to be extended "backwards in time". The currently available evidence leads us to suggest that *P. denitrificans* may be viewed as a free-living, highly adaptable mitochondrion.

ACKNOWLEDGEMENTS

The authors are grateful to Dr Jean Whatley for the electron micrographs, and to Mr D. L. A. Greenway who carried out the experiment reported in Fig. 2. Recent original work from our laboratory reported in this article has been supported by a grant from the Science Research Council. Skilled technical assistance was given by Mrs B. Hicks.

REFERENCES

Ackrell, B. A. C. and Jones, C. W. (1971). *Eur. J. Biochem.* **20**, 22-28.
Aleem, M. I. H. (1966a). *Biochim. biophys. Acta* **113**, 216-224.
Aleem, M. I. H. (1966b). *J Bact.* **91**, 729-736.
Aleem, M. I. H. (1968). *Biochim. biophys. Acta* **162**, 338-347.
Ambler, R. P. (1963). *Biochem. J.* **89**, 349-378.
Andresen, N., Chapman-Andresen, C. and Nilsson, J. R. (1968). *C. r. Trav. Lab. Carlsberg* **36**, 285-317.
Asami, K., Juntti, K. and Ernster, L. (1970). *Biochim. biophys. Acta* **205**, 307-311.
Asano, A. and Brodie, A. F. (1964). *J. biol. Chem.* **239**, 4280-4291.
Asano, A., Imai, K. and Sato, R. (1967a). *Biochim. biophys. Acta* **143**, 477-486.
Asano, A., Imai, K. and Sato, R. (1967b). *J. Biochem., Tokyo* **62**, 210-214.
Ashwell, M. and Work, T. S. (1970). *A. Rev. Biochem.* **39**, 251-290.
Baak, J. M. and Postma, P. W. (1971). *FEBS Lett.* **19**, 189-192.
Beatrice, M. C. and Chappell, J. B. (1974). *Biochem. Soc. Trans.* **2**, 151-153.
Ben Abdelkader, A. and Mazliak, P. (1970). *Eur. J. Biochem.* **15**, 250-262.
Benziman, M. and Goldhamer, H. (1967). *Bact. Proc.* **67**, 103-104.
Beyer, R. E. (1967). *Meth. Enzym.* **10**, 186-194.
Bogin, E., Higashi, T. and Brodie, A. F. (1970). *Biochem. biophys. Res. Commun.* **41**, 995-1001.
Borst, P. (1972). *A. Rev. Biochem.* **41**, 333-376.
Bovee, E. C. and Jahn, T. L. (1973). *In* "The Biology of Amoeba" (Ed. K. W. Jeon), pp. 37-82. Academic Press, New York and London.
Bovell, C. (1957). *Arch. Mikrobiol.* **59**, 13-19.
Bragg, P. D., Davies, P. L. and Hou, C. (1972). *Biochem. biophys. Res. Commun.* **47**, 1248-1255.
Brandon, J. R., Brocklehurst, J. R. and Lee, C. P. (1972). *Biochemistry, N.Y.* **11**, 1150-1154.
Brice, J. M., Law, J. F., Meyer, D. J. and Jones, C. W. (1974). *Biochem. Soc. Trans.* **2**, 523-526.
Brodie, A. F. (1959). *J. Biol. Chem.* **234**, 398-404.
Burgos, J. and Redfearn, E. R. (1965). *Biochim. biophys. Acta* **110**, 473-483.
Burnell, J. N., John, P. and Whatley, F. R. (1975). In preparation.
Butler, W. L. and Hopkins, D. W. (1970). *Photochem. Photobiol.* **12**, 439-450.
Carr, N. G. and Exell, G. (1965). *Biochem. J.* **96**, 688-692.
Chance, B. (1972). *FEBS Lett.* **23**, 3-20.
Chance, B. and Baltscheffsky, M. (1958). *Biochem. J.* **68**, 283-295.
Chance, B. and Williams, G. R. (1955). *Nature, Lond.* **175**, 1120-1121.
Chance, B. and Williams, G. R. (1956). *Adv. Enzymol.* **17**, 65-134.

Chapman, D. and Leslie, R. B. (1970). *In* "Membranes of Mitochondria and Chloroplasts" (Ed. E. Racker), pp. 91-126. Van Nostrand Reinhold Co., New York.

Chapman-Andresen, C. (1971). *A. Rev. Microbiol.* 25, 27-48.

Chappell, J. B. (1968). *Br. med. Bull.* 24, 150-157.

Chappell, J. B. and Haarhoff, K. N. (1967). *In* "Biochemistry of Mitochondria" (Eds E. C. Slater *et al.*), pp. 75-91. Academic Press, New York and London.

Chappell, J. B. and Hansford, R. G. (1972). *In* "Subcellular Components: Preparation and Fractionation" (Ed. G. D. Birnie), 2nd Edn, pp. 77-91. Butterworths, London.

Cheah, K. S. (1969). *Biochim. biophys. Acta* 180, 320-333.

Christianson, R. O., Loyter, A. and Racker, E. (1969). *Biochim. biophys. Acta* 180, 20c-21c.

Clark-Walker, G. D. and Lascelles, J. (1970). *Archs Biochem. Biophys.* 136, 153-159.

Cobley, J. G. and Chappell, J. B. (1974). *Biochem. Soc. Trans.* 2, 146-149.

Cohen, N. S., Bogin, E., Higashi, T. and Brodie, A. F. (1973). *Biochem. biophys. Res. Commun.* 54, 800-807.

Cohen, S. S. (1970). *Am. Scient.* 58, 281-289.

Coleman, J. O. D. and Palmer, J. M. (1972). *Eur. J. Biochem.* 26, 499-509.

Crane, F. L. (1965). *In* "Biochemistry of Quinones" (Ed. R. A. Morton), pp. 183-206. Academic Press, New York and London.

Crompton, M., Palmieri, F., Capano, M. and Quagliariello, E. (1974). *Biochem. J.* 142, 127-137.

Cronan, J. E. Jr. (1968). *J. Bact.* 95, 2054-2061.

Cronan, J. E. Jr. and Vagelos, P. R. (1972). *Biochim. biophys. Acta* 265, 25-60.

Daniel, R. M. (1970). *Biochim. biophys. Acta* 216, 328-341.

Daniels, E. W. (1973). *In* "The Biology of Amoeba" (Ed. K. W. Jeon), pp. 125-169. Academic Press, New York and London.

Daniels, E. W., Breyer, E. P. and Kudo, R. R. (1966). *Z. Zellforsch. mikrosk. Anat.* 73, 367-383.

Davis, D. H., Doudoroff, M., Stanier, R. Y. and Mandel, M. (1969). *Int. J. syst. Bact.* 19, 375-390.

Dayhoff, M. O. (1972). "Atlas of Protein Sequence and Structure". National Biomedical Research Foundation, Maryland.

Dickerson, R. E. (1971). *J. molec. Biol.* 57, 1-15.

Downie, J. A. and Garland, P. B. (1973). *Biochem. J.* 134, 1045-1049.

Downs, A. J. and Jones, C. W. (1974). *Biochem. Soc. Trans.* 2, 526-529.

Drozd, J. W. and Jones, C. W. (1974). *Biochem. Soc. Trans.* 2, 529-531.

Dus, K., Slettin, K. and Kamen, M. D. (1968). *J. biol. Chem.* 243, 5507-5518.

Dutton, P. L. and Storey, B. T. (1971). *Pl. Physiol., Lancaster* 47, 282-288.

Dutton, P. L. and Wilson, D. F. (1974). *Biochim. biophys. Acta* 346, 165-212.

Dutton, P. L., Wilson, D. F. and Lee, C.-P. (1970). *Biochemistry, N.Y.* 9, 5077-5082.

Eilermann, L. J. M., Pandit-Hovenkamp, H. G. and Kolk, A. H. J. (1970). *Biochim. biophys. Acta* 197, 25-30.

Eilermann, L. J. M., Pandit-Hovenkamp, H. G., Van der Meer-van Buren, M., Kolk, A. H. J. and Feenstra, M. (1971). *Biochim. biophys. Acta* 245, 305-312.

Erickson, S. K. (1971). *Biochim. biophys. Acta* 245, 63-69.

Erickson, S. K. and Parker, G. L. (1969). *Biochim. biophys. Acta* **180**, 56-62.

Ernster, L. and Kuylenstierna, B. (1970). *In* "Membranes of Mitochondria and Chloroplasts" (Ed. E. Racker), pp. 172-21. Van Nostrand Reinhold Co., New York.

Ernster, L., Dallner, G. and Azzone, G. F. (1963). *J. biol. Chem.* **238**, 1124-1131.

Ernster, L., Lee, I.-Y., Norling, B. and Persson, B. (1969). *Eur. J. Biochem.* **9**, 299-310.

Ferguson, S. J., John, P., Lloyd, W. J., Radda, G. K. and Whatley, F. R. (1974). *Biochim. biophys. Acta* **357**, 457-461.

Fewson, C. A. and Nicholas, D. J. D. (1961). *Biochim. biophys. Acta* **48**, 208-210.

Flavell, R. (1972). *Biochem. Genet.* **6**, 275-291.

Forget, P. (1968). *Annls Inst. Pasteur, Paris* **115**, 332-342.

Forget, P. and Dervartanian, D. V. (1972). *Biochim. biophys. Acta* **256**, 600-606.

Forget, P. and Pichinoty, F. (1965). *Annls Inst. Pasteur, Paris* **108**, 364-377.

Forrest, W. W. and Walker, D. J. (1971). *Adv. microb. Physiol.* **5**, 213-274.

Frisell, W. R. and Cronin, J. R. (1971). *In* "Electron and Coupled Energy Transfer in Biological Systems" (Eds T. E. King and M. Klingenberg), Vol. 1A, pp. 177-205. Marcel Dekker Inc., New York.

Gel'man, N. S., Lukoyana, M. A. and Ostrovakii, D. N. (1967). "Respiration and Phosphorylation of Bacteria". Plenum Press, New York.

Getz, G. S., Bartley, W., Stirpe, F., Notton, B. M. and Renshaw, A. (1962). *Biochem. J.* **83**, 181-194.

Ghei, O. K. and Kay, W. W. (1972). *FEBS Lett.* **20**, 137-140.

Girard, A. E. (1971). *Can. J. Microbiol.* **17**, 1503-1508.

Golub, E. E. and Bronner, F. (1974). *J. Bact.* **119**, 840-843.

Gray, C. T., Jacobs, N. J. and Ely, S. (1973). *Biochim. biophys. Acta* **325**, 72-80.

Greville, G. D. (1969). *In* "Current Topics in Bioenergetics" (Ed. D. R. Sanadi), Vol. 3, pp. 1-78. Academic Press, London and New York.

Gunsalus, I. C. and Shuster, C. W. (1961). *In* "The Bacteria" (Eds I. C. Gunsalus and R. Y. Stanier), Vol. 2, pp. 1-58. Academic Press, New York and London.

Hamilton, W. A. (1975). *Adv. microb. Physiol.* **12**, 1-53.

Hansford, R. G. and Lehninger, A. L. (1972). *Biochem. J.* **126**, 689-700.

Harms, H. (1969). *Arch. Mikrobiol.* **69**, 180-196.

Harold, F. M. (1972). *Bact. Rev.* **36**, 172-230.

Harold, F. M. (1973). *Ann. N.Y. Acad. Sci.* **227**, 297-311.

Harrison, D. E. F. and Maitra, P. K. (1969). *Biochem. J.* **112**, 647-656.

Hempfling, W. P. (1970a). *Biochim. biophys. Acta* **205**, 169-182.

Hempfling, W. P. (1970b). *Biochim. biophys. Res. Commun.* **41**, 9-15.

Henderson, P. J. F. (1971). *A. Rev. Microbiol.* **25**, 393-428.

Hendler, R. W., Towne, D. W. and Shrager, R. I. (1975). *Biochim. biophys. Acta* **376**, 42-62.

Hersey, D. F. and Ajl, S. J. (1950). *J. gen. Physiol.* **34**, 295-304.

Hill, G. C. and White, D. C. (1968). *J. Bact.* **95**, 2151-2157.

Hinkle, P. C. (1973). *Fedn Proc. Fedn Am. Socs exp. Biol.* **32**, 1988-1992.

Hinkle, P. C. and Horstman, L. L. (1971). *J. biol. Chem.* **246**, 6024-6028.

Hippe, H. (1967). *Arch. Mikrobiol.* **56**, 248-277.
Horio, T., Higashi, T., Sasagawa, M., Kusai, K., Nakai, M. and Okunuki, K. (1960). *Biochem. J.* **77**, 194-201.
Hughes, D. E., Lloyd, D. and Brightwell, R. (1970). *Symp. Soc. gen. Microbiol.* **20**, 295-322.
Imai, K., Asano, A. and Sato, R. (1967). *Biochim. biophys. Acta* **143**, 462-476.
Imai, K., Asano, A. and Sato, R. (1968a). *J. Biochem., Tokyo* **63**, 207-218.
Imai, K., Asano, A. and Sato, R. (1968b). *J. Biochem., Tokyo* **63**, 219-225.
Ishaque, M., Donawa, A. and Aleem, M. I. H. (1973). *Archs Biochem. Biophys.* **159**, 570-579.
Ishikawa, S. (1970). *J. Biochem., Tokyo* **67**, 297-312.
Ishikawa, S. and Lehninger, A. L. (1962). *J. biol. Chem.* **237**, 2401-2408.
Jagendorf, A. T. (1967). *Fedn Proc. Fedn Am. Socs exp. Biol.* **26**, 1361-1369.
John, P. (1973). *J. gen. Microbiol* **75**, xvii.
John, P. and Hamilton, W. A. (1970). *FEBS Lett.* **10**, 246-248.
John, P. and Hamilton, W. A. (1971). *Eur. J. Biochem.* **23**, 528-532.
John, P. and Whatley, F. R. (1970). *Biochim. biophys. Acta* **216**, 342-352.
John, P. and Whatley, F. R. (1975). *Nature, Lond.* (in press).
Jones, C. W. and Redfearn, E. R. (1966). *Biochim. biophys. Acta* **113**, 467-481.
Jones, C. W., Ackrell, B. A. C. and Erickson, S. K. (1971). *Biochim. biophys. Acta* **245**, 54-62.
Jones, C. W., Ackrell, B. A. C. and Erickson, S. K. (1971). *Biochim. biophys. Acta* **245**, 54-62.
Kaback, H. R. (1974). *Science, N.Y.* **186**, 882-892.
Kaltwasser, H., Vogt, G. and Schlegel, H. G. (1962). *Arch. Mikrobiol.* **44**, 259-265.
Kamen, M. D. and Horio, T. (1970). *A. Rev. Biochem.* **39**, 673-700.
Kamen, M. D. and Vernon, L. P. (1955). *Biochim. biophys. Acta* **17**, 10-22.
Kashket, E. R. and Brodie, A. F. (1963). *Biochim. biophys. Acta* **78**, 52-65.
Kirk, B. I. and Hanson, J. B. (1973). *Pl. Physiol., Lancaster* **51**, 357-362.
Klingenberg, M. (1968). *In* "Biological Oxidations" (Ed. T. P. Singer), pp. 3-54. Interscience, New York.
Klingenberg, M. (1970). *In* "Essays in Biochemistry" (Eds P. N. Campbell and F. Dickens), Vol. 6, pp. 119-159. Academic Press, London and New York.
Klingenberg, M. and Kröger, A. (1967). *In* "Biochemistry of Mitochondria" (Eds E. C. Slater *et al.*), pp. 11-27. Academic Press, London and New York.
Kluyver, A. J. (1956). *In* "The Microbe's Contribution to Biology" (Eds A. J. Kluyver and C. B. van Niel), pp. 93-129. Harvard University Press, Cambridge, U.S.A.
Kluyver, A. J. and Verhoeven, W. (1954a). *Antonie van Leeuwenhoek* **20**, 214-262.
Kluyver, A. J. and Verhoeven, W. (1954b). *Antonie van Leeuwenhoek* **20**, 337-358.
Knivett, V. A. and Cullen, J. (1965). *Biochem. J.* **96**, 771-776.
Knobloch, K., Ishaque, M. and Aleem, M. I. H. (1971). *Arch. Mikrobiol.* **76**, 114-125.
Knook, D. L., Van't Riet, J. and Planta, R. J. (1973). *Biochim. biophys. Acta* **292**, 237-245.
Kocur, M., Martinec, T. and Mazanec, K. (1968). *Antonie van Leeuwenhoek* **34**, 19-26.

Kornberg, H. L., Collins, J. F. and Bigley, D. (1960). *Biochim. biophys. Acta* **39**, 9-24.

Kornberg, H. L. and Morris, J. G. (1965). *Biochem. J.* **95**, 577-586.

Kröger, A. and Dadák, V. (1969). *Eur. J. Biochem.* **11**, 328-340.

Lam, Y. and Nicholas, D. J. D. (1969a). *Biochim. biophys. Acta* **172**, 450-461.

Lam, Y. and Nicholas, D. J. D. (1969b). *Biochim. biophys. Acta* **180**, 459-472.

Lance, C. and Bonner, W. D. Jr. (1968). *Pl. Physiol., Lancaster* **43**, 756-766.

Lawford, H. G. and Garland, P. B. (1972). *Biochem. J.* **130**, 1029-1044.

Lawford, H. G. and Haddock, B. A. (1973). *Biochem. J.* **136**, 217-220.

Lawford, H. G. and Williams, G. R. (1971). *Biochem. J.* **123**, 571-577.

Lee, C.-P. and Ernster, L. (1966). *In* "Regulation of Metabolic Processes in Mitochondria" (Eds J. M. Tager *et al.*), pp. 218-234. Elsevier Publishing Co. Amsterdam.

Lee, C.-P., Ernster, L. and Chance, B. (1969). *Eur. J. Biochem.* **8**, 153-163.

Lehninger, A. L. (1964). "The Mitochondrion". W. A. Benjamin Inc., New York.

Lehninger, A. L., Carafoli, E. and Rossi, C. S. (1967). *Adv. Enzymol.* **29**, 259-320.

Leiner, M., Schweikhardt, F., Blaschke, G., Konig, K. and Fischer, M. (1968). *Biol. Zbl.* **87**, 568-591.

Lemberg, R. and Barrett, J. (1973). "Cytochromes". Academic Press, London and New York.

Lyons, J. M., Wheaton, T. A. and Pratt, H. K. (1964). *Pl. Physiol., Lancaster* **39**, 262-268.

Margoliash, E. and Shejter, A. (1966). *Adv. Protein Chem.* **21**, 113-286.

Margulis, L. (1970). "Origin of Eukaryotic Cells". Yale University Press, New Haven and London.

Meijer, A. J. and Van Dam, K. (1974). *Biochim. biophys. Acta* **346**, 213-244.

Meyer, D. J. and Jones, C. W. (1973). *Eur. J. Biochem.* **36**, 144-151.

Miki, K. and Okunuki, K. (1969a). *J. Biochem., Tokyo* **66**, 831-843.

Miki, K. and Okunuki, K. (1969b). *J. Biochem., Tokyo* **66**, 845-854.

Mitchell, P. (1966). *Biol. Rev.* **41**, 445-502.

Mitchell, P. (1968). "Chemiosmotic Coupling and Energy Transduction". Glynn Research Ltd, Bodmin, Cornwall.

Mitchell, P. (1970). *Symp. Soc. gen. Microbiol.* **20**, 121-166.

Mitchell, P. (1972). *FEBS Symp.* **28**, 353-370.

Mitchell, P. and Moyle, J. (1965). *Nature, Lond.* **208**, 1205-1206.

Mitchell, P. and Moyle, J. (1967a). *In* "Biochemistry of Mitochondria" (Eds E. C. Slater *et al.*), pp. 53-74. Academic Press, New York and London.

Mitchell, P. and Moyle, J. (1967b). *Biochem. J.* **105**, 1147-1162.

Mitchell, P. and Moyle, J. (1969). *Eur. J. Biochem.* **9**, 149-155.

Mok, T. C. K., Rickard, P. A. D. and Moss, F. J. (1969). *Biochim. biophys. Acta* **172**, 438-449.

Montal, M., Chance, B. and Lee, C.-P. (1970). *J. memb. Biol.* **2**, 201-234.

Moyle, J. and Mitchell, P. (1973a). *Biochem. J.* **132**, 571-585.

Moyle, J. and Mitchell, P. (1973b). *FEBS Lett.* **30**, 317-320.

Murthy, P. S. and Brodie, A. F. (1964). *J. biol. Chem.* **239**, 4292-4297.

Nass, S. (1969). *Int. Rev. Cytol.* **25**, 55-129.

Needleman, S. B. and Blair, T. T. (1969). *Proc. natn. Acad. Sci. U.S.A.* **63**, 1227-1233.

Newton, N. (1969). *Biochim. biophys. Acta* **185**, 316-331.

Oesterhelt, D. and Stoeckenius, W. (1973). *Proc. natn. Acad. Sci. U.S.A.* **70**, 2853-2857.

Ohnishi, T. (1963). *J. Biochem., Tokyo* **53**, 71-79.
Oishi, K. and Aida, K. (1970). *J. gen. appl. Microbiol., Tokyo* **16**, 393-407.
Oishi, K., Kim, R., Aida, K. and Uemura, T. (1970). *J. gen. appl. Microbiol., Tokyo* **16**, 301-314.
Orme, T. W., Revsin, B. and Brodie, A. F. (1969). *Archs Biochem. Biophys.* **134**, 172-179.
Parsons, D. F., Williams, G. R., Thompson, W., Wilson, D. and Chance, B. (1967). *In* "Mitochondrial Structure and Compartmentation" (Eds E. Quagliariello *et al.*), pp. 29-70. Adriatica Editrice, Bari.
Pascal, M.-C., Pichinoty, F. and Bruno, V. (1965). *Biochim. biophys. Acta* **99**, 543-546.
Passam, H. C. and Palmer, J. M. (1971). *J. exp. Bot.* **22**, 304-313.
Passam, H. C., Berden, J. A. and Slater, E. C. (1973). *Biochim. biophys. Acta*, **325**, 54-61.
Pichinoty, F. (1965). *Annls Inst. Pasteur, Paris* **109**, 248-255.
Pichinoty, F. and d'Ornano, L. (1961). *Biochim. biophys. Acta* **52**, 386-389.
Porra, R. J. and Lascelles, J. (1965). *Biochem. J.* **94**, 120-126.
Postma, P. W. and Van Dam, K. (1971). *Biochim. biophys. Acta* **249**, 515-527.
Pullman, M. E. and Monroy, G. C. (1963). *J. biol. Chem.* **238**, 3762-3769.
Racker, E. (1970). *In* "Essays in Biochemistry" (Eds P. N. Campbell and F. Dickens), Vol. 6, pp. 1-22. Academic Press, London and New York.
Racker, E. (1972). *In* "Membrane Research" (Ed. C. F. Fox), pp. 97-114. Academic Press, New York and London.
Raff, R. A. and Mahler, H. R. (1972). *Science, N.Y.* **177**, 575-582.
Raven, P. H. (1970). *Science, N.Y.* **169**, 641-646.
Ray, S. K. and Cross, G. A. M. (1972). *Nature New Biology*, **237**, 174-175.
Reusch, V. M. Jr. and Burger, M. M. (1973). *Biochim. biophys. Acta* **300**, 79-104.
Revsin, B. and Brodie, A. F. (1967). *Biochem. biophys. Res. Commun.* **28**, 635-640.
Romslo, I. and Flatmark, T. (1973). *Biochim. biophys. Acta* **305**, 29-40.
Roodyn, D. B. (1967). *In* "Enzyme Cytology" (Ed. D. B. Roodyn), pp. 103-180. Academic Press, New York and London.
Rosen, B. P. and McClees, J. S. (1974). *Proc. natn. Acad. Sci. U.S.A.* **71**, 5042-5046.
Rosenberger, R. F. and Elsden, S. R. (1960). *J. gen. Microbiol.* **22**, 726-739.
Rydström, J., Teixeira Da Cruz, A. and Ernster, L. (1971). *Eur. J. Biochem.* **23**, 212-219.
Sapshead, L. M. and Wimpenny, J. W. T. (1972). *Biochem. biophys. Acta* **267**, 388-397.
Sato, R. (1956). *In* "Inorganic Nitrogen Metabolism" (Eds W. D. McElroy and B. Glass), pp. 163-175. Johns Hopkins Press, Baltimore.
Schatz, G. (1967). *Meth. Enzym.* **10**, 197-202.
Schlegel, H. G. and Eberhardt, U. (1972). *Adv. microb. Physiol.* **7**, 205-242.
Schnepf, E. and Brown, R. M. Jr. (1971). *In* "Origin and Continuity of Cell Organelles" (Eds J. Reinert and H. Ursprung), pp. 299-322. Springer Verlag, Berlin.
Scholes, P. B. and Mitchell, P. (1970a). *J. Bioenerg.* **1**, 61-72.
Scholes, P. B. and Mitchell, P. (1970b). *J. Bioenerg.* **1**, 309-323.
Scholes, P. B. and Smith, L. (1968a). *Biochim. biophys. Acta* **153**, 350-362.
Scholes, P. B. and Smith, L. (1968b). *Biochim. biophys. Acta* **153**, 363-375.

Scholes, P. B., McLain, G. and Smith, L. (1971). *Biochemistry, N.Y.* **10**, 2072-2075.
Scocca, J. J. and Pinchot, G. B. (1968). *Archs Biochem. Biophys.* **124**, 206-217.
Senez, J. C. (1962). *Bact. Rev.* **26**, 95-107.
Senior, A. E. (1973). *Biochim. biophys. Acta* **301**, 249-277.
Shipp, W. S. (1972a). *Archs Biochem. Biophys.* **150**, 459-472.
Shipp, W. S. (1972b). *Archs Biochem. Biophys.* **150**, 482-488.
Singer, T. P. (1963). *In* "The Enzymes" (Eds P. D. Boyer *et al.*), Vol. 7, pp. 345-381. Academic Press, New York and London.
Skulachev, V. P. (1971). *In* "Current Topics in Bioenergetics" (Ed. D. R. Sanadi), Vol. 4, pp. 127-190. Academic Press, New York and London.
Skulachev, V. P. (1972). *J. Bioenerg.* **3**, 25-38.
Slabas, A. R. and Whatley, F. R. (1974). *Biochem. Soc. Trans.* **2**, 929-930.
Slater, E. C. (1971). *Q. Rev. Biophys.* **4**, 35-71.
Smith, L. (1968). *In* "Biological Oxidations" (Ed. T. P. Singer), pp. 55-122. Interscience, New York.
Smith, L., Newton, N. and Scholes, P. B. (1966). *In* "Hemes and Hemoproteins" (Eds B. Chance *et al.*), pp. 395-403. Academic Press, New York and London.
Stanier, R. Y. (1970). *Symp. Soc. gen. Microbiol.* **20**, 1-38.
Stoffel, W. and Schieffer, H.-G. (1968). *Hoppe-Seyler's Z. physiol. Chem.* **349**, 1017-1026.
Stouthamer, A. H. (1969). *Meth. Microbiol.* **1**, 629-663.
Stouthamer, A. H. and Bettenhaussen, C. (1973). *Biochim. biophys. Acta* **301**, 53-70.
Swank, R. T. and Burris, R. H. (1969). *Biochim. biophys. Acta* **180**, 473-489.
Taber, H. W. and Morrison, M. (1964). *Archs Biochem. Biophys.* **105**, 367-379.
Tait, G. H. (1975). *Biochem. J.* **146**, 191-204.
Taniguchi, S. and Kamen, M. D. (1965). *Biochim. biophys. Acta* **96**, 395-428.
Taylor, J. F. R. (1974). *Taxon* **23**, 229-258.
Thayer, W. S. and Hinkle, P. C. (1973). *J. biol. Chem.* **248**, 5395-5402.
Timkovich, R. and Dickerson, R. E. (1972). *J. molec. Biol.* **72**, 199-203.
Timkovich, R. and Dickerson, R. E. (1973). *J. molec. Biol.* **79**, 39-56.
Tucker, A. N. and Lillich, T. T. (1974). *Antimicrob. Agents Chemother.* **6**, 572-578.
Vallin, I. (1968). *Biochim. biophys. Acta* **162**, 477-486.
Van der Beek, E. G. and Stouthamer, A. H. (1973). *Arch. Mikrobiol.* **89**, 327-339.
Van de Stadt, R. J., De Boer, B. L. and Van Dam, K. (1973). *Biochim. biophys. Acta* **292**, 338-349.
Van de Stadt, R. J., Nieuwenhuis, F. J. R. M. and Van Dam, K. (1971). *Biochim. biophys. Acta* **324**, 173-176.
Verhoeven, W. (1957). *In* "Bergey's Manual of Determinative Bacteriology" (Eds R. S. Breed *et al.*), 7th Edn, p. 463. Williams and Wilkins Co., Baltimore.
Vernon, L. P. (1956). *J. biol. Chem.* **222**, 1035-1044.
Vernon, L. P. and Mangum, J. H. (1960). *Archs. Biochem. Biophys.* **90**, 103-104.
Vernon, L. P. and White, F. G. (1957). *Biochim. biophys. Acta* **25**, 321-328.
Visser, A. S. and Postma, P. W. (1973). *Biochim. biophys. Acta* **298**, 333-340.
Vogt, M. (1965). *Arch. Mikrobiol.* **50**, 256-281.
Walter, P., Lardy, H. A. and Johsnon, D. (1967). *J. biol. Chem.* **242**, 5014-5018.

Watanuki, M., Oishi, K., Aida, K. and Uemura, T. (1972). *J. gen. appl. Microbiol., Tokyo* **18**, 29-42.
West, I. C. and Mitchell, P. (1974). *Biochem. J.* **144**, 87-90.
Whistance, G. R., Dillon, J. F. and Threlfall, D. R. (1969). *Biochem. J.* **111**, 461-472.
White, D. A., Lennarz, W. J. and Schnaitman, C. A. (1972). *J. Bact.* **109**, 686-690.
White, D. C. and Sinclair, P. R. (1971). *Adv. microb. Physiol.* **5**, 173-211.
White, D. C. and Smith, L. (1962). *J. biol. Chem.* **237**, 1332-1336.
White, D. C., Tucker, A. M. and Kaback, H. R. (1974). *Archs Biochem. Biophys.* **165**, 672-680.
Whiteside, T. L., De Siervo, A. J. and Salton, M. R. J. (1971). *J. Bact.* **105**, 957-967.
Wikström, M. K. F. (1973). *Biochim. biophys. Acta* **301**, 155-193.
Wilkinson, B. J., Morman, M. R. and White, D. C. (1972). *J. Bact.* **112**, 1288-1294.
Yamanaka, T. (1973). *Space Life Sci.* **4**, 490-504.
Yamanaka, T. and Okunuki, K. (1964). *J. biol. Chem.* **239**, 1813-1817.
Yamanaka, T., Nishimura, T. and Okunuki, K. (1963). *J. Biochem., Tokyo* **54**, 161-165.

NOTE ADDED IN PROOF

The literature search on which this article was based ended in May 1974. Since submission of this article the following relevant publications have appeared:

(i) *Paracoccus denitrificans* and the evolutionary origin of mitochondrion (John, P. and Whatley, F. R. (1975). *Nature, Lond.* **254**, 495).

(ii) Phosphate transport in membrane vesicles of *Paracoccus denitrificans* (Burnell, J. N., John, P. and Whatley, F. R. (1975). *FEBS Lett.* **58**, 215).

(iii) Amino acid sequence of *Paracoccus denitrificans* cytochrome c_{550} (Timkovich, R., Dickerson, R. E. and Margoliash, E. (1976). *J. biol. Chem.* **251**, 2197).

(iv) The cytochrome fold and the evolution of bacterial energy metabolism (Dickerson, R. E., Timkovich, R. and Almassy, R. J. (1976). *J. molec. Biol.* **100**, 473).

(v) Oxidative phosphorylation in *Micrococcus denitrificans* (van Verseveld, H. W. and Stouthamer, A. H. (1976). *Arch. Mikrobiol.* **107**, 241).

(vi) Electron transport in aerobically grown *Paracoccus denitrificans*: kinetic characterization of the membrane-bound cytochromes and the stoichiometry of respiration-driven proton translocation (Lawford, H. G., Cox, J. C., Garland, P. B. and Haddock, B. A. (1976). *FEBS Lett.* **64**, 369).

(vii) Bacteria and nuclei in *Pelomyxa palustris*: comments on the theory of serial endosymbiosis (Whatley, J. M., *New Phytol.* (1976). **76**, 111).

Stomatal Behaviour and
Environment

I. R. COWAN

*Department of Environmental Biology, Research School of Biological Sciences,
Australian National University, Canberra, Australia*

I. INTRODUCTION

Evolution has not produced a membrane which has the property of allowing
carbon-dioxide to pass freely, from a gas phase on one side into aqueous solution
on the other, but at the same time impeding the transfer of vapour in the

opposite direction. This fact is crucial to the existence of autotrophic life in the terrestrial atmosphere. The distribution of green plants on the earth's surface is largely determined by the adequacy of soil water to meet the loss by evaporation which is an inevitable concomitant of assimilation of CO_2. Consider the "water-use efficiency" of a hypothetical assimilating surface exposed to the atmosphere. It is

$$\frac{A}{E} = \frac{p - p_s}{e_s - e} \cdot \frac{g^\dagger}{g} \tag{1}$$

where A and E are the molar fluxes of CO_2 and water vapour, p, e and p_s, e_s are the partial pressures of CO_2 and water vapour in the ambient air and at the surface respectively, and $g^\dagger$ and g are conductances to CO_2 and vapour transfer in the boundary layer between the surface and ambient atmosphere. In still air the conductances are proportional to coefficients of molecular diffusion; thus $g^\dagger/g = D_{CO_2}/D_{H_2O} = 0.6$. In moving air the conductances are more nearly equal. The properties of the atmosphere determine an upper limit to water-use efficiency. The partial pressure of CO_2 at the surface cannot be less than zero, and the temperature of the surface cannot be less than that of a fully ventilated wet-bulb thermometer shielded from radiation. Therefore

$$\frac{A}{E} < \frac{p}{e'(T_w) - e} = \frac{p}{\gamma(T - T_w)} \tag{2}$$

where $e'(T_w)$ is saturation vapour pressure at wet-bulb temperature T_w, T the temperature of the atmosphere, and $\gamma \approx 0.67$ mbar $°C^{-1}$, the psychrometric constant. The inequality represents only one of a set of constraints on physiological performance, the others relating to the transfer of CO_2 in the liquid phase and the biochemistry of CO_2 fixation, and the extent to which absorbed radiation is used in photosynthesis rather than dissipated as heat. Some of the constraints can be combined to produce a more complex expression than that above. But the fact remains that there is an overriding physical limitation on water-use efficiency. What strategies are available to an organism to ameliorate the effect of this limitation on physiological performance? In the broadest sense there is only one: to assimilate rapidly when the condition of the atmosphere is such as to promote a relatively small rate of evaporation or when the supply of water to the assimilating surface is adequate to meet the demand, and to assimilate slowly or not at all at other times. The ways in which these strategies are realized are diverse. In higher plants they include what have been termed "drought-escaping" mechanisms; such as leaf abscission, the production of drought-resistant seed and so on. The one mechanism which is available to all higher plants without exception involves the stomata in the epidermis of leaves and other aerial green parts. The cuticle of the outer walls of the epidermis is rather impermeable to both CO_2 and water vapour, and assimilation and transpiration take place primarily through the stomatal pores. Thus the assimilating parts of the plant are surrounded by a porous septum which has the effect of decreasing the conductances, $g^{\dagger}$ and g, to assimilation and evaporation. Stomatal aperture responds to changes in the state of water in the leaf and, directly or indirectly, to changes in all those environmental factors which affect rate of assimilation and rate of evaporation. Therefore, the stomatal mechanism not only acts as a device which prevents desiccation of plant tissue but has, in principle, the capability of varying the conductances to diffusion so as to maximize the ratio of the mean rate of assimilation to mean rate of evaporation, $\bar{A}/\bar{E}$, in a fluctuating environment. There is quite compelling evidence that this is sometimes what the mechanism tends to do; the stomata open with increase in light which promotes assimilation and close with decrease in ambient humidity and other changes in environment which promote evaporation. Yet it would be a mistake to assume that the operation of stomata is programmed in a readily recognizable way. As with all other aspects of physiological functioning stomatal movement is adapted towards the survival and propagation of the individual in the environment in which the species has evolved; only in particular circumstances and periods of ontogenetic development is the requirement likely to be met by anything so simplistically precise as maximization of $\bar{A}/\bar{E}$. The rates of growth and amounts of water used by crop plants do not suggest that water-use efficiency is always a primary consideration. One would expect there to be a distinction between stomatal behaviour in species which exist as isolated perennial plants in arid environments, and that in species—including most crop species—which exist and have evolved in communities where success of the individual is not only a matter of adaptation to the physical environment, but of competition for light and water with its neighbours. Therefore, while we can anticipate some of the behavioural characteristics that evolution has imposed on stomata, the integration of these in a precise function is a difficult task, and varies from species to species. We tend to regard the stomatal mechanism as a control device, but are unable exactly to specify what its "goal" is. It is different with some aspects of physiological functioning in higher animals to which the theory of control systems has been applied. Homeostasis can be identified with the "goal" of the physiological mechanism involved; the control of body temperature and CO_2 concentration in tissue are obvious examples. The difference is largely associated with energy resource. To explore this somewhat peripheral but interesting matter a little further, let us express the maximum water use efficiency of plants in terms of energy. Setting $p = 0.3$ mbar (i.e. 300 p.p.m. by volume) and $T - T_w = 10^{\circ}C$ in eqn (2) and taking the free energy requirement for CO_2 conversion to carbohydrate as 500 kJ mol^{-1}, then, with the latent heat of vapourization of water at 44 kJ mol^{-1}, the maximum efficiency is about 0.5. In agricultural crops, the efficiency is less than 1/20 of this; and the efficiency of individual plants does not exceed 0.1 in the most favourable conditions of temperature, humidity and radiation. An analogous calculation can be applied to respiration in animals. The air expired from the human lung is saturated with water vapour at body temperature; i.e. the vapour pressure is 62 mbar and the temperature 37°C. Assuming the inspired air is at 20°C and contains vapour at a pressure of 12 mbar (corresponding to 10°C wet-bulb depression) then 22 kJ latent heat and 0.5 kJ sensible heat are dissipated per mole air expired. As the concentration of O_2 in air is 21%, and approximately 500 kJ metabolic energy is released per mole O_2 consumed in respiration, the "water-use efficiency" of the lung is potentially about 40. In fact only about 1/5 of the oxygen inspired is removed and therefore the real efficiency is more nearly 8.

That water-use efficiencies are much less than unity in autotrophic organisms and much greater than unity in heterotrophs—a result of the particular concentrations of CO_2 and O_2 in the atmosphere rather than any innate difference in biochemical efficiency—is, in my view, a most significant factor in the evolution of the higher life-forms in the terrestrial environment. A favourable water-use efficiency in animals has allowed the development of homeostasis in body temperature and the operation of a respiratory system based on homeostasis in CO_2 concentration in blood. In contrast, the constraint of a small water-use efficiency in plants is primarily responsible for the close coupling of leaf temperature with its environment* and a stomatal mechanism the function of which is a compromise between conflicting requirements. This chapter is to do with the nature of that compromise.

I shall discuss the influences of hydraulic environment, ambient CO_2 concentration, and irradiance on the behaviour of stomata. The discussion is in three parts, representing, roughly, three levels of biological organization. The first is about the stomatal machinery. Most of the information is drawn from work with epidermal strips, or leaf segments in which the stomatal apparatus is exposed to well-defined, readily controlled, but entirely artificial environments. In the second I describe and analyse stomatal behaviour in intact plants. Here the environment of the stomatal apparatus is a milieu, being affected by the physical properties of the ambient environment, the metabolism of the plant and the functioning of the apparatus itself; much of the discussion is to do with feed-back. Some mathematical theory is used. It is familiar to analysts of man-made control systems but not to most botanists and I have therefore appended an introduction to it. Finally I speculate on the way in which stomatal behaviour in individual plants may be attuned to variation in the physical environment, it being assumed that economy in water use is of predominant importance in the regulation of gas exchange.

---

* Of course, plants may regulate their temperature during limited periods by use of metabolic energy accumulated over longer periods. The spadix of *Symplocarpus foetidus* L. Eastern Skunk Cabbage, maintains an internal temperature 15 to 35°C above ambient air temperature for at least 2 weeks by consuming oxygen at a rate comparable to that of homeothermic animals of equivalent size (Knutson, 1974).

I am conscious of giving scant attention to two topics of importance. The effects of temperature on stomatal functioning are mentioned only in relation to other, more precisely identifiable, effects; they are so ubiquitous that I have shirked the task of attempting a fuller account. And I have confined attention to rather short-term responses of stomata to environment, as opposed to those which take place with time scales of a few days or more and merge with changes associated with the ontogenetic development of plants.

II. THE STOMATAL MECHANISM

A. WATER RELATIONS OF THE EPIDERMIS

Stomatal movement is a manifestation of strain in the epidermis, associated with change in the hydraulic pressure in the epidermal cells. Substantial changes in osmotic pressure take place in guard cells, and may be caused, *inter alia*, by changes in light intensity and concentration of CO_2; variation in osmotic pressure in the neighbouring epidermal cells seems to be relatively small in most species (Meidner and Mansfield, 1968). Therefore change in the hydraulic pressure in guard cells may occur almost independently of change in turgor in the remainder of the epidermis. The distribution of turgor in the whole epidermis, including the guard cells, is affected by the potential of water in the vascular system in the leaf and loss of water by evaporation from the cuticle and the walls of the guard cells and epidermal cells bordering the stomatal and substomatal cavities. Those with interests primarily in guard cell metabolism are likely to think in terms of increase in osmotic pressure in an initially deflated guard cell as being the positive, active process in stomatal movement. But it is by no means certain that decrease in osmotic pressure is not active, also, in energetic terms and, as Raschke (1975a) puts it, the open state of the stoma is as likely as the closed one. My own interests in stomata arose from more general ones in micrometeorology and plant water relations, and I am inclined to think in terms of hydraulic processes in the sense that tends to deflate the guard cell and close the stomatal pore. It is perhaps for this reason I begin by discussing the hydraulic microenvironment of the stomatal apparatus, the assumption being that the pore is at least partially open and therefore permits the escape of water vapour.

Some textbooks imply that loss of water by evaporation from epidermal cells. including those of the stomatal apparatus is negligible, a trivial consequence of imperfections in the cuticle. The rate of loss of water from leaves with closed stomata is usually no more, and often much less, than 1/20 of that with open stomata. There is no way of quantifying the rate of evaporation from the inner walls of guard cells and subsidiary cells, but it has been suggested that a thin layer of cutin on the walls of the guard cell bordering the substomatal cavity would be effective in preventing water loss (Meidner and Mansfield, 1968).

Recently Meidner (1975) has come to a different conclusion, and has shown that the hydraulic conductivity of the epidermis is sufficient to maintain what is, in terms of the functioning of the intact leaf, a substantial rate of evaporation. His discussion of the implications of this has much in common with what I have written here.

It is clear that evaporation from the epidermis, probably in the region of the guard cells, is an important factor in stomatal functioning in some species at least. Lange *et al.* (1971) caused stomata to close and open by decreasing and increasing the humidity of air flowing over the outer surfaces of strips of the lower epidermis of *Polypodium vulgare* and *Valerianella locusta*. They were able to induce movements in single stomata or groups of stomata by the use of small jets of air of varying humidity. They found also that the stomata were sensitive to the humidity of the air to which the inner side of an epidermis was exposed. When the inner side of the epidermis was put in contact with liquid water over its whole length then the stomata did not respond to variation in ambient humidity; when a small sub-epidermal air space was created by introducing a bubble only 2 mm in diameter, then the stomata above reacted to change in ambient humidity. It should be mentioned that the species used in these experiments are unusual in that, in intact leaves, the lower epidermis is attached to the remainder of the leaf only at the margins and main veins; the stomatal apparatus is particularly well situated to act as a humidity sensor because there is minimal hydraulic continuity with the mesophyll. In most species the contact between epidermis and mesophyll is greater (however it seems significant that the guard cell has direct contact with the mesophyll in very few species). Sheriff and Meidner (1974) have shown that in *Tradescantia virginiana*, vein and bundle sheath extensions, characteristic of mesomorphic leaves, are effective in maintaining a close hydraulic communication between epidermis and vascular tissue; but that in *Hedera helix*, in which, as in most xeromorphic species, extensions are absent, there is a relatively poor hydraulic communication between epidermis and vascular tissue. A hydraulic perturbation in the epidermis of *H. helix* was propagated almost immediately to stomata some 100 mm distant, but if applied to the mesophyll and vascular tissue the response of the stomata was delayed some minutes. There is evidence from experiments using isotopic and chemical tracers of water movement in leaves (Maercker, 1965; Tanton and Crowdy, 1972) that the rate of evaporation from guard cells is greater per unit area of wall exposed to the atmosphere than from other tissue. It is important to know whether the loss is sustained primarily by the walls bordering the stomatal pore and substomatal cavity (in which case the rate is modulated, as with evaporation from the mesophyll tissue, by the behaviour of the stomata) or from the cuticle exposed to the external atmosphere. Unfortunately convection of a tracer can occur tangentially to an interface at which evaporation takes place, diffusion tends to blur the results of convection, and, with chemical tracers, other complications due to physicochemical processes may be present. Tanton and Crowdy (1972) found that lead chelate supplied to the roots or cut stems of *Triticum aestivum, Vicia faba, Plantago major* and *Prunus laurocerasus* accumulated beneath and on the surface of the cuticle, being partly associated with the anticlinal walls of the epidermal cells, and in large quantities in the walls of the guard cells. The amounts deposited decreased with decreasing stomatal aperture. Tanton and Crowdy were led to the conclusion that the cuticle of the epidermis is the primary site of water loss in the leaf and the bizarre corollary that cuticular transpiration decreases with decrease in the size of the stomatal pores. A more likely explanation of their observations is that water evaporating from the guard cells did so mainly from the walls bordering the substomatal cavities, and that it was supplied by conduction along the epidermis where lead chelate was trapped in the anticlinal walls and conveyed to the cuticle in association with a relatively small rate of cuticular transpiration. M. Jones (personal communication) has examined, by use of an electron microprobe, the pattern of accumulation of monosilicic acid taken up by transpiring leaves of *Avena sativa*. There was a marked deposition in those regions of the guard cell walls which border the substomatal cavities and extend slightly into the stomatal pore, and in the anticlinal walls of the epidermis. A small concentration occurred in the walls of mesophyll cells. None at all was found in the cuticle. That neither Tanton and Crowdy or Jones detected appreciable amounts of tracer in the mesophyll does not necessarily imply that the total rate of evaporation in the mesophyll was small. The surface area of mesophyll tissue in contact with air-spaces is of the order 10 times the superficial area of a leaf and 100 times the surface area of the guard cells. Nevertheless, it is probable that the guard cells in some species may sustain from their inner walls an appreciable proportion of the loss of water by evaporation from leaves. If they were to do so, it would be a most effective means of engendering rapid and large changes in water potential in the stomatal apparatus in response to changes in ambient humidity, and other changes in environment which tend to affect rate of evaporation, without being uneconomical in terms of total water-use. It may be demonstrated that if the inner walls of the guard cells were wet then they would support a major proportion of the total flux of water from the leaf, but if they were not, the total flux would not, because of that, be much diminished—the reason being that the width of the stomatal pore rather than the extent of the internal evaporating surface is the dominant factor in determining the rate of water loss. Figure 1 shows an electrical analogue of vapour diffusion from the tissue surrounding a substomatal cavity. The electrodes representing the inner walls of the guard and subsidiary cells were isolated from, but maintained at the same potential as, the electrode representing the mesophyll walls, and therefore the current representing evaporation from the former, I_s, could be determined separately from that from the mesophyll, I_m. A further measurement was made with the electrode representing

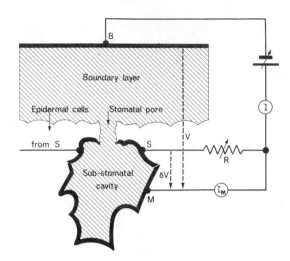

Fig. 1. Electrical analogue of transfer of vapour from a substomatal cavity to the external atmosphere. Hatched area is graphite coated conducting paper representing the space through which vapour diffuses. Thick, black lines are silver paint electrodes representing, M, walls of mesophyll cells, S, inner walls of guard cells and subsidiary cells and, B, ambient atmosphere.

the guard and subsidiary cell walls excised. With the total current initially flowing being arbitrarily taken as 100, the measurements were as follows:

TABLE I

	I_s	I_m	I
Guard cell and subsidiary cell electrode intact	77	23	100
Electrode representing guard and subsidiary cells removed	–	88	88

Thus, although 77% of the total "evaporation" took place from the guard and subsidiary cells, suppression of evaporation from these cells reduced the total amount of evaporation by only 12%. The analogue represented a rather widely open pore (approximately 20 μm); had the pore been more narrow, and had the model been a three-dimensional one, and therefore more realistic, I believe the reduction would have been still less.

    It seems unlikely that the epidermis could support a substantial proportion of the total flux of water in the leaf when the rate of evaporation is large without there being an appreciable reduction in hydraulic pressure (or matrix potential if one prefers to describe pressure in the interstices of cell walls in this way) at the sites at which evaporation takes place. This "drawdown" in pressure, so to speak, and the further reduction in water potential which might take place in the vicinity of the stomatal apparatus due to enhancement of solute concentration by convection with the liquid flow, could, in effect, provide a signal causing the stomata to respond to rate of evaporation. The analogue shown in Fig. 1 does not further discussion of this possibility as it pertains only to the process of vapour diffusion in the substomatal cavity. But it is relevant to a related problem: the effect of small temperature differences within the leaf on the potential of water in the epidermis.

Before the electrodes representing the guard cells and subsidiary cells were removed, some measurements were carried out with the potential on these electrodes made slightly less than that on the mesophyll electrode, in order to simulate a decrease in vapour pressure at the walls of the subsidiary cells and guard cells. Let us take the voltage, V, between mesophyll and atmosphere to represent a vapour pressure difference of 10 mbar. Then, with a reduction in voltage, δV, at the walls of the subsidiary cells at the equivalent of 0.14 mbar, the total "evaporation" was diminished by less than 1%, but the rates of "evaporation" from the individual electrodes were $I_s = 52$ and $I_m = 48\%$ (compared with 77 and 23%, respectively, when there was no reduction in vapour pressure). Variation in vapour pressure with variation in water potential is given by the relation $\partial e/\partial \psi = \rho_v/\rho_l$ where ρ_v and ρ_l are the densites of vapour and liquid water respectively; at $25°C$ a difference in vapour pressure of 0.14 mbar is equivalent to a difference in water potential of about -6 bar. On this basis, the new distribution of vapour flux is consistent with a reduction in water potential of 6 bar in the walls of the subsidiary cells and guard cells—a magnitude which might well cause dimunition in stomatal aperture, to an extent depending on the base level of water potential in the leaf as a whole. Parenthetically, it might be asked whether it is really feasible that approximately half the total rate of evaporation from a leaf with open stomata, and with the quite substantial difference in vapour pressure across the epidermis that has been assumed, could be sustained from the subsidiary cells and guard cells with a reduction in water potential of only 6 bar. An elegant experiment by Meidner (1975) has now shown that it probably is. So far, it has been assumed that the analogue represents an isothermal system. In point of fact a difference in vapour pressure of 0.14 mbar between mesophyll and epidermis could be ascribed to a temperature difference of about $0.07°C$. That such a small difference in temperature is equivalent to what is, in physiological terms, a large difference in water potential indicates the nature of an important interaction. Returning, once again, to the electrical analogue, let us suppose that the voltage on the electrode representing the mesophyll tissue is raised by the equivalent of 0.14 mbar, so that the difference between that and the voltage on the electrodes representing the walls of subsidiary cells and guard cells is now 0.28 mbar. The current from the latter, I_s, is reduced to 27% of the total; that is to say the rate of "evaporation" from the epidermis is halved and it must therefore be supposed that the drawdown in water potential is reduced to -3 bars. We are left with a vapour pressure difference of 0.21 mbar between the mesophyll and epidermis to account for. If it is ascribed to difference in temperature it is the equivalent of about $0.11°C$. How could such a temperature difference occur? If a leaf is exposed to a dry atmosphere and small intensity of radiation, then the latent heat associated with evaporation is drawn from the ambient air with the result that the temperature of the mesophyll tissue is less than that of the epidermis. If, on the other hand, the atmosphere is humid but the irradiance is large, the latent energy derives from the radiation absorbed by the leaf tissue, the excess of sensible heat available is transferred from the leaf to the atmosphere, and the temperature of the mesophyll tissue is greater than that of the epidermis. On the basis of some calculations I believe that the differences may well be in the range $±0.1°C$. Therefore it seems probable that, for a given total rate of evaporation, the rate of evaporation from, and depression of water potential in, the epidermis may be substantially greater or smaller than that which would occur if the leaf were isothermal.

The analogue that has been described is crude, but serves to indicate that there is only a tenuous relationship between the potential of water in leaf tissue in bulk and the local potential of water to which an individual stoma responds in ways now to be discussed.

B. HYDROACTIVATED MOVEMENT

Any explanation of the nature of the stomatal response to change in water potential must involve the mechanical and hydraulic attributes of the stomatal apparatus of course; but explanations vary in the emphasis which is placed on the role of the chemical metabolism of the guard cell. I shall deal first with mechanical and hydraulic attributes.

Because of their specialized geometry and elastic properties, a pair of guard cells is deformed with increase in hydraulic pressure in such a way that the aperture of the stomatal pore between them is increased. Aylor *et al.* (1973) emphasize the importance of two constraints which they believe are relevant for both elliptical and grass-type stomata. The radial micellae of the guard walls (Fig. 2) are relatively inextensible, and serve to link the movement of the ventral wall of each cell with that of the dorsal wall. Also, the length of the entire stomatal apparatus or the length of the common wall between the polar segments of the guard cells must be constrained during expansion. Aylor *et al.* do not agree with the suggestion that is often made that thickening of the ventral wall of the guard cell assists the opening process. But there is a distinction between elliptical stomata which have thick ventral walls and open as a result of lengthening and bending of the guard cells in the way described by Aylor *et al.*, and those, the less common "bellows" type, which have thin ventral walls and open as a result of change in shape of the cross-section of the guard cells, expansion in the direction normal to the epidermal plane being accompanied by a flattening of the ventral walls (Meidner and Mansfield, 1968).

There is not much reliable information about internal pressures required to inflate guard cells and open stomatal pores. With one exception, various

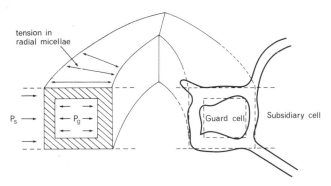

Fig. 2. Representation of a pair of guard cells as rectangular beams, loaded as a result of the stresses exerted by the hydraulic pressure inside the guard cells, P_g, acting on the ventral and dorsal walls of the guard cells, and the hydraulic pressure inside the subsidiary cell, P_s, acting on the dorsal walls of the guard cells.

hydraulic pressures are obtained either by exposing leaf tissue to differing light regimes to bring about differences in osmotic pressure, or by exposing tissue, usually in the form of epidermal strips, to solutions of differing concentration. In either case it is necessary to determine the internal osmotic pressure in order to estimate hydraulic pressure and this is usually done by a plasmolytic method. But most such determinations are of doubtful value because the guard cells have been exposed to plasmolysing solution for so long that significant leakage of solutes may have occurred (Fischer, 1973; Raschke, 1975a). The periods have typically been between 20 and 60 min—of the same order of magnitude as the response of stomata to light intensity. Fischer (1973) measured the width of stomatal aperture in illuminated epidermal strips of *Vicia faba* as a function of the concentrations of external solutions of sucrose. He assumed that leakage of solute was the same for all samples, and that changes in guard cell volume were negligible; on this basis he identified differences in internal hydraulic pressure of the guard cells with differences in the osmotic pressures of the solutions and concluded that stomatal aperture varied with pressure by about 0.5 μm bar^{-1}. It is probable that both assumptions are incorrect but that they have the virtue of engendering errors in opposite senses. Fischer's paper contains a wealth of careful investigation and analysis of factors which influence leakage of solutes and hydraulic equilibration in guard cells in solution; of particular importance, he showed that the half-time for quasiequilibration of water potential is only 20-30 s, provided conditions are such as to minimize the influence of unstirred layers. Raschke (1975a) refers to investigations (Raschke and Dickerson, 1973; Raschke *et al.*, 1973a, b) of the interrelationships between stomatal aperture, guard cell volume, and guard cell pressure in *V. faba*, in which estimates of osmotic pressure were based on observation of incipient plasmolysis after 1 min exposure to plasmolysing solution, the time constant for hydraulic equilibration being about 8 s. In brief, the volume of a guard cell lumen approximately doubles, from 2.5 to 5×10^{-12} l, and the stomatal pore opens from a nearly closed condition to an aperture of about 18 μm, as the hydraulic pressure in the guard cell increases from about 10 to 45 bar—these being the pressures which exist in turgid leaves in the dark, and after prolonged exposure to light, respectively. The pressure with open stomata is greatly in excess of those estimated by Stålfelt (1967) and Meidner (see Meidner and Mansfield, 1968) based on the usual plasmolytic technique. In so far as it is permissible to compare the characteristics of stomata in different species, it is worth noting that Bearce and Kohl (1970) measured osmotic pressures in inflated guard cells of *Chrysanthemum morifolium* and *Pelargonium hortorum* of 22 to 50 bar using a cryoscopic technique.

Meidner and Edwards (1975) have succeeded in directly manipulating the pressure within guard cells and subsidiary cells in *Tradescantia virginiana* by use of microcapillaries inserted into the cells. When a capillary was inserted into a guard cell of an open stomate, turgor was lost and the pore closed partially, apparently in association with penetration of cytoplasm into the capillary. On application of hydraulic pressure in the capillary the stomatal pore could be made to reopen up to, and beyond, its original width. Pore width increased from 4 to 26 μm with increase in pressure from about 1 to 7 bar; that is to say by 4 μm bar^{-1}—very much greater than the results for *Vicia faba* quoted above. However, it was impossible to open an initially closed stomate with an applied pressure of 10 bar—the maximum attainable with the technique used. The latter fact might be thought to indicate that the initial expansion of the guard cell is essentially an irreversible process—irreversible in the thermodynamic sense, that is—the internal pressure decreasing with increase in volume as with the inflation of a rubber balloon. However, the analogy is clearly not satisfactory because the release of pressure in guard cells in open stomata did not lead to complete deflation. Meidner and Edwards (1975) suggest that a temporal change in the elasticity of the guard cell walls associated with ion exchange may precede (during what Stålfelt (1929) termed the *Spannungsphase*) the natural opening of stomata, and that the rigidity of the guard cells in fully closed stomata might be attributable to the fact that this process of relaxation had not taken place. There is a second, rather puzzling aspect of the findings of Meidner and Edwards: that in *Tradescantia* the pressures in guard cells are somewhat less than expected from estimates of internal osmotic pressure using the plasmolytic method. This seems to be inconsistent with the criticism of the traditional plasmolytic method by Fischer (1973) and Raschke (1975a). There is a hint in the report by Edwards and Meidner (1975) that the pressure in guard cells of open stomata in *Vicia faba* may be less than that quoted by Raschke. I shall not attempt to review the multiplicity of factors which could be invoked in an attempt to explain an apparent disparity between direct and indirect measurements of pressure in guard cells—a disparity which might, in any event, prove to be less apparent if various techniques were applied to tissue of the same species and with similar pretreatment.

Meidner and Edwards (1975) caused stomata in *Tradescantia virginiana* to close by applying pressure through microcapillaries inserted into the subsidiary cells. The pressure required was the same as that found to exist in the guard cells of stomata of similar aperture. It has long been known that the aperture of a stoma is not determined by the pressure existing in the guard cells alone, but is affected by the pressure in the subsidiary* cells also. It was directly demonstrated by Heath (1938), who observed that open stomata in *T. zebrina* and *Cyclamen persicum* were made to enlarge when the pressure in the subsidiary cells was released by puncturing. DeMichele and Sharpe (1973) have analysed the mechanics of movement in elliptical stomata of the common type which has thick-walled guard cells. One conclusion is that the subsidiary cells have a mechanical advantage over the guard cells in that uniform increase in pressure in the guard cells and subsidiary cells causes a *decrease* in the aperture of a stomatal pore. Assuming that the relationship between aperture and pressure is linear for small changes, the result may be written

$$\Delta a = \left(\frac{\partial a}{\partial P_g} \right)_{P_s} (\Delta P_g - m \Delta P_s) \tag{3}$$

where a is aperture, P_g and P_s are the pressures in guard cell and subsidiary cell respectively, and the coefficient m is the mechanical advantage of the subsidiary cells. Values of m range between 1.7 and 4.1 for illustrative examples given by DeMichele and Sharpe. DeMichele and Sharpe likened guard cells to rectangular beams caused to bend in the lateral plane under a loading due to the pressures P_s and P_g acting on opposite sides of the dorsal wall of each guard cell, and the pressure P_g acting on the internal surface of the ventral walls (see Fig. 2). The simplest of considerations shows that the mechanical advantage m is greater than unity if this definition of the stresses experienced by the guard cells is accepted. The work done when the pore is caused to open is $P_g \Delta V_g - P_s(\Delta V_g + \Delta V_p)$, V_g being the internal volume of the guard cell and V_p the volume of the stomatal pore. As V_p and V_g are uniquely related, it immediately follows that the mechanical advantage of the pressure in the subsidiary cells is $m = 1 + dV_p/dV_g$. However, no account has been taken of the forces exerted on the guard cells by the walls of the subsidiary cells. In effect the mechanical relationship of the guard cell to the subsidiary cell has been taken as being that of a piston to a cylinder.

A more realistic appreciation of the anatomical arrangement of guard cell and subsidiary cell does not discover any *a priori* reason for asserting $m > 1$. Indeed, if the subsidiary cell tended to change its shape with increase in pressure as does the guard cell in what has been referred to as the bellows type of stomate then m could be negative. Equally, if the subsidiary cell tended to elongate in the

---

* I shall use this term to distinguish cells neighbouring the guard cells whether differentiated from the other epidermal cells or not.

epidermal plane with increase in pressure m would be greater than that calculated from the oversimplified relationship. Even if swelling of an isolated subsidiary cell tended to be isotropic one might expect m to vary with stomatal aperture, being relatively small, and perhaps negative, when deflation of the guard cell causes the subsidiary cell to be distended in the epidermal plane and relatively large when inflation of the guard cell causes the subsidiary cell to be compressed in the epidermal plane. It is worth noting that Heath (1938) did not observe an opening movement in closed or slightly open stomata in *Tradescantia zebrina* when subsidiary cells were punctured; an opening movement only occurred with stomata already widely open. It is probable that m is a function of the variables in eqn (3). The complexity of the relationship potentially provides a rich source of variation in the response characteristics of stomata in different species. The experiments reported by Meidner and Edwards (1975) are difficult to interpret in terms of mechanical advantage, because manipulation of pressure in the subsidiary cell must cause changes in volume, osmotic pressure, and hydraulic pressure in the adjacent guard cell. I am grateful to Professor Meidner for informing me that further experiments, some of them involving simultaneous manipulation of pressures in both cells, have indicated that the mechanical advantage in *T. zebrina* is about 1.6. This is an average over a finite range of pore aperture.

DeMichele and Sharpe discuss the steady state response of stomata to change in leaf water potential, ψ, on the basis of their evaluation of the coefficients in equation (3) and the assumption that $\Delta P_g = \Delta P_s = \Delta \psi$ when the subsidiary cells are turgid, and $\Delta P_g = \Delta \psi$ and $\Delta P_s = 0$ when the subsidiary cells are flaccid. I am inclined to think that this assumption, implying as it does that the amounts of solute in the turgid cells adjust to the changes in volume in such a way that osmotic pressures are constant, is a reasonable one. However, to accept it without discussion certainly short-circuits what many would regard as the central problem in our understanding of stomatal response to change in water potential. Qualitatively, the analysis of DeMichele and Sharpe seems to be supported by the measurements of Glinka (1971) of stomatal aperture in illuminated leaf discs of *Vicia faba* as a function of water potential in solutions of mannitol in which the discs were immersed (Fig. 3). Aperture increased with decrease in water potential in the range 0 to −6.5 bar, but diminished with further decrease in potential because, associated with plasmolysis of the subsidiary cells, $\Delta P_s = 0$. As the increase in aperture between 0 and −6.5 bar is about the same as the decrease between −6.5 and −9 bar, the average mechanical advantage may be estimated as $1 + 2.5/6.5 \approx 1.4$. However, uncertainty about the changes in solute concentration that may have occurred complicates the interpretation of Glinka's experiment. The leaf discs were immersed in solution for 4 h before the observations shown in Fig. 3 were made, and it is probable that translocation of solutes took place between the epidermis and the bathing solution. Osmotic pressures were determined from observations

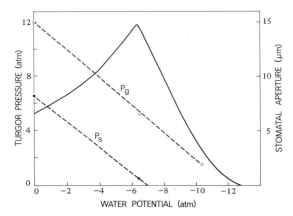

Fig. 3. Width of stomatal pores, and estimates of pressures in guard cells and subsidiary cells, P_g and P_s respectively, in illuminated leaf discs of *Vicia faba* as functions of the potential of water in the solution in which the discs were immersed. (Redrawn from Glinka, 1971.)

of plasmolysis in samples of epidermis floated on mannitol solutions for 45 min in the dark, a procedure which hardly seems likely to reflect the osmotic pressure, particularly that in the guard cells, which obtained in the illuminated discs from which the samples were taken. In using the magnitude of osmotic pressure at incipient plasmolysis to estimate turgor pressures in the leaf discs, no allowance was made for change in solute concentration with change in cell volume. Problems associated with change in solute concentration in guard cells will be taken up in the later parts of this section. But before this, I shall discuss the mechanical properties of guard cells in relation to dynamic attributes of stomata.

The transient characteristics of the response of stomata to changes which affect the hydrology of the epidermis have intrigued stomatal physiologists for almost a century. The reaction of stomata in turgid illuminated leaves to sudden change in rate of evaporation (Darwin 1898), to change in pressure of water supplied to the vascular system (Raschke, 1970a), and to release of "suction" in the vascular system associated with severance of the petiole (Meidner and Mansfield, 1968), are all consistent with the supposition that increase in the potential of water in the cell walls of the epidermis causes a closing movement followed by an opening movement, the final resultant change in aperture often being relatively small. Decrease in potential causes the opposite sequence (see Fig. 4). It has generally be thought that the movements are a manifestation of the opposing effects of changes in turgor in the subsidiary cells and guard cells on stomatal aperture and a difference in the rates at which the turgor in these cells adjusts to perturbation—the subsidiary cells adjusting more rapidly than do the guard cells. I myself incorporated this hypothesis in a model of plant-water relations and stomatal behaviour (Cowan, 1972a), but have since been concerned

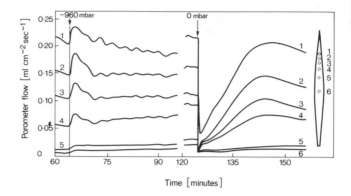

Fig. 4. Stomatal response of stomata in maize to the reduction, and subsequent increase, of pressure (with respect to atmospheric pressure) in the water supplied to the base of a leaf. Flow of air across the leaf due to a small difference in pneumatic pressure was recorded at six locations as indicated. (Redrawn from Raschke, 1970a.)

with two objections to it. One of them now seems to me to be illusory, but the reason for believing it to be so is itself of interest.

The model of DeMichele and Sharpe suggests that there can be very little, if any, change in stomatal aperture without change in guard cell volume; that is to say, there is a close relationship between aperture and volume. That is an implicit assumption in much of what is written about stomatal behaviour. If the relationship were unique, then the immediate effect of a change in pressure in the subsidiary cell would be, not a change in aperture, but rather a change in pressure in the guard cell of magnitude, from eqn (3), $\Delta P_g = m\Delta P_s$. Change in aperture would occur subsequently, but at a rate associated with uptake or release of water by the guard cell. In other words it would take place no faster than the rate of adjustment of the guard cell to a disturbance in the potential of water supplied to it. Thus the whole basis for ascribing transient movements of stomata to a difference in the time constants for hydraulic equilibration in guard cells and subsidiary cells would break down. But there can be no doubt that a change in pressure in the subsidiary cell will bring about *some*, virtually instantaneous, deformation of the stomatal apparatus which is reflected in a change in the aperture of the pore. I suspect that the effect may be quite large. The most useful way of expressing it is to write for change in guard cell volume

$$\Delta V_g = \frac{V_g}{\epsilon_g} [\Delta P_g - m(1 - \sigma)\Delta P_s] \tag{4}$$

where ϵ_g is an elastic modulus depending on the properties not only of the walls of the guard cell, but also the cell walls of the other epidermal cells which may be strained in guard cell movement. The coefficent σ, as comparison of eqn (3) and (4) shows, is

$$\sigma = \frac{(\partial a/\partial P_s)_{V_g}}{(\partial a/\partial P_s)_{P_g}} \tag{5}$$

It is a measure of the extent to which stomatal aperture may be changed as a result of deformation of the guard cells at constant volume. We may expect that σ is greater than zero and less than unity. It will be helpful, later, to have a relationship between volume and pressure in subsidiary cells. Strictly this should be analagous in form to eqn (4)–that is to say V_s should depend on both P_s and P_g–but I shall assume that $(\partial V_s/\partial P_g)_{P_s}$ is negligible compared with $(\partial V_s/\partial P_s)_{P_g}$. Therefore

$$\Delta V_s = \frac{V_s}{\epsilon_s} \Delta P_s \tag{6}$$

where ϵ_s is the elastic modulus of the subsidiary cells.

The second objection to the common explanation of transient responses of stomata to changes in water potential is to do with time constants. It has been mentioned that the time constant for hydraulic equilibration of guard cells in *Vicia faba* with an external solution is of the order 10 s. There is no corresponding information for subsidiary cells in *V. faba*, or about time constants in other species, but provisionally one must suppose that the difference between the time constants of subsidiary cells and guard cells may be no more than a matter of seconds. If this supposition is correct, any transient would be over and done with very quickly–whether the supply of water to the guard cell is mainly through the cell walls of the subsidiary cell, as I assumed (Cowan, 1972a), or mainly throught the protoplast of the subsidiary cell as suggested by Raschke (1975a). The transients which I wish to discuss here are of much longer duration. The right-hand part of Fig. 4, though it relates to *Zea mays*, is typical of what may be observed with other species including, in my own experience, cotton. The initial movement and its reversal take place within a minute or two, but the second movement continues for about 20 min. Raschke (1970a) remarks of this slower movement that it took place "in a manner typical for an active movement after closure similar to that caused by temporary darkening or an increase in the CO_2 concentration". I shall assume that it is due to readjustment of osmotic pressure in the guard cells. The first movement I take to be associated with the establishment of a quasiequilibrium in water potential. There are some complexities in the behaviour of the stomata following application of suction (Fig. 4), but nevertheless the events which occur are more nearly the inverse of those which follow release of suction than they appear to be, because what is plotted is the flow of air across the leaf due to a pressure difference, a quantity which is roughly proportional to the cube of stomatal aperture (Raschke, 1965a). The topic of oscillations will be discussed in a later section.

These various complications suggest the following model. The initial hydraulic adjustment, associated with the "first" stomatal movement, of the guard cells and subsidiary cells to change in water potential, $\Delta\psi$, in the epidermis takes place so rapidly that the amounts of solute in the cells are conserved. Therefore

$$\Delta_1 P_g = \Delta\psi - \frac{\Pi_g}{V_g} \Delta_1 V_g; \, \Delta_1 P_s = \Delta\psi - \frac{\Pi_s}{V_s} \Delta_1 V_s \qquad (7)$$

in which the terms relating to volume account for decrease in osmotic pressure with increase in concentration of solute. The characteristic time is of the order 10 s, but of course relative to a disturbance in the pressure of water applied to the leaf vascular system, as in Raschke's experiment, or a change in rate of evaporation, the change in potential in the vicinity of the stomatal apparatus may itself be delayed and attenuated in a way depending on the hydraulic properties of the epidermis as a whole. Subsequently, the processes which regulate solute concentration in the guard cells operate to restore osmotic pressure to its original magnitude. It is probable that this readjustment, associated with the "second" stomatal movement, involves net exchange of solute by the subsidiary cells also, particularly in those species in which the subsidiary cells are differentiated from other epidermal cells, but, because the volume of a subsidiary cell is usually much greater than that of the guard cell, change in osmotic pressure in the subsidiary cells may be negligible. If so

$$\Delta_2 P_g = +\frac{\Pi_g}{V_g} \Delta_1 V_g; \, \Delta_2 P_s = 0 \qquad (8)$$

The characteristic time for the change is similar to that for response of stomata to change in light intensity or concentration of carbon dioxide, that is to say of the order 10^3 s.

The solutions of eqns (3) to (8) for the two successive stomatal movements in terms of change in water potential are

$$\frac{\Delta_1 a}{\Delta\psi} = \left(\frac{\partial a}{\partial P_g}\right)_{P_s} \frac{\epsilon_g}{\epsilon_g + \Pi_g} \left(1 - m' - m'\sigma \frac{\Pi_g}{\epsilon_g}\right) \qquad (9)$$

$$\frac{\Delta_2 a}{\Delta\psi} = \left(\frac{\partial a}{\partial P_g}\right)_{P_s} \frac{\Pi_g}{\epsilon_g + \Pi_g} (1 - m' + m'\sigma) \qquad (10)$$

where $m' = m\epsilon_s/(\epsilon_s + \Pi_s)$ is the "effective" mechanical advantage. The sum of the two movements is

$$\frac{\Delta a}{\Delta \psi} = \left(\frac{\partial a}{\partial P_g}\right)_{P_s} (1 - m')$$ (11)

The important conclusions to be drawn are as follows. With a *decrease* in potential of water, the first stomatal movement will be an opening one if $m' > 1/(1 + \sigma\Pi_g/\epsilon_g)$. It will be followed by a closing movement only if $m' < 1/(1 - \sigma)$. Thus the sequence which commonly occurs corresponds to a range of positive m' values, extending above and below unity by amounts which increase with increasing σ. The relationship is illustrated in Fig. 5, which shows that two other sequences of movement are possible, but closure followed by opening is not. Figure 5 also indicates how guard cell volume changes: the two changes are always in the same sense, volume being increased if $m' > 1/(1 - \sigma)$ and decreased otherwise. One notes that if $m' = 1$, opening before closure occurs provided σ is finite. I am inclined to suppose $m' \approx 1$ in fully turgid leaves

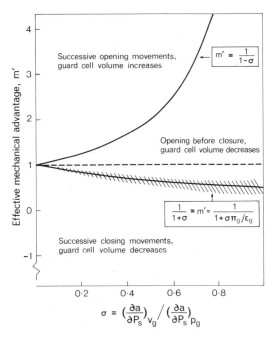

Fig. 5. Delineation of sequence of changes in stomatal aperture and guard cell volume following reduction of water potential in terms of effective mechanical advantage, m, and the elastic parameter, σ, defined by eqn (5).

(corresponding to a magnitude of the mechanical advantage, $m = m'(1 + \Pi_s/\epsilon_s)$, of perhaps about 1.5). Following the suggestion of DeMichele and Sharpe, we might take $m' \approx 0$ in leaves near zero turgor. Then successive closing movements would take place in response to decrease in potential, however near σ may be to unity.

The magnitudes of the first and second changes which take place in both aperture and guard cell volume are summarized in Table II, for the particular cases $m' = 1$ and $m' = 0$. Because the essential assumptions, apart from that about time constants, are to do with the roles of change in pressure in the

TABLE II

Variable	$m' = 1$		$m' = 0$	
	First response	Second response	First response	Second response
$-\dfrac{\Delta a/\Delta \psi}{(\partial a/\partial \psi_g)_{P_s,\, \text{solute}}}$	$+\sigma\dfrac{\Pi_g}{\epsilon_g}$	$-\sigma\dfrac{\Pi_g}{\epsilon_g}$	-1	$-\dfrac{\Pi_g}{\epsilon_g}$
$-\dfrac{\Delta V_g/\Delta \psi}{(\partial V_g/\partial \psi_g)_{P_s,\, \text{solute}}}$	$-\sigma$	$-\sigma\dfrac{\Pi_g}{\epsilon_g}$	-1	$-\dfrac{\Pi_g}{\epsilon_g}$

subsidiary cells and osmotic adjustment in the guard cell, changes have been expressed in terms of those which would occur if P_s and the amount of solute in the guard cell were constant. It is readily shown that

$$\left(\frac{\partial a}{\partial \psi_g}\right)_{P_s,\, \text{solute}} = \left(\frac{\partial a}{\partial P_g}\right)_{P_s} \frac{\epsilon_g}{\epsilon_g + \Pi_g} \tag{12}$$

$$\left(\frac{\partial V_g}{\partial \psi_g}\right)_{P_s,\, \text{solute}} = \frac{V_g}{\epsilon_g + \Pi_g} \tag{13}$$

These are coefficients which may be determined by exposure of isolated pairs of guard cells to external solutions having differing osmotic pressures for just sufficiently long for hydraulic equilibrium to be established, as has been done by Raschke *et al.* (1973b). We see from the last line in Table II that there is a diminution in guard cell volume with decrease in water potential even when the net change in aperture is zero. If this turns out not to be so, or if the magnitude of σ indicated in other ways is shown to be negligible, then it would be strong evidence that much more complicated processes are at work. In any event, the classic explanation based solely on the concept of mechanical advantage and a disparity in time constants for hydraulic adjustment simply will not do.

Had I written this section a few years ago, I might have concluded it here with a modest degree of satisfaction. But to do so now would be to ignore some important findings, of a kind presaged several decades ago, that are currently receiving a great deal of attention. The preceding analysis might prove to be the last, or nearly the last, attempt to explain complex changes in stomatal aperture in terms of elementary physical processes. The classic work is that of Stålfelt (1929, 1955) which led him to define the terms "hydropassive" and "hydro-active" to distinguish movements due to forces outside the guard cell from those due to forces inside the guard cell, hydroactive movements being associated with loss of solute. Is a "hydroactive" movement one which is associated with osmotic adjustment in the guard cell in the way I have described it? If so, it is an esoteric term for what is conceptually a very simple process. Although it is implied in equation (8) that the steady-state change in osmotic pressure is zero, the feature which is essential to the analysis is that there should be a *tendency* for osmotic pressure to revert to that which obtained prior to change in cell volume. That is to say the change in total amount of solute should *be initiated by, and be in the same sense as*, the volume change. Its magnitude would be in the range

$$0 < \frac{\Delta(\Pi V)}{\Delta V} \leqslant \Pi \qquad (14)$$

in which the lower limit represents absence of readjustment and the upper limit complete readjustment. In terms of mechanisms the whole range can be represented, to take a far-fetched example, by a classic osmometer containing a solution of two species, one of them in equilibrium with its own solid phase. But it may equally well encompass cases where membranes are permeable and energy is continuously dissipated—the concentration of diffusible species being maintained by active uptake. Then, decrease in volume would engender a net loss of solutes to the external medium.* But, of course, when the water potential in a cell is caused to change there may be effects on the balance of fluxes and reactions which are not attributable to changes in concentration, but are the cause of them. The example of *Valonia*, in which imposition of an increase in internal pressure causes a decrease in the amount of solute by inhibiting active uptake (Gutknecht, 1968), is well known: the ratio defined in eqn (14) is negative. Raschke (1975a) has evidence that loss of solutes takes place in guard cells exposed to solutions having osmotic pressures greater than 7 bar. I am not clear whether the changes exceed the upper limit of eqn

---

* If active uptake is $k_1 c_0$ and passive loss is $k_2(c_i - c_0)$, where c_0 and c_i are concentrations outside and inside a cell, and k_1 and k_2 are constants, then $c_i = (1 + k_1/k_2)c_0$ at equilibrium. Change in cell volume causes a change $c_i \Delta V$ in the amount of solute which the cell contains. It is this kind of readjustment, which must of necessity take place in cells in which solute concentration is a balance of concentration-dependent fluxes and chemical reactions, that I have in mind.

(14)—which may be written $\Delta\Pi/\Delta V = 0$, Π being internal osmotic pressure of course—and in any event phenomena of this kind with guard cells in osmotic solutions might not reflect what happens in the functioning epidermis. Stålfelt (1955) describes an experiment, with illuminated leaf segments which were being allowed slowly to lose water, in which the amount of solute in guard cells began to decrease while stomatal aperture was increasing. Shortly afterwards the stomata began to close. Stålfelt uses this example to emphasize what is meant by hydroactive movement: a movement caused by, rather than causing, loss of solute. However, the interpretation is valid only if it is assumed that guard cell volume was increasing in association with stomatal aperture. I have suggested that stomatal aperture may be an unreliable indicator of guard cell volume, not only in unsteady, but in quasisteady conditions. For example, equilibrium changes in volume and aperture due to change in water potential may be shown to take place in the opposite sense if $1 < m' < 1/(1 - \sigma)$. However, there is an increasing amount of evidence that Stålfelt's interpretation of this and similar observations may be substantially correct (Raschke, 1975a). It stems from the spate of work begun by Little and Eidt (1968) and Mittelheuser and van Steveninck (1969), who demonstrated that exogenous abscisic acid reduced transpiration and closed stomata, and that of Wright (1969) and Wright and Hiron (1969), who showed that ABA was formed in wilting leaves. These separate observations were focused by what had been a long-standing interest in the "aftereffect" of decrease in water potential on stomatal aperture (Stålfelt, 1955; Milthorpe and Spencer, 1957; Fischer et al., 1970). Hsiao (1973), Milborrow (1974) and Raschke (1975a) have reviewed the role of endogenously created abscisic acid as a modulator of stomatal behaviour, and here I will mention only a few results of importance. Loveys and Kriedemann (1973) found that concentrations of ABA-like compounds in leaves of Vitis vinifera doubled within 15 min of leaf excision, the water potential having fallen to −15 bars at that stage and a stomatal closing movement having become evident. Kriedemann et al. (1972) found that stomatal closure occurred when the ABA concentrations in transpiring leaves of Phaseolus vulgaris and Rosa spp. were approximately doubled by uptake of exogenous ABA through the severed petioles. Stomatal closure began 8 min after application in the first species and 32 min after application in the second. When ABA was applied to the cut base of leaves of Zea mays, closure began 3 min later. While it may be that the speed of ABA synthesis in leaf tissue is too slow to initiate stomatal movement within one or two minutes of a sudden change in leaf water potential, it certainly is not too slow to be involved in the later stages of the movement. Cummins et al. (1971) and Horton (1971) showed that recovery from the effects of ABA taken up in the vascular system is rapid if concentrations are not too large. If it is variation in the concentration of ABA, or ABA-like hormones, associated with changes in water potential in leaf tissue which are responsible for what I have called the "second" stomatal movement, then, clearly, it is a simple matter to explain the sequence of movements without invoking the property defined as σ in eqn (5). It is only necessary that the effective mechanical advantage, m', exceeds unity so that the stomata first open in response to decrease in water potential: the change in volume of the guard cell is no longer of primary importance. If ABA synthesized in response to decrease in turgor in leaf tissue will engender decrease in turgor associated with loss of solute in the guard cells, then might not loss of turgor in the guard cells be reinforced by synthesis of ABA, or other compounds within the guard cells themselves? The effect would be what Raschke (1975a) calls "hydroactive feedback in guard cells". The implications would depend on what triggers synthesis of ABA. ABA seems to be synthesized in chloroplasts (Milborrow, 1974) and if it were triggered by conformational changes resulting from loss of water from a cell then most of what I have said about stomatal mechanics would remain relevant, because the "first" change in guard cell volume would initiate the "second" stomatal movement.

There is no evidence as yet that synthesis of ABA is caused by small changes of water potential in turgid tissue. Abscisic acid and related compounds are very strong contenders for the major role in stomatal closure within wilting or near-wilting plants (indeed it is impossible to see how a guard cell could be deflated if osmotic pressure remained at, say, 40 bar), but their involvement in stomatal movement in turgid leaves is speculative.

We have discussed stomatal responses which are *activated* by change in the state of water—hence the title of this section. Whether, and in what way, they are *active* is the central problem.

C. ACTIVE MOVEMENT

I shall give only a short account of processes involved in stomatal movements which can indubitably be called active in the sense that guard cell metabolism is directly involved, the emphasis being on matters which are most relevant to later sections of this work. To attempt more would be beyond my competence, and superfluous because of the excellent review by Raschke (1975a).

The fact that one only in about 20 cells in a leaf, and five cells in the epidermis, is a guard cell, that the guard cell is small and cannot readily be separated makes it an unattractive system for the biochemist and biophysicist. Yet progress in our understanding of the processes underlying changes in osmotic pressure has been quite rapid in the last decade, and shows promise of continuing. It is now clear that increase in osmotic pressure in guard cells is largely associated with accumulation of potassium salts (Fujino, 1967; Fischer, 1968; Fischer and Hsiao, 1968; Sawnhey and Zelitch, 1969; Humble and Raschke, 1971) and that the sources of potassium are the neighbouring epidermal cells (Raschke and Fellows, 1971; Willmer and Pallas, 1973). The anions seem to be mainly organic, with a predominence of malate (Allaway, 1973; Pallas and Wright, 1973), but chloride is also involved to a variable extent (Humble and Raschke, 1971; Raschke and Fellows, 1971). Recent estimates of large changes in osmotic pressure in *Vicia faba*, already referred to, may be accounted for by the results of measurements of potassium with an electron microprobe (Humble and Raschke, 1971) and the assumption that the associated anion derives from malic acid (Raschke *et al.*, 1973a; Raschke, 1975a).

Combining long established facts with new information, Willmer *et al.* (1973) pointed out that guard cells have some of the characteristics of leaf tissue in plants having Crassulacean acid metabolism. Stomatal opening is accompanied by decrease in starch, increase in pH, and the production of malic acid through the carboxylation of phosphoenolpyruvate; the relative activities of carboxylating enzymes are similar to those in CAM. But the extent of the parallel is uncertain in terms of metabolic functioning. The association between synthesis of malate and accumulation of potassium in the vacuole, and the inhibition of the production and consumption of malate by CO_2 at concentrations lower than those which normally exist in the atmosphere, are essential characteristics of metabolism in guard cells which do not have counterparts in CAM. The theory of guard cell metabolism and stomatal movement by Levitt (1974) attributes major functions in the control of the system to the guard cell chloroplasts. The primary stimuli to stomatal opening and closure are changes in pH in the cytoplasm brought about by the establishment and cessation of photoelectron transport. It is these changes in pH which influence the rate at which malic acid is produced, or decarboxylated, and the exchange of hydrogen for potassium ions across the plasmalemma. Whatever the likelihood of each of several individual hypotheses (some of them contentious) being correct, the major criticism of the theory as a model of stomatal functioning is that it does not explain why stomata in the dark can be open and respond to changes in CO_2 concentration in much the same way as in the light. Levitt assumes that opening in the dark is due to anaerobic respiration, and increase in pH in the cytoplasm, associated with decreased concentration of oxygen in the leaf. That the concentration of oxygen could decrease sufficiently seems most unlikely and, for obvious reasons, any stomatal opening it might cause would be ephemeral.

It is a puzzling feature of guard cells that there is no specific function that can clearly be attributed to the chloroplasts. It is at least arguable that energy for opening and maintenance of ion fluxes is obtained by oxidative phosphorylation of substrate imported into the guard cell (Raschke, 1975a). Despite the fact that plasmadesmata between guard cells and subsidiary cells do not seem to be frequent and may be absent in some species, sugars can be translocated along the epidermis to the guard cell (Pallas, 1964). Several authorities (Ketallaper, 1963; Meidner and Mansfield, 1968; Raschke, 1975a) have emphasized the predominant role of concentration of CO_2 in the intercellular air-spaces in affecting stomatal aperture through a direct effect on guard cell metabolism. Scarth (1932) first suggested that opening of stomata in light and closure during darkness was due to decrease and increase, respectively, in intercellular concentration of CO_2 associated with change in rates of photosynthesis and respiration in the leaf as a whole. Raschke (1970b) showed that the kinetics of the stomatal response to CO_2 in *Zea mays* is independent of illumination, and has reviewed evidence which indicates that the relationship between stomatal aperture and concentration of CO_2 is hardly affected by light (Raschke, 1975a). As the formation of malate requires CO_2, the problem of explaining the inhibitory effects of CO_2 is a difficult one. Raschke (1975a) speculates that malate is required for both opening and closing, but in different parts of the cell, that in the cytoplasm determining the direction and magnitude of ion fluxes into and out of the vacuole. He states that "PEP carboxylase and the malic dehydrogenases appear to form a push-pull system which adjusts the intracellular malate level and pH in some relationship to the intercellular $[CO_2]$ ". In some species, maximum stomatal aperture does not occur in CO_2-free air, but at a concentration of about one-third that of the normal atmosphere (Drake and Raschke, 1974; Raschke, 1975a).

There are, however, effects of illumination on the steady-state stomatal aperture in intact leaves which cannot be related to intercellular CO_2 concentration (Meidner and Mansfield, 1968) and there is sometimes a marked sensitivity of stomata to light in isolated epidermal strips (e.g. Humble and Hsiao, 1970; Fischer, 1971). The action spectrum for stomatal opening is dissimilar to that for photosynthesis, blue light having a relatively greater effect on stomatal opening (Meidner and Mansfield, 1968; Hsiao *et al.*, 1973). Ketallaper (1963) pointed out that apparent sensitivity of stomata to light might be explained on the basis that CO_2 concentration within the guard cell is related not only to the intercellular concentration, but to the net rates of CO_2 fixation within the guard cell itself; Shaw and MacLachlan (1954) measured a small light-induced uptake of radioactive CO_2 in *Tulipa gesneriana*, about 0.1 to 0.2 the potential capacity estimated on the basis of chlorophyll content. But there is also the possibility that blue light stimulates the activity of enzymes and respiratory processes involved in the stomatal mechanism (Raschke, 1975a). Although the evidence tends to confirm the sense of Scarth's suggestion on the whole it relies heavily on observations with a very small number of species, the work of Raschke and his colleagues with *Zea mays* forming a major contribution. It is possible that direct effects of light may be much more marked in some species than others. Certainly the sensitivity of stomata to CO_2 concentration varies greatly, being relatively very small in tomato for example (Hurd, 1969) and smaller in C_3 than in C_4 species (Pallas, 1965). Raschke (1975b) has recently shown that there is a marked interaction between carbon-dioxide and ABA concentration in their effect on stomatal aperture in *Xanthium strumarium*, and is led to speculate on the way in which stomatal sensitivity to CO_2 concentration may change with change in the potential of water in the leaf. The interaction was present, but not as strong, in *Gossypium hirsutum* and *Commelina communis*.

The characteristic times for "photoactive" movements of stomata seem to be very different in different species. The half-time for light-stimulated uptake of potassium by guard cells and increase in stomatal aperture in epidermal strips is about 20 min for *Zea mays* (Raschke and Fellows, 1971) and about 1 h for *Vicia faba* (Humble and Hsiao, 1970). Closure is generally quicker; the half-time appears to be about 5 min in *Z. mays* (Raschke and Fellows, 1971). Neither opening nor closing is strictly first-order in time and therefore half-times are a rough indication only of the speeds of the processes.

When stomata are caused to open or close in intact leaves, they are affected by the changes in rate of transpiration which take place. The effects of steady-state change in osmotic pressure may be readily superimposed on those due to change in water potential in the epidermis according to the theory outlined in the previous section. The first of eqns (8) becomes

$$\Delta_2 P_g = \Delta_\infty \Pi + \frac{\Pi_g}{V_g} \Delta_1 V_g \tag{15}$$

in which $\Delta_\infty \Pi$ represents the change, due to change in CO_2 concentration or light intensity, of the reference signal for the osmotic regulatory system.

The "second" stomatal movement, previously described by eqn (10), becomes

$$\Delta_2 a = \left(\frac{\partial a}{\partial P_g}\right)_{P_s} \left\{ \Delta_\infty \Pi + \frac{\Pi_g}{\epsilon_g + \Pi_g} (1 - m' + m'a)\Delta\psi \right\} \tag{16}$$

The first movement is unchanged. The potential, ψ, may be expected to decrease if the stomata have been made to open. If the cause is an increase in solar radiation, out-of-doors, then there will have been also a direct effect on rate of evaporation associated with increase in temperature of the leaf. One might ask whether the decrease in ψ could exactly balance the increase in osmotic pressure. If it were to do so, the stomata would have been made to open very rapidly in accordance with eqn (9), but in such a way that no further movement would take place. It seems apparent that this could not happen. If $m' \approx 1$, as I have suggested might be the case in turgid leaves, then the decrease in ψ would have to be greater, probably several times greater, than the increase in osmotic pressure. On the other hand if we adopt the "hydroactive" hypothesis of transient stomatal response to water potential in turgid leaves and suppose that magnitudes of m' of 3 or 4 are feasible, then a balance could be achieved with a modest decrease in water potential. This matter will be taken up later in the chapter.

D. SUMMARY
The main points which emerge from what has been written, and are relevant to what is to come, are these.

The potential of water in the epidermis may differ from that in the mesophyll by an amount depending not only on rate of evaporation from a leaf as a whole, but on characteristics of the physical environment which affect the micro-hydrology of the leaf. The response of stomata to change in the supply or loss of water in turgid leaves has complex dynamic characteristics which are not adequately explained by previous notions based on the mechanical and hydraulic properties only of the stomatal apparatus. Some readjustment of the solute concentration in the guard cells is involved. The longer of the two time constants which characterize the major features of the response is similar to that for stomatal movement induced by change in light intensity or CO_2 concentration. It is not clear whether the response is a function of the intrinsic properties of the stomatal apparatus or whether it is affected by stimuli transmitted from other parts of the leaf. It seems certain that stomatal closure in wilting leaves is due, in part at least, to a response to ABA, or ABA-like hormones, synthesized in the leaf. It is becoming apparent that the stomatal apparatus is more intimately connected with the metabolism of the plant as a whole than has previously been thought. The effects of water potential in the plant and intercellular concentration of CO_2 are interrelated in a way which may involve ABA. There is a web of interactions the complexity of which we probably do not yet envisage.

III. STOMATAL BEHAVIOUR IN THE INTACT PLANT

A. QUASISTEADY BEHAVIOUR AND PLANT-WATER RELATIONS

The role of stomata in the hydrology of the soil-plant-atmosphere system is illustrated by the block diagram in Fig. 6. Leaf conductance g_l is that property of the leaf epidermis, dominated by the number and dimensions of the stomatal

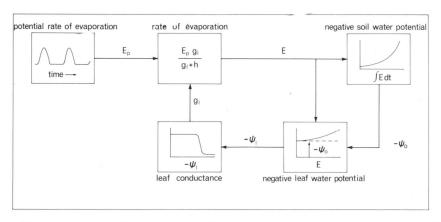

Fig. 6. Interrelationship between leaf conductance to vapour transfer, g_l, rate of evaporation, E and potential of water in the leaf, ψ_l. The quantities E_p, h and ψ_0 are potential rate of evaporation, combined conductance of leaf boundary layers to heat and vapour transfer, and potential of water in the soil, respectively.

pores, which pertains to vapour transfer. It may be defined as the rate of evaporation per unit area of leaf, E, divided by the difference in humidity across the epidermis. In general, its influence on rate of evaporation may be represented by the relation $E/E_p = g_l/(g_l + h)$ where E_p is potential rate of evaporation (that which would occur if the leaves behaved as though saturated with water, i.e. g_l were infinite), and h is a combination of the conductances of the leaf boundary layers to heat and vapour transfer. Because E_p outdoors is dominated, directly or indirectly, by the flux of solar radiation it tends to vary diurnally being at its greatest near midday and relatively very small at night. It is unnecessary to go into detail yet because the precise influence of g_l on rate of evaporation is not critical to the present discussion. Rate of evaporation affects plant water relations in two ways. In the long run it is responsible for depletion of the water content in the soil and a decrease in the potential of water available for absorption by the root system, ψ_0. It also causes a "drawdown", so to speak, of water potential in the leaf, ψ_l, with respect to that in the soil, according to what is often called the "van den Honert hypothesis" of water transport. The loop in Fig. 6 is completed by the function representing the response of leaf conductance to leaf water potential. An example of the way in which such a system might work is shown in Fig. 7. When leaf water potential exceeds a certain critical level the stomata are fully open during the day and leaf conductance is a maximum. Then, variation in rate of evaporation is closely related to variation in

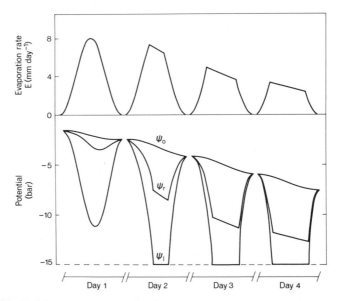

Fig. 7. Hypothetical time course of rate of evaporation, leaf water potential, ψ_l, potential of water at the root surfaces, ψ_r, and potential of water in the soil, ψ_0, on successive days (Cowan, 1965).

the physical environment; rate of evaporation fluctuates diurnally with a maximum about midday because the most influential climatic factors, radiation and temperature, do so in this way. The potential of water in the leaf depends on the potential of water in the soil and the rate of evaporation. As the amount of water in the soil is depleted there occurs each day a period of successively increasing length during which leaf water potential tends to fall below the critical level. But the stomata respond so sensitively that they, in effect, limit the rate of evaporation to an extent which prevents the potential of water in the leaf decreasing any further. The rate of evaporation relative to the rate of evaporation which would occur if the leaf conductance were maximum is an indication of stomatal closure.

It is obvious that the block diagram in Fig. 6 and the temporal sequence illustrated in Fig. 7 make no allowance for complexities associated with the spatial distribution of leaves and roots in plants and, for that reason alone, are unrealistic. But together they summarize the interrelationship and implications of the particular processes I shall discuss in this section.

Hsiao (1973) has reviewed data which demonstrate a threshold level of water potential or relative water content above which leaf resistance (reciprocal of leaf conductance) remains constant and below which resistance increases perhaps 20- to 30-fold with a further decrease in potential of about 5 bar; threshold values are −7 to −9 bar in tomato (Duniway, 1971) and the adaxial surface in bean (Kanemasu and Tanner, 1969), −10 to −12 bar in soybean (Boyer, 1969) and the abaxial surface in bean (Kanemasu and Tanner, 1969), and −12 to −16 bar in grape (Kriedemann and Smart, 1971). The observations of Ehlig and Gardner (1964) with cotton, sunflower, trefoil and pepper; Millar, Duysen and Wilkinson (1968) with wheat; Kassam (1973) with *Vicia faba*; Beadle *et al.* (1973) with corn and sorghum; and other data summarized by Turner (1974a)—see Fig. 8a—lead to the same generalization. There are some exceptions to the rule, e.g. Boyer (1970) found that leaf resistance in corn increased with decrease in leaf water potential below −3.5 bar. There is evidence that the threshold values and sensitivities of stomata in the adaxial and abaxial surfaces in amphistomatous leaves differ (Shimshi, 1963; Kanemasu and Tanner, 1969; Kassam, 1973; Sharpe, 1973).

Remarkably, there seems to have been little investigation of the interacting effects of irradiance and leaf water potential on stomatal behaviour. Kassam (1973) found that the nature of the stomatal response in *Vicia faba* to change in leaf water potential depended whether irradiance was greater or less than that required to cause maximum opening in the turgid plant. At the higher irradiance stomatal resistance *decreased* slightly with decrease in potential in the range −2 to −7 bar before increasing greatly at lower potentials. At the lower irradiance, on the other hand, resistance increased with decrease in leaf water content within the range corresponding to 0 to −7 bar water potential. In contrast, Beadle *et al.* (1973) found very little evidence, if any, for an interaction between

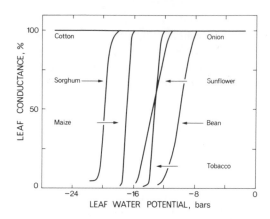

Fig. 8(a). Leaf conductances in several species, at similar irradiances, as functions of leaf water potential. (Redrawn from Turner, 1974a. Data for sunflower, cotton, bean and onion due, respectively, to Berger, 1973; Jordan and Ritchie, 1971; Kanemasu and Tanner, 1969; Millar *et al.*, 1971.)

irradiance and leaf water potential in their effects on stomatal aperture in corn and sorghum.

Threshold leaf water potentials vary with age and growth conditions of plants. Cotton plants grown in a growth room had a critical water potential of −16 bar while leaf resistance of cotton in the field remained small at a leaf water potential of less than −27 bar (Jordan and Ritchie, 1971). McCree (1974) and Turner (1974a) describe similar but smaller effects with other species. Millar *et al.* (1971), Turner and Begg (1973) and Turner (1974a) suggest that differences may be related to a greater capability of plants in field conditions to maintain leaf turgor by increase of osmotic pressure. Diurnal changes in osmotic pressure in maize and sorghum take place in such a way as to partially offset the effect of diminishing water potential on turgor (Turner, 1974b). There is an indication that the characteristics in different species differ less when stomatal response is related to turgor pressure rather than water potential (Fig. 8b). In general terms it is not altogether surprising if stomata act to maintain positive turgor, rather than an apparently arbitrary level of water potential. But, on the basis of what is known about leaf hydrology, there is no *a priori* reason for supposing that stomatal aperture is more closely related to leaf turgor pressure than to leaf water potential. If it is so, it implies that osmotic pressure in guard cells is in some way correlated with osmotic or turgor pressure in the leaf in bulk. It is tempting to invoke the role of ABA in a hypothetical linkage of this kind. However, there are observations which indicate that the role of ABA in its effect on stomatal aperture and water loss is one of integral control rather than proportional control; increase in ABA in leaves of *Vitis labrusca* was in part a function of the time over which diminished water potential occurred (Liu *et al.*, 1975).

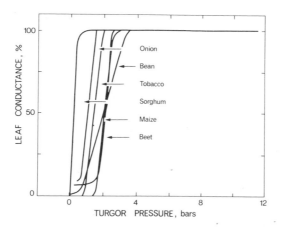

Fig. 8(b). Leaf conductances in several species, at similar irradiances, as functions of turgor pressure in leaf tissue. (Redrawn from Turner, 1974a. Data for beet, bean and onion due respectively, to Biscoe, 1972; Kanemasu and Tanner, 1969; Millar *et al.*, 1971.)

All the observations that have been referred to, except the last, relate to mesophytic, annual, crop plants and it is to these that the generalization of threshold potential must be confined. It can be taken as certain that no such generalization would apply to all plants in the whole range of natural environments. For example, *Acacia harpophylla* F. Muell., brigalow, exhibits a continuous, though steepening, decline in leaf conductance over a range of leaf water potentials extending from −5 to −65 bar (van den Driessche *et al.*, 1971).

What now of the other part of the system: the way in which leaf water potential varies with rate of evaporation? It is unnecessary to review how, beginning with the ideas of Gradmann as publicized by van den Honert (1948), the soil-plant-atmosphere system has been treated as a catena of resistances each of which sustains a decrease in water potential dependent on the flux of water supported (Philip, 1957; Gardner, 1960; Cowan, 1965). If these resistances were constant, then the potential in any part of the catena would decrease linearly with increasing flux of water. In particular we should have

$$\psi_l - \psi_0 = E(R_s + R_p) \tag{17}$$

where R_s and R_p are the resistances associated with transport in the soil and in the plant. But in many experiments ψ_l is found to vary with E in a complex way, even with plants growing in nutrient solution or in a medium which can be expected to keep the roots well supplied with water. The only characteristic which is common to all the observations is that $\partial \psi_l / \partial E \leqslant 0$. There is evidence that resistance varies diurnally, being least in the morning and early afternoon, and increasing towards the end of the day (Barrs and Klepper, 1968). A summary of published observations has been provided by Hailey *et al.* (1973) and could now be supplemented from the work of Boyer (1974), Kaufmann and

Hall (1974) and Neumann *et al.* (1973). Those that have been obtained with *Helianthus annuus* are illustrated in Fig. 9. Why is the resistance so variable? In general terms it may partly be related to the fact that what is measured when we determine what is called leaf water potential is not, whatever else it may be, a potential in the flow pathway. It is assumed, of course, that there is thermodynamic equilibrium between the water in the protoplasts of plant tissue and that in the vessels and cell walls which permeate the tissue; that is to say the cells act as "exquisite gauges" (Dixon, 1938; Weatherley, 1970) of the state of water in the conducting system. But Boyer (1974) argues, from the results of a series of well-conceived experiments, that there is a disequilibrium in growing tissue between water in the protoplasts and in the conducting tissue which supports the transpiration stream. We may write for the increase in leaf water content associated with growth

$$\frac{dW}{dt} = \text{function } (\psi_l + \Pi_l) = \frac{\psi_c - \psi_l}{R} \tag{18}$$

where W is volume of water per unit area of leaf, ψ_l and Π_l the water potential and osmotic pressure in the tissue, ψ_c the water potential in the conducting system, and R is the resistance to water uptake by the protoplasts. The first equality represents the finding that cell expansion associated with growth is a function of internal turgor pressure (Cleland, 1967; Boyer, 1968; Green, 1968; Acevedo *et al.* 1971); the second assumes that it is related to the difference in potentials of water in the conducting system and in the protoplasts. If, now, equation (17) is corrected for the effects described, we find:

$$\psi_l = \psi_o - R_p E - (R + R_p)\frac{dW}{dt} \tag{19}$$

so that $\psi_l < \psi_0$ when $E = 0$. Boyer assumed that $dW/dt = 0$ for $\psi_l < -4$ bar (corresponding to rapid rates of evaporation) and found, from measurements of ψ_l and E, that $R_p = 3.3 \times 10^7$ bar s m^{-1}. He also found that, when $E = 0$, $\psi_l \approx -2$ bar and $dW/dt = 2 \times 10^{-9}$ m s^{-1}. From the estimate of $R + R_p$ which results, he obtained $R = 10^9$ bar s m^{-1}.

The magnitude of R raises doubts in my mind about Boyer's analysis. Presumably it is the inverse of the sum, in parallel, of the conductances to water uptake by the individual cells in the leaf. With a total cell surface area of about 10 m^2 per m^2 leaf having a permeability of 10^{-7} m s^{-1} bar^{-1} (a rather small value—see Cowan and Milthorpe, 1971) one estimates $R = 10^8$ bar s m^{-1}. In other words Boyer's measurement yields a value which is at least an order of magnitude greater than one might expect. Furthermore, although equation (18) might explain why $\psi_l < \psi_0$ at $E = 0$, it does not explain why ψ_l varies with E in the way Boyer observed. The equation suggests that ψ_l would diminish relatively slowly with increase in E until dW/dt becomes zero at -4 bar, and would then

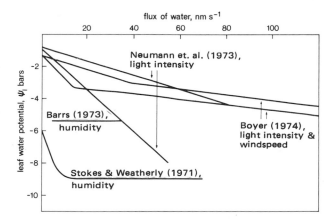

Fig. 9. Variation of leaf water potential due to variation of rate of evaporation in sunflower plants. The means by which rate of evaporation was caused to change is indicated.

diminish more rapidly. That is the opposite of what occurred. However, I would not suggest that Boyer's argument is not as important and relevant as others to do with non-linearity in the potential : flux relationship in plants; for example the treatment of coupled solute and water flow in roots by Fiscus (1975). What does seem certain is that no one hypothesis will explain all the observations of varying plant resistance or even the variation in one experiment. Among many possible complexities, there is one worth noting here. It stems from comparisons of four experiments with *Helianthus annuus*, and two with *Zea mays*. The results of the former are those drawn in Fig. 9.

Neumann *et al.* (1973) provide almost the only data indicating that plants might behave like tubes. Water potential in leaves of sunflower plants decreased linearly with evaporation rate, the latter being varied by varying light intensity. Ambient humidity was kept constant. The resistance of individual plants varied from 5.13×10^7 bar s m^{-1}, the maximum volume flux being about 10^{-7} m s^{-1}. Similar results were obtained with *Zea mays* and *Glycine max*. The observations are in marked contrast with those of Stokes and Weatherley (1971). The water potential in leaves of sunflower was −6 bar in the dark; in the light it decreased to −9 bar as rate of evaporation was increased to about 1.5×10^{-8} m s^{-1}, and then remained steady while rate of evaporation was increased to 7×10^{-8} m s^{-1}. Rate of evaporation was varied by varying ambient humidity. Barrs (1973) caused evaporation rate in sunflower and maize to vary by varying ambient humidity and found that leaf water potentials were constant, being −5 and −3 bar respectively. Finally, Boyer (1974) caused evaporation rate in sunflower to vary by varying a combination of light intensity and windspeed. His results are intermediate between those of Neumann *et al.* on the one hand, and Stokes and Weatherley, and Barrs, on the other. Leaf water potential decreased as evaporation rate was increased but the resistance was small, about 2×10^7 bar s m^{-1}. There are also some remarkable data (Camacho-B. *et al.*, 1974) showing no significant variation in leaf water potential in sunflower with rates of evaporation varying from almost zero to 1.5×10^{-7} m s^{-1}. Mean potential was -2 bar. Ambient humidity and temperature were varied.

The differences in techniques used to change rate of evaporation in these experiments cause me to wonder whether there might be an interrelationship at work between differences in temperature, rate of evaporation, and water potential in different parts of the leaf of a kind I have described in a section on the water relations of the epidermis. It could have nothing to do with depression of water potential in the dark, of course. Nor could it explain a complete absence of any observable effect of rate of evaporation on the potential of water in the leaf. But it might have something to do with changes and differences in the slope of the flux : potential curves. If the location of the flow path in the leaf changed with rate of evaporation, and was different in different environments then the potential of water in the leaf tissue in bulk would not be uniquely related to the flux of water. It would be premature to speculate further. The basic practical difficulty is this: plant cells are "exquisite gauges" of their own microenvironment but we do not see what individually they show.

Let us consider next the relationship between leaf water potential and evaporation rate with plants in soil—that is to say with R_s in equation (17) retained. The term R_s, "rhizosphere resistance" as it has been called by Etherington (1967), accounts for the gradient in soil water potential in the vicinity of a root which takes up water. Because the hydraulic conductivity of soil diminishes with decrease in water content, the gradient for a given rate of uptake, in other words the resistance R_s, increases as the soil becomes drier and the potential of water in it is decreased. However, Newman (1969a, b) and Andrews and Newman (1969) have shown, almost conclusively to my mind, that the magnitude of R_s is not sufficiently great to matter very much unless the soil becomes exceedingly dry. They argue that theoretical treatments of the physical basis of R_s (Philip, 1957; Gardner, 1960; Visser, 1964; Cowan, 1965) have assumed unrealistically sparse root systems or—what amounts to the same thing—unrealistically large rates of water uptake by roots, and that the experimental evidence for the existence of an appreciable R_s is equivocal.

Nevertheless there may be an appreciable resistance in the water transport pathway with plants in soil in excess of that which obtains when the roots are immersed in solution. Even in moist soil, roots can make contact with free water only over discrete small areas of the epidermis which diminish in size and number as the soil dries. Because only a limited part of the epidermis functions in taking up water, the effective resistance of the root epidermis as a whole is proportionately increased. Thus there is a resistance, increasing with decrease in soil water content and therefore having the same effect as a "rhizosphere resistance", but which is a physiological resistance conditioned by the amount of water in the soil in what is, in principle, but not in detail, a rather simple way. The argument is set out elsewhere (Cowan and Milthorpe, 1968).

We have travelled, against the stream so to speak, from one end of the plant to the other, and one may wonder whether discussion of root environment is in place in a chapter primarily about stomata. However, stomatal physiology cannot arbitrarily be separated from other aspects of plant physiology, and what happens in the soil, though remote from the stomata, is probably as significant to their functioning as events in the atmosphere. A great many experiments on stomatal behaviour are performed on plants with their roots in nutrient solution and their leaves in rather dry air. If I am correct about the importance of root-water contact, then this condition is one which differs radically from the real world environment, at least for the species which are usually used, and the significance of stomatal functioning (as distinct to the mechanics of it) is the less likely to be divulged by such experiments.

Finally let us return to the subject of diurnal variation of stomatal aperture in relation to leaf water potential in plants out-of-doors. The phenomenon of "midday closure" is not as well documented as the frequency of reference to it might imply; and often, because of a lack of other physiological data, its occurrence is susceptible to various interpretations. The observations of Loftfield (1921) and others have been tentatively explained (Heath and Orchard, 1957) as an effect of increased intercellular concentration of CO_2 due to respiration (or in the context of recent work one might suggest, photorespiration), enhanced by temperature in the middle part of the day. Recent investigations include that of Schulze et al. (1975), who showed that midday closure in Prunus armeniaca was not due to increased intercellular CO_2 concentration and was concurrent with increase, rather than decrease, in leaf water content. Despite the apparent capacity of the hydraulic system in plants sometimes to sustain large fluxes of water without substantial increase in gradient potential, there are many examples in which leaf water potential exhibits a pronounced diurnal variation attributable to variation in rate of evaporation. But, often, this is not associated with a tendency for stomata to close in the middle part of the day. Stomata did not close in cotton when leaf water potential decreased from -10 to -27 bar during the course of the morning and early afternoon (Jordan and Ritchie, 1971). But with other crop plants there is sound evidence that stomatal closure does occur as a result of the interplay of climate and the hydraulic properties of soil and plant as it affects leaf water potential. Begg et al. (1964) observed a closure of stomata shortly after midday in Pennisetum typhoides followed by a reopening during the afternoon, the closing and opening movements being preceded by decrease and increase in the relative water content of the leaves. Turner and Begg (1973) and Turner (1974b) found that stomatal conductance to vapour diffusion in maize, sorghum and tobacco followed a time course closely correlated with intensity of radiation when the soil was wet, but with dry soils it decreased rapidly during the morning at a time when the turgor pressure of the leaves was reduced nearly to zero. Turgor pressure began to increase in the late afternoon but the stomata did not reopen until the following morning. Sharpe (1973) found that stomata in cotton in wet soil did not close during the day even under severe light and temperature regimes, but closed early in the morning with a tendency for those in the abaxial surfaces to reopen late in the afternoon with plants in dry soil. Leaf water potential was not measured.

In this brief review of stomatal behaviour and plant water relations I have concentrated on what seems to be an attribute of most crop plants: the absence of any positive action of the stomata to reduce rate of evaporation until the turgor pressure of water approaches zero, the tendency to increase osmotic pressure and therefore maintain turgor as water potential is diminished, the ability to support large fluxes of water with a small difference in potential between roots and leaves, and sufficiency in the length of roots below ground to prevent a substantial "rhizosphere" resistance developing as the soil dries. All these are characteristics which tend to preserve turgor and not to conserve water. They have been accentuated to a degree by the plant breeder, and there is now interest in removing some of them by breeding (Passioura, 1972). Means of causing stomata to close have been widely investigated. But the characteristics are probably present in most species which grow in regions sufficiently moist to support fairly dense communities of plants, and which have a growth habit which does not expose them to influences tending to promote a greater rate of evaporation than that sustained by their neighbours. The matter is relevant to a later discussion of the efficiency of water use in plants. First, though, I shall discuss two aspects of stomatal behaviour which are in marked contrast to that which has been described; the response of stomata to changes in environment which cause rapid changes in rate of evaporation, and the evidence that is accumulating that change in ambient humidity can have a direct influence on stomatal aperture.

B. DYNAMIC BEHAVIOUR AND THE HYDRAULIC LOOP

Figure 10 illustrates a phenomenon which, had he observed it, might have caused Stephen Hales (1727) to amend his comment: "But as plants have not a dilating and contracting *Thorax*, their inspirations and expirations will not be so frequent as those of animals, but depend wholly on the alternate changes from hot to cold, for inspiration, and vice versa for expiration." What is shown are three of 70 successive oscillations in stomatal conductance, rate of evaporation, and leaf water content that Dr Troughton and I observed in a cotton plant maintained in a constant environment (Cowan, 1972a). The measurements relate to a single leaf on an intact plant (fluxes being expressed as averages per unit area of leaf), but oscillations in the other leaves occurred in synchrony. The rate of change in water content of the leaf, dW/dt, was obtained from the difference

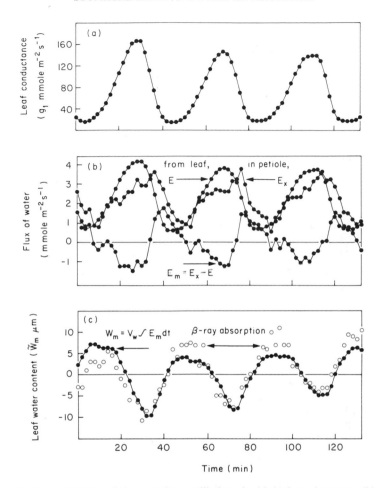

Fig. 10. Characteristics of free-running oscillations in (a) leaf conductance, (b) water flux, and (c) water content, in cotton. Data relate to a single leaf on an intact plant. The rate of change of water content of the leaf, E_m, was obtained as the difference of water flux in the petiole, E_x, and rate of evaporation, E. The alternating component of leaf water content, W_m, is the integral of this quantity with respect to time, multiplied by the molar volume of water. The scale of the measurements of β-ray absorption by the leaf is arbitrary. (Data of Troughton and Cowan; Cowan, 1972a.)

between the flux in the petiole, E_x, and the rate of evaporation, E. The alternating component of the leaf water content, $\widetilde{W}$, is the integral of this quantity with respect to time. Leaf conductance, g_l and its reciprocal leaf resistance, r_l, are defined by the equations

$$E = \frac{g_l(w_i - w)}{1 + g_l/g_b} = \frac{w_i - w}{r_l + r_b} \tag{20}$$

where w_i is the humidity in the intercellular air spaces in the leaf (assumed to be saturated), w is the humidity of the ambient air, and $g_b = 1/r_b$ is the conductance of the external boundary layer. The boundary layer, in the present context, is a concept, related to gas transfer in the region of a solid surface, which is rather like the "unstirred layer" which bedevils measurement of the permeability of membranes (see Dainty, 1963). Approximately, g_b varies with $(u/b)^{1/2}$ where u is windspeed and b is breadth of the leaf, but is best determined in any given conditions by measurement of evaporation from, and temperature of, a wet leaf replica.

A note on the units I shall use for conductance is appropriate here, because they are not those generally employed. The reasons for changing a well-entrenched practice are given in Appendix A. E will be expressed as molar flux of water vapour per unit area of leaf and w as mole fraction of water vapour in air. Therefore g has the same units as E. For all practical purposes w may be taken as equivalent to both the ratio of vapour pressure to total air pressure, and to the volume fraction of vapour in air. Therefore g is 10^3 times the flux density that would be sustained by a vapour pressure difference of 1 mbar in air at 1 bar total pressure. The analogous conductance for CO_2, g^+, is 10^6 times the molar flux density that would be sustained by a concentration difference of 1 $\mu l/l$, i.e. 1 p.p.m. The conversion from conductance as defined here to that which is generally used varies as the absolute temperature; at 25°C a conductance of 1 mole m^{-2} s^{-1} is equivalent to 2.5 cm s^{-1}. Leaf conductances throughout this chapter relate to transfer from both surfaces of a leaf in parallel. Strictly, this is not valid except in the unlikely instance that the transport properties on the two sides of the leaf are identical (see Gale and Poljakoff-Mayber, 1968; Moreshet et al., 1968), but is less open to objection if the boundary layer conductance is large compared with the leaf conductance as it was with the observations in Fig. 10 ($g_l/g_b < 0.1$).

It is more common to use resistance rather than conductance in describing the properties of the leaf pertaining to gas exchange, primarily because the processes of vapour transfer in the epidermis and boundary layer are in series. But, conceptually, there is much to recommend the use of conductance. When g_b is large compared with g_l, flux of vapour (and sometimes CO_2 also) is nearly linear with leaf conductance. And, sometimes, leaf conductance is more nearly a linear function of those internal physiological conditions which are affected by the fluxes of vapour and CO_2 than is resistance. To use resistance can create a misleading impression of non-linearity in the interrelationships between gas exchange, stomatal behaviour and associated aspects of plant functioning which is very inconvenient. Nevertheless, resistance is often the more convenient quantity to employ in formal expressions and calculations, and I shall use it when appropriate.

Behaviour of the kind shown in Fig. 10 has been observed in some dozen or more species, with periods varying from 15 to 50 min, but generally close to 35 min, and it is now not unreasonable to suppose that it can occur in most mesophytes (see Barrs, 1971; Cowan, 1972a). It is known to take place in both mono- and dicotyledonous plants, the former including species having the Hatch-Slack C_4-dicarboxylic pathway of assimilation. Several models of the mechanism responsible for oscillations have been described (Karmanov et al., 1966; Lang et al., 1969; Raschke, 1970b; Hopmans, 1971; Cowan, 1972a). They have in common one basic assumption: that the oscillations are due to the properties of the loop in which rate of evaporation affects, through physiological processes, the aperture of stomata and stomatal aperture in turn affects rate of evaporation. Whatever the precise nature of the physiological elements, the way in which oscillations can occur in such a loop may be described, with reference to Fig. 11, as follows.

Suppose a small oscillation $\tilde{w}$, having angular frequency ω, is superimposed on the mean ambient humidity. Then rate of evaporation and leaf conductance will also oscillate with frequency ω. Had leaf conductance remained constant the oscillation in rate of evaporation would have been $\tilde{E} = \tilde{w} \cdot \partial E/\partial w$ but, because the oscillation in leaf conductance is fed back into the system, the actual oscillation in rate of evaporation will be modified both in amplitude and phase. Let us set $\tilde{E} = \cos(\omega t)$—after all we are at liberty to choose units and time scale to make it so. Now the amplitude of the oscillation in g_l relative to that in E is the gain, B, of the physiological component of the loop; and the phase of the oscillation in g_l with respect to the oscillation in E is the phase shift, ζ, of the physiological component of the loop. That is to say the gain and

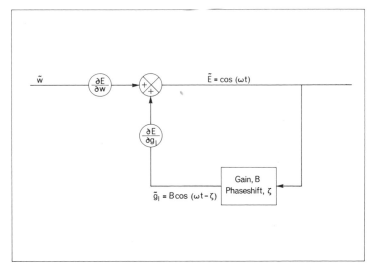

Fig. 11. Schematic diagram showing interrelationship between oscillations in rate of evaporation, E, and leaf conductance, g_l, caused by an oscillation in ambient humidity, w. The gain, B, and phase shift, ζ, are frequency-dependent properties of the plant.

the phase characteristics are defined by writing $\tilde{g}_l = B \cos(\omega t + \zeta)$. Both B and ζ are functions of the frequency, ω. What may be loosely termed the physical component of the loop (because it relates primarily to the process of gas diffusion in the stomatal pores) is the effect of leaf conductance on rate of evaporation. It does not impose a phase shift, or, more correctly, the phase shift it imposes is negligibly small compared with that in the physiological component. Its gain—what will be called "environmental gain" (Farquhar, 1973; Farquhar and Cowan, 1974)—is $\partial E/\partial g_l$ and is independent of ω. Therefore the total phase shift in the loop is ζ and the total gain is $B\partial E/\partial g_l$.

It may now be shown that the amplitude of the output of the system, $\tilde{E}$, is greater than that of the input of the system, $\tilde{w} \cdot \partial E/\partial w$, if $2 \cos \zeta > B\partial E/\partial g_l$. And it is immediately apparent that if the phase shift is zero at some particular frequency, and the corresponding total loop gain is equal to unity, then the output is sustained by the feedback. If the gain at this frequency is made greater than unity, the oscillations will successively increase in amplitude until they are limited by the non-linear characteristics of the system. That is to say, the oscillations are self-perpetuating and take place when ambient humidity is maintained constant. This, then, is the way in which spontaneous oscillations occur in a closed loop. Dr Farquhar and I conceived the notion of testing the supposition that oscillations were associated with the properties of the hydraulic feedback loop by manipulating the environmental gain which, from eqn (20), is

$$\frac{\partial E}{\partial g_l} = \frac{(w_i - w)}{(1 + g_l/g_b)^2} \tag{21}$$

The simplest way of doing so is to change the ambient humidity. However, the results might be ambiguous because the rate of evaporation would be caused to change and the repercussions of that could not be predicted. Instead, Dr Farquhar set up a system of humidity control which could maintain rate of evaporation constant despite changes in leaf conductance (provided leaf conductance was not excessively small). In other words, an artificial feedback system was imposed on the plant which had the effect of making $\partial E/\partial g_l$ zero. When it was used to maintain rate of evaporation in a cotton plant constant (Fig. 12), the oscillations which otherwise occurred were inhibited. The technique is somewhat similar to the use of the "current clamp" in membrane physiology (see Cole, 1968). Apart from the context in which I have described it, it may also be used to explore stomatal response to environmental variables such as light and CO_2 concentration at constant evaporation rate (Verfaille, 1972). In many experiments of course, including that shown in Fig. 12, leaf temperature is "clamped" by varying ambient temperature with respect to a thermometric device attached to the leaves. That is another example of negating a feedback loop intrinsic to the functioning of a plant by imposing an external one.

I shall not discuss all the various observations which have been made with

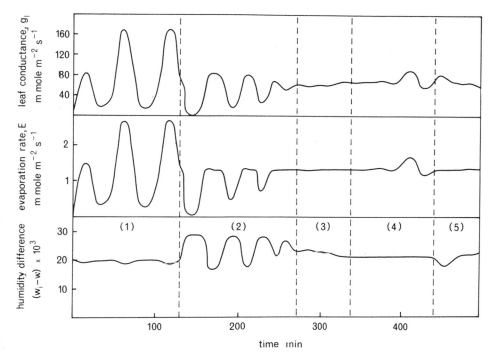

Fig. 12. The effects of controlling ambient humidity, w, and rate of evaporation, E, on stomatal functioning in a cotton plant. During stages (1) and (4) humidity is constant and leaf conductance, g_l oscillates; during stages (3) and (5) rate of evaporation is maintained constant and the oscillations are suppressed. (Farquhar and Cowan, 1974.)

oscillations of the type shown in Fig. 10. Of particular interest are the experiments of Ehrler *et al.* (1965), Barrs and Klepper (1968) and Lang *et al.* (1969) with cotton, Karmanov and Savin (1964) with beans of a Mexican variety, Barrs and Klepper (1968) and Cox (1968) with sunflower and Hopmans (1969, 1971) with *Phaseolus vulgaris.* Barrs (1971) has provided a comprehensive review, and shorter reviews may be found in Hopmans (1971) and Cowan (1972). More recently a series of elegant experiments with *Avena sativa* has been reported (Johnsson, 1973; Brogårdh and Johnsson, 1973, 1974; Brogårdh *et al.*, 1974). There is a strong concensus that the coupling which tends to maintain synchrony in the oscillations in the different leaves of a plant is due to the rapid propagation of changes of water potential in the vascular system. Experiments by Sheriff (1973, 1974) confirm that phase changes at the relevant frequency in the vascular system are quite small. In some instances the oscillations in differing leaves are not in phase however (Hopmans, 1971; Teoh and Palmer, 1971), presumably because the hydraulic communication is not sufficiently great. Treatment of the roots or lower stem which reduces the hydraulic conductance to uptake of water tends to enhance the probability of oscillations (Barrs and

Klepper, 1968; Brogårdh and Johnsson, 1973); excision of the root system usually causes oscillations to cease (Lang *et al.*, 1969). Oscillations usually occur in environments tending to promote a large rate of evaporation: this is probably an example of the role of environmental gain rather than actual rate of evaporation because oscillations can occur in the dark (Hopmans, 1971) when actual rate of evaporation is small due to small leaf conductance. A model which combines the gross hydraulic properties of a plant with "hydropassive" opening and closing movements of stomates associated with change in water potential (Cowan, 1972a; Farquhar and Cowan, 1974) is successful in simulating these various characteristics of stomatal behaviour (see also Brogårdh *et al.*, 1974). I now think, however, that the movements involve adjustment of osmotic pressure in the guard cells, and might be associated with changes in concentration of abscisic acid, or abscisic acid-like compounds in the leaf, in the way I have discussed.

The experiment described in Fig. 12 has associated a particular aspect of stomatal behaviour with the properties of a particular feedback loop. How can the properties of that loop be quantified in such a way that they can, on the one hand, be compared with relevant information about the stomatal mechanism, and, on the other, be used to predict how stomata will behave in environments different to those in which observations have been made? The study of free running oscillations is of limited usefulness because their characteristics are greatly influenced by the non-linear properties of the system involved—in a truly linear (and therefore unreal) system, the amplitude of the oscillations would successively increase without limit. It is unfortunate that limitations of mathematical theory are such that we often have to treat non-linearity as an aberration of second-order importance. In order to apply linear analysis one deals with small deviations from the steady state. The feedback loop becomes stable if the environmental gain is reduced to about 19×10^{-3} or less by decreasing the humidity difference across the leaf epidermis (Farquhar and Cowan, 1974) and the measurements to be described were carried out with this condition applied. I am grateful to Miss C. F. Tabor for making them available to me prior to the publication of a fuller account.

Figure 13 shows variations in rate of evaporation and leaf conductance in cotton following a step change in ambient humidity. An increase in rate of evaporation occurs immediately, followed by an opening of the stomata. Because of feedback, the subsequent behaviour of both rate of evaporation and leaf conductance is quite complex. If we assume that change in humidity has no *direct* effect on stomata, then it is the effect of rate of evaporation on leaf conductance that is of physiological interest; the change in humidity is no more than a convenient means of altering rate of evaporation. But the change in rate of evaporation is not at all the simple type of input that, by choice, one would use in an experiment. There is the problem of relating output to input so as to retain all fundamental information about the characteristics of the physiological mechanisms involved, but to discard trivial information about the particular

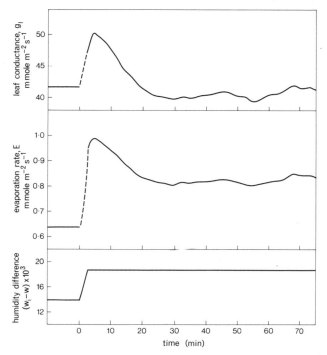

Fig. 13. The effects of decreasing ambient humidity, w, on rate of evaporation, E, and leaf conductance, g_l, in a cotton plant. (Data of Miss C. F. Tabor.)

form of the input. It can be tackled by use of the Laplace transformation. For the reader who is unfamiliar with the technique an outline of the immediately relevant aspects is given in Appendix B. It is no substitute for the treatments provided, for example, by Jaeger (1961) or Spiegel (1965), but I have it in mind that these do not make easy introductory reading. On the subject of transfer functions one cannot do better than recommend the delightfully lucid account by Machin (1964) together with a commentary (which should be glanced at first) by Wilkins in the same volume. Texts on regulation in biological systems (e.g. Grodins, 1963) are also likely to prove helpful.

Let Δg_l be the difference between the conductance at any time, t, after the initial perturbation in humidity and the steady-state conductance that obtained at $t < 0$. Let ΔE be the corresponding difference in rate of evaporation. Then the transfer function relating change in conductance to change in rate of evaporation is defined as

$$G(p) = \frac{\int_0^\infty e^{-pt}\, \Delta g_l\, \mathrm{d}t}{\int_0^\infty e^{-pt}\, \Delta E\, \mathrm{d}t} = \frac{L(\Delta g_l)}{L(\Delta E)} = \frac{\overline{\Delta g_l}}{\overline{\Delta E}} \qquad (22)$$

where the integrals are the Laplace transforms of Δg_l and ΔE, and the subsequent expressions have alternative shorthand notations for the same quantities. The Laplace operator p is a weighting variable. The greater p, the more weight is given to the short-term or high-frequency characteristics of both g and E, and vice versa. As $p \to \infty$, $G \to 0$ because, due to the time lag always present in a real system, $\Delta g_l = 0$ when $t = 0$. As $p \to 0$, G approaches the final steady-state magnitude of $\Delta g_l / \Delta E$. The transfer function implicitly contains all the information about the dynamic characteristics of the response in g_l to change in E in a form which is independent of the way in which E happened to have varied in the experiment from which G was derived—although in practice the way in which E varies affects the accuracy with which G may be determined. The uses of transfer functions are several. By a process of inversion, the variation of the output (Δg_l) may be predicted for any prescribed variation of the input (ΔE). Transfer functions may be manipulated algebraicly to combine the properties of a number of interacting feedback loops. Employed in this way they have conceptual value even when the results are not applied quantitatively. We shall later consider the variation in conductance when rate of evaporation, intercellular concentration of CO_2, c_i, and light intensity, I, all vary. Distinguishing the transfer function presently being discussed by the subscript E, the effect of all three influences on leaf conductance is written.

$$\overline{\Delta g_l} = G_E \overline{\Delta E} + G_{c_i} \overline{\Delta c_i} + G_I \overline{\Delta I} \tag{23}$$

and may be combined with expressions for the change in both rate of evaporation and intercellular concentration of CO_2 due to the change in conductance. In the particular observations illustrated in Fig. 13, light intensity was constant and the magnitude of G_{c_i} was found to be negligibly small; therefore only the first term on the right-hand side will be considered now. Figure 14a shows $G(p)$ for the observations in Fig. 13. I find it convenient to plot transfer functions against $1/p$ rather than p, because the short-term characteristics of the system relate to the properties of the transfer function at small magnitude of $1/p$ and vice versa. However, the important properties of a transfer function often do not yield readily to visual inspection. The algebraic function which fits the computed curve well is

$$G = \frac{a}{1 + p\tau_1} - \frac{b \exp(-p\tau_0)}{1 + p\tau_2} \tag{24}$$

To obtain the response to any prescribed variation in E one must find the inverse transform $L^{-1}(G\overline{\Delta E})$ as described in Appendix B. When $\Delta E = 1(t)$, the unit step function,* the then $\Delta g_l / \Delta E = L^{-1}(G/p)$. This is the unit response, U say.

---

* $1(t) = 0$ for $t < 0$, and 1 for $t \geqslant 0$.

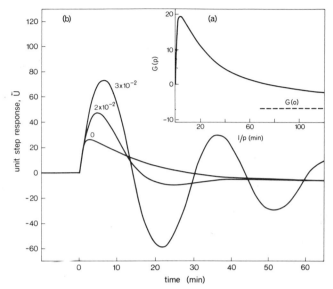

Fig. 14. (a) transfer function, $G(p)$, relating the response in leaf conductance to change in rate of evaporation shown in Fig. 13 as a function of the reciprocal of the Laplace operator, p; (b) computed responses, $\overleftrightarrow{U}$, of leaf conductance to unit step increase $(\Delta E)g_l$ for three magnitudes of environmental gain, $\partial E/\partial g_l$. For $\partial E/\partial g_l = 0$ then $\overleftrightarrow{U} = U$, the response to unit step increase in actual rate of evaporation.

For G given by equation (24),

$$U = a\,\{1 - \exp[-t/\tau_1]\} - 1(t - \tau_0)b\,\{1 - \exp[-(t - \tau_0)/\tau_2]\} \qquad (25)$$

If the reponse of g_l to a more complicated variation in E is required, the inverse transformation of $G\overline{\Delta E}$ may be hard to find. However, it is always possible to determine the response, provided the unit step response is known, by use of the principle of superposition. As shown in Appendix B

$$\Delta g_l = L^{-1}(G\overline{\Delta E}) = U * \frac{\partial E}{\partial t} \qquad (26)$$

The convolution, denoted by the symbol $*$, may be evaluated numerically.

The unit step response is plotted in Fig. 14b. It consists of an opening movement and a delayed closing movement, each of which occurs at a rate decaying exponentially with time. The time constants are $\tau_1 \approx 2$, $\tau_2 \approx 16$ min, and the delay is $\tau_0 \approx 1$ min. The two movements are similar in magnitude and the net result is that the steady-state response is small. It is hardly necessary to point out that the step response is consistent with what has been said in an earlier discussion about the "first" and "second" responses of stomata to a sudden perturbation of the water relations of the leaf epidermis; and the rapidity of the initial movement lends support to the supposition that evaporation may take place from sites rather close to the stomatal apparatus. Is it also consistent with the occurrence of spontaneous oscillations at increased environmental gain? To investigate this, first split the change in rate of evaporation into two parts, i.e.

$$\Delta E = (\Delta E)_{g_l} + \frac{\partial E}{\partial g_l} \Delta g_l \tag{27}$$

where $(\Delta E)_{g_l}$ is the change which would have occurred had there been no variation in conductance: in the context of the present discussion it is equivalent to $(\partial E/\partial w) \Delta w$, where w is ambient humidity but it allows also for changes due to variation in other external conditions directly influencing rate of evaporation, e.g. rate of ventilation. The other part is the change due to variation in leaf conductance and its magnitude depends on the environmental gain, $\partial E/\partial g_l$. Considering, now, the Laplace transform of equation (27), and making use of the definition of the hydraulic transfer function G, it follows that

$$\frac{\overline{\Delta g}}{(\overline{\Delta E})_{g_l}} = \frac{G}{1 - \dfrac{\partial E}{\partial g_l} \cdot G} = \overleftarrow{G}, \text{say} \tag{28}$$

The function $\overleftarrow{G}$ incorporates the effect of feedback, as the superscript implies; it is the "closed loop" transfer function describing the dynamic properties of the intact loop (see Fig. 15a). It would reduce to the "open loop" transfer function, G, if the environmental gain, $\partial E/\partial g_l$, were to be made effectively zero, i.e. if the loop were broken. In fact the loop cannot be physically broken (as, for example, may be done with an electrical circuit by breaking a connection, or in biochemistry by the use of a blocking inhibitor). However, as we have seen, G may be determined directly and thinking of it as an open-loop transfer function is a useful unifying concept.

Just as the function G enables one to predict the variation in g for any prescribed variation ΔE, the function $\overleftarrow{G}$ enables one to predict the variation in g_l for any prescribed variation $(\Delta E)_{g_l}$. As indicated in Appendix B, the roots, p_1, p_2 etc. of the denominator $1 - (\partial E/\partial g_l)G$ in equation (28) determine some important characteristics of the closed loop system. If they are real the system is non-oscillatory; if they occur in conjugate complex pairs it is oscillatory. If any of the real roots, or real parts of the complex roots, are positive then the system is unstable. The roots are readily found if G may be expressed as the ratio of two polynomials in p as would be the case if $\tau_0 = 0$ in equation (24). If $\tau_0 \ll \tau_2$, as it is in the particular function we have been considering, it is permissible to replace the exponential term representing delay by $1 - p\tau_0$ for most purposes. But in general the presence of non-linear terms complicates the analysis. An alternative approach is to determine the response of the system numerically by recourse to

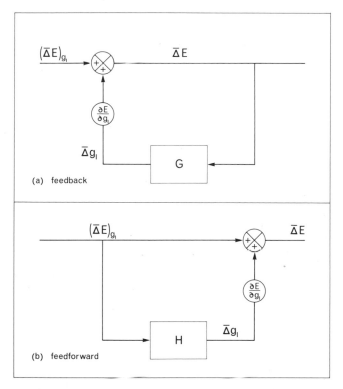

Fig. 15. Schematic diagrams of (a) feedback, and (b) feedforward, hypotheses of stomatal responses to changes in environment which tend to promote change in rate of evaporation. In (a) leaf conductance, g_l, responds to change in the actual rate of evaporation from the leaf, E; in (b) leaf conductance responds to the change in rate of evaporation that would have occurred had there been no response in leaf conductance, $(\Delta E)g_l$. The physiological transfer functions G and H relate the transformed changes in leaf conductance, $\overline{\Delta}g_l$, to the transformed "inputs", $\overline{\Delta}E$ and $(\overline{\Delta}E)_{g_l}$ respectively.

the convolution theorem. For $(\Delta E)_{g_l} = 1(t)$ it follows from equation (28) that the unit step response of the closed loop is

$$\overleftarrow{U} = U + \frac{\partial E}{\partial g_l} U \star \frac{\partial \overleftarrow{U}}{\partial t} \tag{29}$$

The numerical solution of this equation is carried out by a process which is directly analogous to the operation of the closed loop. It involves forward integration with the increase in $\overleftarrow{U}$ occurring during each successive step in time being fed back into the integral represented by the convolution.

Computations of $\overleftarrow{U}$ for two different magnitudes of environmental gain are shown in Fig. 14b. As $\partial E/\partial g_l$ is increased the system tends towards instability. The frequency of the oscillations agrees quite well with that observed in plants exhibiting spontaneous oscillations. The magnitude of $\partial E/\partial g_l$ at which instability occurs is 33×10^{-3}. In practice we find that instability actually occurs at about 19×10^{-3}. Therefore the analysis, though it seems to be qualitatively satisfying, is quantitatively inaccurate.

There are several reasons why the results of an analysis of this kind go astray. If the system is not linear, then extrapolation of the observations is not valid. It becomes necessary to determine the transfer function G at various mean levels of the input signal, E, in order to characterise its behaviour. The properties of G may vary with time. Figure 13 indicates that the system fluctuates in a way which is not readily related to its preceding history. Often it is not possible to determine a transfer function because leaf conductance does not become sufficiently steady for the integrands in equation (22) to converge. We find a great deal of unexplained variation in G, not so much in its general form, but in its magnitude. These are problems which may become less obtrusive when techniques are improved, and more measurements are made, but are likely to remain to some extent. There is also a quite different, and fundamentally interesting, potential source of error. The assumption that stomata respond to change in humidity only insofar as they are sensitive to change in rate of evaporation may be incorrect. I shall discuss the implications of this in a separate section.

The method of quantifying the properties of a feedback loop which has been outlined is not the only one available. Another approach is to determine the frequency response to periodic perturbation of the loop; that is to say, find the gain B and the phase shift ζ as functions of the perturbing frequency ω. The form of the transfer function is readily obtained from such information by means of the well-known relation

$$Be^{i\zeta} = G_E(i\omega) \tag{30}$$

The method is, in principle, a better one than that which has been described here because it is less affected by the presence of non-linearity and noise in the system. However, because some of the frequencies of particular interest correspond to periods of 30 min, and more, the method is extremely time-consuming, and does not lend itself to replication. It has been used by Farquhar (1973). In many ways the visual presentation of the properties of a transfer function in terms of gain and phase shift, i.e. in the form of a Bode diagram, is more immediately informative than the plot of the function itself. Figure 16 shows $B(\omega)$ and $\zeta(\omega)$ computed from the transfer function described by equation (26) (parenthetically it may be worth remarking that β and ζ would be more accurately obtained directly from the original measurements of evaporation and conductance by use of the integral Fourier transform). The phase shift passes through zero at the natural undamped frequency of the system, corresponding to a period of 30 min. With increasing environmental

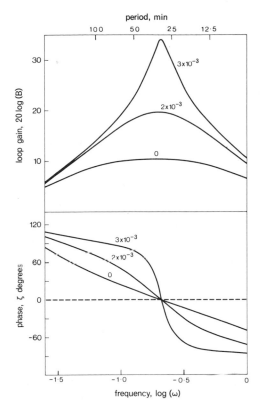

Fig. 16. Bode diagram of the response of leaf conductance in terms of gain, B, in decibels, and phase shift, ζ, to sinusoidal variation of evaporation rate at frequency ω. The definition of B and ζ is illustrated in Fig. 11. The response has been computed from the transfer function in Fig. 14.

gain, resonance in the closed loop at a frequency close to the natural frequency becomes greater and more sharply defined.

In this section we have discussed how the characteristics of the hydraulic transfer function, G, together with the magnitude of environmental gain, $\partial E/\partial g_l$, might determine the behaviour of the hydraulic loop. The action of the stomata, because their initial response tends to reinforce any environmental change causing change in rate of evaporation, is one of positive feedback and it is this which can cause continuous oscillations to take place. It is not known whether continuous cycling often occurs out-of-doors or if it is associated with particular conditions in which physiological experiments are conducted. The way in which $\partial E/\partial g_l$ is affected by the characteristics of the environment is important in this respect, and that will be discussed next. But I do not want to overemphasize the importance of continuous cycling. It is one manifestation, only, of the properties of the hydraulic feedback loop and one which may have little significance in its own right. Later I shall explore the notion that the functional role of positive feedback in the hydraulic loop is to enhance the rapidity with which stomata respond to changes in irradiance.

C. THE CONCEPT OF ENVIRONMENTAL GAIN

Farquhar (1973) has analysed the factors influencing environmental gain. He points out that one must be cautious in attributing significance to the form of its dependence on leaf conductance. It has been assumed that leaf conductance responds linearly to change in evaporation rate. Then it is appropriate to express environmental gain as $\partial E/\partial g_l$. With leaf temperature held constant, $\partial E/\partial g_l$ is given by eqn (21) and one is led to conclude that the gain increases with decreasing conductance. Therefore any disturbance, such as decrease in light intensity or increase in ambient concentration of CO_2, which causes stomatal closure would increase gain and increase the tendency of the stomatal system to oscillate (Cowan, 1972a). However, if leaf resistance, r_l, rather than conductance, were to respond linearly to change in rate of evaporation, then environmental gain would be more appropriately expressed as $-\partial E/\partial r_l = g_l^2 \, \partial E/\partial g_l$ and the implications would be exactly the opposite. On the whole it seems that the response of stomata to rate of evaporation is more nearly linear when expressed in terms of conductance rather than resistance, but it is by no means certain.

The effects of environment on gain are less equivocal but are quite complex. The correct basis on which to make comparisons is to assume similar rates of evaporation and similar leaf conductances. Making use of equation (20), equation (21) becomes

$$\frac{\partial E}{\partial g_l} = \frac{E/g_l}{1 + g_l/g_b} \tag{31}$$

Therefore, the influence of environment on the hydraulic loop is not completely accounted for by measuring rate of evaporation and leaf conductance; the stability of the loop depends also on the conductance of the boundary layer. As g_b is usually much greater than g_l the effect seems to be small. But generally, equations (20) and (21) are inadequate to describe the variation of E and $\partial E/\partial g_l$ with environment because leaf temperature is not normally held constant and therefore w_l may vary. This is illustrated in Fig. 17. Rate of evaporation is determined by the intersection of the line (a), which represents equation (21), with a second line, (b), which represents the influence of the energy balance of the leaf on rate of evaporation and is derived as follows.

The flux of sensible heat from unit area of leaf is

$$\phi - \lambda E = C_p g_b^{\ddagger}(T_l - T) \approx \lambda g_b^{\ddagger}(w_l' - w')/\epsilon \tag{32}$$

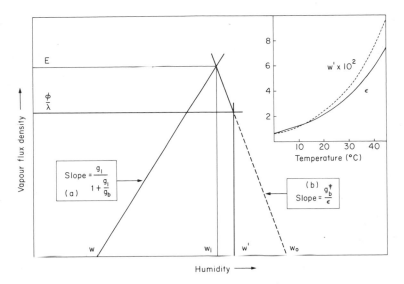

Fig. 17. Rate of evaporation, E, and internal humidity, w_i, in a leaf as the intersection of line (a) representing the influence of w_i on rate of evaporation with line (b) representing the influence of w_i on heat transfer, $E - \phi/\lambda$, ϕ being net radiation absorbed per unit area of leaf and λ being latent heat of vaporization of water. g_l, g_b, $g_b^\ddagger$ are leaf conductance, boundary layer conductance to vapour transfer, and conductance to heat transfer, respectively. w is ambient humidity, and w' humidity of air saturated at ambient temperature. The insert shows w' and $\epsilon = 1.51\ dw'/dT$ at normal atmospheric pressure as functions of temperature.

where ϕ is net flux of radiation absorbed per unit area of leaf, λ molar latent heat of vaporization of water, C_p molar heat capacity of air at constant pressure, $g_b^\ddagger$ boundary layer conductance to heat transfer*, w_l' saturation humidity of air at leaf temperature T_l, and w' saturation humidity of air at ambient temperature T. The quantity $\epsilon = \lambda/C_p \cdot dw'/dT \approx 1.51\ de'/dT$ is the rate of increase in the latent heat content of saturated air with increase in sensible heat content. Equation (32) is discussed in Appendix A. Making the usual assumption that the intercellular spaces in a leaf are saturated with water vapour it becomes

$$E = \phi/\lambda - g_b^\ddagger(w_i - w')/\epsilon \tag{33}$$

This is the line, (b), of negative slope in Fig. 17. Strictly, it is not straight because ϵ varies with temperature; but provided the difference between leaf and air temperature does not exceed a few degrees it can be taken as approximately straight. The point at which it intersects the line (a) of positive slope determines E and w_i, the point at which it intersects the horizontal line, $E = \phi/\lambda$,

* Variation in the emission of thermal radiation with variation in leaf temperature may be taken into account in the conductance $g_b^\ddagger$, ϕ then being taken as the net flux of radiation that would obtain if $T_l = T$ (Jones, 1976). When this is done $g_b^\ddagger$ is somewhat greater than g_b.

corresponds to w', the humidity of air saturated at ambient temperature. When g_l, and therefore the slope of (a), varies, w_i may be kept constant by varying w'. It is equivalent to shifting the line (b) laterally, without change of slope, and the variation in w' defines the variation in ambient temperature. This represents the experimental device of maintaining leaf temperature constant. With constant ambient conditions, the line (b) is fixed and it is readily derived from the geometry of the diagram that

$$E = \frac{g_l(w_0 - w)}{1 + g_l(\epsilon/g_b^{\ddagger} + 1/g_b)} \tag{34}$$

where w_0 is the intercept of the line (b) on the abscissa. Therefore environmental gain is now

$$\frac{\partial E}{\partial g_l} = \frac{E/g_l}{1 + g_l(\epsilon/g_b^{\ddagger} + 1/g_b)} \tag{35}$$

It is the counterpart of equation (31). At a given rate of evaporation and given leaf conductance, environmental gain is less if ambient temperature rather than leaf temperature is held constant. In the context of the data in Fig. 13 the difference is small; the denominator in equation (35) is 1.15. But the maximum conductance in cotton and many other species is about an order of magnitude greater than that in Fig. 13, and in many experiments the boundary conductance is somewhat smaller (due to smaller rates of ventilation). Therefore the difference between equations (31) and (35) may become important. Then, the device of keeping leaf temperature constant is one which increases the likelihood of instability in stomatal functioning. On the other hand many of the chambers in which physiological experiments are carried out may engender stability, not only because the rate of ventilation is small, but because it is the humidity and temperature of the ingoing, rather than the ambient, air which is controlled. The combination is such that change in rate of evaporation is resisted by the changes in ambient humidity and temperature which are caused by it; that is to say, environmental gain is effectively decreased. It is perhaps not generally realized that stability in stomatal functioning may be an *artifact of the way in which the environment is controlled.* We tend to think that the properties of an environment are adequately defined if ambient humidity, temperature and CO_2 concentration (w, T and c, respectively) are measured. But it may be equally important to determine quantities such as $\partial w/\partial E$, $\partial T/\partial E$, $\partial c/\partial A$ (A being rate of assimilation) which define the dynamic properties of an experimental environment. And, if the dynamic properties of the environment affect the stability of the stomatal system, may it not be that they affect stomatal functioning in other, more subtle ways?

The ecological significance of environmental gain will be touched on later. Use will also be made of the quantity defined as w_0 in Fig. 17. It is

$$w_0 = \epsilon\phi/(\lambda g_b^{\pm}) + w' \qquad (36)$$

and is an estimate of the humidity which would obtain in the leaf if leaf conductance were zero (and ϵ were constant). Formally, it is somewhat like a quantity entering into the equation for carbon-dioxide exchange in leaves, the compensation concentration Γ , and that analogy will be developed. If w_0 is substituted in eqn (34) the resulting expression is recognizable as a form of the "combination equation" first derived for evaporation from vegetation by Penman (1953) and expressed and applied in many ways since.

D. DIRECT RESPONSE TO HUMIDITY

Until recently, there was a consensus that stomatal regulation of plant water relations is achieved entirely by a process of feedback. That is to say, the stomata respond not to those external factors which tend to influence the rate of water loss from the leaf, but to components of the internal physiological state which are affected by the rate of water loss. As Meidner and Mansfield (1968) put it, "It is the actual rate of transpiration which is important." My discussion of both quasisteady and dynamic stomatal behaviour in the intact plant presumes that this is so. However, I have mentioned that Lange et al. (1971) demonstrated a direct influence of change in ambient humidity on the aperture of stomata in epidermal strips of *Valeriana locusta* and *Polypodium vulgare*. Raschke (1970a) obtained evidence of it with detached leaves of *Zea mays*. Several investigations with intact plants also point to the existence of such an effect. They are cited in Schulze et al. (1972) and Hall et al. (1976). Here I will concentrate attention on the measurements of Schulze and his co-workers who have not only studied the phenomenon but have discussed its significance in an ecological context (Schulze et al., 1972, 1974, 1975a, b).

Figure 18 shows the reaction of *Prunus armeniaca*, growing in the Negev Desert, to change in ambient humidity. The measurements relate to single twigs enclosed in a naturally illuminated cuvette in which ambient temperature was controlled so as to keep leaf temperature constant. As the difference in humidity between leaves and air was successively increased by decreasing ambient humidity, the resistance of the stomata to vapour diffusion was caused to increase and the net rate of assimilation to decrease. In this respect the results are unremarkable. What is remarkable is that the increase in resistance on each occasion was so great that the rate of evaporation decreased despite the increase in humidity difference. Reduction in rate of evaporation was accompanied by an increase in the water content of the leaves. When, finally, the ambient humidity was increased, then rate of evaporation increased and the water content of the

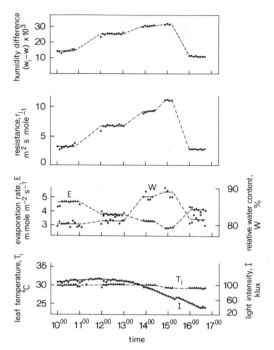

Fig. 18. Changes in resistance to vapour transfer, r, evaporation rate, E, and relative leaf water content, W, in *Prunus armeniaca* due to changes in ambient humidity, w. Leaf temperature, T, was kept constant. Resistance is $r = (w_i - w)/E$, where w_i is leaf internal humidity. Lighting was natural. (Redrawn from Schulze *et al.*, 1972.)

leaves declined. Similar results were obtained with the wild desert plants *Hammada scoparia* and *Zygophyllum dumosum*. It is worth quoting from the paper of Schulze *et al.* (1972): "When the stomata close at low air humidity the water content of the leaves increases. The stomata open at high air humidity in spite of a decrease in leaf water content. This excludes a reaction via the water potential in the leaf tissue and proves that the stomatal aperture has a direct response to the evaporative conditions in the atmosphere." They imply that the mechanism may be associated with peristomatal transpiration; that is to say, due to loss of water not under stomatal control which takes place at a sufficient rate to affect the turgor of the stomatal apparatus. The argument is a compelling one. I shall try to formalise it and discuss some of its implications.

The feedback hypothesis of stomatal response to changes in environment which tend to alter rate of evaporation is illustrated in Fig. 15a. The closed loop transfer function for response in rate of evaporation rather than leaf conductance is

$$\frac{\overline{\Delta E}}{(\overline{\Delta E})g_l} = \frac{1}{1 - \dfrac{\partial E}{\partial g_l} G} \tag{37}$$

In the present context $(\Delta E)g_l = (\partial E/\partial w) \cdot \Delta w$, where w is humidity. Stability demands that the denominator of the closed loop transfer function in eqn (37) be positive for all real positive values of the Laplace operator p. Therefore change in rate of evaporation will always take place in the same sense as $(\Delta E)g_l$, i.e. E will increase if w decreases. In an extreme case ΔE may approach zero following a change in humidity of finite duration. This would occur if, in effect, the system were a device employing integral control to maintain evaporation rate constant. Then $G \to -\infty$ as $p \to 0$. What is quite clear is that the system, as it has been defined, cannot explain the observations in Fig. 18 because E decreases with increase of w, and vice versa. The argument is a little different from the one put forward by Schulze et al. (1972), in that it does not depend on the assumption that leaf conductance increases with leaf water content. There is no absolute impediment to assuming the opposite, as was indicated in an earlier discussion of change in stomatal aperture with change in water potential.

A "direct" response of stomata to change in environment tending to cause change in rate of evaporation may be expressed as

$$\bar{\Delta}g_l = H(\bar{\Delta}E)_{g_l} \qquad (38)$$

where H is a new transfer function. Using equation (27), i.e. $\Delta E = (\Delta E)_{g_l} + (\partial E/\partial g_l)\Delta g_l$, it follows that

$$\frac{\bar{\Delta}E}{(\bar{\Delta}E)_{g_l}} = 1 + \frac{\partial E}{\partial g_l} H \qquad (39)$$

The system represented by this equation is shown in Fig. 15b. It is appropriate to call it "feedforward" as a comparison with Fig. 15a makes clear. There is no formal difficulty in explaining the nature of the observations in Fig. 18 as a manifestation of a feedforward response. It is only necessary that $H(0)$ should be negative and sufficiently large in magnitude for the right-hand side of eqn (39) also to be negative.

At this point a practical difficulty makes itself obvious. Let us accept that observations showing $\Delta E/(\Delta E)_{g_l} < 0$ indicate a direct effect of humidity on the stomatal mechanism. It is apparent, however, that the converse is not true. One has no way of knowing, from measurements of rate of evaporation, leaf conductance and leaf water content, whether there is, or is not, a direct effect of humidity on the stomatal mechanism unless the stomata give the game away, so to speak, by reacting to the extent that $\Delta E/(\Delta E)_{g_l} < 0$. It might be that stomata in all species are to some extent sensitive to external humidity. If that were indeed so, then the analysis of transient stomatal behaviour in cotton that I have outlined in a previous section would be incomplete. To combine elements of feedforward and feedback in a formal expression involving two transfer functions is an easy matter, but the result is cumbersome and of little practical value as it is not clear how the transfer functions can be determined. Instead, therefore, I shall make some initial assumptions which lead to an expression which is less general but relates to previous discussion of the water relations of the leaf epidermis. Suppose that

$$\overline{\Delta g_l} = G^+(\overline{\Delta E_c} + \alpha\overline{\Delta E_s}) \tag{40}$$

where E_c is the rate of evaporation from the external cuticle of the epidermis, and E_s is that which takes place from within the leaf and emerges from the stomatal pores. What is being assumed is that the feedback and feedforward transfer functions, G and H, differ only in magnitude and may be combined in the one function, G^+. The interpretation of α will be discussed shortly. We have, as before, $\Delta E = (\Delta E)_{g_l} + (\partial E/\partial g_l)\Delta g_l$. But now there are the additional relations

$$\Delta E = \Delta E_c + \Delta E_s$$

$$\Delta E_c = \frac{g_c}{g_l}(\Delta E)_{g_l} \tag{41}$$

where g_c is cuticular conductance. The basis of the first is plain; the second supposes that interference between the vapour streaming from the stomatal pores and that emanating from the cuticle is negligible. Combining these relations with equation (40) it may be shown that

$$\frac{\overline{\Delta E}}{(\overline{\Delta E})_{g_l}} = \frac{1 + \dfrac{g_c}{g_l}\dfrac{\partial E}{\partial g_l}(1 - \alpha)G^+}{1 - \alpha\dfrac{\partial E}{\partial g_l}G^+} \tag{42}$$

Consider the interpretations of α. If $\alpha = 1$ then the stomata respond to the total rate of evaporation rather than the individual components of it. It implies that the epidermis is in hydraulic equilibrium with the mesophyll tissue, because the stomata are proportionally no more sensitive to evaporation from the cuticle than to evaporation from the leaf as a whole. With $\alpha = 1$ equation (42) reduces to equation (37), the feedback equation. If $\alpha = 0$, on the other hand, the stomata are sensitive only to cuticular evaporation. There must be minimal hydraulic communication between the stomatal apparatus and the mesophyll tissue, and there can be no evaporation from the walls of the guard cells and subsidiary cells bordering the substomatal cavities. With $\alpha = 0$ equation (42) becomes similar to equation (39), the feedforward equation, and, depending on the magnitudes of the terms in the numerator, it is possible that the direct response of the stomata to cuticular evaporation will manifest itself in the way that has been discussed.

It would seem from this analysis that the characteristics which promote "direct" sensitivity of the stomata to ambient humidity tend to preclude sensitivity of the stomata to the potential of water in the leaf, and vice versa.

But of course, a response of stomata to decrease in leaf water potential may be brought about by other means, such as those associated with ABA synthesis. Schulze *et al.* (1972) suggestd that "an additional hydroactive component brought about by increased water stress in the whole plant" probably accounted for certain aspects of their observations with irrigated and non-irrigated *Prunus armeniaca* at two levels of ambient humidity. At each level of humidity the mean rate of evaporation was least in the unirrigated plants. In both irrigated and unirrigated plants the rate of evaporation was greatest at the higher level of humidity, the effect being most marked in the unirrigated plants—to the extent that the unirrigated plants transpired about as rapidly in the more humid atmosphere as did the irrigated plants in the less humid atmosphere. The way in which stomatal conductance varied during the course of the day depended both on soil water content and ambient humidity. The stomata tended to close in the middle part of the day in unirrigated plants in the more humid atmosphere; that is to say midday closure was associated with low water content and large rate of evaporation. There have been few experiments which have as clearly demonstrated the diversity of the effects of hydraulic environment on stomatal behaviour.

Of all changes in environment, a change in ambient humidity with leaf temperature held constant is likely to be least equivocal in its effects. It is difficult to imagine that it does anything other than cause a change in rate of evaporation. Combined with the conclusion that the responses shown in Fig. 18 are feedforward responses, the case that the physiological mechanism is associated with what has been termed peristomatal transpiration is very strong. Of course all one might strictly conclude is that there is a loss of water from the leaf to which the stomata are sensitive and which is not modulated by the reaction of the stomata to it. For example (for illustrative purposes only), if some of the stomata did not react directly to an increase in rate of evaporation from inside the substomatal cavities, but transmitted a signal which caused *other* stomata to close, this would be consistent with the definition of a feedforward response to water loss not under stomatal control.

However, there are other possibilities that should be borne in mind. Not all controls systems are adequately described as "feedback", "feedforward", or a combination of the two. They may have capabilities of adaptation and learning. Imagine what is feasible if control of stomatal aperture were supervised by a systems engineer, of micrometric proportions, inside the leaf. Knowledge of the internal flux of liquid water and of stomatal aperture would enable him to estimate the gradient in humidity across the epidermis. This provides a rough measure of environmental gain and he could adjust stomatal aperture accordingly. As the adjustment would depend on criteria of his own choosing, he could arrange that rate of water loss from the leaf be least when the external environment tended to promote a rapid loss, and vice versa. With certain additional measurements available, his knowledge of environmental gain and therefore his ability to predict the results of adjustment could be very much improved. But there is an alternative strategy at his disposal: observations of the consequences of trial variations in stomatal aperture could be used to determine environmental gain directly, and the results could be used to set an appropriate mean level of stomatal aperture. I have mentioned the possibility that fluctuations in leaf conductance may serve an exploratory purpose elsewhere (Cowan, 1972a). To an observer outside the leaf the question which may be most interesting is not so much the technique used by the systems engineer to obtain information, but the rationale of the decisions he makes. The "direct" response of stomata to change in humidity seems eminently sensible because it is one which conserves water. But its implications cannot be assessed without taking into account the effect of change in stomatal aperture on assimilation of CO_2. This is the subject of the next section.

E. QUASISTEADY RESPONSE TO LIGHT AND CO_2

An obvious feature of stomatal response to change in light intensity is its variability. Figure 19a shows responses for several species determined in field conditions. It is a popular misconception that stomata generally attain maximum aperture when exposed to light intensity which is a small fraction of full sunlight. It arises because responses are usually plotted in terms of resistance rather than conductance, and because methods of determining resistance tend to be insensitive (and perhaps the signal itself is noisy) when leaf resistance is small. Turner (1970) found that leaf conductance in sorghum grown in a plant chamber increased linearly with light intensity up to the equivalent of full sunlight; when plotted in terms of resistance such a response is, of course,

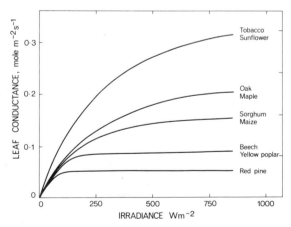

Fig. 19(a). Leaf conductance in several species as functions of irradiance. (Redrawn from Turner, 1974a. Data due to Waggoner and Turner, 1971; Woods and Turner, 1971; Stephens *et al.*, 1972; Berger, 1973; Turner and Begg, 1973.)

hyperbolic, and this may easily lead to statements about 'saturation', 'threshold values' and the like. Nevertheless there is a tendency for saturation to occur in some species, and the slope of the conductance: intensity curve diminishes with increasing light intensity in most. The characteristics vary with leaf age (Fig. 19b) and the intensity of light to which a plant has been accustomed (Björkman *et al.*, 1972a; Björkman, 1973). There often is a difference between the behaviour of the adaxial and abaxial surfaces; Turner (1970) found that the conductances of the two epidermes in sorghum were about the same when the illumination of the abaxial was about 1/5 that of the adaxial surface. How do all these differences and changes relate to differences and changes in the carbon and water metabolism in leaves? I shall discuss certain, very limited, aspects of the problem. If we exclude those species, having Crassulacean acid metabolism, with the capability of assimilating CO_2 at night, the most marked genotypic differences in stomatal behaviour are found between species which employ the Calvin cycle, only, to fix carbon and those which have the Hatch-Slack C_4-dicarboxylic pathway (C_3 and C_4 species, as they are commonly called). Maximum leaf conductance tends to be greater, (Ludlow, 1970; Slatyer, 1970; Downes, 1971) and sensitivity to CO_2 concentration smaller (Pallas, 1965), in C_3 than in C_4 species. The differences are such that maximum net rate of photosynthesis, and the response of net rate of photosynthesis to change in light intensity, differ less between the two categories than the intrinsic characteristics of the metabolic pathways would suggest (Akita and Moss, 1973; Gifford, 1974). Nevertheless there is some justification for distinguishing C_3 and C_4 species as "inefficient" and "efficient", respectively (Black *et al.*, 1969). The latter tend to grow more quickly, and use less water. I shall use the same subjective terminology here, but in relation to the way in which stomatal

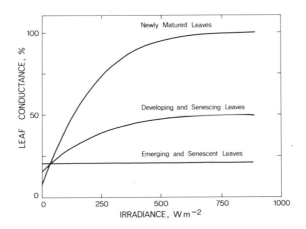

Fig. 19(b). Leaf conductance in leaves of different ages as functions of irradiance. (Redrawn from Turner, 1974a.)

behaviour in C_3 and C_4 species is attuned to the characteristics of leaf metabolism.

Raschke (1975a) holds that response of stomata to light intensity is primarily a manifestation of their sensitivity to concentration of CO_2 in the intercellular air spaces in the leaf. Intercellular concentration of CO_2 is an internal physiological parameter linking the response of stomata to light induced changes in rate of assimilation in a rather analogous way to that in which leaf water potential links the response of stomata to changes in rate of evaporation. However, there is an important difference between this system and the hydraulic loop. In the latter, rate of evaporation is a function of the environment and the physics of diffusion in stomata; change in leaf water potential has very little direct effect on rate of evaporation. In the system involving CO_2 transfer, rate of assimilation is as much a function of internal CO_2 concentration as CO_2 concentration is a function of rate of assimilation; both of them depend not only on stomatal behaviour but on metabolic characteristics of leaf mesophyll tissue. The interrelationships are illustrated in Fig. 20. Leaf conductance to CO_2 transfer, $g_l^\dagger$, is defined by the relation

$$A = \frac{g_l^\dagger(c - c_i)}{1 + g_l^\dagger/g_b^\dagger} = g^\dagger(c - c_i), \text{ say} \qquad (43)$$

where A is net rate of assimilation, c_i and c are concentrations of CO_2 in the intercellular spaces in the leaf and in the ambient air, respectively, and $g_b^\dagger$ is the conductance of the external boundary layer. The equation is the counterpart of equation (20) for rate of evaporation, the conversions between the conductances being $g_b^\dagger = 0.63\, g_l$ and $g_b^\dagger = 0.74\, g_b$ (see Appendix A). In a previous section it was shown how the influence of the boundary layer on rate of evaporation is magnified due to the coupling of heat and vapour transfer in leaves. The effect is responsible for the presence of the term ϵ in the denominator of equation (34). No such effect takes place with CO_2 transfer, and the boundary layer usually has a relatively minor influence. For the purposes of discussion here, I shall assume $g_l^\dagger \approx g^\dagger$, the conductance to transfer of CO_2 in the gas phase.* Rearranging equation (43) we have what is represented in one of the boxes in Fig. 20: intercellular concentration of CO_2 expressed as a function of ambient CO_2 concentration, rate of assimilation and conductance to CO_2 transfer. It in turn influences rate of assimilation (lower box) in a way which depends on light intensity; and leaf conductance in a way which *might* depend on light intensity.

Figure 21 illustrates some features of CO_2 fixation in C_4 and C_3 species. It is assumed that there is no systematic difference in the response of assimilation to light intensity when the carboxylating system is "saturated" with CO_2. The variation that exists amongst species having the same metabolic pathway

---

* This approximation is not acceptable when the magnitude of intercellular CO_2 concentration is computed from measurements of the other terms in equation (43).

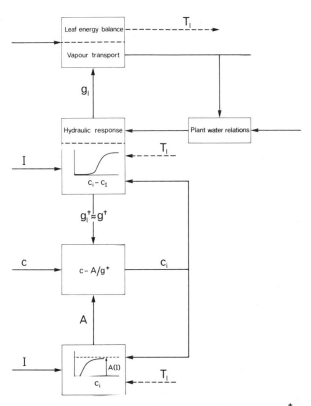

Fig. 20. Interrelationship between leaf conductance to CO_2 transfer, $g_l^\dagger$, intercellular CO_2 concentration, c_i, and rate of assimilation, A. The "inputs" c and I are ambient CO_2 concentration and light intensity respectively. Also indicated is the coupling of the metabolic system with the hydraulic system, g_l being leaf conductance to vapour transfer, and T_l being leaf temperature.

(Hesketh, 1963), and amongst plants of one species adapted to different light regimes (Björkman, 1973), is sufficiently great to obscure any general difference that may exist. The slope dA_I/dI at small intensity is about 0.06 mol E^{-1} (McCree, 1972; Björkman, 1973). Extrapolation to full sunlight ($I_{max} \approx 2mE\ m^{-2}\ s^{-1}$) yields $A_I = 120\ \mu mol\ m^{-2}\ s^{-1}$. However, the curve is not linear and approaches different maxima in different species. The functional dependence of rate of assimilation on intercellular CO_2 concentration is rather more simple in C_4 than in C_3 species. Due to the apparent absence of photorespiration, rate of assimilation is positive at all but very small concentrations of CO_2. With sufficient light and at small intercellular concentrations of CO_2, rate of assimilation increases rapidly with increase in c_i, the "internal conductance", $\partial A/\partial c_i = k$ say*, being in the range of 0.4 to 2.0 mol $m^{-2}\ s^{-1}$ in

---

* I use this distinguishing symbol in order to avoid the impression that the quantity has anything more than a superficial resemblance to conductance in the gas phase, $g^\dagger$.

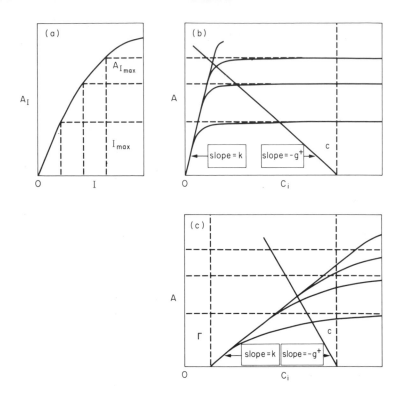

Fig. 21. Hypothetical diagram showing rate of assimilation, A, as a function of light intensity, I, and intercellular CO_2 concentration, c_i: (a) rate of assimilation with saturating concentration of CO_2, A_I, as function of I; (b) $A(I,c_i)$ in C_4 species; (c) $A(I,c_i)$ in C_3 species. The quantities k and g^+ are internal and gas phase conductances to CO_2 transfer, respectively; Γ and c are the compensation and ambient concentrations of CO_2. Actual rate of assimilation and actual intercellular CO_2 concentration correspond to the intersection of the straight line, slope $-g^+$, with the $A(I,c_i)$ characteristic for the particular light intensity.

tropical grasses, corresponding to internal resistances, in conventional units, of 1 to 0.2 s cm^{-1} (Bull, 1969; Gifford and Musgrave, 1970; Ludlow, 1970; Downes, 1971; McPherson and Slatyer, 1973). Conductances in *Amaranthus viridis* (Bull, 1969) and *Tidestroma oblongifolia* (Pearcy *et al.*, 1971) are similar; those in *Atriplex rosea* (Björkman *et al.*, 1971) and *Atriplex spongiosa* (Osmond *et al.*, 1969) are about 0.25 mol m^{-2} s^{-1}. All of these values relate to mature leaves and near optimum temperature for photosynthesis. Internal conductance may vary markedly with leaf age (Bull, 1969; Slatyer, 1970), and be very much smaller in plants grown in deleterious conditions, e.g. with limited water supply (Bull, 1969). In plants having large internal conductance the transition between the "light-saturated" ($A \approx kc_i$) and "CO_2-saturated ($A \approx A_I$) parts of the $A(c_i,I)$ characteristics is centred on an intercellular concentration of CO_2 considerably less than ambient concentration, even when the light intensity is equivalent to full sunlight. Taking $k = 1.3$ mol m^{-2} s^{-1} and $A_I = 80$ μmol m^{-2} s^{-1} at full sunlight, one finds $c_i \approx 60$ p.p.m.

I shall not comment yet on the characteristics of C$_3$ species, illustrated in Fig. 21c, for I shall discuss stomatal behaviour in C$_4$ species first.

The influence of a given magnitude of the gas phase conductance, $g^\dagger$, on rate of assimilation and intercellular concentration of CO_2 may be obtained by a construction shown in Fig. 21; A and c_i correspond to the intersection of $A(c_i,I)$ with the line of slope $-g^\dagger$ intercepting the abscissa at ambient concentration of CO_2. The maximum value of $g^\dagger$, $g^\dagger{}_{max}$ say, associated with fully open stomata and rapid ventilation, places a constraint on the rate of assimilation that may be attained at any given concentration of CO_2. However, it is about 0.25 mol m^{-2} s^{-1} in the most rapidly assimilating C$_4$ species, corresponding to a resistance in conventional units of 1.6 s cm^{-1} (Gifford, 1974), and this is sufficiently large to allow assimilation at a light intensity equivalent to full sunlight to proceed at a rate not very much smaller than the maximum, $A_{Im\,max}$, that the carboxylating system is capable of. With C$_4$ plants not experiencing an immediate deficiency in the supply of water but not being unduly liberal in the use of it, the paradigm of efficiency in stomatal behaviour suggests itself. As light intensity varies, the conductance $g^\dagger$ should be manipulated so that intercellular CO_2 concentration is just to the right of the line, $A = kc_i$. That is to say, rate of assimilation should be limited by light intensity rather than supply of CO_2, but intercellular concentration should not greatly exceed the magnitude, $c_i = A_I/k$, which corresponds to the "breakpoint" of the $A(c_i\,I)$ characteristic at any given light intensity. There is evidence that this is sometimes what occurs. But it is not clear how it is achieved.

If the sensitivity of conductance to intercellular concentration is written in the finite difference form

$$\Delta g^\dagger = -\beta \Delta c_i \qquad (44)$$

and combined with equation (43) it is found that

$$\Delta c_i = \frac{\Delta c - \Delta A/g^\dagger}{1 + c_i'\beta} \qquad (45)$$

where $c_i' = (\partial c_i/\partial g^\dagger)_{c,A}$ plays the same role in this system that environmental gain plays in the response of stomata to change in rate of evaporation. The term c_i' is by no means constant, of course, being equal to $A/g^{\dagger^2} = (c - c_i)g^\dagger$ Because of this, and the probability that β is not constant either, the differences in equation (45) must be taken as small.

We also have, by combining equations (44) and (45), the response in conductance

$$\Delta g^\dagger = \frac{-\beta(\Delta c - \Delta A/g^\dagger)}{1 + c_i'\beta} = -\overleftarrow{\beta}(\Delta c - \Delta A/g^\dagger) \qquad (46)$$

where $-\overleftarrow{\beta} = (\partial g^\dagger/\partial c)_A$ is the closed-loop response of stomata to change in concentration of CO_2, in contrast to $-\beta$ which is the open-loop response. If β is very large, $\overleftarrow{\beta} \approx 1/c_i'$.

Provided β is sufficiently large, change in c_i associated with a change in ambient concentration of CO_2, or (more realistically when we are concerned with functioning out-of-doors) change in rate of assimilation, becomes vanishingly small. The stomatal apparatus acts as a high-gain amplifier which, by means of negative feedback, ensures that the difference between c_i and some constant, internally generated, reference signal is zero. With C_4 plants having large internal conductance, maintenance of $c_i = A_{Imax}/k$, perhaps about 60 p.p.m., is a rough approximation to what we have defined as the paradigm of efficiency. As the difference in CO_2 concentration across the epidermis would be about 260 p.p.m., $g^\dagger$ would be no more than 23% greater than necessary to maintain $A = A_I$ even at the very smallest light intensity. But is the sensitivity of stomata to CO_2 concentration sufficiently great in real plants to maintain c_i constant, or almost constant? The measurements of Gifford and Musgrave (1970) with *Zea mays* showed that intercellular CO_2 concentration was very tightly controlled by stomatal adjustment as ambient concentration of CO_2 was altered. In contrast, other experiments, e.g. those of Raschke (1965b) with *Z. mays*, have shown that stomata in C_4 plants can remain partially open when c_i is caused to be much greater than that normally occurring in the light. Rate of assimilation in three C_4 grasses almost doubled as ambient CO_2 concentration was increased from its normal level to about 1000 p.p.m. (Ludlow and Wilson, 1971): clearly, intercellular concentration had more than doubled. Figure 22 shows data obtained by Downes (1971) with *Sorghum sudanense*. They provide an almost immediate indication of the gain of the CO_2-based feedback loop. As A is almost constant with increase in ambient CO_2 concentration above 300 p.p.m. we can deduce from eqn (43) that if c_i also is to be constant it must be that

$$\Delta\left(\frac{1}{g^\dagger}\right) = \frac{\Delta c}{A} \tag{47}$$

That is to say the increase in resistance to CO_2 transfer with increase in ambient CO_2 concentration must be inversely proportional to the rate of assimilation. It appears that the data in Fig. 22 might conform to eqn (47). However, a calculation shows that the slopes are considerably less than the corresponding magnitudes of $1/A$. Therefore the open-loop gain is insufficient to maintain c_i approximately constant, a conclusion which is confirmed in Fig. 23a. Increase in c_i is about half the increase in c in the range 300 to 700 p.p.m. This is a measure of the ratio of the closed- and open-loop responses of the system, for equation (45) may be written more succinctly as

$$\Delta c_i = \frac{\overleftarrow{\beta}}{\beta}(\Delta c - \Delta A/g^\dagger) \tag{48}$$

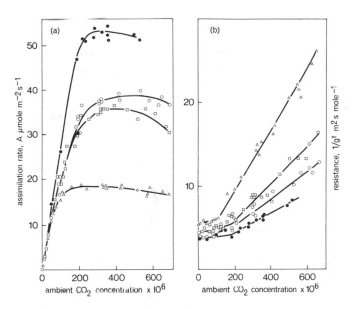

Fig. 22. Rate of assimilation, A, and resistance to CO_2 transfer in the gas phase, $1/g^\dagger$, in attached leaves of *Sorghum sudanense* as functions of ambient concentration of CO_2; open symbols, $\triangle$, $\square$, and $\circ$, relate to plants grown at small light intensity and tested at 62, 186, and 310 W m^{-2} photosynthetically active radiation, respectively, and the symbol $\bullet$ relates to a plant grown at large light intensity and tested at 310 W m^{-2} photosynthetically active radiation. (Redrawn from Downes, 1971.)

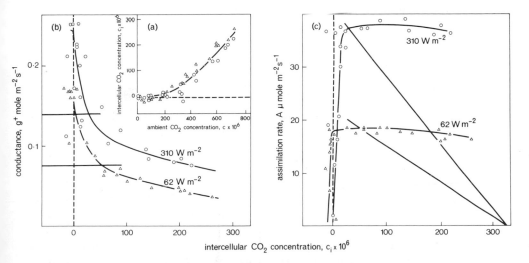

Fig. 23. (a) intercellular CO_2 concentration, c_i, as a function of ambient concentration, (b) conductance, $g^\dagger$, as a function of c_i, and (c) rate of assimilation, A, as a function of c_i, in *Sorghum sudanense* at two light intensities. Magnitudes of $g^\dagger$ at normal ambient concentration of CO_2 are indicated in (b) and are represented by the negative slope of the straight lines in (c). (Computed from data of Downes, 1971, shown in Fig. 22.)

However, the plot of $g^\dagger$ in Fig. 23a shows that β increased markedly when c_i decreased below 50 or 60 p.p.m., and that $g^\dagger$ increased in association with increase in light intensity. These features did not prevent c_i decreasing from about 50 to about 25 p.p.m. with increase in light intensity at normal ambient concentration of CO_2, but 25 p.p.m. was apparently sufficient to maintain $A = A_I$ at the higher light intensity. It would not be reasonable to attempt to quantify the magnitude of β at small c_i. There are always doubts about the accuracy with which $g^\dagger$ may be estimated from measurements of rate of transpiration. If $g^\dagger$ is overestimated c_i is underestimated, and vice versa. In determining β, inaccuracy is compounded by the same process of feedback that is being considered. One may be more confident of the magnitude of change in $g^\dagger$ associated with change in light intensity for it persists over the whole range of intercellular concentrations.

A disadvantage of CO_2-based control of stomatal functioning is that it cannot cause an increase in the availability of CO_2 when the demand for it by the mesophyll tissue is increased. As the stomata open with decrease in c_i, they cannot then be responsible for increase in c_i. This limitation of feedback is similar to the one discussed in the previous section. It is removed if the stomata respond directly to the change in environment which initiates activity in the feedback loop; in the present context, to change in light intensity. This "feedforward" action may be taken into account by rewriting equation (44) as

$$\Delta g^\dagger = -\beta \Delta c_i + \Delta g_I^\dagger \tag{49}$$

where $g_I^\dagger$ is a function of light intensity. Equations (46) and (48) become

$$\Delta g^\dagger = -\overleftarrow{\beta}(\Delta c - \Delta A/g^\dagger) + \frac{\overleftarrow{\beta}}{\beta} \Delta g_I^\dagger \tag{50}$$

and

$$\Delta c_i = \frac{\overleftarrow{\beta}}{\beta} (\Delta c - \Delta A/g^\dagger + c_i'\Delta g_I^\dagger) \tag{51}$$

The second relation shows that c_i may now increase with increase in light intensity. Assuming $A = A_I$, c_i increases provided

$$g^\dagger c_i'\Delta g_I^\dagger = (c - c_i)\Delta g_I^\dagger > \Delta A_I \tag{53}$$

The change $\Delta g_I^\dagger$ in Fig. 23 is a little more than half what is necessary to maintain c_i constant.

But what if rate of assimilation fails to respond to light intensity, for reasons which need not be specified but might be associated with malfunction in the leaf mesophyll tissue? From equation (50) the change in conductance which would then take place relative to that which would occur if the leaf functioned normally is

$$\frac{(\Delta g^\dagger)_{c,A}}{(\Delta g^\dagger)_{c,\Delta A = \Delta A_I}} = \frac{dg_I^\dagger/dA_I}{\beta/g^\dagger + dg_I^\dagger/dA_I} \tag{54}$$

Unless β is very large, the stomata respond to light intensity in a way which is not sensitive to the real requirements of leaf function. Yet if β is very large the direct response to light intensity is redundant as equation (51) shows. It is not difficult to envisage why this incompatibility occurs; the feedforward and feedback responses oppose each other. The necessity for compromise is removed if the influence of light intensity is to modulate the reference signal for the CO_2-based feedback system, rather than superimpose an additional change in conductance. This may be expressed by rewriting eqn (49) as

$$\Delta g^\dagger = -\beta(\Delta c_i - \Delta c_I) \tag{55}$$

It is consistent with the suggestion (Ketallaper, 1963; Raschke, 1975) that direct sensitivity of stomata to light intensity is associated with diminution in concentration of CO_2 in guard cells relative to that in the substomatal cavity. Equation (51) becomes

$$\Delta c_i = \frac{\overleftarrow{\beta}}{\beta}(\Delta c - A/g^\dagger) + \overleftarrow{\beta} c_i' \Delta c_I \tag{56}$$

As β becomes large, $\overleftarrow{\beta} c_i'$ approaches unity and $\Delta c_i \approx \Delta c_I$. The paradigm of efficient stomatal behaviour is then fulfilled if $dc_I/dA_I = 1/k$. Equation (54) becomes

$$\frac{(\Delta g^\dagger)_{c,A}}{(\Delta g^\dagger)_{c,\Delta A = \Delta A_I}} = \frac{dc_I/dA_I}{1/g^\dagger + dc_I/dA_I} \tag{57}$$

There remains some inflexibility in the system, but it is small if dc_I/dA_I is about $1/k$. It would be surprising indeed if evolution did not prefer equation (55) to equation (49). Yet the data in Fig. 23 do not demonstrate it. The two curves are similar in shape insofar as one can tell, but are displaced vertically, consistent with equation (49), rather than horizontally as eqn (55) would imply. It is to be hoped that experiments will soon clarify what is an important distinction.

There are other data with C_4 species which seem to indicate that light intensity influences leaf conductance directly. I find evidence for it in the data of McPherson and Slatyer (1973) for *Pennisetum typhoides*. The measurements of Ludlow and Wilson (1971) show that c_i increased in *P. purpureum* as the stomata opened in response to increase in light intensity. The quantity c_i/A (which is best termed "apparent internal resistance" rather than "internal" or "mesophyll resistance", as it is usually called) remained constant as illumination was increased from 2000 to 10 000 ft-candles, despite the fact that rate of assimilation doubled. However, observations such as these, and those of Downes that have been analysed here, are susceptible to another interpretation. Perhaps the stomata respond not to light intensity, but to rate of assimilation. That is to say, there may be a metabolite, synthesized in the mesophyll tissue at a rate which increases with rate of assimilation, and translocated to the epidermis where it causes the stomata to open. We have no *a priori* reason to suppose that change in intercellular CO_2 concentration as such is the *only* means by which the mesophyll tissue signals its short-term requirements for CO_2 to the stomata. Hanebuth and Raschke (1973) appreciated this possibility and investigated it with *Zea mays* (see also Raschke, 1975a). They showed, by means which included reversing the direction from which a leaf was illuminated, that stomatal aperture was independent of light intensity provided the quantum flux, and therefore rate of assimilation, in the mesophyll was constant. But in other experiments they did find a small difference in the resistances, $1/g^+$, in leaves in the light and in leaves in the dark, which was not attributable to difference in intercellular CO_2 concentration. To my mind their observations tend to support the notion that it might have been associated with the process of carbon fixation in the mesophyll tissue rather than a direct effect of light in the guard cells. However, they conclude otherwise. Their experiments are a model for investigations with species in which the role of intercellular CO_2 concentration might be less dominant than in *Z. mays*. If stomata receive signals from the mesophyll via the epidermis then the analysis of stomatal responses that I have essayed in this section should be extended appropriately. The additional loop would have properties of positive feedback, in that increase in rate of assimilation would cause the stomata to open, and opening of the stomata, by facilitating the transfer of CO_2, might increase rate of assimilation. As with a direct influence of light intensity on stomata, it would be most effective if the stimulus were to modulate the reference signal for the feedback system based on stomatal sensitivity to intercellular CO_2 concentration. Evidence is accumulating (Raschke, 1975a, b) that there is a relationship between intercellular CO_2 concentration and abscisic acid in their effect on stomatal aperture: each modifies the influence of the other. It seems probable to me that this interaction, or one involving compounds similar to abscisic acid (Loveys and Kriedemann, 1973, 1974), will be found to involve a third component: rate of assimilation in the mesophyll tissue.

Because intercellular CO_2 concentration is usually small in illuminated C_4 plants and, due to the response of the stomata, tends to remain so as illumination is varied, it may appear that the stomata control rate of assimilation. McPherson and Slatyer (1973) say, of their observations with *Pennisetum typhoides*, that the stomata "exerted the greatest single influence on rate of net photosynthesis, accounting for 60 to 80% of the total resistance to net photosynthesis". The remark is potentially misleading. With the marked exception of the observations of Ludlow and Wilson (1971) noted earlier, most experiments show that rate of assimilation is not increased by increasing ambient CO_2 concentration above its normal magnitude because the photosynthetic system is already working at full capacity. Provided they are not unduly affected by other influences such as decrease in the potential of water in the plant, the stomata tread a delicate path appearing to cause, but not quite causing, a shortage in CO_2 available to the mesophyll tissue. It is the reverse in C_3 plants. Rate of assimilation is always influenced by stomatal aperture, although the constraint is often small. It is a result of the shape of the $A(c_i, I)$ characteristics, illustrated in Fig. 21c. These are based on observations with *Simmondsia chinensis* by G. J. Collatz (unpublished) and are consistent with the models of assimilation in C_3 species of Laisk (1970), Hall (1971), and Peisker (1974). The obvious differences between these characteristics and those for C_4 species are largely attributable to photorespiratory release of carbon in the glycolate pathway and the competitive inhibition of the activity of RuDP carboxylase by oxygen. Net rate of assimilation is zero at $c_i = \Gamma$, the compensation concentration. It increases relatively slowly with increase in concentration of CO_2, internal conductance, k, being about 0.16 mol m^{-2} s^{-1} or less (see Gifford, 1974). The carboxylating system is far from being "saturated" at normal physiological concentrations of CO_2 and therefore rate of assimilation is never entirely dominated by intensity of light.

At light intensities less than about 1/4 full sunlight, the stomata tend to be so widely open that the apparent internal conductance, A/c_i, is considerably greater than k. That is to say, the system operates in a domain of Fig. 21c in which rate of assimilation is sensitive to both c_i and I. The potential increase in rate of assimilation with increase in leaf conductance is small, but is always positive. Hall and Björkman (1975) give data for *Gossypium hirsutum* at low irradiance which illustrates the point well.

In Table III the conductance $g^\dagger$ is an order of magnitude greater than the apparent internal conductance $A/(c_i - \Gamma)$. Yet A increases linearly with c_i as the final two columns show, and therefore by implication A would increase with increase in $g^\dagger$ also. Parenthetically I will anticipate some later discussion by remarking that Table III probably describes a physiological state not greatly different from that obtaining in experiments of Jones and Mansfield (1972) with barley at light intensity 34 W m^{-2}. It was found, not surprisingly, that closure of

TABLE III
Influence of CO_2 concentration on rate of assimilation in cotton (data from Hall and Björkman, 1975)

Plant no.	I (W m^{-2})	C (p.p.m.)	A (μmol m^{-2} s^{-1})	$g^\dagger$ (mmol m^{-2} s^{-1})	c_i (p.p.m.)	$A/(c_i - \Gamma)^*$ (mmol m^{-2} s^{-1})	$\Delta A/\Delta c_i$ (mmol m^{-2} s^{-1})
1	16.7	255	1.78	137	242	9.3	8.9
		353	2.48	78	321	9.2	
2	20.9	257	2.17	121	239	11.5	13.4
		348	3.27	121	321	12.1	

* Assuming Γ = 50 p.p.m.

the stomata by application of ABA caused a decrease in the transpiration ratio. The result, other than that the stomata were caused to close, is irrelevant to any supposition about the usefulness of antitranspirants in field conditions. It does serve to make the point, however, that it is not easy to envisage, in a heuristic way, what the strategy of stomatal behaviour should be in relation to C_3 metabolism at small light intensity. The $A(c_i, I)$ characteristics do not present us with an obvious paradigm as do those for C_4 plants.

At light intensity greater than about 1/4 full sunlight, stomata in C_3 plants tend to be at near maximum aperture. Leaf conductance is usually about 0.4 mol m^{-2} s^{-1}, somewhat greater than in C_4 plants. A. E. Hall (personal communication) has measured a leaf resistance to water vapour in sunflower of 0.2 s cm^{-1} in conventional units, equivalent to $g_l^\dagger \approx 2$ mol m^{-2} s^{-1}. The stomata continue to exert an influence, but a minor one only, on rate of assimilation; c_i is typically about 250 p.p.m. Rate of assimilation responds to increase in ambient CO_2. For example, it increased threefold in each of two legumes as c was increased from 320 to 1400 p.p.m., and was by no means saturated at the higher concentration (Ludlow and Wilson, 1971).

It may be concluded that stomata in most C_3 species only have a marked effect on rate of assimilation when they are caused to close by factors not directly related to the carbon metabolism of the leaf, for example diminution of leaf water potential (Troughton, 1969). Exceptional are those species in which stomata are directly affected by change in ambient humidity (or other attributes of the environment which influence rate of evaporation) in the way which has been discussed earlier. In these the stomata may be partially closed with ambient humidity deficits which are typical of those to which the plants are normally exposed during the day (Camacho-B et al., 1974; Hall et al., 1975). The condition of partial closure is normal.

It is an interesting aspect of stomatal behaviour that it often gives the impression of hyperactivity. Even in plants which commonly grow in environments tending to promote a very small rate of evaporation only, stomata usually close at night and may respond rapidly to small changes in light intensity during the day. A good example is provided by the measurements of Björkman et al. (1972b) with Alocasia macrorrhiza growing in extreme shade on the floor of the rainforest in Lamington National Park, Queensland (Fig. 24). The variation in conductance appears to be particularly well attuned to the metabolic requirements of the leaf, but one may wonder why the stomata do not remain open continuously in such a continuously humid environment—in any event the difference in CO_2 concentration across the leaf epidermis was about 10 p.p.m. only. Is stomatal behaviour of this kind vestigial, is it present because it is functionally significant in exceptional circumstances, or does it serve an immediate function which is not apparent? In the next section I shall analyse some aspects of the dynamic response of stomata to light.

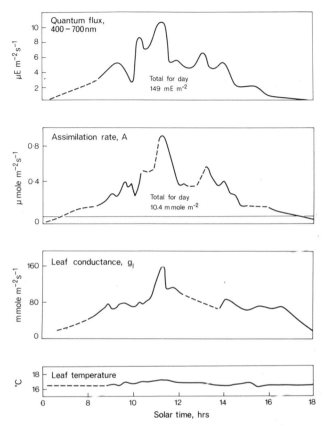

Fig. 24. Variation during the course of an overcast day of quantum flux, rate of assimilation, leaf conductance, and temperature, in a leaf of *Alocasia macrorrhiza* growing in rain forest in Lamington National Park, Queensland. (Redrawn from Björkman *et al.*, 1972.)

F. DYNAMIC RESPONSE TO LIGHT AND CO_2

In the previous section stomatal behaviour in relation to the carbon metabolism of leaves was discussed with no more than passing reference to activity in the hydraulic feedback loop. This may be justified in discussing quasisteady interrelationships in turgid leaves but transient responses to changes in irradiance or CO_2 concentration cannot be treated in this way. Figure 25 is a generalized version of Fig. 20. Stomatal and mesophyll characteristics are represented by transfer functions, and inputs and outputs by Laplace transforms of deviations from the steady state. The transfer function representing stomatal response may be defined by the relation

$$\bar{\Delta}g_l = G_E \,\bar{\Delta}E + G_{c_i}\bar{\Delta}c_i + G_I\bar{\Delta}I + \dots \qquad (58)$$

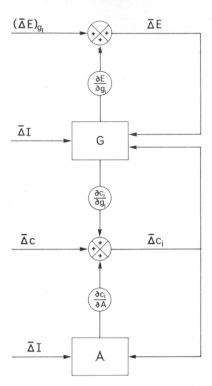

Fig. 25. Interrelationship between the transfer function G which describes response of leaf conductance, g_l, and transfer function $\mathscr{A}$ which describes response of rate of assimilation, A, to the various inputs shown. E is rate of evaporation, I light intensity, and c ambient concentration of CO_2.

The function may be extended to encompass responses to ambient humidity and leaf temperature if desirable, but only the terms written explicitly, and represented in Fig. 25, will be needed here. In the steady state, the second and third components of G relate to quantities used in the previous section, i.e.

$$G_{c_i}(0) = -1.6\beta; \qquad G_I(0) = 1.6\beta dc_I/dI \qquad (59)$$

in which the numerical factor allows for the differing coefficients of diffusion of water vapour and carbon-dioxide in the air in the stomatal pores. The transfer function A represents response of assimilation, so that

$$\bar{\Delta}A = \mathscr{A}_{c_i}\bar{\Delta}c_i + \mathscr{A}_I\bar{\Delta}I + \dots \qquad (60)$$

Here, also, additional terms may be necessary, e.g. to take account of the response of rate of assimilation to temperature. The function $\mathscr{A}$ undoubtedly has dynamic properties of great complexity. But it is possible that the time constants of some of them are sufficiently short, and those of the others sufficiently long, for these properties to be neglected in the interpretation of particular studies of stomatal behaviour. Then the components $\mathscr{A}_{c_i}$ and $\mathscr{A}_I$ may be replaced by

$$\mathscr{A}_{c_i}(o) = \frac{\partial A(c_i, I)}{\partial c_i} \; ; \qquad \mathscr{A}_I(o) = \frac{\partial A(c_i, I)}{\partial I} \qquad (61)$$

These, too, are characteristics which were discussed in the previous section; with C_4 plants the first was identified with internal conductance, k, when the supply of CO_2 is "limiting", and the second with the slope of the light response curve, dA_I/dI, when light is "limiting".

When one or other of the inputs on the left-hand side of Fig. 25 is changed, activity is transmitted throughout the system. For example, Aubert and Čatský (1970) showed that stomata in banana responded to light more rapidly the drier the atmosphere. It probably occurred because the activity generated in the hydraulic part of the system is a function of environmental gain, $\partial E/\partial g_l$, and therefore increases with decrease in humidity (Cowan, 1972a). That this kind of interaction occurs is obvious, indeed axiomatic; but it is often neglected. It is an intrinsic part of the investigation of stomatal physiology to determine separately the characteristics of the various loops in the system of which stomata are a part; it is then a problem in ecophysiology to put them together again in ways which relate to stomatal functioning in the natural, rather than experimental, environment. I shall describe a simple example of this sequence.

Figure 26 shows the effects of step changes in irradiance and ambient carbon dioxide concentration on leaf conductance and rate of evaporation in cotton. The one curve in each case, (marked g_l, E), represents both leaf conductance and evaporation rate. The plant is the one that provided the data in Fig. 12 and as with the data in that figure leaf temperature was maintained constant at $25°C$. Clearly, the stomatal system verges on instability, as a result, we suspect, of feedback in the hydraulic loop. We do not believe that the characteristics of the CO_2 feedback loop are such that they could give rise to oscillatory behaviour of the kind shown in Fig. 9. In any case, analysis shows that the gain of that loop was negligible. Rate of assimilation was very small, about 3 μmol m^{-2} s^{-1}, and so the difference in CO_2 concentration across the boundary layer and leaf epidermis was only 30-40 p.p.m. Therefore the concentration of CO_2 in the intercellular air spaces, c_i, was coupled very closely to that in the atmosphere; when the latter was increased by 100 p.p.m., rate of assimilation increased by 15% only, c_i increased by 90 p.p.m. and the variation that subsequently occurred in association with the fluctuations in leaf conductance was less than 10 p.p.m. For the purpose of determining G_{c_i} in eqn (58), little error is engendered by taking $\Delta c_i = 90 \times 10^{-6}$. When irradiance was increased rate of assimilation was doubled, from 1.5 to 3 μmol m^{-2} s^{-1} and there was a rapid decrease in c_i of about 20 p.p.m. However, the increase in c_i was insufficient to account for the opening of the stomata that then took place; in fact c_i was almost restored to its initial magnitude as a result of the increase in leaf

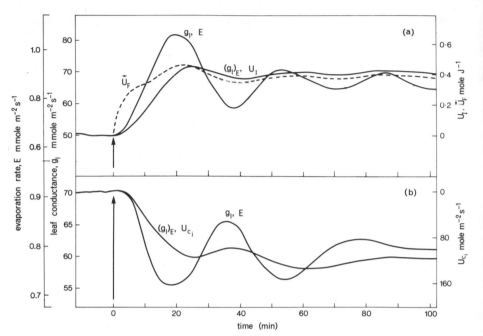

Fig. 26. Response of leaf conductance in cotton to (a) increase in light intensity from 37 to 85 W m^{-2}, and (b) increase in ambient concentration of CO_2 from 250 to 350 p.p.m. The lines marked g_l, E represent both conductance and rate of evaporation; they relate to observations with ambient humidity and leaf temperature maintained constant. The lines marked $(g_l)_E$ are estimates of the variation in conductance that would have occurred had rate of evaporation been maintained constant; they also represent the unit step responses, U_I and U_{c_i} to change in light intensity and intercellular concentration of CO_2. The broken curve $\check{U}_F$ is an estimate of the variation in leaf conductance that would have occurred due to change in light intensity had ambient temperature, rather than leaf temperature, been maintained constant.

conductance. For the purpose of determining G_I in eqn (58), Δc_i could be neglected. These are circumstances which greatly simplify further analysis, implying as they do that feedback due to the interrelationship of g_l and c_i is negligible, but they are likely to be peculiar to C_3 plants growing at low irradiance.

In principle the effects of activity in the hydraulic loop may be removed from the observed responses of stomata to irradiance and CO_2 concentration by the following method. First, change in conductance is expressed in the form

$$\bar{\Delta}g_l = G_E\bar{\Delta}E + (\bar{\Delta}g_l)_E \tag{63}$$

in which $(\Delta g_l)_E$ is the change in conductance not attributable to change in rate of evaporation. Making use of the convolution theorem to invert the equation, it follows that

$$(\Delta g_l)_E = \Delta g_l - U_E * \frac{\partial E}{\partial t} \tag{64}$$

where U_E is the unit step response of g_l to change in rate of evaporation. If U_E is known, the convolution can be evaluated numerically. When it is subtracted from the observations in Fig. 26 eqn (64) shows that the components which remain are estimates of responses to light and CO_2 unencumbered by activity in the hydraulic feedback loop. In practice, though, we find that in order to remove the major part of the oscillatory component of the observations, it has to be assumed that the magnitude of U_E is about twice that which we determined on the basis of stomatal response to ambient humidity. It seems, therefore, that the stomatal system may contain a mechanism engendering oscillations in conductance with a period of about 30 min which is not adequately represented by our concept of the hydraulic feedback loop. On the other hand, as was previously mentioned, we find a great deal of variation in determinations of G_E. It is possible that the magnitude relevant to the observations in Fig. 26 was indeed about twice that found from the responses of leaf conductance and rate of evaporation to change in humidity. For the purposes of example, the latter is taken to be the correct explanation. The "corrected" responses are those marked $(g_l)_E$ in Fig. 26. When expressed relative to the changes ΔI and Δc_i they are the unit step responses, U_I and U_{ci}, of leaf conductance to change in irradiance and intercellular concentration of CO_2. They are very roughly similar in shape, suggesting that a common mechanism is involved. The result is consistent with the view that the light response is essentially a response to change in CO_2 concentration inside the guard cell associated with change in rate of uptake. It is interesting to note that the magnitudes of the responses relative to the changes in irradiance and CO_2 concentration are similar. However, relative to changes in I and c_i which actually take place in natural conditions, the response to irradiance is much the more significant.

The response times indicated by U_I and U_{ci} are considerably longer than that, for example, in *Zea mays* (Raschke, 1972) but are not atypical of other species (see Meidner and Mansfield, 1968). They are longer than the response time of the intact system; that is to say of the observed changes in leaf conductance. Perhaps the characteristics of the hydraulic feedback loop are designed to enhance the speed of the response to light (Cowan, 1972a). It is not uncommon in manmade control systems to reduce the time constant and increase the gain of a system by superimposing a loop having characteristics of positive feedback (Ogata, 1970). Activity *initiated* in the hydraulic loop as a result of the metabolic response of stomata to light is not the only way in which the hydraulic loop may be involved. Because only a small proportion of the radiant energy incident on a leaf is fixed in chemical bonding, increase in irradiance normally causes increase in leaf temperature, internal humidity, and (provided the stomata are partially open) rate of evaporation. The only exception is with an artificially controlled environment in which leaf temperature is held constant, as with the observations shown in Fig. 26. The increase in rate of evaporation may be calculated from eqns (34) and (36), or directly from Fig. 17, as

$$\frac{\partial E}{\partial I} = \frac{\partial \phi}{\partial I} \cdot \frac{\partial E}{\partial \phi} = \frac{\partial \phi}{\partial I} \cdot \frac{\epsilon g_l/(\lambda g_b^\ddagger)}{1 + g_l(\epsilon/g_b^\ddagger + 1/g_b)} \tag{65}$$

The quantity $\partial \phi/\partial I$ is approximately unity if I is expressed in terms of energy flux. The long-wave radiation in the solar spectrum is rather efficiently reflected and transmitted by leaves and so does not contribute greatly to ϕ; what it does contribute roughly makes up for the visible radiation which is reflected or transmitted. If I is expressed in terms of photon flux then $\partial \phi/\partial I \approx 210$ kJ E^{-1}. The third expression in eqn (65) yields $\partial E/\partial I \approx 3$ μmol J^{-1} for the conditions pertaining to Fig. 26. As the change in irradiance was 48 W m^{-2}, the change in rate of evaporation that would have occurred had leaf temperature not been held constant is estimated as 0.14 mmol m^{-2} s^{-1}. The estimate relates to the solar spectrum, but it happens to agree quite well with a second one based on the fact that, when irradiance was increased, ambient air temperature in the experimental chamber was automatically reduced about 0.6°C by the system which controlled leaf temperature.

Setting $\bar{\Delta}E = (\partial E/\partial I)\bar{\Delta}I + (\partial E/\partial g_l)\bar{\Delta}g_l$ in eqn (58), and taking $G_{c_i}\bar{\Delta}c_i \approx 0$ as before, it follows that

$$\frac{\bar{\Delta}g_l}{\bar{\Delta}I} = \frac{G_I + \dfrac{\partial E}{\partial I} G_E}{1 - \dfrac{\partial E}{\partial g_l} G_E} = \overleftarrow{G}_F, \text{ say} \tag{66}$$

The transfer function $\overleftarrow{G}_F$ describes the response of leaf conductance to change in irradiance, taking into account the effects of change in net radiation on rate of evaporation, and feedback in the hydraulic loop. If it were possible to break the hydraulic loop, i.e. make $\partial E/\partial g_l$ zero, without affecting $\partial E/\partial I$ then $\overleftarrow{G}_F$ would become the open-loop transfer function $G_F = G_I + (\partial E/\partial I)G_E$. The corresponding unit step response to irradiance is

$$U_F = U_I + \frac{\partial E}{\partial I} U_E \tag{67}$$

It is plotted in Fig. 27, $\partial E/\partial I$ having been taken as 3 μmol J^{-1}. To find the closed-loop unit step response, use may be made of the convolution theorem. Thus

$$\overleftarrow{U}_F = U_F + \frac{\partial E}{\partial g_l} U_E * \frac{\partial \overleftarrow{U}_F}{\partial t} \tag{68}$$

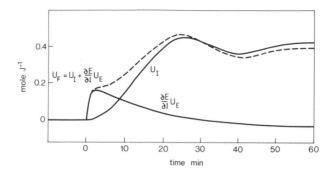

Fig. 27. Unit step responses, U_I and U_E, of leaf conductance in cotton to light intensity and rate of evaporation, respectively, together with a combination of the two, U_F, in proportions relating to the influence of radiation on rate of evaporation.

the relationship between this equation and eqn (66) being similar to that between eqns (29) and (28).

Equation (68) has been used to evaluate the closed-loop response $\overline{U}_F$, with the magnitude of environmental gain being taken as that which would have obtained if ambient temperature, rather than leaf temperature, had been maintained constant (i.e. $\partial E/\partial g_l = 9 \times 10^{-3}$, rather than 13×10^{-3}). The result is plotted in Fig. 26; in terms of the scale in the left of the diagram it represents an estimate of g_l which may be compared with the original observations. The opening movement is much quicker and, subsequently, conductance is much more nearly steady. Though we have not conducted a systematic experiment, we do have some evidence that this prediction is qualitatively consistent with what happens when leaf temperature is not controlled.

Two particular points emerge from this analysis. The first concerns the physiology of the stomatal apparatus, and relates to the discussion following eqn (16), in which the combined effects of change in external water potential and internal osmotic pressure on stomatal aperture are combined. The argument there suggests that what is shown in Fig. 26 is indicative of a "hydroactive" explanation of transient response to change in rate of evaporation. The initial increase in leaf conductance following increase in irradiance would be a hydromechanical response; but subsequent movement would be relatively small because the metabolic processes associated with "photoactive" stomatal opening would be partially offset by those associated with "hydroactive" stomatal closure. The second point concerns ecology. At first sight, the characteristics of the hydraulic loop, as indicated in Fig. 13, seemed paradoxical; the stomata open in response to decrease in ambient humidity and therefore exacerbate the increase in rate of water loss. In the present context they no longer seem so. Rapid changes in humidity do not take place out-of-doors but rapid changes in photon flux density do—as a result of passing cloud, variation of leaf orientation and shade in moving foliage, and, in forest canopies, the passage of sunflecks. Of course it remains to be shown that the analysis I have outlined here is relevant to plants in climatic and edaphic conditions in which rapid response to irradiance is likely to be an advantage. They would be conditions in which intercellular concentration of CO_2 is tightly controlled by the stomata; and a thorough analysis would have to take account of all the interactions indicated in Fig. 25.

G. SUMMARY

In the majority of crop plants, and perhaps other species also, stomata operate in the light in such a way as to maintain positive turgor in the leaves. There is no evidence that they react to quasisteady changes in internal water relations, or to environmental changes which cause quasisteady changes in internal water relations, provided leaf turgor pressure exceeds 1-2 bar. However, the transient response of stomata to changes in environment which cause rapid changes in plant-water relations is quite dramatic. On the face of it it seems to imply that control of water loss is inefficient, the stomata opening when the changes are such that rate of evaporation is enhanced. The behaviour is consistent with the hypothesis that stomata react to changes in actual rate of evaporation rather than the perturbations which cause them. Thus stomata both affect and are affected by rate of evaporation from the leaf. Positive feedback in this relationship is sometimes sufficient to cause instability in stomatal functioning. In some species, particularly shrubs or trees in arid and semi-arid conditions, the stomata appear to respond to changes in ambient humidity. By implication the same response would occur with change in other factors which tend to cause changes in rate of evaporation, such as irradiance and ambient temperature, but the effect would be confounded with the responses of the stomata to irradiance and temperature *per se*. The explanation of the "direct" response to humidity is obscure; it seems most likely to be associated with peristomatal transpiration; certainly it is inconsistent with the simple feedback hypothesis of stomatal action.

Stomatal behaviour differs markedly between plants which employ the Calvin cycle, only, to fix carbon and those which have the Hatch-Slack C_4-dicarboxylic pathway. Because the carboxylating system in the latter becomes saturated with CO_2 at concentrations much less than ambient concentration, the stomata are able to respond to intercellular CO_2 concentration, c_i, and irradiance in such a way that regulation of c_i is well attuned to the characteristics of carbon metabolism in the mesophyll. Usually c_i is less than 100 p.p.m. and decreases slightly with increasing irradiance, but there is evidence that the response of the stomata to light is sometimes sufficiently large to allow c_i to increase, despite increasing rate of assimilation.

Leaf conductance is generally greater in C_3 plants than in C_4 plants and c_i is usually about 250 p.p.m., but the carboxylating system is not saturated unless c_i is artificially increased to a level above normal ambient concentration. Therefore the stomata always place some constraint on rate of assimilation. The constraint is a maximum at full sunlight when the potential capacity of the carboxylating system may be several times greater than the rate of assimilation which actually occurs. Rate of assimilation is not closely related to conductance in the gas phase unless the stomata are caused to close by influences other than light.

Transient responses of leaf conductance in cotton to increase in irradiance and decrease in ambient concentration of CO_2 are similar in shape. They contain an oscillatory component which is probably attributable to activity in what has been termed the hydraulic loop. The responses are almost exactly out of phase relative to a response to increase in rate of evaporation. If the light and hydraulic responses are superimposed in suitable proportions, the combined response describes a rapid increase in conductance succeeded by a relatively steady state. It is suggested that the transient response of stomata to change in rate of evaporation may be a device which is designed to enhance the speed of the response to light.

IV. ECOLOGICAL CONSIDERATIONS

A. STOMATAL BEHAVIOUR AND ECONOMY IN THE USE OF WATER

The notion that natural selection leads to leaves having a form and function tending to minimize loss of water relative to the amount of CO_2 taken up is an appealing one. It has been discussed, and used with some degree of success to "explain" the variation of leaf size with climate, by Parkhurst and Loucks (1972). Let me say at once, though, that the arguments on which it is based seem to me to be shaky. It is not obvious that plants which are economical in their use of water will prove to be the "fittest" when they are in continuous competition with plants which are less economical in their use of water. I would suppose that economy is most likely to confer selective advantage in sparse communities in arid environments in which individual plants are only occasionally subjected to direct competition with others for water resources. Cohen (1970) suggests that "efficient" plants, i.e. those which economize in their use of water, are large, widely spaced, deeply rooted trees or shrubs which, because of that spacing, have root systems which do not overlap to a large extent; plants which have the capacity to root themselves in a specific soil or rock element in the soil complex; and summer annuals growing in unirrigated soils. He surmises that "inefficient plants" are those having root systems which are either very shallow or are considerably overlapped by adjacent root systems, for these depend on supplies of water which, if not used, are lost by evaporation direct from the soil or by absorption by the roots of neighbouring plants. The category would include ephemeral desert species which are efficient (as are all species which survive) in a broad sense, but may be "inefficient" in the restricted sense used here.

If a plant is to be efficient in terms of water use, how should it vary its leaf conductance with time in relation to the fluctuations which occur in ambient environment? Parkhurst and Loucks suggest, in effect, that stomatal aperture would vary during the course of each day so that the average rate of evaporation divided by the average rate of assimilation is a minimum. That, roughly, is the proposition that will be explored here. But I prefer to phrase it in a slightly modified way: stomatal aperture would vary so that the average rate of evaporation is a minimum for the particular average rate of assimilation. Thus the danger of finding that the stomata should remain continuously closed will be avoided. Incidently, the discussion that follows could equally well be based on the converse proposition: stomatal aperture would vary so that the average rate of assimilation is a maximum for the particular average rate of evaporation.

Let us imagine that we had been able to impose a series of small perturbations, sometimes positive sometimes negative, on stomatal aperture in a leaf during the course of a day in such a way that there was no change in the total amount of CO_2 taken up, i.e. $\int \delta A \, dt = 0$, where δA is the disturbance in rate of assimilation due to a perturbation in stomatal aperture. Clearly the number of ways in which this could be achieved is without limit. If the unperturbed time course of stomatal aperture were the most efficient possible then it must be that any such sequence of imposed perturbations would cause $\int \delta E \, dt > 0$, where E is the disturbance in rate of evaporation. Assuming that A increases monotonically with increase in leaf conductance we may rewrite the inequality as

$$\int \left[\frac{\partial E}{\partial A} \delta A + \tfrac{1}{2} \frac{\partial^2 E}{\partial A^2} (\delta A)^2 + \cdots \right] dt > 0 \qquad (69)$$

Provided δA is small further terms may be neglected, which is equivalent to assuming that, within the range of interest, E is a quadratic function of A. It is important to note that the differentials with respect to A are partial differentials at constant time. In general both $\partial E/\partial A$ and $\partial^2 E/\partial A^2$ may be expected to vary with time, as a result of variation in stomatal aperture and environment, although I shall now show that efficiency may in fact require that stomatal aperture should continually adjust so that the former does not do so. If δA is sufficiently small then the linear term will be very much greater than the quadratic term at any given instant of time. But if $\partial E/\partial A$ varies during the course of the day then we can always envisage a sequence of perturbations in stomatal aperture such that δA is positive when $\partial E/\partial A$ is small, and negative when $\partial E/\partial A$ is large, the net result of this negative correlation being that the integral of $(\partial E/\partial A)\delta A$ is itself negative. Therefore a necessary condition that the observed time course of stomatal aperture is efficient is that $\partial E/\partial A$ is invariant with time. Then the first component of the integral in eqn (68) is zero and it becomes necessary to consider the nature of the term in $(\delta A)^2$. Now it is widely believed that, irrespective of environmental conditions and the magnitude of leaf conductance, rate of evaporation is relatively much more sensitive to change in conductance than is rate of assimilation. That is to say, if stomatal aperture is caused to vary while environmental variables are held constant, the sense of the curvature in the relationship between E and A is always such that $\partial^2 E/\partial A^2 > 0$. If this were so, obviously the second component of the integral in eqn (69) is invariably positive, and we conclude that the average rate of evaporation is least for a given average rate of assimilation if

$$\frac{\partial E}{\partial A} = \frac{\partial E/\partial g_l}{\partial A/\partial g_l} = R', \text{say} \qquad (70)$$

is a constant. The argument may readily be extended to encompass the efficiency of a whole plant having leaves which experience differing local microenvironments at any given instant of time. Provided $\partial^2 E/\partial A^2 > 0$ for all leaves, then minimisation of water lost for a given amount of growth requires that R' as defined in eqn (70) should be constant not only in time, but should be uniform amongst the different leaves. In other words stomata should adjust so that what has been termed environmental gain, $\partial E/\partial g_l$, is in constant and uniform proportion to the equivalent quantity relating to assimilation, $\partial A/\partial g_l$. It is encouraging that these two quantities, so intimately involved in the dynamics of the stomatal mechanism, should appear, *ab initio*, in a theorem concerning efficiency of stomatal behaviour.

The criterion in eqn (70) was first derived by Farquhar (1973), who also pointed out that if the curvature $\partial^2 E/\partial A^2 > 0$ then constancy in $\partial E/\partial A$ defines not the most efficient, but the most inefficient mode of stomatal behaviour. That rate of evaporation may sometimes be less sensitive to change in stomatal aperture than is rate of assimilation, and that this has implications for optimization of stomatal behaviour had been suggested by Cowan and Troughton (1971). The implications can best be appreciated by considering the effect on the argument which leads to eqn (70) of the constraint $0 \leqslant A \leqslant A_{max}$, where A_{max} is the rate of assimilation that would be sustained at any given instant of time if the stomata were fully open. Sometimes eqn (70), depending on the variation of environmental conditions, may imply a magnitude of A which contravenes one or other of the limits. Then the actual magnitude of A will be 0 or A_{max}, whichever is the particular limit. It is easily shown that the constrained variation in A, and the associated magnitude of E, is still in conformity with the proposition concerning efficiency. Provided $\partial^2 E/\partial A^2$ is everywhere positive then $\partial E/\partial A$ at $A = 0$ is greater than R' when the magnitude of A implied by eqn (70) is less than zero, and $\partial E/\partial A$ at $A = A_{max}$ is less than R' when the implied magnitude of A exceeds A_{max}. Because a perturbation of A at $A = 0$ must of necessity be positive, and a perturbation of A at $A = A_{max}$ must of necessity be negative, it is clear that the imposition of the limits causes the first component of the integral in eqn (69) to be finite and positive; the inequality is still satisfied. Now the smaller the curvature $\partial^2 E/\partial A^2$ the wider will be the fluctuations in E and A, required to maintain $\partial E/\partial A$ constant in a changing environment, and the greater the periods of time for which $A = 0$ or A_{max}. In the extreme, if the curvature is virtually zero, the system operates only at the limits $A = 0$ or A_{max}. What now happens if $\partial^2 E/\partial A^2$ is negative? The system also operates at the limits, with $A = 0$ or $A = A_{max}$, depending whether E/A at A_{max} is greater or less than a critical magnitude. Thus efficient stomatal behaviour has an "on-off" character determined entirely by the variation in the transpiration ratio that would obtain if the stomata remained fully open.

B. ECONOMY IN RELATION TO METABOLISM AND ENVIRONMENT

In order to progress further, an interrelationship between E and A is needed, and the conditions pertaining to its curvature must be established. Equation (36) may be rewritten, using resistances rather than conductances,

$$E = \frac{w_0 - w}{r_l + er_b^{\ddagger} + r_b} \tag{71}$$

in which

$$w_0 = er_b^{\ddagger}\phi/\lambda + w' \tag{72}$$

With regard to the process of assimilation it will be supposed that the stomata operate so that irradiance is not limiting, i.e. the carboxylating system functions on the linearly rising parts of the characteristics in Fig. 21b or 21c and therefore $A = k(c_i - \Gamma)$. Then

$$A = \frac{c - \Gamma}{r_b^{\dagger} + r_l^{\dagger} + 1/k} \tag{73}$$

Making use of the relations $r_l^{\dagger} = 1.6\,r_l$ and $r_b^{\dagger} = 1.37\,r_b$ it follows that

$$\frac{c - \Gamma}{A} - \frac{1.6(w_0 - w)}{E} = \frac{1}{k^*} \tag{74}$$

with

$$\frac{1}{k^*} = \frac{1}{k} - 1.6er_b^{\ddagger} - 0.2r_b \approx \frac{1}{k} - 1.6er_b^{\ddagger} \tag{75}$$

By differentiation it may readily be verified that $\partial^2 E/\partial A^2$ is positive or negative accordingly as k^* is positive or negative. The quantity $1/k^*$ might appropriately be called "supraresistance (to transfer of CO_2)" for the following reason. It is frequently suggested that rate of assimilation is relatively less sensitive than rate of evaporation to change in stomatal aperture because CO_2 encounters an additional resistance, the "internal resistance" $1/k$, over and above those encountered by CO_2 and water vapour in common. The argument is imprecise primarily because it ignores what is, in effect, a resistance to transfer of water vapour that has no counterpart in CO_2 transfer: that associated with the term ϵ

in eqn (71). It becomes precise if $1/k^*$ is regarded as the additional resistance to CO_2 transfer. Writing $R = E/A$ for the transpiration ratio, eqn (74) may be rearranged as

$$R = R_0 + \frac{E}{k^*(c - \Gamma)} = \frac{R_0}{1 - A/[k^*(c - \Gamma)]} \tag{76}$$

in which

$$R_0 = \frac{1.6(w_0 - w)}{c - \Gamma} \tag{77}$$

is the limit of R as leaf conductance approaches zero. Equation (76) demonstrates, in a particularly simple way, how transpiration ratio tends to vary with leaf conductance. With environment constant, increase in E or A can be taken as an indication of an implicit increase in conductance; if $1/k^*$ is positive then transpiration ratio increases, and vice versa.

Table IV shows estimates of $1.6\ er_p^{\pm} = 1/k - 1/k^*$, the difference between internal resistance and "supraresistance". With plants having C_3 metabolism $1/k$ may be taken as having a minimum magnitude of about $6\ m^2\ s\ mol^{-1}$ (an internal resistance in conventional units of $2.4\ s\ cm^{-1}$). In most C_3 species in most conditions it probably exceeds $10\ m^2\ s\ mol^{-1}$. One concludes from an examination of Table IV that $1/k^*$ is positive unless boundary layer conductance is very small and temperature exceptionally high. Therefore transpiration ratio

TABLE IV
Estimates of $1.6\ er_b^{\pm}$ in $m^2\ s\ mol^{-1}$ for various combinations of windspeed, u, leaf breadth, b, and temperature

Temperature (°C)	ϵ	$u/b\ s^{-1}$								
		2	5	10	20	50	100	200	500	1000
16	1.75	3.5	2.7	2.1	1.7	1.1	0.8	0.6	0.4	0.3
32	4.06	7.6	5.9	4.7	3.7	2.5	1.9	1.4	0.9	0.7
48	8.49	14.0	11.5	9.4	7.4	5.2	3.9	2.9	1.9	1.4

usually increases with increase in leaf conductance in C_3 plants. It is the reverse with C_4 plants, for which $1/k \approx 1\ m^2\ s\ mol^{-1}$. Unless boundary layer conductance is large and temperature relatively low, $1/k^*$ is negative and transpiration ratio decreases with increase in leaf conductance.

When $1/k^*$ is positive the criterion of efficiency stated in eqn (70) may be applied. From eqn (74) it then follows that

$$\frac{\partial E}{\partial A} = \frac{c - \Gamma}{1.6(w_0 - w)} \cdot \frac{E^2}{A^2} = R' \tag{81}$$

and therefore

$$R = (R_0 R')^{1/2} \tag{82}$$

Further, by resubstituting eqn (74), or making use of eqns (76) and (77),

$$E = k^*(c - \Gamma)(R_0 R')^{1/2} \left[1 - \left(\frac{R_0}{R'} \right)^{1/2} \right] \tag{83}$$

and

$$A = k^*(c - \Gamma) \left[1 - \left(\frac{R_0}{R'} \right)^{1/2} \right] \tag{84}$$

Thus we have functions which purport to represent real magnitudes of E and A in terms of three variables, Γ, k^* and R_0, the variation in these variables being associated with variation in the intrinsic characteristics of leaves and the physical nature of their microenvironments. Of course, they are realistic only if irradiance is sufficiently great and leaf conductance is sufficiently small for equation (73) to be an appropriate description of assimilation. The constant term, R', is, in effect, a measure of a plant's need for growth or, conversly, its need to conserve water. If R' is infinite the need for growth is paramount; rate of assimilation is the maximum the plant is capable of, whatever the conditions of its environment relating to water loss. If R' is zero then conservation of water is paramount and the stomata are closed all the time. The magnitude of R' might be expected to increase with increase in the amount and potential of water available to a plant, and to be relatively large in species which grow in environments in which the immediate expectation of rainfall is relatively great. For a leaf having given metabolic characteristics and experiencing a given microclimate the ratio of mean rate of evaporation to mean rate of assimilation is uniquely related to R'. The greater R', the greater $\bar{E}/\bar{A}$. The quantity $R_0 = 1.6(w_0 - w)/(c - \Gamma)$ is predominantly a measure of the influence of atmospheric conditions on transpiration ratio. Its role in eqns (83) and (84) may be understood by referring to Fig. 28 and assuming Γ and k^* are constant. Rate of assimilation decreases with increase in R_0; that is to say the plant minimizes its water use by reduction of its activity in those leaves and during those times when transpiration ratio is potentially greatest. Rate of evaporation is zero when $R_0 - 0$ not because the stomata are closed but because $w_0 = w$. It increases with increase in R_0 to a maximum at $R_0 = R'/4$ and decreases thereafter as the influence of diminishing leaf conductance overrides the effect of increasing $w_0 - w$. It may readily be shown that the maximum corresponds to $E = 1.6\, k^*(w_0 - w)$.

Figure 29 illustrates some solutions of equations (82) to (84) for a hypothetical diurnal variation in R_0. Because Γ has been assumed constant the broken curve in the upper part of the diagram represents both R_0 and $w_0 - w$. I have taken it to be out of phase with solar time because saturation humidity at ambient temperature, w', which is a large component of w_0 in dry climates (see

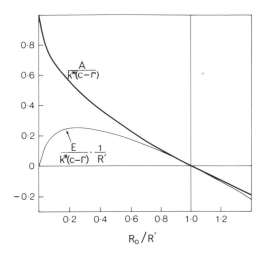

Fig. 28. Parametric representation of eqns (83) and (84) relating to the variation of rate of evaporation, E, and rate of assimilation, A, in "efficient" plants.

eqn (72)), normally lags behind the diurnal variation in solar radiation. Thus R_0 is greater at dusk, and for some while afterwards, than it is at dawn. The curves representing transpiration ratio, $R = E/A$, are derived from eqn (82), three magnitudes of R' having been assumed. The magnitude of R corresponding to $R' = 300$ is less than that of R_0 during the middle part of the day. Leaf conductance, rate of evaporation and rate of assimilation are then zero. At other times E and A are given by eqns (83) and (84), with $c - \Gamma = 200 \times 10^{-6}$, and $1/k^* \approx 18$ m^2 s mol^{-1}—consistent with a C$_3$ plant having a large internal resistance to CO$_2$ transfer. I have continued these curves not only throughout the day but during part of the night also, because their nocturnal magnitudes are not without relevance to the interrelationship between environment and the occurrence of Crassulacean acid metabolism (Osmond, 1975). However, in plants having C$_3$ or C$_4$ metabolism, irradiance must limit rate of assimilation during the earliest and latest periods of the day and that is indicated by the curve $A = A_I$. The corresponding limits of rate of evaporation and transpiration ratio have been computed by setting $A = A_I$ in eqn (74). In the interests of indicating the way in which estimates of E and A derive from the concept of efficiency, some generality has been sacrificed in Fig. 29. It is worth noting that eqns (83) and (84), and Fig. 28, imply that the one set of curves can represent high rates of evaporation and assimilation corresponding to a small magnitude of the $1/k^*$, or low rates of evaporation and assimilation associated with large $1/k^*$. It then becomes evident from Fig. 29, that for the same average rate of assimilation the variance in E and A would be greatest when $1/k^*$ is small.

If plants did behave in the way predicted in Fig. 29 what would be the mechanism? It might fall into one, or more, of three categories. The shape of an

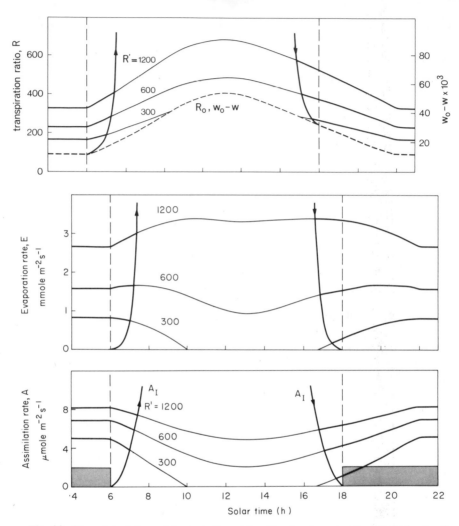

Fig. 29. Diurnal variation of transpiration ratio, T, rate of evaporation, E, and rate of assimilation, A, based on eqns (83) and (84) for the three magnitudes of R shown. The line R_0 represents the limit of T as leaf conductance approaches zero; it also represents the difference in humidity, $w_0 - w$, between leaf and air that would obtain if leaf conductance were zero. The lines marked A_I, and their counterparts in relation to E and T, represent limitations associated with the influence of light intensity on rate of assimilation and leaf conductance.

endogenous rhythm in leaf conductance might be modulated by some particular circumstance of the environment, e.g. the potential of water in the soil. But this would provide a first approximation only, to the required behaviour because it would take no account of stochastic components in the variation of atmospheric properties which influence rate of evaporation and rate of assimilation. In effect the stomata would be attuned to climate, but not sensitive to meteorology. Alternatively leaf conductance might respond in a direct open-loop fashion to changes in those environmental factors which tend to affect rate of evaporation and rate of assimilation, the responses being such that the criterion of efficiency is satisfied. The functional expression corresponding to the previous analysis is readily found, but is so cumbersome as to be rather unilluminating. However, if boundary layer conductance is very much larger than both leaf conductance and internal conductance to CO_2 transfer (to the extent $k^* \approx k$ and $w_0 \approx w_i$), then it may be shown that

$$g_l = 1.6k\left\{\left(\frac{R'}{R_0}\right)^{1/2} - 1\right\} = 1.6k\left\{\left[\frac{R'(c - \Gamma)}{1.6(w_i - w)}\right]^{1/2} - 1\right\} \quad (85)$$

Leaf conductance decreases with increase in humidity difference between leaf and air and varies with leaf temperature in the way implied by the temperature dependence of k and Γ. Of course, one might readily have predicted this from eqns (71) and (73). If boundary layer conductance is not large then g_l is a complex function of all the variables in these equations. One might reasonably doubt that the open-loop responses of stomata could meet the requirement.

The third possibility is that stomata may operate in such a way as to maintain $\bar{E}/\bar{A}$ constant not through a direct response to environmental conditions but through sensitivity to the environmental gain $\partial E/\partial g_l$ and its counterpart $\partial A/\partial g_l$. The idea is an appealing one because it is a simple concept and, at the same time, is consistent with the presence of complex dynamic characteristics in the stomatal apparatus.

Schulze et al. (1974, 1975a) consistently found that leaf conductance in Prunus armeniaca in the Negev Desert decreased in the middle part of the day provided the environmental conditions were such as to promote rapid evaporation (Fig. 30), and they have examined the relation between this trend and a number of external and plant internal factors: it is not due to an intrinsic endogenous rhythm; nor is it associated, at least not to a major extent, with depression of water potential in the plant or increase in intercellular CO_2 concentration—the traditional explanations of midday stomatal closure. Most of the variation is attributable to a negative response of leaf conductance to increase in humidity difference between leaves and air, and a positive response to increase in leaf temperature. The nature of the former was the subject of a previous section in this chapter. The extent to which these two responses accounted for the time course of leaf resistance to vapour diffusion on a particular day is indicated in Fig. 30 In another paper, Schulze et al. (1975b) investigated the influence of the responses on transpiration ratio in Prunus armeniaca. They show that the humidity response causes $\bar{E}/\bar{A}$ to be reduced. But the temperature response seems to be anomalous because it causes leaf conductance to increase when ambient temperature is large, and the capacity of the leaf to assimilate, as influenced by the parameters k and Γ, is greatly reduced. It

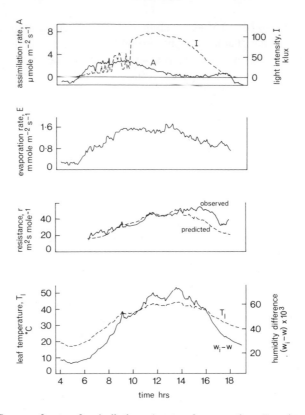

Fig. 30. Course of rate of assimilation, A, rate of evaporation, E, resistance to vapour diffusion, r, in *Prunus armeniaca* on 17 August 1971, in the Negev Desert, together with relevant environmental variables. Observed resistance is $r = (w_i - w)/E$ where w_i and w are the leaf internal and ambient humidities, respectively. (Redrawn from Schulze *et al.*, 1974.)

suggests either that the notion of efficiency, as defined in eqn (70), is inappropriate or that the subsequent analysis based on eqns (71) and (73) is inadequate. I shall comment on the likelihood of the latter possibility now, and reserve comment on the former until later.

Equation (73) is an oversimplified and incomplete description of the dependence of rate of assimilation on leaf conductance in C_3 plants. It seems reasonable to suppose that stomata in "efficient" plants generally do not open so far that irradiance rather than the supply of CO_2 alone has a direct effect on rate of assimilation. However, this is not valid for C_3 plants at low irradiance. It will be an interesting exercise to apply the criterion $\partial E/\partial A$ = constant, using a function for rate of assimilation of the kind depicted in Fig. 21c, and to modify the curves in Fig. 29 appropriately. But it is only the shape of the curves very early and very late in the day that would be affected, and the matter need not detain us here. The influence of leaf temperature on rate of assimilation has more complex implications. It may readily be seen from Fig. 17 that the internal humidity in a leaf, that is to say, saturation humidity at leaf temperature, is

$$w_i = w_0 - er_b^{\ddagger}E \tag{86}$$

Roughly, saturation humidity increases by a factor of 1.7 for a $10°C$ increase in temperature within the range of physiological temperatures. The compensation concentration of CO_2 in C_3 plants, Γ, has a Q_{10} somewhat greater than this, but the general sense of the metabolic influence of temperature in eqn (74) can be appreciated by assuming that Γ is a linear function of w_i over the range of leaf temperatures associated with change in leaf conductance at constant environment. Therefore

$$c - \Gamma = c - \Gamma(w_0) + \Gamma'er_b^{\ddagger}E \tag{87}$$

where Γ' is the gradient of $\Gamma(w_0)$. Thus the influence of temperature on the compensation point may be incorporated in eqn (74) without much formal complication. However, if $\Gamma(w_0)$ is close to c then the additional term in E becomes very significant and could cause the curvature $\partial^2 E/\partial A^2$ to be negative. Transpiration ratio would then decrease with increase in leaf conductance. Of course, many C_3 species in hot arid environments exhibit one or more characteristics which will minimize the effects discussed; e.g. small leaves which engender small $r_b^{\ddagger}$ and thereby reduce both w_0 and the magnitude of the term in E in eqn (87), and seasonal changes in the photosynthetic characteristics such that the temperature at which $\Gamma = c$, the "upper temperature compensation point", may be as high as $55°C$ when mean daily temperatures are greatest (Lange et al., 1975). However, Prunus armeniaca is a broad-leaved mesophytic species in which the optimum temperature for photosynthesis is about $30°C$ and the upper temperature compensation point is about $50°C$ (Lange et al., 1975). To analyse observations such as those in Fig. 30 in terms of stomatal efficiency it would be essential to take the last term in eqn (87) into account. It is probable that the notion of efficiency in terms of economy of water use would be found to be inadequate, for we may anticipate that stomatal behaviour might at times be more closely attuned to the need to prevent irreversible effects of excessive leaf temperature.

Let us now consider efficiency of water use with $1/k^* < 0$. Plants with negative supraresistance might be said to be "superconductive". "Superconductivity" is a state depending on a rather fine balance of physiological and environmental factors as Table IV indicates. The environmental component, $1.6er_b^{\ddagger}$, is likely to vary greatly during the course of a day, temperature being the dominant influence, with the variation in windspeed providing some compensation. Insofar as "superconductivity" is mainly associated with C_4 plants, the physiological component, $1/k$, is less open to question than it is when applied to problems with C_3 plants. Also, the apparent absence of photorespiration greatly simplifies discussion.

In some C_4 plants the ability to make use of high irradiances and adapt to high temperature is quite remarkable. Consider the performance of the C_4 species *Tidestroma oblongifolia* growing in Death Valley, California (Fig. 31), (Pearcy *et al.*, 1971). This plant is active at the hottest period of the year when maximum temperature is $48°C$ and the difference in humidity across the epidermis, $w_i - w$, is 112×10^{-3}—equivalent to about 112 mbar. It functions with its stomata wide open at midday, leaf conductance being about 0.2 mol m^{-2} s^{-1}. It then achieves an assimilation rate of 35 μmol m^{-2} s^{-1}. Pearcy *et al.* showed that rate of assimilation increases linearly with increase in irradiance up to full sunlight. One may readily calculate that the difference in CO_2 concentration across the leaf epidermis at the maximum rate of assimilation is about 280 p.p.m. Therefore leaf conductance can be very little more than sufficient to maintain the rate that the photosynthetic system is capable of at that irradiance, and the internal resistance, $1/k$, is about 1 m^{2}s mole^{-1}. No information is provided by Pearcy *et al.* about leaf boundary layer resistance, and in any case that which obtained in the experimental chamber may have differed from that of the exposed leaves outside the chamber. Let us assume for the purposes of an example that $u/b = 100$ s^{-1}. Then, with $k = 1$ m^2 s mol^{-1}, $A = 35$ μmol m^{-2} s^{-1}, $c = 320 \times 10^{-6}$ and $\Gamma = 0$, we first find using Table IV that $1/k^* = -2.9$ m^{2}s mol^{-1}, and then from eqn (76) that $(R - R_0)/R = -0.32$. That is to say transpiration ratio would be *increased* by about 32% if the stomata were caused to close almost entirely. If they were caused to close to the extent that rate of assimilation were reduced by a factor of 2 then the

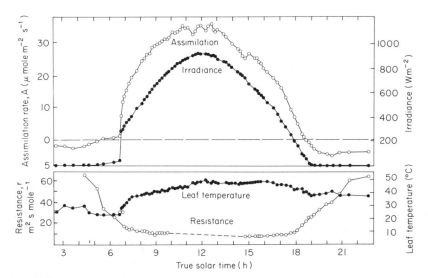

Fig. 31. Course of rate of assimilation, irradiance, leaf temperature, and total resistance to vapour transfer in *Tidestromia oblongifolia* on 1 July 1970, in Death Valley, California. (Redrawn from Pearcy *et al.*, 1971.)

transpiration ratio would be increased by 16%. Certain conclusions follow from this example. Given that the soil and plant hydraulic system is capable of delivering water to the leaves at the required rate without excessive drawdown in water potential, then there is no instantaneous benefit in terms of water conservation to be gained by a plant such as *T. oblongifolia* by diminution in stomatal aperture. If there were a need to reduce the rate of loss of water it would be done more effectively in terms of maximizing rate of assimilation by closing the stomata in some leaves entirely, than by reducing stomatal activity in all the leaves to a uniform extent.

Thus we see that "efficient" stomatal behaviour in superconductive plants, C_4 plants in the main, demands that the stomata in any one leaf should either be open to the extent that $A = A_I$ or should be fully shut. Which leaves should have open stomata and which should have closed stomata and how should the proportion vary during the course of the day? Irrespective of whether we are concerned with the distribution of activity amongst leaves or in time, the criterion, deriving from eqn (77), that the stomata should be open is that

$$R(A_I) = \frac{1.6(w_0 - w)}{c - \Gamma - A_I/k^*} \tag{88}$$

should be less than a critical magnitude which would be smaller the greater the need to conserve water. In arid environments the maximum magnitude of $w_0 - w$, occurring in the early afternoon is usually at least twice as great as the magnitude late in the afternoon. The denominator of the expression for R in eqn (88) also varies diurnally; if $1/k^*$ is negative, it has a maximum roughly in phase with solar time but the magnitude of the fluctuation is not relatively large. With the example for *Tidestromia oblongifolia*, $-A_I/k^* = 105 \times 10^{-6}$ at midday while, of course, the constant component $c - \Gamma \approx 320 \times 10^{-6}$. Therefore R as given by eqn (88) is likely to be greatest at times when $w_0 - w$ is greatest and it is then that "efficiency" demands that stomatal closure should occur. Nevertheless it is interesting that negative $1/k^*$ tends to cause R to be more nearly constant throughout the day than it would otherwise be. It has still greater repercussions in terms of differences in transpiration ratio associated with differences in light microclimate of the leaves in a plant at a given instant of time. The component of $w_0 - w$ directly attributable to solar radiation is not relatively large in very hot, very dry climates (perhaps contrary to expectation); the long-wave component of the radiation term in eqn (72) is often positive (the net loss of thermal radiation to the atmosphere being less than the net gain of thermal radiation from the ground), w' is very large and w relatively very small. Therefore the numerator in equation (88) may not be as sensitive to differences in irradiance associated with differences in leaf orientation, or shading, as is the denominator. In certain conditions then it seems probable that not only is transpiration ratio least when stomata are fully open but transpiration ratio is least in those leaves most fully exposed to the sun.

How the stomatal behaviour suggested by eqn (88) might be achieved in terms of mechanism, how particularly it could relate to the CO_2 feedback system which appears to be primarily responsible for the response of stomata to irradiance in C_4 plants, is a subject for speculation. It would be more profitable to find out whether stomata in C_4 plants do exhibit a much greater response to changes in the difference in humidity between leaf air than do C_3 plants, whether the response is much faster—whether in fact they exhibit indications of "on-off" behaviour that the criterion of water-use efficiency suggests they should. I know of no evidence to support the supposition that they do. *Tidestromia oblongifolia* is not an efficient plant in the limited sense of the criterion that I have used in this chapter; at least not in the conditions in which Pearcy *et al.* studied it. It appears to occupy a rather unique niche in arid zone ecology, in that its water relations are such that stomatal limitation of gas exchange does not occur even when rate of evaporation is as high as 20 mmol m^{-2} s^{-1}. But in the study of another C_4 species which inhabits Death Valley, *Atriplex hymenelytra*, Pearcy *et al.* (1971, 1974) found that rate of assimilation was limited by the stomata; yet there was no evidence of stomatal closure at midday. There is some evidence with cultivated C_4 plants which is relevant. Stomata in *Zea mays* seem to exhibit a very marked direct response to ambient humidity (Raschke, 1970a), and also a very rapid response to light and intercellular concentration of CO_2 (Raschke, 1975a).

I suspect that the primary reason why the proposition to do with economy of water use may prove inconsistent with the observed characteristics of short-term stomatal behaviour, particularly in C_4 species, is because it takes no account of intrinsic costs, e.g. of establishing and maintaining foliage. It would be inconsistent that foliage carrying a large investment in terms of photosynthetic capacity should then be employed discontinuously in the interests of conserving water, unless the cost-benefit ratio of developing a root and vascular system capable of maintaining a water supply to the leaves was excessive. Indeed one would expect the root-shoot ratio of a plant would be designed to prevent such a conflict of short-term and long-term interests. Stomatal behaviour bears the same relation to longer term phenomena in plants as tactics do to strategy. It is the nature of tactics to be flexible but if we seek to see a general principle underlying them we should expect it to reflect the strategy. It is for this reason that it would be surprising to find the criterion of efficiency in stomatal behaviour applying in a plant which grows, as does *Tidestromia* for example, at a period of the year when transpiration ratio, other things being equal, is bound to be greatest. It is more reasonable to assume that there are circumstances surrounding the ecology of the species that make our criterion redundant.

Clearly, the difficulty in applying optimisation theory to stomatal control of gas exchange in plants arises in defining the criterion of performance. In its general form the criterion must be that the partial derivative of the fitness of a species with respect to any feasible variation in stomatal characteristics should be zero. The problem we encounter is not so much that the performance criterion may be imperfectly satisfied, but that we do not know how to reexpress the criterion in terms of the physiological variables directly affected by stomatal functioning; that is to say rate of transpiration and rate of assimilation. I have suggested that, in the short term, and in certain circumstances, the criterion may degenerate to $(\partial E/\partial g_l)/(\partial A/\partial g_l) = $ constant. It is worth remarking that, although this particular relationship may be oversimplistic, there can be little doubt that the two quantities $\partial E/\partial g_l$ and $\partial A/\partial g_l$ are components of the general criterion of performance in a particular species. Any answer to the question "why is leaf conductance what it is?" will demand, at the very least, that we know what the rates of transpiration and assimilation would be if leaf conductance were slightly different from what it is. In investigations of gas exchange in natural conditions, sufficient information is usually obtained to calculate $\partial E/\partial g_l$. It is to be hoped that techniques will be developed to find the other, much more elusive, component, $\partial A/\partial g_l$, in a way which does not depend on assumptions about internal resistance to CO_2 uptake, and the like. Then we shall be much more likely to discover broad principles underlying stomatal behaviour, its relationship with environment, and its diversity amongst species.

C. SUMMARY

We have explored some implications of the hypothesis that stomata control gas exchange so that the total diurnal loss of water by the plant is the minimum possible for whatever total amount of carbon is taken up. The hypothesis is satisfied if the stomata maintain the ratio $(\partial E/\partial g_l)/(\partial A/\partial g_l) = \partial E/\partial A$ constant, provided the curvature $\partial^2 E/\partial A^2$ is positive. The latter proviso generally obtains in C_3 plants, rate of evaporation being relatively more sensitive to change in leaf conductance than rate of assimilation.

By applying the criterion $\partial E/\partial A = $ constant to the equations describing rate of evaporation and rate of assimilation in a leaf, it is possible to determine optimal time courses of leaf conductance, and the corresponding rates of gas exchange, in terms of the metabolic characteristics of the leaf, and the environment to which it is exposed. The analysis and computations outlined in this section are based on a simplified model of CO_2 uptake, appropriate only at large irradiance and small leaf conductance. However the solutions obtained serve to indicate that the theory is in qualitative agreement with certain aspects of stomatal behaviour observed in arid regions. Optimisation requires that leaf conductance diminish in the middle part of the day, sometimes to the extent that rate of transpiration is less than that occurring in the early morning and late afternoon. Behaviour of this kind might be brought about in several ways, but it is inconsistent with feedback hypotheses of stomatal action; it is consistent with recent observations that stomata respond directly to the difference in humidity between leaf and air.

In some circumstances the relative sensitivity of rate of transpiration to change in leaf conductance may be less than that of rate of assimilation, that is to say $\partial^2 E/\partial A^2$ may be negative. It is most likely to be so in C_4 plants, by virtue of their characteristically small internal resistance to assimilation, but could also occur in C_3 plants when windspeed is unusually small and ambient temperature is very high. The hypothesis of optimisation then suggests that stomatal behaviour should have an "on-off" character. However, there is no evidence that such behaviour takes place.

V. DISCUSSION

In this article I have endeavoured to develop a concept of the stomatal mechanism and other mechanisms associated with gas exchange in leaves as elements in a system of interacting loops. I have used terms drawn from control theory to discuss the behaviour of the system under the influence of perturbations in environment. There are several formidable impediments to the use of such theory however. Firstly we have no coherent appreciation of the internal mechanism of the elements in the system. To suggest that the stomatal apparatus is a "black box" seems hardly to do justice to many excellent investigations of particular processes that take place in guard cells and subsidiary cells. But it is very plain that what knowledge exists of the internal structure and functioning of these cells does not amount to a basis for predictions of the properties of the apparatus as a whole. Similar remarks apply to the other "elements" in the system, such as those pertaining to assimilation in the mesophyll. This being so, we are not in a position to compute the behaviour of the system on the basis of *a priori* knowledge of the characteristics of the elements within it. The role of the stomatal mechanism must be defined entirely in terms of what can empirically be observed of its properties as a transducer. I have shown how certain experimental techniques and analyses may be used to compute a "stomatal transfer function". However, because the relationships of the output to the various inputs are non-linear and interactive, the analysis is valid for small perturbations, only, about a particular steady-state. We cannot hope to obtain a complete description of the dynamic characteristics of the stomatal apparatus over the whole range of interest. Also, there is the difficulty of identifying and measuring the inputs of the stomatal mechanism. Indeed the concept of the stomatal apparatus as a discrete element is open to criticism. As was indicated in an early section of this article the hydrology of the leaf and epidermis is very complex, and bulk leaf water potential may bear little relation to the potential of water in the vicinity of the stomata. Unfortunately we are not able to monitor the latter. Also, we do not know where abscisic acid is synthesised in response to water stress or how it is translocated. Measurement of ABA requires the destruction of rather large amounts of leaf. In view of these difficulties the practical course at present may be to regard rate of evaporation in the leaves, and the state of water supplied to the roots as the hydraulic inputs to the system. Then, of course, the transducer being considered is not the stomatal apparatus; it is nothing less than the entire plant. There is the additional complication that stomata are sometimes, perhaps always, directly sensitive to the humidity of the ambient air. That this has been generally recognised only recently is doubtless because the influence of change in humidity is readily confounded with the influence of change in rate of water loss from the leaf as a whole. Similar problems arise when we consider stomatal behaviour in relation to the carbon metabolism of the leaf, and there is an argument for treating the plant holistically and taking rate of assimilation as the input. This would be particularly appropriate if stomatal aperture were influenced, as I have suggested may occur, by a product of photosynthesis. But the most widely accepted view is that change in intercellular CO_2 concentration is the most important, if not the only, signal related to photosynthetic activity in the mesophyll tissue. Whether or not change in intercellular CO_2 concentration is always the predominant influence underlying stomatal movement associated with change in light intensity is an open question. The assertion that light intensity has only a small direct influence on stomata may turn out to be as misplaced as a similar one, not infrequently stated a few years ago, about the influence of ambient humidity.

Clearly, then, the theory of control systems is not a panacea for ignorance of fundamental processes or a substitute for experiments in the classical tradition of physiology. Perhaps it is most useful in showing whether or not any particular hypothesis concerning the structure of the system is consistent with observed performance. Several examples have been discussed in this article. Here I will emphasize two of them. If stomatal control of the flux and state of water were entirely a matter of feedback, then it would not be possible for a change in environment *tending* to increase rate of transpiration to in fact result, as is sometimes observed, in a steady-state decrease in rate of transpiration. A decrease in rate of transpiration, unless transient, implies that the stomata directly sense and respond to the change in external environment. Such a response constitutes "feedforward" control of plant water relations. The second example pertains to stomatal control of rate of assimilation. If control were based entirely on negative feedback associated with stomatal response to intercellular concentration of CO_2, then it would not be possible for a change in environment which directly stimulated rate of CO_2 uptake in the mesophyll to cause also a steady-state increase in intercellular concentration of CO_2. An increase in the latter, unless transient, would imply either a direct response of the stomata to the change in external environment, that is to say feedforward control; or, alternatively, positive feedback control based on a response of the stomata to a signal generated by the process of CO_2 fixation in the mesophyll. The concept of a closed loop transfer function and a knowledge of the conditions pertaining to stability provide a sound general basis for conclusions such as these.

I emphasize the distinctions between feedback and feedforward control of gas exchange because it is a central theme of what I have written. The activity of stomata in many species appears to be consistent with the classical hypothesis that stomata are primarily sensitive to plant internal water relations and intercellular concentration of CO_2. However, I have argued, in the latter parts of this article, that optimisation of gas exchange in some circumstances may demand that stomata behave in a quite different way, one which could not be achieved by negative feedback alone; and have shown that the predicted behaviour is, in one important respect at least, in accord with that observed in some cultivated and wild species in arid conditions. The topic of optimisation seems to bring us much closer to the kernel of control theory: the design of systems which satisfy required criteria of performance. But in biological systems it is fitness which is the criterion of performance, and natural selection which results in design. If the reader has been patient enough to follow me thus far, some comments on the difficulty of expressing the criterion of performance in terms which relate directly to the function of stomata in controlling gas exchange will be fresh in the mind, and need not be reiterated. But I would now add to them these final remarks. The best empirical estimate we have of criteria of stomatal performance is in fact represented by actual performance. What makes the study of stomatal performance, and its relationship with environment, intrinsically exciting is that it reflects more directly than any other aspect of plant functioning the resolution of what Raschke (1975a) termed the dilemma of land plants: how to accumulate carbon without excessive loss of water.

VI. APPENDIXES

APPENDIX A. NOTES ON EQUATIONS FOR HEAT AND MASS TRANSFER

Insofar as transfer of vapour in a leaf is by diffusion, the quantities used in this chapter to describe the process relate to the equation

$$J = -\frac{D}{PV}\frac{de}{dz} = -\frac{D}{V}\frac{dw}{dz} \tag{89}$$

where J is the molar flux density, D the diffusion coefficient, e the partial pressure, w the mole fraction, of vapour in air, and P and V are the total pressure and the molar volume of the moist air. In equating the two expressions for J it is assumed that deviations from perfect gas laws are negligible and that gradient in air pressure, P, is negligible.

There are two alternative ways of expressing diffusion of vapour that are in common use. The mass fraction of water vapour in air or specific humidity is

$$q = \frac{M_w}{M_a}w = \frac{M_w}{M_a}\cdot\frac{e}{P} \tag{90}$$

in which M_w is the molecular weight of water vapour and $M_a = 29.0\,(1 - 0.38w)$

is the apparent molecular weight of the moist air. Therefore eqn (90) rewritten in terms of mass flux density and mass fraction is

$$M_w J = -D\rho \, \frac{dq}{dz} \qquad (91)$$

with $\rho = M_a/V$ the density of the moist air. The absolute humidity is

$$\chi = \rho q \qquad (92)$$

Differentiating, making use of eqn (91), and remembering that ρ is inversely proportional to absolute temperature T, it follows that

$$M_w J = -D\left(\frac{d\chi}{dz} - \frac{\chi}{T}\frac{dT}{dz}\right) \approx -D\frac{d\chi}{dz} \qquad (93)$$

The term involving the temperature gradient is probably negligible for diffusion through stomata because transfer of heat across the epidermal cells "short circuits" the gas phase in the pores. Its influence in the boundary layer of leaves is usually small also; the worst case is with evaporation from a shaded leaf into warm, humid, slowly moving air and then it might exceed -10% of the gradient in humidity. Despite the fact that it may be acceptable for most working purposes, the approximate form of eqn (93) should nevertheless not be used, as it sometimes is, as the fundamental basis for the treatment of diffusion problems pertaining to leaves. Rate of diffusion depends not only on the gradient of the number of molecules per unit volume but the gradient of the speed with which they move. This fact is taken into account in eqns (89) and (91).

Penman and Schofield (1951) provided the initial treatment of vapour and CO_2 transfer in leaves from which the use of resistance terminology derives. Their definitions of the "equivalent lengths", L_s and L_a, of the stomatal array and the external atmosphere (of the leaf epidermis and the boundary layer, as we now prefer) were based on an analogue with diffusion in one dimension. Using the expressions for diffusion in eqns (89), (91), and (93) we may write for the rate of vapour transfer in moles per unit area of leaf

$$E = \frac{D}{PV} \cdot \frac{(e_i - e)}{L_s + L_a} = \frac{D}{V} \cdot \frac{(w_i - w)}{L_s + L_a} = \frac{D\rho}{M_w} \cdot \frac{(q_i - q)}{L_s + L_a} = \frac{D}{M_w} \cdot \frac{(\chi_i - \chi)}{L_s + L_a} \qquad (94)$$

where e_i and e are partial pressures inside the leaf and in the ambient air; and so on,* the analogous equations for CO_2 will be obvious.

---

* Of course, eqn (94) glosses over many complexities which have been treated since the paper of Penman and Schofield. Some of these relate to the role of convection in vapour transfer. Equation (89) should strictly be written $J = -D/V$ grad $w + (U/V)w$, where J is the vector flux of water vapour and U is the velocity of flow. In the boundary layer of a leaf a pattern of flow is imposed due to the inertia of the moving ambient air and the result is that transfer is only in part due to diffusion. The length L_a varies approximately as the third power of the diffusion coefficient. As J is still a linear function of w we may expect E still to be related to the difference in w across the epidermis and boundary layer. But in fact

The only advantage in defining resistances, or conductances, to vapour or CO_2 transfer is to make the expressions in eqn (94) formally simple. The way in which it is best done depends on which of the expressions are most useful and whether transfer is to be expressed in terms of mass, or in moles. There can be no doubt of the value of the first expression; vapour pressure is the practical quantity used to describe the state of vapour in equilibrium with a liquid or solid phase. Tables of vapour pressure are almost always employed in problems concerning water vapour in leaves. Mole fraction is the most closely related measure of the concentration of water vapour or CO_2 in air, as the conversion involves only the total pressure. For practical purposes it is identical with volume fraction, the traditional measure of CO_2 concentration. In the laboratory, CO_2-air mixtures are usually produced by combining measured volumes. If one employs mole fraction as the measure of concentration it is logical to express the flux in moles. It is a practice favoured by biochemists, but likely to be less acceptable to micrometeorologists who prefer to use mass fraction and mass flux density. However, mass fraction is not presently used in relation to CO_2 and the conversion from mass fraction to vapour pressure is different for different gases. The use of χ does not seem to me to have any general advantages.

It is on the basis of these observations that I have preferred to define leaf and boundary layer resistances, r_l and r_b, and the corresponding conductances, so that they relate conveniently to differences in partial pressure and mole fraction, and flux density in moles. Equation (94) becomes

$$E = \frac{1}{P} \cdot \frac{(e_i - e)}{r_l + r_b} = \frac{(w_i - w)}{r_l + r_b} = \frac{M_a}{M_w} \frac{(q_i - q)}{r_l + r_b} = \frac{V}{M_w} \cdot \frac{(\chi_i - \chi)}{r_l + r_b} \tag{95}$$

in which $r_l = L_s V/D$ and $r_b = L_a V/D$. These resistances differ from the analogous resistances for CO_2, but they do not depend very strongly on other properties of the air. The ratio V/D is independent of pressure and inversely proportional to absolute temperature. This may have advantages. For example, the effect of change in total pressure on gas exchange and stomatal aperture is not without physiological interest (Gale, 1973), and it is better in this context not to use an implied measure of stomatal aperture which is itself dependent on pressure. The molar flux of CO_2 to unit area of leaf is

$$A = \frac{1}{P} \frac{(p - p_i)}{r_l^\dagger + r_b^\dagger} = \frac{(c - c_i)}{r_l^\dagger + r_b^\dagger} \tag{96}$$

---

vapour diffusion itself causes convection and it has recently been pointed out to me that the error involved in disregarding this is roughly of the same magnitude as the maximum error that can occur due to the approximation in eqn (93) (G. S. Campbell, personal communication). When water diffuses away from a wet surface into air, there being no flow of air to the surface, the convection stream is $U/V = J$. Therefore the flux equation becomes $J = -D/V \cdot (1/1 - w)(dw/dz)$ which is non-linear. Roughly, the effect is to cause $L_s + L_a$ to vary as $1 - \bar{w}$, where $\bar{w}$ is the mean of w and w_i.

in which p and c are partial pressure and mole fraction, respectively, of CO_2 in air, and $r_l^\dagger = 1.60\, r_l$ and $r_b^\dagger = 1.37\, r_b$ are the leaf and boundary layer resistance to transfer of CO_2.

For comparison with eqn (95), mass flux density expressed with the resistances that are in common usage is

$$M_w E = \frac{M_w \rho}{M_a P} \frac{(e_i - e)}{r_l^0 + r_l^0} = \frac{M_w \rho (w_i - w)}{M_a} \frac{}{r_l^0 + r_b^0} = \frac{\rho(q_i - q)}{r_l^0 + r_b^0} = \frac{(\chi_i - \chi)}{r_l^0 + r_b^0} \tag{97}$$

with

$$r^0 = L_s/D = r_l/V$$
$$r_b^0 = L_a/D = r_b/V \tag{98}$$

These resistances are proportional to pressure and inversely proportional to the square of the absolute temperature. With $V = 25$ l at 1 bar and $25°C$ then

$$r\ (\mathrm{m^2\ s\ mol^{-1}}) = 2.5\, r^° \ (\mathrm{s\ cm^{-1}}) \tag{99}$$

Table V shows typical magnitudes of various resistances, $r^°$ in s cm^{-1}, to gas transfer in leaves, and the corresponding resistances, r, and conductances, $g = 1/r$, in the units used in this chapter.

For the flux of sensible heat per unit area of leaf

$$H = C_p \frac{T_l - T}{r_b^\dagger} \tag{100}$$

in which $C_p = 29.2$ J mole^{-1} is the molar heat capacity of moist air at constant pressure, T_l and T are the temperatures of leaf and ambient atmosphere, and $r_b^\dagger$ is the boundary layer resistance to heat transfer. Assuming that the internal vapour pressure, e_i, in a leaf corresponds to saturation vapour pressure at leaf temperature, eqn (100) may be rewritten

$$H = \frac{\lambda}{P} \frac{e_i - e'}{er_b^\dagger} = \lambda \cdot \frac{w_i - w'}{er_b^\dagger} \tag{101}$$

where $\lambda = 44$ kJ mol^{-1} is the molar heat of vapourization of water, e' and w' are saturation vapour pressure and saturation humidity, respectively, at the temperature of the ambient air, and

$$\epsilon = \frac{\lambda}{C_p P} \cdot \frac{de'}{dT} \approx 1.51 \frac{de'}{dT} \tag{102}$$

is the increase in latent heat content with increase in sensible heat content of saturated air.

Very often in the use of eqn (100) in the context of problems of evaporation, the psychrometric "constant" γ is introduced. The substitution that is made is

$$\frac{C_p}{r_b^\dagger} = \frac{\gamma \lambda}{P r_b} \tag{103}$$

TABLE V

Component	Resistance, r^0 (s cm^{-1})		Resistance, r (m^2 s mol^{-1})		Conductance, g (mol m^{-2} s^{-1})	
	Water vapour	CO$_2$	Water vapour	CO$_2$	Water vapour	CO$_2$
Leaf epidermis						
Stomata wide open	1.0	1.6	2.5	4.0	0.40	0.25
Stomata closed (cuticle)	40	–	100	–	0.01	–
Internal (CO$_2$ limiting photosynthesis)						
C$_3$ species	–	4.0	–	10.0	–	0.10
C$_4$ species	–	0.4	–	1.0	–	1.0
Boundary layer						
Broad leaves, low windspeed	0.40	0.55	1.00	1.38	1.00	0.73
Narrow leaves, high windspeed	0.10	0.14	0.25	0.35	4.0	2.9

Equations (95) and (100) then reduce to the psychrometric equation if $H = -\lambda E$ and $r_l = 0$. If one sets $r_b^\ddagger = 1.12\, r_b$, as mass and heat transfer theory suggests (Cowan, 1972b), eqn (103) yields

$$\gamma = 0.59 \text{ mbar } C^{-1} \text{ with } P = 1 \text{ bar} \tag{104}$$

which is quite close to the best theoretical estimate of γ. But in fact the magnitude employed is $\gamma = 0.67$ mbar C^{-1}, the empirical magnitude for use with ventilated psychrometers. It so happens that this magnitude is close to that derived by setting $r_b^\ddagger = r_b$ in eqn (103). That this is so seems to be fortuitous.

It is sometimes convenient to modify the resistance, or conductance, to heat transfer to account for the effect of change in leaf temperature on the emission of thermal radiation (Peisker, 1973; Jones, 1976). Assuming unit emissivity, and considering transfer from both sides of a leaf in parallel, the conductance, $1/r_b^\ddagger$, becomes

$$g_b^\ddagger = 0.89\, g_b + 8\, \sigma T^3/C_p \tag{105}$$

where σ is Stefan's constant and T is absolute temperature. The additional conductance, taken into account in Table IV, is 0.32, 0.44 and 0.52 mol m^{-2} s^{-1} at temperatures 16, 32 and 48°C, respectively.

APPENDIX B. LAPLACE TRANSFORMS, CONVOLUTIONS, AND TRANSFER FUNCTIONS

To increase our understanding of the dynamic characteristics of a biological system we may perturb its environment and observe the changes in physiological functioning which are brought about. Usually we perturb one characteristic of environment at a time, keeping all others constant. The variation in the environment is the "input" or "signal" and the variation in the functioning of the system is the "output" or "response". It is convenient if the input is a simple one. An example is the unit step function $1(t)$, which is, by definition, zero for $t < 0$, and unity for $t \geq 0$. The corresponding output is called the "unit step response" of the system and will be assigned the symbol $U(t)$. The generalized step input, of size A rather than unity, and occurring at time τ rather than zero, is $A \times 1(t - \tau)$. Provided the system is "linear" and its properties do not change with time, the corresponding output is $A \times U(t - \tau)$.

Often the input is not varied in a simple fashion. There are two common reasons for this. Sometimes what we wish to regard as the input, $X(t)$ say, can only be manipulated indirectly; and sometimes we may wish to regard some natural environmental fluctuation as the input. However, it is always possible to regard whatever input occurs as due to the superposition of a number of step functions. Expressed in this way it is

$$X_m = \sum_1^m (X_n - X_{n-1}) \cdot 1(m - n) \tag{106}$$

and the output can be deduced by superimposing step responses in the same way, i.e.

$$Y_m = \sum_1^m (X_n - X_{n-1}) \cdot U(t - \tau_n) \tag{107}$$

If greater accuracy in the representation of the input is required, one can regard it as the sum of an infinite number of infinitely small steps (together, sometimes, with the sum of a finite number of finite steps). The output becomes

$$Y = \int_0^t \overset{\circ}{X}(\tau) \cdot U(t - \tau)\, \mathrm{d}\tau \tag{108}$$

where $\overset{\circ}{X} = \mathrm{d}X/\mathrm{d}t$. This integral is an example of a convolution integral. In conventional notation it is written

$$Y = \overset{\circ}{X} * U = U * \overset{\circ}{X} \tag{109}$$

By setting $\tau' = t - \tau$ in eqn (108) it is readily confirmed that it makes no difference in which order the convolution is expressed. There are two ways in which we may want to make use of this representation. We may wish to determine Y given that we know X and U. Or we may want to determine U given that we know X and Y. So far we have concentrated on, and indicated, the mechanism for tackling the first problem. Let us now think of the second.

The function U can be regarded as a characteristic of the system. It is the response corresponding to a simply defined input. The function Y is not a characteristic of the system alone. It depends also on the nature of the input X, which we shall assume is complicated—that is to say not the type of input to which one would relate output by choice. Thus, in the integral defining Y, there is mixed, in what seems to be an inextricable way, valuable information about the system characteristics and trivial information about the form of the input function. The problem is to extract the valuable information.

Suppose we measured the area under the output curve and the area under the input curve and formed the ratio, i.e. found

$$G(0) = \frac{\displaystyle\int_0^\infty Y\mathrm{d}t}{\displaystyle\int_0^\infty X\mathrm{d}t} \tag{110}$$

the reason for assigning the ratio this particular symbol being deferred for the moment. By measuring the areas we have eliminated all information about variation in time and simply obtained an impression of the size of the output related to the size of the input. One might expect the ratio to be the same for all forms of input; that is to say, that the ratio is a characteristic of the system. But although it fits some of our requirements, it clearly lacks all information about the dynamic characteristics of the system—it does not tell us how rapidly, for example, the response increases or dies away. Suppose now we modify the ratio slightly and write

$$G(p) = \frac{\int_0^\infty Ye^{-pt}Y dt}{\int_0^\infty Xe^{-pt}X dt} \qquad (111)$$

$G(p)$ is now a function of p. The parameter p is a weighting variable, if it is large, the magnitudes of the integrals are determined primarily by the transient features of X and Y; if it is small the integrals depend on the more enduring features of X and Y. Intuitively one might suppose that, in the spectrum of G, all the essential information that is to be learnt about the system by examining X and Y is retained. We now have to justify this expectation. In order to do this, we examine the nature of the integrals. Incidentally, the integrals are the Laplace transforms of X and Y, p is the Laplace operator, and $G(p)$ is the transfer function of the system. The Laplace transform of a variable Y, say, is often written $L(Y)$.

Making use of eqn (108) we have

$$L(Y) = \int_0^\infty e^{-pt} \left\{ \int_0^t \overset{\circ}{X}(\tau) \cdot U(t-\tau)\, d\tau \right\} dt \qquad (112)$$

Because (in a stable system) there is no output preceding an input, it must be that $U(t-\tau) = 0$ for $\tau > t$. Therefore the upper limit of the convolution integral may be reset at infinity. Then the order of integration may be reversed so that

$$L(Y) = \int_0^\infty \left\{ \int_0^\infty e^{-pt} U(t-\tau)\, dt \right\} \overset{\circ}{X}(\tau)\, d\tau \qquad (113)$$

Consider now the inner integral. It may progressively be adjusted as follows

$$\int_0^\infty e^{-pt} U(t-\tau)\, dt = \int_\tau^\infty e^{-pt} U(t-\tau)\, dt = e^{-p\tau} \int_0^\infty e^{-pt'} U(t')\, dt' = e^{-p\tau} L(U) \qquad (114)$$

the equivalence of the first and third expressions being illustrated in Fig. 32. Therefore eqn (113) becomes

$$L(Y) = L(U) \cdot \int_0^\infty e^{-p\tau} \overset{\circ}{X}(\tau)\, d\tau = L(U) \cdot L(\overset{\circ}{X}) \qquad (115)$$

We now see that by taking the Laplace transform of Y we have separated two types of information. $L(U)$ is a system characteristic whereas $L(\overset{\circ}{X})$ is entirely concerned with the form of the input. The transform $L(\overset{\circ}{X})$ may be integrated by parts as shown below

$$\int_0^\infty e^{-p\tau} \frac{dX}{d\tau}\, d\tau = \left. e^{-p\tau} X \right|_0^\infty + p \int_0^\infty e^{-p\tau} X\, d\tau = pL(X) \qquad (116)$$

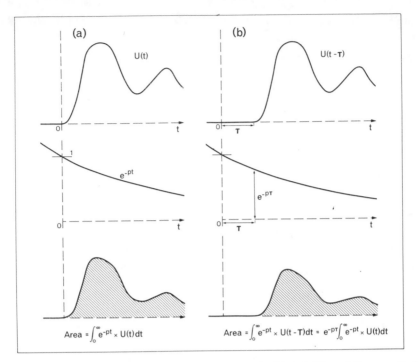

Fig. 32. The relationship between the transforms of $U(t)$ and $U(t - \tau)$ for a particular magnitude of the Laplace operator, p.

as $X(0) = 0$. Finally, from eqns (111) and (115)

$$G(p) = \frac{L(Y)}{L(X)} = \frac{L(U)}{L(1)} = pL(U) \qquad (117)$$

The result is quite general: *whatever* the form of the input, the transfer function, as we have defined it, is identical with the quantity $pL(U)$ which is a characteristic of the system. It is a general result, not proved here, that the Laplace transform of a function is unique; if we know the transform then the function itself is known in principle. If by measurement of the output, Y, for a given input, X, we have determined $G(p)$ for a system according to eqn (111), then the output may be determined for any *other* prescribed input by the inverse process, i.e. finding

$$Y = L^{-1} (G(p) \cdot L(X)) \qquad (118)$$

Assuming that $G(p)$ and $L(X)$ are mathematical functions of p, there is a method, embodied in the "inversion theorem", which can be used to find Y. In practice one may often avoid the necessity of using this method by recourse to a table of transforms. Table VI is a very abbreviated example. The first of the general relations needs no explanation, the second and third have essentially

TABLE VI
Laplace transforms

	$F(t)$	$L(F)$
1	$a\,F_1 + b\,F_2$	$a\,L(F_1) + b\,L(F_2)$
2	$1(t - \tau)F(t - \tau)$	$e^{-p\tau}L(F)$
3	$\dfrac{\mathrm{d}F}{\mathrm{d}t}$	$pL(F) - F(0)$
4	$\displaystyle\int_0^t F_1(\tau) \cdot F_2(t - \tau)\,\mathrm{d}\tau$	$L(F_1) \cdot L(F_2)$
5	δ	1
6	1	$\dfrac{1}{p}$
7	$e^{p_1 t}$	$\dfrac{1}{p - p_1}$
8	$t e^{p_1 t}$	$\dfrac{1}{(p - p_1)^2}$
9	$e^{p_1 t} \sin(\omega t)$	$\dfrac{\omega}{(p - p_1)^2 + \omega^2}$
10	$e^{p_1 t} \cos(\omega t)$	$\dfrac{p - p_1}{(p - p_1)^2 + \omega^2}$

been derived in eqns (114) and (116), and the fourth follows from a comparison of eqns (109) and (115).

The unit impulse function, δ, is a useful mathematical fiction: a rectangular pulse, infinitely narrow and infinitely high and having unit area. It may be defined as

$$\delta = \lim_{\tau \to 0} \frac{1(t) - 1(t - \tau)}{\tau} \tag{119}$$

and it is readily shown that the transform is unity. When the input to a system is $X = \delta$, then the output, from eqn (118), is $Y = L^{-1}(G)$. Thus the most succinct definition of a transfer function is that it is the transform of the output for a unit impulse input. The discussion leading to eqn (117) can be carried through using the concept of δ. However, although inputs sometimes approximate to δ for practical purposes, the function cannot, of course, be strictly realized. That is why I have preferred to deal with step functions.

The other transforms in Table VI, except perhaps the last, are readily verified. As an example of the use of the table, one may find the unit step response with G as described by eqn (24). It is the inversion of G/p that is required, that is to say

$$L^{-1}\left\{ a\left(\frac{1}{p} - \frac{\tau_1}{1 + p\tau_1} \right) - b\,\exp(-p\tau_0)\left(\frac{1}{p} - \frac{\tau_2}{1 + p\tau_2} \right) \right\} \tag{120}$$

The result, shown in eqn (25), is obtained using entries 1, 2, 6, and 7 in Table VI. The inversion of G is the differential of the unit step response, just as δ is the differential of $1(t)$.

Let us consider the inversion of a transfer function, that is to say the output corresponding to a unit impulse input. Most transfer functions that are met with can be written as the ratio of two polynomials, i.e.

$$G = \frac{y_0 + y_1 p + y_2 p^2 + \cdots y_m p^m}{x_0 + x_1 p + x_2 p^2 + \cdots x_n p^n} = \frac{Y(p)}{X(p)} \tag{121}$$

where $m < n$. The condition on the order of the polynomials is determined by the fact that in any real system $G(p) \to 0$ as $p \to \infty$. Generally the ratio can be split into partial fractions of the kind $a/(p - p_1)$, where p_1 is a root of $X(p)$. These fractions may then be inverted according to entry 7 in Table VI; occasionally a multiple root might demand the use of entry 8. The Heaviside inversion theorem provides a convenient means of carrying out this process once the roots are known, but in practice the order of the polynomials in eqn (121) will not be large unless the transfer function derives from a manmade system in which the characteristics of individual components are known, or from a theoretical exercise. Experimental data from biological systems are usually such that one would rarely be justified in fitting a transfer function with $n > 3$. The nature of the roots determines the dynamic properties of the system. If any of the real roots, or real parts of complex roots, is positive, then the system is unstable. If there are complex roots they occur in conjugate pairs, as in entries 9 and 10 in Table VI, and the system is oscillatory. The magnitude of the imaginary parts, $i\omega$ and $-i\omega$, of the roots is the angular frequency of the oscillation, while the real part, p_1, determines the time constant $\tau = -1/p_1$ of its decay. If G as in eqn (24) is substituted in eqn (28) for the closed-loop response, and the term $e^{-p\tau_0}$ is neglected, the resulting expression is readily inverted, but I will describe only the characteristics of the system that may be discerned from the denominator of $\ddot{G}$. It is found that the denominator of $\ddot{G}$ is a quadratic, $p^2 + 2cp + d$, say, in which

$$c = [\tau_1 + \tau_2 - \partial E/\partial g_l(a\tau_2 - b\tau_1)]/(2\tau_1\tau_2) \text{ and}$$
$$d = [1 - \partial E/\partial g_l(a - b)]/(\tau_1\tau_2)$$

With $a = 36$, $b = 43$, $\tau_1 = 1.6$ min, and $\tau_2 = 16$ min it is found that $c = 0$ when $\partial E/\partial g_l = 35 \times 10^{-3}$. This is the magnitude of the environmental gain such that the oscillation is just undamped. The natural undamped frequency is then $\omega_n = d^{1/2} = 0.22$ min^{-1} corresponding to a period of 28 min. It may be verified that these estimates are not much affected by including $\exp(-p\tau_0) \approx 1 - p\tau_0$, with $\tau_0 = 0.7$ min. This short delay term only affects behaviour for short periods of time after an initial disturbance, corresponding to a range of p where the estimate of G was in any case uncertain.

The fitting of expressions of the form of equation (121) to numerical determinations of a transfer function is not an easy task. I am not aware of the proper statistical approach, assuming that one exists. What would, for many purposes, be regarded as an insignificant error in curve fitting can be extremely important, for different functions of time can yield transforms which appear to be very similar in shape. It is preferable to manipulate data numerically as far as possible, and to examine the results in terms of functions of real time. It is for this reason that I have made use of the convolution theorem in the analysis of the observations in Fig. 26, and thus avoided where possible the necessity of transforming the data.

ACKNOWLEDGMENTS

I am grateful to Professor R. O. Slatyer and Dr. C. B. Osmond for encouragement and helpful discussion; to Mrs. C. F. Donnelly, née Tabor, for providing data used in analyses of stomatal response; and especially to Dr. G. D. Farquhar — the influence of my association with him permeates a large proportion of what I have written. I thank also Miss D. Lee and my wife for secretarial help.

REFERENCES

Acevedo, E., Hsiao, T. C. and Henderson, D. W. (1971). *Pl. Physiol.* **48**, 631-636.
Akita, S. and Moss, N. (1973). *Crop Sci.* **13**, 234-236.
Allaway, W. G. (1973). *Planta* **110**, 63-70.
Andrews, R. E. and Newman, E. I. (1969). *New Phytol.* **68**, 1051-1058.
Aubert, B. and Čatský, J. (1970). *Photosynthetica* **4**, 254-256.
Aylor, D. E., Parlange, J.-Y. and Krikorian, A. D. (1973). *Am. J. Bot.* **60**, 163-171.
Barrs, H. D. (1971). *A. Rev. Pl. Physiol.* **22**, 223-236.
Barrs, H. D. (1973). *In* "Plant Response to Climatic Factors", Proc. Uppsala Symp., 1970. (Ecology and Conservation 5) (Ed. R. O. Slatyer), pp. 249-258. UNESCO, Paris.
Barrs, H. D. and Klepper, B. (1968). *Physiologia Pl.* **21**, 711-730.
Beadle, C. L., Stevenson, K. R., Neumann, H. H., Thurtell, G. W. and King, K. M. (1973). *Can. J. Pl. Sci.* **53**, 537-544.
Bearce, B. C. and Kohl, H. C. Jr. (1970). *Pl. Physiol.* **46**, 515-519.
Begg, J. E., Bierhuizen, J. F., Lemon, E. R., Mistra, D. K., Slatyer, R. O. and Stern, W. R. (1964). *Agric. Meteorol.* **1**, 294-312.
Berger, A. (1973). *In* "Plant Response to Climatic Factors", Proc. Uppsala Symp., 1970. (Ecology and Conservation 5) (Ed. R. O. Slatyer), pp. 201-212. UNESCO, Paris.
Biscoe, P. V. (1972). *J. exp. Bot.* **23**, 930-940.
Björkman, O. (1973). *In* "Current Topics in Photophysiology" (Ed. A. Giese), Vol. 8, pp. 1-63. Academic Press, New York and London.
Björkman, O., Nobs, M., Pearcy, R., Boynton, J. and Berry, J. (1971). *In* "Photosynthesis and Photorespiration" (Eds M. D. Hatch, C. B. Osmond and R. O. Slatyer), pp. 105-119. Wiley-Interscience, New York and London.

Björkman, O., Boardman, N. K., Anderson, J. M., Thorne, S. W., Goodchild, D. J. and Pyliotis, N. A. (1972a). *Yearb. Carnegie Instn* **71**, 115-135.
Björkman, O., Ludlow, M. M. and Morrow, P. A. (1972b). *Yearb. Carnegie Instn* **71**, 94-102.
Black, C. C., Chen, T. M. and Brown, R. H. (1969). *Weed Sci.* **17**, 338-344.
Boyer, J. S. (1968). *Pl. Physiol.* **43**, 1056-1062.
Boyer, J. S. (1969). *Science, N.Y.* **16**, 1219-1220.
Boyer, J. S. (1970). *Pl. Physiol.* **46**, 236-239.
Boyer, J. S. (1974). *Planta* **117**, 187-207.
Brogårdh, T. and Johnsson, A. (1973). *Physiologia Pl.* **28**, 341-345.
Brogårdh, T. and Johnsson, A. (1974). *Physiologia Pl.* **31**, 311-322.
Brogårdh, T., Johnsson, A. and Klockare, R. (1974). *Physiologia Pl.* **32**, 258-267.
Bull, T. A. (1969). *Crop Sci.* **9**, 726-728.
Camacho-B, S. E., Hall, A. E. and Kaufmann, M. R. (1974). *Pl. Physiol.* **54**, 169-172.
Cleland, R. (1967). *Planta* **77**, 182-191.
Cohen, D. (1970). *Israel J. Bot.* **19**, 50-54.
Cole, K. S. (1968). "Membranes, Ions and Impulses". University of California Press, Berkeley and Los Angeles.
Cowan, I. R. (1965). *J. appl. Ecol.* **2**, 221-239.
Cowan, I. R. (1972a). *Planta* **106**, 185-219.
Cowan, I. R. (1972b). *Agric. Meteorol.* **10**, 311-329.
Cowan, I. R. and Milthorpe, F. L. (1968). *In* "Water Deficits and Plant Growth" (Ed. T. T. Kozlowski), Vol. 1, pp. 137-193. Academic Press, New York and London.
Cowan, I. R. and Milthorpe, F. L. (1971). *In* "Respiration and Circulation" (Eds P. L. Altman and D. S. Dittmer), pp. 705-709. Fedn Am. Socs exp. Biol. Bethesda, Md.
Cowan, I. R. and Troughton, J. H. (1971). *Planta* **97**, 325-336.
Cox, E. F. (1968). *J. exp. Bot.* **19**, 167-175.
Cummins, W. R., Kende, H. and Raschke, K. (1971). *Planta* **99**, 347-351.
Dainty, J. (1963). *In* "Advances in Botanical Research" (Ed. R. D. Preston), Vol. 1, pp. 279-326. Academic Press, London and New York.
Darwin, F. (1898). *Phil. Trans. R. Soc. Ser. B* **190**, 531-621.
DeMichele, D. W. and Sharpe, P. J. H. (1973). *J. theor. Biol.* **41**, 77-96.
Dixon, H. H. (1938). *Proc. R. Soc. Ser. B* **125**, 1.
Downes, R. W. (1969). *Planta* **88**, 261-273.
Downes, R. W (1971). *In* "Photosynthesis and Photorespiration" (Eds M. D. Hatch, C. B. Osmond and R. O. Slatyer), pp. 57-62. Wiley Interscience, New York, London, Sydney, Toronto.
Drake, B. and Raschke, K. (1974). *Pl. Physiol.* **53**, 808-812.
Duniway, J. M. (1971). *Physiol. Pl. Pathol.* **1**, 537-546.
Edwards, M. and Meidner, H. (1975). *Nature, Lond.* **253**, 114-115.
Ehlig, C. F. and Gardner, W. R. (1964). *Agron. J.* **56**, 127-130.
Ehrler, W. L., Nakayama, F. S. and van Bavel, C. H. M. (1965). *Physiologia Pl.* **18**, 766-775.
Etherington, J. R. (1967). *J. Ecol.* **55**, 373-380.
Farquhar, G. D. (1973). "A study of the responses of stomata to perturbations of environment". Ph.D. Thesis, Australian National University.
Farquhar, G. D. and Cowan, I. R. (1974), *Pl. Physiol.* **54**, 769-772.

Fischer, R. A. (1968). *Science, N.Y.* **160**, 784-785.
Fischer, R. A. (1971). *Pl. Physiol.* **47**, 555-558.
Fischer, R. A. (1973). *J. exp. Bot.* **24**, 387-399.
Fischer, R. A. Hsiao, T. C. (1968). *Pl. Physiol.* **43**, 1953-1958.
Fischer, R. A., Hsiao, T. C. and Hagan, R. M. (1970). *J. exp. Bot.* **21**, 371-385.
Fiscus, E. L. (1975). *Pl. Physiol.* **55**, 917-922.
Fujino, M. (1967). *Sci. Bull. Fac. Educ. Nagasaki Univ.* **18**, 1-47.
Gale, J. and Poljakoff-Mayber, A. (1968). *Physiologia Pl.* **21**, 1170-1176.
Gale, J. (1973) (p. 136).
Gardner, W. R. (1960). *Soil Sci.* **89**, 63-73.
Gifford, R. M. (1974). *Aust. J. Pl. Physiol.* **1**, 107-117.
Gifford, R. M. and Musgrave, R. B. (1970). *Physiologia Pl.* **23**, 1048-1056.
Glinka, Z. (1971). *Physiologia Pl.* **24**, 476-479.
Green, P. B. (1968). *Pl. Physiol.* **43**, 1169-1184.
Grodins, F. S. (1963). "Control Theory and Biological Systems". Columbia University Press, New York.
Gutknecht, J. (1968). *Science, N.Y.* **160**, 68-70.
Hailey, J. L., Hiler, E. A., Jordan, W. R. and van Bavel, C. H. M. (1973). *Crop Sci.* **13**, 264-267.
Hales, Stephen (1727). "Vegetable Statiks". W. Innys and R. Manby and T. Woodward, London; also Oldbourne, London (1961).
Hall, A. E. (1971). *Yearb. Carnegie Instn* **70**, 530-540.
Hall, A. E. and Björkman, O. (1975). *In* "Perspectives of Biophysical Ecology" (Eds D. M. Gates and R. B. Schmerl), pp. 55-72. Springer-Verlag, Berlin.
Hall, A. E. and Kaufmann, M. R. (1975). *Pl. Physiol.* **55**, 455-459.
Hall, A. E., Camacho-B, S. E. and Kaufmann, M. R. (1975). *Physiologia Pl.* **33**, 62-65.
Hall, A. E., Schulze, E.-D. and Lange, O. L. (1976). *In* "Water and Plant Life—Problems and Modern Approaches" (Eds O. L. Lange, L. Kappen and E.-D. Schulze), Ecological Studies Vol. 15, Springer-Verlag, Berlin.
Hanebuth, W. and Raschke, K. (1973). *In* "Plant Research 172" pp. 139-144. Report of the MSU/AEC Commission Plant Research Laboratory for the Year 1972. MSU/AEC Plant Research Laboratory, Michigan State University.
Heath, O. V. S. (1938). *New Phytol.* **37**, 385-395.
Heath, O. V. S. and Orchard, B. (1957). *Nature, Lond.* **180**, 180-181.
Hesketh, J. D. (1963). *Crop Sci.* **3**, 493-496.
Honert, T. H. van den (1948). *Disc. Faraday Soc.* **3**, 146-153.
Hopmans, P. A. M. (1969). *Z. PflPhysiol.* **60**, 242-254.
Hopmans, P. A. M. (1971). *Meded. LandbHoogesch. Wageningen* 71-73.
Horton, R. F. (1971). *Can. J. Bot.* **49**, 583-585.
Hsiao, T. C. (1973). *A. Rev. Pl. Physiol.* **24**, 519-570.
Hsiao, T. C., Allaway, W. G. and Evans, L. T. (1973). *Pl. Physiol.* **51**, 82-88.
Humble, G. D. and Hsiao, T. C. (1970). *Pl. Physiol.* **46**, 483-487.
Humble, G. D. and Raschke, K. (1971). *Pl. Physiol.* **48**, 447-453.
Hurd, R. G. (1969). *New Phytol.* **68**, 265-273.
Jaeger, J. C. (1961). "An Introduction to the Laplace Transformation", 2nd Edn. Methuen, London.
Johnsson, A. (1973). *Physiologia Pl.* **28**, 48-50.
Jones, H. G. (1973). *New Phytol.* **72**, 1089-1094.
Jones, H. G. (1976). *J. appl. Ecol.* **13**, 605-622.

Jones, R. J. and Mansfield, T. A. (1972). *Physiologia Pl.* **26**, 321-327.
Jordan, W. R. and Ritchie, J. T. (1971). *Pl. Physiol.* **48**, 783-788.
Kanemasu, E. T. and Tanner, C. B. (1969). *Pl. Physiol.* **44**, 1547-1552.
Karmanov, V. G. and Savin, V. N. (1964). *Dokl. Akad. Nauk SSSR* **154**, 16-19.
Karmanov, V. G., Meleshchenko, S. N. and Savin, V. N. (1966). *Biofisika* **11**, 147-155.
Kassam, A. H. (1973). *New Phytol.* **72**, 557-570.
Kaufmann, M. R. and Hall, A. E. (1974). *Agric. Meteorol.* **14**, 85-98.
Ketallapper, H. J. (1963). *A. Rev. Pl. Physiol.* **14**, 249-270.
Knutson, R. M. (1974). *Science, N.Y.* **186**, 746-747.
Kriedemann, P. E. and Smart, R. E. (1971). *Photosynthetica* **5**, 6-15.
Kriedemann, P. E., Loveys, B. R., Fuller, G. L. and Leopold, A. C. (1972). *Pl. Physiol.* **49**, 842-847.
Laisk, A. (1970). *In* "Prediction and Measurement of Photosynthetic Pro-ductivity", pp. 295-306. Proc. IBP/PP Tech. Meeting, Trêbon, Pudoc, Wageningen.
Lang, A. R. G., Klepper, B. and Cumming, M. J. (1969). *Pl. Physiol.* **44**, 826-830.
Lange, O. L., Lösch, R., Schulze, E.-D. and Kappen, L. (1971). *Planta* **100**, 76-86.
Lange, O. L., Schulze, E.-D., Evenari, M., Kappen, L. and Buschbom, U. (1975). *Oecologia (Berl.)* **18**, 45-53.
Levitt, J. (1974). *Protoplasma* **82**, 1-17.
Little, C. H. A. and Eidt, D. C. (1968). *Nature, Lond.* **220**, 498-499.
Liu, W. T., Pool, R., Wenkert, W. and Kriedemann, P. E. (1975). *Physiologia Pl.* In press.
Loftfield, J. V. G. (1921). *Publs Carnegie Instn* **314**, 1-104.
Loveys, B. R. and Kriedemann, P. E. (1973). *Physiologia Pl.* **28**, 476-479.
Loveys, B. R. and Kriedemann, P. E. (1974). *Aust. J. Pl. Physiol.* **1**, 407-415.
Ludlow, M. M. (1970). *Planta* **91**, 285-290.
Ludlow, M. M. and Wilson, G. L. (1971). *Aust. J. biol. Sci.* **24**, 449-470.
Machin, K. E. (1964). *Symp. Soc. exp. Biol.* **18**, 421-445.
Maercker, U. von (1965). *Protoplasma* **60**, 61-78.
McCree, K. J. (1972). *Agric. Meteorol.* **9**, 191-216.
McCree, K. J. (1974). *Crop Sci.* **14**, 273-278.
McPherson, H. G. and Slatyer, R. O. (1973). *Aust. J. biol. Sci.* **26**, 329-339.
Meidner, H. (1975). *J. exp. Bot.* **26**, 666-673.
Meidner, H. and Edwards, M. (1975). *J. exp. Bot.* **92**, 319-330.
Meidner, H. and Mansfield, T. A. (1968). "Physiology of Stomata". McGraw-Hill, Maidenhead.
Milborrow, B. V. (1974). *A. Rev. Pl. Physiol.* **25**, 259-307.
Millar, A. A., Duysen, M. E. and Wilkinson, G. E. (1968). *Pl. Physiol.* **43**, 968-972.
Millar, A. A., Gardner, W. R. and Goltz, S. M. (1971). *Agron. J.* **63**, 779-784.
Milthorpe, F. L. and Spencer, E. J. (1957). *J. exp. Bot.* **8**, 414-437.
Mittelheuser, C. J. and van Steveninck, R. F. M. (1969). *Nature, Lond.* **221**, 281-282.
Moreshet, S., Koller, D. and Stanhill, G. (1968). *Ann. Bot.* **32**, 695-701.
Neumann, H. H., Thurtell, G. W. and Stevenson, K. R. (1973). *Can. J. Pl. Sci.* **54**, 175-184.
Newman, E. I. (1969a). *J. appl. Ecol.* **6**, 1-12.
Newman, E. I. (1969b). *J. appl. Ecol.* **6**, 261-272.

Ogata, K. (1970). "Modern Control Engineering". Prentice-Hall, Englewood Cliffs, N.J.

Osmond, C. B. (1975). *In* "Environmental and Biological Control of Photosynthesis" (Ed. W. Junk), pp. 311-321. The Hague.

Osmond, C. B., Troughton, J. H. and Goodchild, D. J. (1969). *Z. Pflphysiol.* **61**, 218-237.

Pallas, J. E. (1964). *Bot. Gaz.* **1964**, 102-107.

Pallas, J. E. (1965). *Science, N.Y.* **147**, 171-173.

Pallas, J. E. Jr., and Wright, B. G. (1973). *Pl. Physiol.* **51**, 588-590.

Parkhurst, D. F. and Loucks, O. L. (1972). *J. Ecol.* **60**, 505-537.

Passioura, J. B. (1972). *Aust. J. agric. Res.* **23**, 745-752.

Pearcy, R. W., Björkman, O., Harrison, A. T. and Mooney, H. A. (1971). *Yearb. Carnegie Instn* **70**, 540-550.

Pearcy, R. W., Harrison, A. T., Mooney, H. A. and Björkman, O. (1974). *Oecologia (Berl.)* **17**, 111-211.

Peisker, M. (1973). *Kulturpflanze* **21**, 97-100.

Peisker, M. (1974). Photosynth. **8**, 47-50.

Penman, H. L. (1950a). *Q. Jl R. met. Soc.* **76**, 330.

Penman, H. L. (1953). *Rep. Inst. Hort. 13th Congr. London* **2**, 913-924.

Penman, H. L. and Schofield, R. K. (1951). *Symp. Soc. exp. Biol.* **5**, 115-129.

Philip, J. R. (1957). *Trans. 3rd Congr. Int. Comm. Irrig. Drainage* Question 8, 8.125.

Raschke, K. (1965a). *Planta* **67**, 225-241.

Raschke, K. (1965b). *Z. Naturf.* **20**, 1261-1270.

Raschke, K. (1970a). *Pl. Physiol.* **45**, 415-423.

Raschke, K. (1970b). *Planta* **91**, 336-363.

Raschke, K. (1972). *Plant Physiol.* **49**, 229-234.

Raschke, K. (1975a). *A. Rev. Pl. Physiol.* **26**, 309-340.

Raschke, K. (1975b). *Planta* **125**, 243-259.

Raschke, K. and Dickerson, M. (1973). *In* "Plant Research '72", pp. 153-154. Report of the MSU/AEC Plant Research Laboratory, Michigan State University.

Raschke, K. and Fellows, M. P. (1971). *Planta* **101**, 296-316.

Raschke, K., Dickerson, M. and Pierce, M. (1973a). *In* "Plant Research '72", pp. 149-153. Report of the MSU/AEC Plant Research Laboratory, Michigan State University.

Raschke, K., Dickerson, M. and Pierce, M. (1973b). *In* "Plant Research '72", pp. 155-157. Report of the MSU/AEC Plant Research Laboratory, Michigan State University.

Sawhney, B. L. and Zelitch, I. (1969). *Pl. Physiol.* **44**, 1350-1354.

Scarth, G. W. (1932). *Pl. Physiol.* **7**, 481-504.

Schulze, E.-D., Lange, O. L., Buschbom, U., Kappen, L. and Evenari, M. (1972). *Planta* **108**, 259-270.

Schulze, E.-D., Lange, O. L., Evenari, M., Kappen, L. and Buschbom, U. (1974). *Oecologia* **17**, 159-170.

Schulze, E.-D., Lange, O. L., Kappen, L., Evenari, M. and Buschbom, U. (1975a). *Oecologia* **18**, 219-233.

Schulze, E.-D., Lange, O. L., Evenari, M., Kappen, L. and Buschbom, U. (1975b). *Oecologia (Berlin)* **19**, 303-314.

Sharpe, P. J. H. (1973). *Agron. J.* **65**, 570-574.

Shaw, M. and MacLachlan, G. A. (1954). *Can. J. Bot.* **32**, 784-794.

Sheriff, D. W. (1973). *J. exp. Bot.* **24**, 796-803.

Sheriff, D. W. (1974). *J. exp. Bot.* **86**, 562-574.
Sheriff, D. W. and Meidner, H. (1974). *J. exp. Bot.* **25**, 1147-1156.
Shimshi, D. (1963). *Pl. Physiol.* **38**, 713-721.
Slatyer, R. O. (1970). *Planta* **93**, 175-189.
Spiegal, M. R. (1965). "Schaums Outline of Theory and Problems of Laplace Transforms". Schaums Outline Series. McGraw-Hill, New York.
Stålfelt, M. G. (1929). *Planta* **8**, 287-340.
Stålfelt, M. G. (1955). *Physiologia Pl.* **8**, 572-593.
Stålfelt, M. G. (1967). *Physiologia Pl.* **20**, 634-642.
Stephens, G. R., Turner, N. C. and de Roo, (1972). *Forest Sci.* **18**, 326-330.
Stokes, R. and Weatherley, P. E. (1971). *New Phytol.* **70**, 547-554.
Tanton, T. W. and Crowdy, S. H. (1972). *J. exp. Bot.* **23**, 600-618.
Teoh, C. T. and Palmer, J. H. (1971). *Pl. Physiol.* **47**, 409-411.
Troughton, J. H. (1969). *Aust. J. biol. Sci.* **22**, 289-302.
Turner, N. C. (1970). *New Phytol.* **69**, 647-653.
Turner, N. C. (1974a). *In* "Mechanisms of Regulation of Plant Growth" (Eds R. L. Bieleski, A. R. Ferguson and M. M. Cresswell), pp. 423-432. Bull. 12. Royal Soc. of N.Z., Wellington. 1974.
Turner, N. C. (1974b). *Pl. Physiol.* **53**, 360-365.
Turner, N. C. and Begg, J. E. (1973). *Pl. Physiol.* **51**, 31-36.
van den Driessche, R., Connor, D. J. and Tunstall, B. R. (1971). *Photosynthetica* **5**, 210-217.
Verfaillie, G. R. M. (1972). *J. exp. Bot.* **23**, 1106-1119.
Visser, W. C. (1964). *Tech. Bull. Inst. Ld Wat. Mgmt Res., Wageningen* **32**, 1-21.
Waggoner, P. E. and Turner, N. C. (1971). *Conn. Agric. exp. Sta. Bull.* **726**.
Weatherley, P. E. (1970). *In* "Advances in Botanical Research" (Ed. R. D. Preston), Vol. 3, pp. 171-206. Academic Press, London and New York.
Willmer, C. M. and Pallas, J. E. Jr. (1973). *Can. J. Bot.* **51**, 37-42.
Willmer, C. M., Pallas, J. E. Jr. and Black, C. C. Jr. (1973). *Pl. Physiol.* **52**, 448-452.
Woods, D. B. and Turner, N. C. (1971). *New Phytol.* **70**, 77-84.
Wright, S. T. C. (1969). *Planta* **86**, 10-20.
Wright, S. T. C. and Hiron, R. W. P. (1969). *Nature, Lond.* **224**, 719-720.

Evolutionary Patterns and Processes in Ferns

J. D. LOVIS

Department of Plant Sciences, University of Leeds, England

I. INTRODUCTION*

In the earlier decades of this century research into the phylogeny and classification of the ferns was dominated by two men, F. O. Bower and Carl Christensen.

Bower's massive trilogy (Bower, 1923, 1926, 1928) on the comparative morphology, phylogeny and classification of the ferns had a profound influence on the development of this field, and even today constitutes a most valuable source, having half a century later still not been superseded in many respects. There is no doubt, however, that Bower's second volume, which includes the eusporangiate and other arguably primitive ferns, is more impressive and has stood the test of time better than his treatment of the leptosporangiate ferns, wherein his concepts are now seen clearly to be erroneous in several important respects.

In his "Index Filicum" (the first part of which was published in 1905/06, to be followed by supplements in 1913, 1917 and 1934) Christensen provided an indispensable basic source for all work on fern taxonomy and nomenclature. The original Index and the first three supplements were compiled by Christensen alone and only more recently (Pichi-Sermolli *et al.*, 1965) has the Index been updated by preparation of a fourth supplement. The value of Christensen's Index is enhanced by his unrivalled accuracy and consistently sound taxonomic judgement. Christensen was, however, content to apply the classification of the leptosporangiate ferns provided by Diels (1899/1902), and not until near the close of his life did he publish a phylogenetic system of his own, as an essay in Verdoorn (1938).

The end of the second world war saw the appearance in rapid succession of three distinguished works and a subsequent upsurge of activity and interest in pteridological studies. Copeland's "Genera Filicum" (1947), a description of all the known genera of ferns, stands as an essential source, second only to "Index Filicum". However, a new classification of the ferns applied in this work by Copeland has failed to gain general acceptance, in spite of the obvious practical advantage of the complete attribution of genera to the various families recognized in the system, principally because the limits of some of the most important families were much too comprehensive and are patently unnatural. On the other hand the scheme of Holttum (1947), published at the same time, is especially noteworthy as the first attempt at a radical reclassification of the ferns, incorporating also a two-dimensional display of phylogenetic relationships, which gained any wide measure of acceptance.

Manton (1950) brilliantly exploited the application of a new cytological technique, the squash method for study of meiosis. This technical advance

* Authorities for names of species are given in section II, concerned with palaeobotany, but are omitted elsewhere except for a few instances where circumstances make their inclusion necessary.

provided a most powerful tool whose application during the past 25 years has created an important new line of evidence of relevance not only to phylogenetic problems but also for analysis of processes and patterns of recent evolution. Both of these topics constitute major sections of this review.

Of more recent developments in studies of fern evolution, there is space here to mention only four publications. Knowledge of fossil ferns has been well served by the appearance of the relevant part of Harris's monograph of the Yorkshire Jurassic flora (Harris, 1961) and the volume relating to ferns in Boureau (1970). Together, these provide a most significant clarification of the state of our knowledge of the fossil evidence relating to the origin of the modern ferns. A discussion of this topic constitutes the first main section of this review. Mention must be made of a conference on the phylogeny and classification of the ferns held at the Linnean Society in London in April 1972. The symposium volume subsequently published (Jermy et al., 1973) provides a valuable survey of recent research and the range of current opinion, and a most useful introduction to the variety of classificatory systems for ferns proposed in recent years. Finally, attention must also be drawn to the 1973 Amherst symposium on the evolution of systematic characters in ferns (Taylor and Mickel, 1974), and in particular to the valuable review by Wagner on spore characters in relation to fern phylogeny.

The diversity of systems of classification constitutes a real obstacle to an appreciation of discussions on fern phylogeny by anyone other than specialists in this field. A principal difficulty is caused by the ambiguity of the name Polypodiaceae. Until relatively recently, this name was generally applied so as to comprise all of the ferns included in Bower's third volume (1928), i.e. those he regarded as truly leptosporangiate ferns. They can well be called the "modern" ferns though to do so is to prejudge the question of their relative age. It may be best to refer to them collectively as the "polypodiaceous" ferns, within quotation marks to avoid confusion with the term Polypodiaceae now restricted to a relatively small group. This terminological difficulty arises because it has been generally agreed, at least since Bower's work, that the "polypodiaceous" ferns are of polyphyletic origin. Different major groups have arisen from quite distinct points in the more ancient ferns. Of these the most numerous and diverse is referred to here as the dennstaedtiaceous or dennstaedtioid radiation, and comprises the majority of those ferns which bear indusiate superficial sori. Other major radiations are: (1) the adiantaceous radiation, ferns characterized by marginal or submarginal sori, and including Adiantum, the cheilanthoid and gymnogrammoid ferns, and, most probably, also Pteris; and (2) the poly-podiaceous radiation, here using the term polypodiaceous in a restricted sense, including only ferns characterized by naked superficial sori. Since modern classifications attempt to be natural, it follows that there can be no formal modern term equivalent to Polypodiaceae in its former sense.

For reasons which will become manifest later, I have found it necessary to provide yet another classification in this paper. In order to make it possible for readers other than pteridologists to follow the phylogenetic arguments involved without recourse to the general literature, I have included a synopsis of the classification of Christensen (1938), which is the most accurate of the systems using the term Polypodiaceae in its old broad sense, set alongside the framework of my own system for ready comparison (Table I).

Finally, it will be noted that the Psilotales (*Psilotum* and *Tmesipteris*) are treated here as ferns. This view has been advocated at length by Bierhorst (1968a,b, 1969, 1971, 1973), who has produced a formidable compilation of detailed morphological evidence in support of his contention that these plants are closely related to *Stromatopteris*, a monotypic genus related to Gleichenaceae. Bierhorst's arguments have received scant consideration (e.g. Sporne, 1970, p. 49; Pichi-Sermolli, 1973, p. 27; Foster and Gifford, 1974, p. 117). There is unfortunately not space enough in this review to discuss this hypothesis at the length it deserves. In any case, the concept is so iconoclastic that it may be that a new generation of pteridologists must arise before it can receive a truly objective evaluation. Certainly, Bierhorst's proposal that the psilotean genera be transferred not merely to the Filicopsida but directly to the Filicales is too radical for me to accept in its entirety at the present time.

However, the evidence assembled by Bierhorst does constitute a strong case indicating not merely that the Psilotales are closer to the ferns than are either group to the Sphenopsida or Lycopsida but also that a real phylogenetic affinity may exist between them and certain primitive ferns, e.g. *Stromatopteris*. Accordingly, the Psilotales are included here as a separate order of ferns.

II. FOSSIL RECORD OF "POLYPODIACEOUS" FERNS

Seward once wrote in these terms: "From Jurassic rocks in various parts of the world numerous fossils have been described under the generic names *Aspidium, Asplenium, Davallia, Polypodium* and *Pteris*. In the great majority of cases such records leave much to be desired from the point of view of students who appreciate the dangers of relying on external similarity between vegetative organs, and on resemblances founded on obscure impressions of sori" (Seward, 1910, p. 377).

Nearly a quarter of a century later Seward could still write, "We know very little of the early history of the Polypodiaceae. The number of well-preserved examples in Mesozoic floras is insignificant" (Seward, 1933, p. 434). Referring specifically to early Cretaceous ferns, he wrote, "The few examples referred to *Onychiopsis, Adiantites*[*] and other genera implying relationship with the

---

* Records of *Adiantites* need to be evaluated with particular circumspection, since this is a form-genus to which are attributed Carboniferous and Permian fossils which are not even ferns at all, but pteridosperms.

TABLE I

Comparison between the classification of the "polypodiaceous" ferns by Christensen (1938) and that proposed in the present paper. The systematic order in which Christensen presented his subfamilies is retained but in order to clarity the comparison the elements of my system are shuffled out of their systematic sequence.

Christensen, 1938 (POLYPODIACEAE)		Lovis, 1977	
Subfamily	Subfamily	Subfamily	Family
DENNSTAEDTIOIDEAE	Dennstaedtiae / Hypolepideae	Dennstaedtioideae / Hypolepidoideae	DENNSTAEDTIACEAE
LINDSAYOIDEAE / DAVALLIOIDEAE / OLEANDROIDEAE		Hypolepidoideae / Lindsaeoideae / Davallioideae / Oleandroideae	DAVALLIACEAE
PTERIDOIDEAE	Chaetopterides / Leptopterides	Pteridoideae	
GYMNOGRAMMOIDEAE / VITTARIOIDEAE		Adiantoideae / Vittarioideae	ADIANTACEAE (and PARKERIACEAE)
ONOCLEOIDEAE / BLECHNOIDEAE		Onocleoideae / Athyrioideae	BLECHNACEAE / ASPLENIACEAE
ASPLENIOIDEAE	Asplenieae / Athyrieae	Athyrioideae	
WOODSIOIDEAE		Dryopteridoideae	DRYOPTERIDACEAE
DRYOPTERIDOIDEAE	Dryopterideae / achrostichoid derivatives / Thelypterideae	Tectarioideae	THELYPTERIDACEAE
DIPTERIDIOIDEAE			DIPTERIDACEAE / CHEIROPLEURIACEAE
POLYPODIOIDEAE			POLYPODIACEAE / GRAMMITIDACEAE
ELAPHOGLOSSOIDEAE			LOMARIOPSIDACEAE

Polypodiaceae . . . are for the most part too imperfectly known to be of much value as trustworthy records" (Seward, 1933, p. 394). The only example amongst Mesozoic records which Seward discusses further here is *Onychiopsis psilotoides* (Stokes and Webb) Oishi, a characteristic Wealden (Lower Cretaceous) fern, ". . . so called because of the close resemblance of the fronds, both fertile and sterile, to those of some living species of *Onychium* especially the Japanese species *Onychium japonicum*. Though no well-preserved sporangia have been discovered, it is very probable that *Onychiopsis* is a representative of the Polypodiaceae" (Seward, 1933, p. 387).

In contrast to Seward's cautious and critical approach, Emberger (1944, p. 254), in his treatise on plant fossils and their equivalents amongst living plants, simply states that *Davallia* (*D. saportana* Racib.) is known since the Lias period, *Adiantum* and *Polypodium* "connu depuis le Jurassique", *Asplenium* (*A. dicksonianum* Heer) since the Lower Cretaceous, and *Onoclea* (*O.* cf. *sensibilis* L.) since the Upper Cretaceous. Later on, in discussing the development of floras during geological time, he gives the bald statement, with respect to the Jurassic floras, that "Les Polypodiacées commencent (*Davallia, Adiantum, Polypodium*)". In a completely reset and very substantially augmented and rewritten second edition (Emberger, 1968) these statements are repeated (pp. 352, 354, 356 and 368), again without qualification or reservation, with the sole exception that with respect to *Polypodium oregonense* Font. Emberger writes that "au sens strict, il n'ya à signaler que le genre *Polypodium* L., du Jurassique". With regard to this fossil Seward had long since (Seward, 1910, p. 377) pointed out that although Fontaine (in Ward *et al.*, 1905, p. 64) stated that "the fructification seems near enough to that of *Polypodium* to justify the placing of the plant in that genus", the fact that sporangia had not been found rendered such an attribution unacceptable. In fact, the illustrations of Fontaine's specimens (op. cit., pl. X, figs. 1, 2 and 4 = fertile examples) show that the whole aspect of this fossil is utterly unlike *Polypodium* or for that matter any known genus of Polypodiaceae (*sensu stricto*, see p. 231), and is more consistent with Cyatheaceae, though without proper study this suggestion has scarcely much more merit than Fontaine's determination. Fontaine's reference to some specimens being preserved sorus downwards suggests that reinvestigation by modern transfer methods might be rewarding.

As noted above, Emberger cites without qualification *Asplenium dicksonianum* Heer as indicative of the presence of the living genus *Asplenium* in the Lower Cretaceous, in spite of the fact that many years previously, in discussing the status of this taxon, Seward (1926, p. 83) had stated that "there is no justification for the use of the generic name *Asplenium*". It is quite clear that Heer referred fossils to living genera with much greater freedom than is now considered acceptable. He also had a narrow view of specific limits; no less than four more species of *Asplenium* named by Heer (*A. johnstrupii, A. nordenskioldii, A. pringelianum* and *A. puilaskense*) are referred by Seward, along with

A. dicksonianum, to the form genus *Sphenopteris* under the name *S. psilotoides* Halle, a name Seward preferred "to names implying more knowledge of affinity than we at present possess", in view of the fact that the Greenland material he was then considering was entirely sterile. Specimens found elsewhere with sterile portions of indistinguishable morphology but also bearing fertile axes have been referred to *Onychiopsis*, a genus so named because in its morphology it is close to the modern genus *Onychium*, a member of the Adiantaceae, not the Aspleniaceae.

In case it is assumed that Emberger's review is an isolated modern example of an uncritical approach to Heer's species, it may be noted that *Asplenium dicksonianum* Heer is also cited by Gnauck and Collett (in Harland *et al.*, 1967) as the first fossil record of this genus, their only qualification being the comment that Berry (1919) doubts that all specimens attributed to this name represent the same species.

Furthermore, no less an authority than Arnold has written, though without further elaboration: "Many of the unclassified ferns of the Mesozoic are undoubtedly polypodiaceous. The literature, however, does contain references to many genera usually placed in the Polypodiaceae, a few being *Davallia* (Early Jurassic), *Adiantum* (Jurassic), and *Asplenium* and *Onoclea* (Cretaceous). *Blechnum, Aspidium, Dryopteris, Pteris, Woodsia* and *Woodwardia* have all been reported from the Tertiary. Our knowledge of the fossil occurrences of these genera is far from exact because most of the determinations were based upon vegetative foliage" (Arnold, 1964, p. 62).

It is therefore not surprising that there is considerable confusion amongst pteridologists, at least those primarily concerned with living plants, regarding the true state of our knowledge of the fossil history of "polypodiaceous" ferns.

A principal difficulty in the evaluation of fossil material lies in the paucity of characteristics diagnostic of a "polypodiaceous" fern, the only features universal to this grouping being the small leptosporangiate sporangium, containing only 64 spores, and the vertical annulus. Certain other features, such as monolete (= monolaesurate) spores, and the presence of a perispore, are restricted to "polypodiaceous" ferns, but are not present in all groups.

However, our understanding has been greatly enhanced by a catalogue recently published by Andrews and Boureau (in Boureau, 1970) and the volume devoted to the pteridophytes produced by Harris (1961) in his series on the flora of the Yorkshire Jurassic.

A. MESOZOIC ERA

1. Yorkshire Jurassic flora

Ferns are "so abundant in the Yorkshire Jurassic flora that they may well have formed the dominant herbs on land" (Harris, 1961, p. 71). It is generally true that ferns are prominent in the Mesozoic until the Lower Cretaceous, whereafter they diminish in importance, no doubt in direct consequence of the diversification and increasing abundance of angiosperms in the latter half of the Cretaceous period. The Yorkshire Jurassic fern flora is not the most abundant known in terms of numbers of species, but this statement requires some qualification, since, as Harris says, "It seems that imperfect knowledge tends to multiply fern species." Thus in the case of the Yorkshire Jurassic, Phillips (1875) recognized no fewer than 44 ferns, a number reduced with due reason by Seward (1900) to 15. An additional 15 species recorded by Harris all constitute new discoveries. Whatever its relative size, there is no doubt that the Yorkshire Jurassic fern flora constitutes the best known and the most critically studied fern flora of this period, and it is appropriate to survey its content here, since to do so will provide a necessary perspective to this enquiry.

Two families are particularly prominent both in numbers of species and in numbers of individuals; these are the Osmundaceae and the Dicksoniaceae. The Osmundaceae is represented by two genera; one genus, *Todites*, with sporangia covering the lower surface of the pinnule lamina in a manner comparable to the living *Todea*, is abundant while the other, *Osmundopsis*, with fertile segments indistinguishable from those of a slender form of modern *Osmunda*, is rare. With respect to the Dicksoniaceae (treated by Harris as distinct from the other great family of tree-ferns, the Cyatheaceae) the great majority of species and individual specimens belong to the genus *Coniopteris*, which, together with the less frequent *Kylikipteris* and *Eboracia*, is placed by Harris in the subfamily Thyrsopterideae.* This group has only one living representative, the small tree-fern *Thyrsopteris elegans* Kunze, which today possesses a bizarre relict distribution, being confined to Juan Fernandez, remote in the south Pacific. Harris refers two more species of dicksonioid ferns to the genus *Dicksonia* itself (*D. mariopteris* Wilson and Yates, and *D. kendallii* Harris), the only ferns in this flora, other than *Marattia anglica*, which are referred to a living genus. Harris explains that the generic name *Dicksonia* "is used here for fossils in which the sorus is broad and protected by an indusium of two equal, robust valves as in *D. antarctica*" (Harris, 1961, p. 176).

Apart from the Osmundaceae, two other families of very ancient fossil history, extending back into the Palaeozoic, are represented, namely Marattiaceae and Schizaeaceae, although the equally old Gleicheniaceae appear, rather curiously, to be absent from these Jurassic rocks. Only one representative of the Marattiaceae is described by Harris, *Marattia anglica* (Thomas) Harris, but a second marattiaceous fern, *Angiopteris neglecta* van Cittert, has been discovered in the Yorkshire Jurassic more recently (van Cittert, 1966; Hill and van Kojninenberg van Cittert, 1973). The Schizaeaceae may also be represented by two genera. The attribution of *Klukia* to the Schizaeaceae is not in doubt, since

---

* As interpreted by Harris, this subfamily is not synonymous with subfamily Thyrsopteridoideae of Holttum and Sen (1961), since *Culcita* is excluded.

it clearly possesses schizaeaceous sporangia with the very characteristic and conspicuous apical annulus. The other genus, *Stachypteris*, present as *S. spicans* Pomel, is a rare and inadequately known fossil. The morphology of the characteristic fertile spikes is clearly suggestive of the living genus *Lygodium*, and the spores are very similar to those of *Klukia exilis* (Phillips) Raciborski, but although the details of the sporangia have not yet been clearly elucidated, what is known of the annulus is not obviously consistent with the Schizaeaceae. It has been suggested that the affinities of *Stachypteris* are with a very isolated extant family of obscure relationship, the Loxsomaceae (Bower, 1926, pp. 258-259), but Harris, referring to the sporangia, states: "It is certain, however, that their arrangement is as in *Lygodium* and has nothing to do with a Loxsomaceous sorus" (Harris, 1961, p. 134). The affinities of *Stachypteris* remain a subject for controversy and require further study.

Two more families, not known earlier than the Trias, but particularly characteristic of Jurassic rocks, though both still surviving today in the form of a very few representatives in the Indo-Malaysian region, are the Dipteridaceae and Matoniaceae. The Dipteridaceae are represented in the Yorkshire Jurassic by three genera, *Dictyophyllum, Clathropteris* and *Haussmannia*, whose affinity to the living genus *Dipteris* is evident. In the case of *H. dichotoma* Dunker the resemblance to narrow-leaved species of *Dipteris*, such as *D. quinquefurcata* Baker, is striking. Matoniaceae are represented in the Yorkshire Jurassic by two genera, *Phlebopteris* and *Matonidium*. Their affinity to the living genera *Matonia* and *Phanerosorus* cannot be doubted, and in this case the resemblance between *Matonidium goeppertii* (Ettingshausen) Schenk and the extant *Matonia pectinata* R. Br. is very close indeed.

2. Aspidistes

Only one more species in the Yorkshire Jurassic flora remains to be considered, but this fossil, *Aspidistes thomasii* Harris, is one of particular interest and importance in the present context, because it is the only one attributed by Harris, albeit with some reservation, to a polypodiaceous group, the family [*sic*] Aspideae. Virtually all of the known specimens of *A. thomasii* are from the famous Gristhorpe Bed in Cayton Bay and were collected by H. Hamshaw Thomas, although not studied by him, and first described by Harris (1961, pp. 181-186). The plane of cleavage of the fossils is such that the sori can be studied only by the transfer method. It is the sori which give this fossil its peculiar interest and importance, since they are of dryopteroid type, complete with a persistent indusium approximately intermediate in form between the peltate indusium of *Polystichum setiferum* (Forskål) Woynar and the heart-shaped indusium of *Dryopteris filix-mas* (L.) Schott. The characteristics of the sporangia are compatible with a "polypodiaceous" fern, these being small (*c.* 200 μm) in comparison with other Mesozoic ferns, and bearing an approximately vertical annulus (*c.* 12 cells) of leptosporangiate form. There is also some evidence that the sporangia ripened at different times. Spore counts suggest a spore number per sporangium of 48, which is less than the number (64) characteristic of "polypodiaceous" ferns. The form of the frond fragments, and the shape of the pinnules, is compatible with the range of modern *Dryopteris*, save only that the mode of branching is katadromic, and not anadromic (see Harris, 1961, p. 145, fig. 50: H-K) although, as Harris suggests, this is probably not important. The spore of *Aspidistes* is, however, not consistent with *Dryopteris*, being triangular in shape and devoid of a perispore, whereas *Dryopteris* and its relatives have wedge-shaped spores, ovoid to reniform in side view, monolete, and possessing a perispore.*

As Harris writes, "The occurrence of an aspidioid sorus as early as the Middle Jurassic is disturbing." *Aspidistes thomasii* thus presents a dilemma, which Harris states succinctly: "It may be that these 'advanced' ferns had already evolved but were living in the background (as is often held for the angiosperms). . . . It may also be that *Aspidistes* is not a member of the Aspideae (as is indeed suggested by its spore) but merely an interesting and early imitation from another stock" (Harris, 1961, p. 182).

Harris draws attention to a distinctive feature of *Aspidistes thomasii*: the lower surface of the lamina bears "numerous scattered, apparently sessile, unicellular glands" (p. 182). He states (p. 186) that he has seen similar glands on living "polypodiaceous" ferns, not only on "the 'golden fern' *Cheilanthes argentea* var.",† which clearly has no close affinity with dryopteroid ferns, but also, thanks to the assistance of R. E. Holttum, in "two less familiar species of the Dryopteridoideae", one of which is *Thelypteris* (*Coryphopteris*) *viscosus* (J.Sm.) Ching. It would seem that at this time Harris may not have appreciated the full possible significance of this observation, since although these same two species are included in the Dryopteridoideae *sensu* Christensen (1938), as thelypteroids they are excluded from Dryopteridoideae *sensu* Holttum (1947, 1949).

A crucial point is Holttum's belief that the *Dryopteris* series of genera is fundamentally distinct from the *Thelypteris* series and is of quite separate evolutionary origin. The distinction between these two series of ferns had already been recognized by Christensen (1938) but Holttum (1947, 1949) placed much greater weight on their separateness, and removed the thelypteroid series into an independent family, the Thelypteridaceae. In this he had been anticipated by Ching (1940) but the significance of Ching's revision was lessened

---

* A species of *Dryopteris* (*D. indica* Sharma *nom. illegit., non D. indica* v. A. v. R. (1909)) has recently been described from Middle Jurassic rocks of the Rajmahal Hills, India (Sharma, 1971). The sori, which are minute (0.5 x 0.6 mm), are described as of reniform shape, but since no mention is made of indusium, sporangia or spores, the evidence is clearly insufficient to justify referral to the Dryopteridaceae, let alone attribution to a living genus.

† Harris probably means to refer here to *Pityrogramme argentea* var. *aurea* (Willd.) Domin, not to *Cheilanthes argentea* (Gmelin.) Kunze.

by his simultaneous fractionation of the Polypodiaceae of Engler and Prantl into some 33 families. As is discussed more fully below (p. 282), Holttum's treatment was subsequently emphatically vindicated by cytological studies (Manton, 1950; Manton and Sledge, 1954) which established a clear discontinuity in chromosome numbers, the true dryopteroids having a base number of $x = 40$ or 41, whereas in Thelypteridaceae the base number is more variable, with $x = 27\text{-}36$.

Holttum has commented briefly on the identity of *Aspidistes thomasii* on several occasions (Holttum, 1963, 1971, 1973b). In a nutshell (Holttum, 1963, p. 67), "*Aspidistes* [*thomasii*] looks like an early *Thelypteris*." This conclusion will be less than obvious to anyone whose experience of thelypteroid ferns is confined to the British flora, for our three representatives look very different, all possessing fronds which are only bipinnatifid, with a characteristic square insertion of the pinnae (oblique in *A. thomasii*) and moreover bearing sori which are either totally exindusiate, or else possess an exceedingly delicate and evanescent indusium.

However, Holttum is uniquely well qualified to make a judgement on this issue, having spent much time in recent years on a revision of the thelypteroid ferns found in the Old World, culminating in the recognition of 23 genera from amongst the components of *Thelypteris s.l.* Moreover, he is well placed to appreciate the practical problems of the palaeobotanist, having himself assisted Seward in collecting Cretaceous fossils in Greenland, and worked with him on Mesozoic or Tertiary floras from Rhodesia, Ceylon and Scotland (Seward and Holttum, 1921, 1922, 1924).

Some species of *Coryphopteris*, e.g. *C. viscosa* (Bak.) Holttum, have spherical glands comparable to those of *Aspidistes thomasii* and also match the fossil in possessing rather robust conspicuous and persistent indusia. Holttum believes that species of *Coryphopteris* "may represent the most primitive extant members of the family" (Holttum, 1973b, p. 177). The spores of this genus are monolete, however, as is generally true throughout the Thelypteridaceae, though there is very considerable variation within the family with respect to the morphology of the perispore (see the very elegant SEM study by Wood (1973)). However, in the present context, it is of the greatest possible interest that in one small group of species, to which Holttum has given the very appropriate name *Trigonospora*, the spores are trilete.

Keeping within the bounds of what can be determined from the existing fossil specimens, Harris described the frond of *Aspidistes thomasii* as "repeatedly pinnate". However, the mode of branching and the form of the leaflets resembles that seen in species of *Pseudophegopteris* and *Macrothelypteris*, two living genera including species with tripinnatifid or even tripinnate fronds. Even if the largest specimens of *A. thomasii* only represent single pinnae, they are well matched by pinnae of *Macrothelypteris polypodioides* (Hook.) Holttum and are also close to those of the better-known *M. torresiana* (Gaud.) Ching (Lovis, 1975). Discussing *Pseudophegopteris* and *Macrothelypteris* Holttum (1973b,

p. 177) has written, "These genera do not look primitive, and might have become bipinnate secondarily." However he himself (Holttum (1971, 1973b)) proposed a very persuasive hypothesis arguing a common origin for Thelypteridaceae and *Cyathea* which would indicate that a bipinnate frond is an ancestral character for the Thelypteridaceae, and in an earlier communication he stated specifically that "if one regards Thelypteridaceae as related to *Cyathea*, the genera *Pseudophegopteris* and *Macrothelypteris* may be regarded as showing the most primitive frond form in the family" (Holttum, 1969, p. 8). Even if the bipinnate character is not strictly primitive in these two genera, it is then nevertheless a reversion to the primitive condition.

The same hypothesis for the origin of the Thelypteridaceae also requires that the trilete spore must be a primitive character in the family, and Holttum (1971, p. 29) comments that *Trigonospora* "in other aspects is not evidently primitive . . .". Elsewhere he comments that ". . . the facts may indicate that evolution in the family has resulted in primitive characters being dispersed among different species-groups of today" (Holttum, 1971, p. 18). Such a seemingly irrational distribution of primitive characters amongst members of a group, wherein the one species may display a conservative condition for one character, but a highly evolved state for one or more other features, is a relatively familiar phenomenon. It is, for instance, well recognized in the flowering plants amongst members of the Ranales and Magnoliales. Species of *Drimys*, sect. *Tasmannia*, show as primitive a condition of the carpel as is known in any genus (Bailey and Swamy, 1951), but this is combined in the same flower with a relatively highly evolved stamen.

Returning to consideration of the identity of *Aspidistes thomasii*, it is now clear that there is no character of this fossil which cannot be matched somewhere amongst living members of the Thelypteridaceae, and there appears to be no good reason why it should not be accepted as an ancient representative of that family. If *A. thomasii* is accepted as a thelypteroid fern, then it is apparent that, since those characteristics it possesses which are unusual but not unknown amongst living thelypteroid ferns are exactly those which would be considered primitive for thelypteroids if they originated from a common stock with *Cyathea, A. thomasii* constitutes strong support for this hypothesis for the origin of the Thelypteridaceae.

If Holttum's view concerning (1) the essential separateness of the dryopteroid and thelypteroid lineages and (2) the derivation of the thelypteroids from a cyatheoid stock are basically correct, then one rather piquant conclusion follows from the acceptance of *Aspidistes thomasii* as a member of the Thelypteridaceae, since it then tells us nothing about the main (dennstaedtioid) radiation of "polypodiaceous" ferns. Whereas it is generally accepted that *Dryopteris* and its near relatives occupy an approximately central and advanced position in the main development of "polypodiaceous" ferns, according to both of Holttum's schemes of phylogeny (1949, 1973a) *Thelypteris* is far removed from the main source of these ferns (Figs 1, 2).

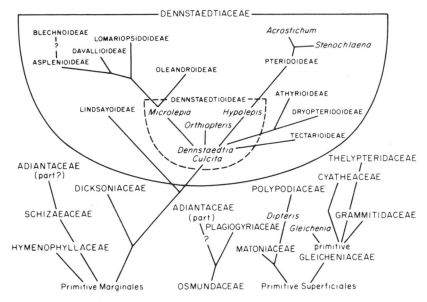

Fig. 1. Diagram showing the interrelations of the various groups of ferns according to the scheme of classification of Holttum (1949). (Reproduced from Holttum, op. cit., p. 275.)

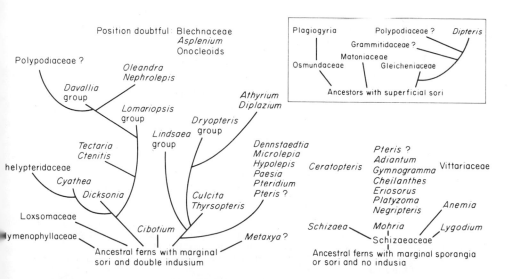

Fig. 2. "A tentative scheme showing possible interrelationships of groups of ferns, according to the author's present understanding; he has little first-hand knowledge of the genera placed above Schizaeaceae, and intends no implication as to relationships among them. Where generic names are entered, related genera are assumed to be included" (Holttum, 1973a, p. 5). (Reproduced from Holttum, loc. cit.)

Before leaving *Aspidistes thomasii*, two further points deserve note. Recognition of the true nature of this fossil enables us to add the Thelypteridaceae to the list of families present in the Yorkshire Jurassic flora. Also occurring in the same plant bed in which *A. thomasii* was found are Matoniaceae and Dipteridaceae, which have left undoubted descendants in the form of *Matonia* and *Dipteris* growing today, sometimes together, in the Malaysian region. It may not be entirely coincidental that some of the more primitive living thelypteroid genera, such as *Coryphopteris* and *Macrothelypteris* as well as *Trigonospora*, appear to have their present centre of distribution in the same region,* with some species growing in habitats very similar to those favoured by *Matonia* and *Dipteris.*

More recently, two more species of *Aspidistes* have been described, *A. beckerii* Lorch from Jurassic rocks in Israel (Lorch, 1967) and *A. sewardii* Watson from Wealden (Lower Cretaceous) rocks in England (Watson, 1969).

Only a few fragments of *Aspidistes beckerii* have been discovered, and neither sporangia nor spores have been preserved. Conspicuous sori with circular indusia are present. As Lorch recognizes, "*Aspidistes thomasii* . . . is different in the shape of its pinnules as well as in their venation" (Lorch, 1967, p. 143). The venation of *A. beckerii* shows distinctive anastomoses, resembling those seen in *Weichselia reticulata* (Stokes and Webb) Fontaine (see Alvin, 1971, pl. VI),† and is in general very unlike the venation of thelypteroid ferns. It is unfortunate that this fossil should have been placed in the same genus as *A. thomasii*, since these two species are almost certainly unrelated. The true affinity of *A. beckerii* is problematical, but it is surely not thelypteridaceous, and it need not necessarily be a "polypodiaceous" fern. On the contrary, the curious venation suggests to me the possibility that it may be a member of the Matoniaceae with highly reduced pinnules, somewhat after the manner of *Matonidium americanum* Berry (see Brown, 1950, p. 10, fig. 8) or even of some specimens of *M. goeppertii* (cf. Harris, 1961, p. 114, fig. 37F), but possessing a more developed indusium than is characteristic of *Matonidium*, more like that of the extant *Matonia pectinata*. One sorus near the base of the specimen illustrated by Lorch (pl. IVF) shows an impression with the appearance of a ring of six or seven large sporangia, after the fashion of *M. pectinata*, though this feature is not mentioned by Lorch.

---

* In this context it is interesting that a second species of *Todea*, *T. papuana* Hennipman (1968), is now known. *Todea* was formerly believed to be a monotypic genus with one species, *T. barbara* (L.) Moore, confined to South Africa, Australia and New Zealand. *Todea papuana* is appreciably closer at least in the form of its fertile pinnae than is *T. barbara* to their Jurassic antecedents *Todites* spp., e.g. *T. williamsonii* (Brongn.) Seward and *T. denticulatus* (Brongn.) Krasser. It would seem that the Malaysian region has been peculiarly favoured as a refuge for the survival of the more immediate descendants of some elements of the Jurassic fern flora.

† Although not necessarily of significance in the present context, it is nevertheless of interest that Alvin, who has made a detailed study of *Weichselia reticulata* (Alvin, 1968, 1971, 1975), concludes that this remarkable fern "may be envisaged as a highly specialized derivative of an early Matoniaceous stock" (Alvin, 1971, p. 26).

Aspidistes sewardii is known only as a single indifferently preserved pinna fragment, long misidentified as a specimen of *Weichselia mantellii* (Brongniart) Seward. Although substantially smaller, *A. sewardii* has otherwise an evident resemblance to *A. beckerii*, and is just as distinct from *A. thomasii*. In complete contrast to our knowledge of *A. beckerii*, nothing is known of the venation of *A. sewardii* but, on the other hand, information is available concerning the sporangia and spores. There is a "... thick peltate indusium covering about 12 sporangia arranged in a circle. Sporangium containing 32 spores (sporangial wall unknown). Spores trilete ..." (Watson, 1969, p. 236). Watson refers *A. sewardii* to the Polypodiaceae, and states (loc cit.) that "amongst living genera *A. sewardii* is most similar to *Polystichum* agreeing in the form of the indusium and the spore output". However, these features are equally consistent with *Matonia* and, moreover, not only the trilete spores (unknown in Dryopteridaceae) and the arrangement of the sporangia, but also the general morphology of the fossil (utterly unlike any species of *Polystichum* known to me), are all compatible with Matoniaceae. The indusium of *Matonia* is shed at maturity. That the unique specimen of *A. sewardii* was fossilized in an immature state is clearly indicated by the remarkably consistent spore counts obtained by Watson (nine of the 12 sporangia macerated yielded exactly 32 spores, the others 29-31 spores).

If *Aspidistes sewardii* and *A. beckerii* are indeed related, as seems probable, then the totality of the evidence known from these two fossil taxa, being complementary in character, clearly suggests very strongly that both are matoniaceous ferns.

3. Gleicheniopsis

In his recent brief survey of the evidence for the existence of "polypodiaceous" ferns in the Mesozoic, Harris (1973, pp. 42/43) states that he knows of no "... Mesozoic ferns with a strong claim to the Polypodiaceae ..." other than *Aspidistes* and *Gleicheniopsis*.

Gleicheniopsis was first described by Tutin (1932) from the Middle Cretaceous of Greenland. The fossils (which consist only of small fragments) had earlier been attributed to species of *Gleichenia* ($\equiv$ *Gleichenites*), but "... Tutin found, by transfers, that the sori had not just a few large sporangia as in *Gleichenia* but a great many small ones producing 32 spores. ... Tutin called it *Gleicheniopsis* and linked it with *Gleichenia* but I suppose he might easily have called it *Polypodium* (in the broad sense)" (Harris, 1973, p. 42).

Tutin in fact distinguished two species of *Gleicheniopsis, G. fecunda* (Heer) Tutin, with $5 < 7$ sori per pinnule and $20 < 40$ sporangia per sorus, and *G. sewardii* Tutin, with $2 < 4$ sori per pinnule and only $12 < 20$ sporangia per sorus. Probably both species produced 32 spores in each sporangium. Tutin had, I believe, three reasons for associating his *Gleicheniopsis* with the Gleicheniaceae. Firstly, one fragment shows the attachment of the bases of two consecutive pinnules to the rachis, displaying the crowded rectangular insertion characteristic of many gleicheniaceous ferns. Secondly, associated with the fragments of *Gleicheniopsis fecunda* "... are numerous dichotomising axes and a few circinate buds of a Gleicheniaceous type. These were not found in actual connexion with the pinnae, but it is highly probable that they are all parts of the same plant, particularly as no pieces of any other fern occur in association with the Gleicheniopsis" (Tutin, 1932, p. 504). It must be admitted that here Tutin is in some danger of falling victim to the classic trap that haunts palaeobotanists, namely the hazard involved in the assumption of organic relationship from suggestive evidence of association, without evidence of physical continuity. It should be noted that Tutin's study was based solely on museum material, without field experience of the actual localities. Furthermore, according to Seward (1926, p. 68) "*Gleichenites* is the most abundant genus in the Greenland Cretaceous vegetation", and at least one species, *G. porsildii* Seward, is known (Seward, op. cit., p. 77) from the locality concerned, Ritenbenk's coal-mine. The case based on association is therefore not strong.

However, Tutin's third reason seems to me to be more secure. Not all Greenland species of *Gleichenites* have a low number of sporangia per sorus. Thus, according to Tutin, *Gleichenites nordenskioldii* (Heer) Seward has 2 < 3 sporangia per sorus, *G. nitida* Harris 1 < 12 sporangia per sorus, but *G. porsildii* has 40 < 50 sporangia per sorus. In each case, the number of spores produced by each sporangium is large, of the order of 100 in the two first-named species and about 250 in *G. porsildii*. Tutin's argument is that *G. porsildii* occupies an intermediate position between *Gleichenites* and *Gleicheniopsis*. In other words, what we have here is not simply a contrast between one fern with sori containing a few sporangia each producing many spores, and another with sori with many sporangia producing relatively few spores, but a series: (1) few sporangia/many spores, (2) many sporangia/many spores, and (3) many sporangia/few spores.

Another species of *Gleicheniopsis*, *G. erlansonii* Miner, intermediate between *G. fecunda* and *G. sewardii*, has been described, also from Greenland, but from Upper Cretaceous rocks (Miner, 1934). The material on which this species is based is somewhat less fragmentary than that available to Tutin, and its gleicheniaceous character was not doubted by Miner.

Arnold goes to the extreme, with regard to *Gleicheniopsis*, of stating: "Its three species, however, form a series that intergrades with *Gleichenia*, so the living and fossil forms may merely represent an expanded genus" (Arnold, 1964, p. 61).

Nevertheless, the relationship between *Gleichenites* and *Gleicheniopsis* remains open to question. Even if the series does represent stages in a natural sequence, it is not clear whether we should interpret it as a recapitulation of the evolutionary events actually involved in the formation of, say, for argument's sake, a pro-*Grammitis* form from a gleicheniaceous ancestor, or a late and abortive essay by Cretaceous Gleichenaceae which mimics one of the steps involved in the establishment of a leptosporangiate condition. I use the expression "recapitulation" advisedly. The fragmentary nature of the fossils does not permit much more than an intuitive judgement, but my opinion is that although the sorus of *Gleicheniopsis* can be interpreted as approximating to the "polypodiaceous" condition, Tutin was correct in regarding the morphology of *Gleicheniopsis* as essentially gleicheniaceous. If this evaluation is correct, surely the Middle Cretaceous is too late a point in time for the occurrence of so early a stage in the evolution of the Grammitidaceae.

4. Other relevant Mesozoic fossils

If *Aspidistes thomasii*, although acceptable as a genuine "polypodiaceous" fern, nevertheless tells us nothing regarding the development of the main (dennstaed-tioid) line and radiation of "polypodiaceous" ferns, is there any other point in the Jurassic flora where information regarding the origin of the main line of evolution might be found? The only possible source is the group of ferns classified by Harris under the Dicksoniaceae, whose identity has excited considerable discussion because, as Harris (1961, p. 140) remarks, "Several species . . . included in the genus *Coniopteris* look very like such 'Polypodiaceous' genera as *Davallia* and *Odontosoria.*" It is, however, difficult to form confident judgements on the basis of comparison with living genera on account of the fragmentary nature of the fossils, and authorities are not in agreement on their interpretation. Holttum, commenting briefly on Harris's (1961) account of these fossils, makes no specific comment on the two species attributed by Harris to *Dicksonia*, but states that in his judgement "the fossil most like living *Dicksonia* is *Coniopteris hymenophylloides, C. murrayana* is perhaps more like *Culcita*" (Holttum, 1963, p. 65), whereas Harris writes (1961, p. 142) that ". . . the sori of *C. hymenophylloides* are like those of *Thyrsopteris*". Harris draws attention (1961, p. 158) to the fact that two fossils attributed to *Davallia* by Teixeira (1945, 1948), *D. alfeizeranensis* and *D. delgadoi*, from the Upper Jurassic and Lower Cretaceous of Portugal respectively, look just like *C. hymenophylloides* (Brongniart) Seward (a few specimens may be nearer to *C. burejensis* (Zalessky) Seward), and points out that *Davallia* could readily be distinguished from this group of fossils by its spores, which are bilateral in form, not triradiate as in *Coniopteris*. In fact, *C. hymenophylloides* has relatively massive sporangia, and is perhaps the least likely candidate amongst Yorkshire species of *Coniopteris* to represent a "polypodiaceous" fern. On the other hand, a better case can be made out with respect to other species. Thus Harris (1961, p. 142) pointed out that the sori of *C. murrayana* Brongniart and *C. burejensis* resemble those of lindsaeoid genera, e.g. *Lindsaea, Odontosoria* and *Schizoloma*, rather than those of *Thyrsopteris*. The spores of *Schizoloma* and of most species of *Lindsaea* are trilete, so no simple character, such as exists for *Davallia*, is available to distinguish them from *Coniopteris*. The character which could serve to distinguish these genera from fragments of dicksonioid ferns are only rather subtle features of the sporangium, indusium and soral organization, all of which will be difficult to evaluate in fossil material. Recently Harris has written, "I have realised that some of the ferns I ascribed (Harris, 1961) to the Mesozoic genus *Coniopteris* . . . should have been compared at least with *Dennstaedtia* and possibly other genera" (Harris, 1973, p. 43). It is of crucial importance that it is exactly these groups, the Lindsaeoideae and Dennstaedtioideae, which lie in Holttum's (1949) phylogenetic scheme at the very base of the main "poly-podiaceous" radiation, the stem of which is derived from the direction of the Dicksoniaceae. Their position on Holttum's more recent scheme (Holttum, 1973a) is one of lesser importance, but they are still derived from relatively early points on the *Culcita/Thyrsopteris* branch. In either scheme, it is exactly these groups which one would expect to be evolving at an early stage from a dicksonioid or thyrsopteroid source. The uncomfortable fact must be faced that the intermediate stages which must have existed before forms referable to the modern genera *Lindsaea* or *Dennstaedtia* evolved may well be in practice virtually indistinguishable in a fossil state from dicksonioid or thyrsopteroid ferns. Thus, though there is at present no possible justification for referring *Coniopteris*-type fossils to the modern genus *Davallia*, the possibility that some species of *Coniopteris* represent genuine pro-*Lindsaea* or pro-*Dennstaedtia* forms deserves critical study.

Andrews and Boureau (in Boureau, 1970) discuss, under the heading Polypodiaceae, a number of ferns which have not yet been mentioned here. We should, however, first draw attention to their treatment of Davalliaceae in isolation from the Polypodiaceae, which is otherwise interpreted by them in the traditional inclusive sense of Engler and Prantl, and of Christensen. Under *Davallia*, Teixeria's two species (see p. 245 above) are listed (though Harris's reservation regarding their identity is made clear) together with *D. niehhutzuensis* Tutida, described from the Lower Cretaceous of Manchuria. Four more Cretaceous species of *Davallia* are listed, all from the USSR, but these appear to be founded on spore records alone.

Andrews and Boureau have taken a very critical and exclusive approach to the content of their chapter on the Polypodiaceae. They may have unconsciously allowed their standard of scepticism to drop somewhat, when considering the Davalliaceae, below the level they would have adopted to these fossil records had they, like the great majority of authors, treated the Davalliaceae as "polypodia-ceous" ferns. This reflects a potential attitude of mind against which we should all be on our guard. It is appropriate to take a highly critical approach to any records of "polypodiaceous" genera from Jurassic or Cretaceous rocks. However, if we allow our scepticism to become so deeply entrenched that it takes on the character of dogma, we could be led to the firm conclusion that true "polypodiaceous" ferns did not occur in Jurassic time. This dogma may subsequently cause us to be disposed to reject other Jurassic candidates as not "polypodiaceous" without an adequately objective assessment. There is in this clearly the possibility of there developing a dangerous circularity of argument, i.e. this fossil is of Jurassic age therefore it cannot be truly "polypodiaceous", and, conversely, this genus (e.g. *Davallia* or *Lindsaea*) is a "polypodiaceous" fern therefore it will not be found fossil in Jurassic rocks and, ergo! the Jurassic records must be errors.

Returning to the survey of the fossil Polypodiaceae presented by Andrews and Boureau, it is interesting that for the Jurassic period they include only two of the species of *Aspidistes* already discussed, and a fossil described some considerable time ago, *Adiantites lindsayoides* Seward (1904), from Jurassic rocks of Victoria. Seward himself writes of the resemblance of the pinnules of this fossil to those of recent species of *Lindsaea*, which is indeed striking, and states that it "may be a true Polypodiaceous fern, but in this case, as in many similar instances, nothing is known of the structure of the sporangia" (Seward, 1910, p. 377). This fossil comes from a quite different part of the world from that whence most of the fossils discussed here originate and it is very desirable that this fossil be reinvestigated.

Turning now to Cretaceous records of "Polypodiaceae" other than davallioid ferns, Andrews and Boureau consider only three fossils worthy of inclusion, and one of these, *Dennstaedtia tschuktschorum* Kryshtofovich (1958), from the USSR, is mentioned only by name which, in the context of their publication, I interpret as indicating that insufficient information is available for them to be able to form an opinion regarding the identity of this fossil.

Australopteris coloradica (Brown) Tidwell, Rushforth and Reveal (1967) from Lower Cretaceous rocks in, as its specific name suggests, the mid-West of the United States, was first placed by Brown (1950), on the basis of sterile material only, in the living genus *Bolbitis* (Lomariopsidaceae). Tidwell *et al.* obtained abundant new material, which included fertile specimens (Tidwell *et al.*, 1967, pl. 3, figs 1 and 3), in which the distribution of the sori is entirely unlike that obtaining in *Bolbitis*. They quote the opinion of D. B. Lellinger (personal communication) that this fossil can be compared with species of the living genus *Drynaria*, which is of quite different affinity, being a member of the Polypodiaceae *sensu stricto* (i.e. *sensu* Holttum, 1947). In particular, the pinna shape and soral distribution is comparable to that of *D. rigidula* (Schwartz) Bebb, although the venation pattern of this species is different. Tidwell *et al.* recognized that this comparison should not be pushed too far, and therefore preferred to create a new genus to receive this unique fossil.

Lastly, there is *Onoclea inquirenda* Hollick (1906) from New York and Massachusetts. These fossils have a conceivable likeness to small and very shattered fragments of the fertile frond of *Onoclea*, a genus belonging to a small and isolated group, the Onocleoideae of Christensen, of problematical affinity. Holttum (1947, p. 125; 1973a, p. 5) is unable to place the relationship of these ferns, but in a recent monograph Lloyd (1971) regards them as related to the aspidiaceous (dryopteridaceous) ferns. The quality of these fossil fragments is very poor (unlike the sterile specimens of *Onoclea* from the Palaeocene of

Montana and Mull – see p. 251 below), and the determination has to be very
tentative. Although this fossil has been obtained from four different locations, it
seems that no trace of the sterile foliage (which is highly characteristic, and
could be confused only with that of *Woodwardia*) has been found at any one of
the sites.

Andrews and Boureau consider two more potentially relevant genera of
Mesozoic ferns under the category "*incertae sedis*". One of these, *Odontosorites*,
represented by *O. heerianus* (Yokoyama) Kobayashi and Yosida (1944), from
Jurassic rocks of north Manchuria, has a strong resemblance to the modern genus
Odontosoria (Lindsaeoideae) both in form of the pinnules and the indusia.
However, details of the sporangia have not been observed, and this fossil is
subject to the same reservations regarding its true nature as apply to *Adiantites
lindsayoides* and to those species of *Coniopteris* which approach lindsaeoid ferns
in their morphology.

The second is *Onychiopsis*, a genus widespread in Lower Cretaceous rocks, so
named on account of its close resemblance to the living *Onychium* (Adiantaceae), and long familiar as the species *Onychiopsis elongata* (Geyler) Yokoyama
and *O. psilotoides* (Stokes and Webb) Oishi (incl. *O. mantellii* (Brongniart)
Seward)). It seems that for Seward this genus constituted the most likely
candidate for a Jurassic "polypodiaceous" fern (see p. 234 above). Two more
species of *Onychiopsis* have been described in recent years, *O. tenuiloba* Lorch
(1967), from Jurassic rocks in Israel, and *O. paradoxus* Bose and Sukh Dev
(1961), from Gondwana rocks in India. Furthermore, some of Seward's original
English fertile material of *O. psilotoides* has been re-examined by Tattersall
(1961), who was able to observe a layer of elongated cells which she suggested
can be interpreted as a membranous indusium. She also extracted spore-masses
from the specimen, but could find no trace of sporangia, even though the surface
of the fertile pinnule has a lumpy appearance consistent with individual
sporangia of "polypodiaceous" size lying underneath a thin indusium. Bose and
Sukh Dev (1961; see also Surange, 1966) could also find no trace of sporangia
inside the fructification of *O. paradoxus* and felt compelled, in the present state
of our knowledge, to regard the fertile pinnule of their species as an "exannulate
monosporangiate fructification". According to this interpretation, the layer of
elongated cells observed by Tattersall would constitute the sporangium wall, and
not an indusium. If Bose and Sukh Dev's interpretation is correct, then the
resemblance to *Onychium* is purely superficial, and *Onychiopsis* can have no
relationship whatsoever to the Adiantaceae.

A different approach to the search for ancient "polypodiaceous" ferns is
provided by palynological studies. The attribution of Cretaceous spores to the
genus *Davallia* by Russian authors has already been mentioned. Of greater
interest is the description by Couper (1953) of *Leptolepidites verrucatus* from
Upper Jurassic rocks in New Zealand, a spore which resembles, in size and shape
and in its type of sculpture, that of *Leptolepia novae-zealandiae* (Col.) Kuhn

(Dennstaedtioideae), although the verrucate projections are considerably larger than those of the Recent species. Couper (1958) has also recorded from British Cretaceous rock the genus *Peromonolites*, which he considers can be attributed either to the Blechnaceae or the Aspleniaceae. In spite of such contributions as those of Harris (1955) and Erdtman and Sorsa (1971), the taxonomy of fern spores is still in a relatively undeveloped state.

Before leaving the subject of Mesozoic fossils, mention should be made of the existence of a number of poorly understood fossils in various floras which collectively suggest that the Cyatheoideae were more numerous and diverse in the later Mesozoic than is yet recognized. Some may have been thelypterida-ceous ferns. Apart from *Polypodium oregonense* Fontaine, mentioned above (p. 234), other fossils in this category known to me to exist in a fertile state include *Cladophlebis parva** Fontaine from the Potomac flora of Lower Cretaceous age (Berry, in Clark *et al.*, 1911; pl. XXXI, fig. 1), and *C. hukuiensis* Oishi from the Upper Jurassic of Japan (Oishi, 1940, pl. XVI, fig. 2).

B. TERTIARY PERIOD

The evaluation of records of "polypodiaceous" ferns from Tertiary rocks presents a different but not less difficult problem. From this period a great many fragmentary fern fossils are known, a high proportion of which have been attributed to living genera. In most cases it may be accepted that these fossils are "polypodiaceous" ferns, but beyond that their attributions are generally highly questionable, often being based on superficial study of rather inadequate specimens. Recently both Harris (1973) and Andrews (in Boureau, 1970) have written with decided emphasis on this theme. Thus Harris has written (1973, p. 43), "I believe the reason why few of these specimens have been studied in the manner of the best work on Mesozoic and Palaeozoic plants is a different tradition based on a different outlook for the more recent fossils. Tertiary floras are dominated by the leaves of dicotyledonous trees so they look familiar, but the older floras are full of plants which are plainly strange so nothing can be assumed. . . . Apparently, the Tertiary dicotyledon leaves belong in the main to living genera, so the problem is to spot the right genus which is done by a process of matching. . . . The Tertiary species of *Asplenium* and *Pteris* have been matched in this way, but on the whole glibly and with much less careful interest. Good Tertiary ferns deserve as close a study as the older ones, so such glib determinations deserve and get little attention."

Andrews (in Boureau, 1970, p. 361) is equally forthright in his justification of the critical and exclusive policy taken in determining which fossil records should qualify for inclusion in the chapter on Polypodiaceae, his approach being

---

* Berry believed that *Cladophlebis parva*, together with some other species in the Potomac flora "are all to be included in the subfamily Dryopterideae" (op. cit., p. 242) but the spores of *C. parva* are trilete, and the indusium and sporangia are as yet unknown.

not less sceptical and selective in relation to Tertiary fossils than for Mesozoic fossils. His comments are too extensive to quote at length here even in paraphrase but he points out that, when one considers the difficulty that botanists can experience in identifying with certainty some modern ferns even when supplied with good fertile specimens, it makes no sense to give a specific name to a minute fragment of a sterile imprint. The attribution of sterile debris to living genera such as *Asplenium, Dennstaedtia* and *Dryopteris* can only cause confusion.* He admits that certain important fossils have perhaps been omitted, but maintains that the great majority of those excluded are not of any significance and have only served to prejudice the cause of palaeobotany.

Andrews and Boureau include six fossils from Palaeocene rocks, all from the USA. *Allantodiopsis erosa* (Lesquereux) Knowlton and Maxon is so named for the remarkable resemblance of its fructification to that of the athyrioid fern *Diplaziopsis* (= *Allantodia* Wallich, *non* R.Br.) *javanica* (Blume) C.Chr., but its authors recognize that it cannot be associated with *D. javanica* because its venation is entirely different. The veins of *Diplaziopsis* form conspicuous anastomoses over the inframarginal half of the lamina, whereas those of the fossil remain free and more or less parallel throughout their length. The real affinity of *Allantodiopsis* remains uncertain. No details are known of the structure of the sorus and it would seem that the available evidence does not exclude the possibility that this fossil is a marattiaceous fern. As Andrews (op. cit., p. 361) points out, the synonymy of this fern illustrates very well how little reliance can be placed on the earliest studies of these fossils. Both Brown and Knowlton regard as identical with *Allantodiopsis erosa* several fossil taxa described by their original authors as species of *Asplenium, Osmunda* or *Pteris*!

It is interesting, in view of the great phylogenetic importance claimed by Holttum (1949) and others (e.g. Mickel, 1973) for *Dennstaedtia* and its immediate relatives, that a fossil attributed to this genus, *Dennstaedtia americana* Knowlton (1910), is included. The fossils clearly have a resemblance to modern *Dennstaedtia* (cf. Brown, 1962, pl. 6, figs 1-2, 5-7); Knowlton, 1930, pl. 2, figs 7-9 (as *D. crossiana* Knowlton); Knowlton, 1910, pl. 63, fig. 4, pl. 64, figs 3-5), but Knowlton himself, writing about *D. crossiana* Knowlton (justifiably treated by Brown as conspecific with *D. americana*), states that "it is referred to *Dennstaedtia* with considerable hesitation, for to distinguish between this genus and *Hymenophyllum* or *Davallia* is next to impossible with material so scanty and obscure" (Knowlton, 1930, p. 21). However, Arnold and Daugherty (1964) have studied excellently preserved mineralized rhizomes and petiole bases

---

* "L'attribution de ces débris stériles à des genres vivants tels que *Dennstaedtia, Asplenium, Aspidium, Dryopteris*, etc., ne peut entrainer que des confusions et des maltendus dans le receusement des fossiles. Si l'on considère les difficultés quel'on a à identifier certaines des espèces vivantes les plus communes lorsqu'on dispose de bons spécimens fertiles, il semble insensé d'attribuer un nom d'espèce a de miniscules fragments d'empreintes stériles."

from Upper Eocene strata whose anatomy approaches most closely to that of the Dennstaedtioideae, and which they therefore name *Dennstaedtiopsis aerenchymata*. Moreover, concerning this fossil Mickel (1973, p. 138) writes, "I have seen the type, and I am sure that, based on its anatomy and hairs, it certainly belongs near *Dennstaedtia* if not in it."

Dryopteris meeteetseana Brown (1962) has the position of its sori clearly preserved, and this fossil most probably does belong to the Dryopteridaceae. The general aspect of the fossil and in particular the dissection of the pinnules is not characteristic of the Thelypteridaceae, but is suggestive of *Ctenitis* or *Lastreopsis* rather than *Dryopteris*. It is therefore more likely that this form belongs to the Tectarioideae than to the Dryopteridoideae.

Onoclea hesperia Brown (1962) from Montana is exceptional amongst these fossils in that it is remarkably similar to living *Onoclea* and it is indeed doubtful whether the fossil can be separated from the modern species *O. sensibilis* L. *Onoclea* is also known fossil from rocks in the Old World, from Mull, as *O. hebraidica* Forbes (Gardner and Ettingshausen, 1879/1882, pl. XIII, figs 5, 6; Seward, 1910).

Salpichlaena anceps (Lesquereux) Knowlton (1930) was so called by Knowlton because of its close resemblance to *Salpichlaena volubilis* J.Sm., an otherwise monotypic Recent American blechnoid genus. Since no fertile specimens are known, attribution to this living genus is unjustified, notwithstanding the considerable authority of the opinion of W. R. Maxon (see Knowlton, 1930, pp. 27-28). Again, it is not clear from the available evidence whether a marattiaceous affinity can be excluded, and I am furthermore unconvinced that this fossil is truly distinct from *Allantodiopsis erosa*, which it clearly resembles very closely indeed, so much so that Brown (1962, p. 41) states that all but one of the earlier records cited in synonymy under *S. anceps* by Knowlton are indeed *Allantodiopsis erosa*.

However, there is no reason to doubt that the fossil named *Woodwardia arctica* (Heer) Brown (1962), also belonging to Blechnaceae, is correctly so named, in view of the fact that well-preserved fertile fragments (Brown, 1962, pl. 7, fig. 5) as well as sterile portions have been found. (Sterile specimens can be remarkably difficult to distinguish from *Onoclea*.)

As well as *Dennstaedtiopsis aerenchymata*, Arnold and Daugherty (1963) have described another mineralized fern from the Clarno formation (Upper Eocene) of Oregon, *Acrostichum preaureum*, of which petioles, fragments of rachis and portions of fertile fronds are all preserved. The anatomy of the rachis is particularly close to that of the Recent species *A. aureum* (Adiantaceae). This discovery adds confidence to the earlier determination of sterile fragments from English Eocene rocks as a species of *Acrostichum, A. lanzeanum* (Gardner and Ettingshausen) Seward (Gardner and Ettingshausen, 1879/1822, pl. I and II; Seward, 1910, pp. 350, 379; Chandler, 1962, pl. 5, figs 4 and 5).

Apart from a bare mention of two species placed in the fossil genus

Polypodites, the only other fossils included which are stated to be of Eocene age are two species attributed to *Adiantum*, *A. francisii* (Ball, 1931) and *A. anastomosum* (Brown, 1940). Of these, the first is only a fragment, and the material figured of the second is not adequate to justify the generic attribution.

It is noteworthy that certain genera (e.g. *Asplenium, Pteris*) most frequently quoted in the early literature have not so far been mentioned. In fact, *Pteris* is not included at all by Andrews and Boureau, though the analogous *Pteridium* appears as *P. calabazensis* (Dorf) Graham (1963) from Miocene rocks of the western USA, a fossil equivalent of the almost ubiquitous living fern *P. aquilinum.*

Andrews and Boureau do include *Asplenium alaskanum* Hollick (1936), from Alaska. The best specimens of this fossil are of substantial size, and the definition of the impressions is good. The fertile fragments bear linear sori (Hollick, 1936, pl. 4, figs 5 and 6), which Hollick states "are clearly asplenoid in character", but it is not certain in what exact sense he uses the term "asplenoid". It is unlikely that it would be possible to determine with certainty whether these sori are truly of asplenioid rather than athyrioid or diplazioid type. The Aspleniaceae and Athyrioideae are not always easily separated, even when studying pressed rather than fossil material. Indeed, in much of the early literature (e.g. Hooker and Baker, 1865/68) athyrioid ferns were classified in *Asplenium.* However, the general aspect of the fossil, in particular the broad midrib and the shape and dissection of the pinnae, is more typical of the Athyrioideae than the Aspleniaceae, and in my opinion *Asplenium alaskanum* is more likely to be an *Athyrium* or *Diplazium* rather than a true *Asplenium.* Furthermore, the age of this fossil is uncertain. Andrews and Boureau simply state that it is of Tertiary age. Hollick was confident, on floristic grounds, that the bulk of the Alaskan Tertiary flora is of Eocene age. However, *Asplenium alaskanum* is known from three sites (Yakutat Bay; Kuiu Island, Alexander Archipelago; Berg Lake, Controller Bay), but from only one of these (Berg Lake) is any considerable number of species (18) recorded by Hollick, and only two of these appear in the list of that part (approximately half) of the Alaskan Tertiary flora which he considers characteristic of the widely recognized "Arctic Miocene" flora (now generally accepted as of Eocene age). Fertile specimens of *A. alaskanum* are known only from Yakutat Bay. Only two of the eight species found at this locality belong to the so-called "Arctic Miocene" flora. This area was the subject of a detailed early survey by Tarr and Butler (1909), who wrote (p. 163) of the plant fossil bed series that "their position in the late Tertiary seems highly probable . . .", and elsewhere (p. 26) that they ". . . are tentatively assigned to Pliocene age". It seems probable therefore not merely that *Asplenium alaskanum* is more likely to be a member of the Athyrioideae than the Aspleniaceae, but also that this fossil is restricted to rocks of relatively late Tertiary age.

Finally, I should like to draw attention to the fact that the rich Palaeocene flora of western North America contains some other distinctive fern fossils which though not mentioned by Andrews and Boureau are nevertheless of evident interest, since they are already represented by good and potentially informative fertile material which surely deserves investigation by modern techniques. Most obvious of these is *Saccoloma gardnerii* (Lesquereux) Knowlton, a fossil which in sterile condition is not easily separated from *Allantodiopsis*, but which possesses a very distinctive fructification (Brown, 1962, pl. 5, fig. 19) superficially very similar to that of the living *Saccoloma elegans* Kaulfuss, but its structure has apparently not been investigated. Knowlton (1930, p. 27) writes of *Saccoloma* as "davallioid polypodiaceous ferns", but this genus is now usually regarded as a member of the Dennstaedtiaceae (*sensu stricto*; see Mickel, 1973).

C. CONCLUSION

Before we attempt to summarize what the fossil record can tell us of the origins of "polypodiaceous" ferns, we should again remind ourselves that the problem is a complex one. The "Polypodiaceae" *sensu* Engler and Prantl is not a natural group. Since at least the time of Bower, opinion has generally favoured the hypothesis that the "polypodiaceous" ferns are to a greater or lesser extent of polyphyletic origin. The problem is therefore not comparable with that of the origin of the angiosperms, which most authorities now believe to have been monophyletic. The universality of the characteristic embryo sac and the remarkable double fertilization mechanism amongst angiosperms appears to settle the argument (Parkin, 1923; Takhtajan, 1969, p. 6).

The problem of the angiosperms is not so acutely a question of when they arose, though indeed precisely when is still a substantial problem, but of from where? A remarkable transformation in the fossil floras of the world takes place about the Middle Cretaceous; in the Lower Cretaceous the angiosperms are still numerically a minor element in the fossil record, but in the Mid-Cretaceous the angiosperms come rushing on to the stage with many living families, including some supposedly primitive and others regarded as advanced, already recognizable (Chesters *et al.*, 1967; Takhtajan, 1969). By the Upper Cretaceous they have become the dominant element in the floras of the world, a position they have of course maintained to the present day. By contrast, at the same time the ferns become of lesser numerical importance in the fossil record, never again being as prominent as they were in the Jurassic and Lower Cretaceous. It is tempting to seek a relationship between the advent of the angiosperms and the transformation in the taxonomic structure of the fern floras, between the radiation and expansion of the angiosperms and the rise of "polypodiaceous" ferns. In other words, the development and radiation of "polypodiaceous" ferns occurred as a response to the enormous changes that colonization by angiosperms wrought on the ecological structure of the plant communities of the world.

To recapitulate our assessment of the evidence: very, very few confident "polypodiaceous" fossils are known from any part of the Mesozoic period. Undoubtedly the best example is *Aspidistes thomasii*, of Middle Jurassic age, which, following Holttum's lead, I believe to have been a member of the Thelypteridaceae. No known feature of the fossil is inconsistent with this identification. One must remember, however, that, according to Holttum, the Thelypteridaceae lie somewhat apart from the main line of "polypodiaceous" evolution. Other than *Aspidistes thomasii*, the Jurassic period affords only the difficult problem of the possibly lindsaeoid or dennstaedtioid nature of some species of *Coniopteris*, at present regarded as members of the Thyrsopteridioideae or Dicksonioideae. Technically this question will be extremely difficult to resolve, which is all the more frustrating since the evolution of dennstaedtiaceous stock from a source like *Coniopteris* would be an expected development in relation to contemporary concepts of fern phylogeny (e.g. Holttum, 1949, 1973a). The Cretaceous period has remarkably little to offer us. I have given reasons why I believe that *Gleicheniopsis* is probably no more than an interesting but abortive gleicheniaceous experiment, even though it could be interpreted as a possible pro-Grammitidaceae or pro-Polypodiaceae state. However, it does seem that *Australopteris coloradica* may well be close to *Drynaria*, in the Polypodiaceae (*sensu stricto*). Lastly, *Onoclea inquirenda* might after all indeed be an *Onoclea*, though the evidence is very slim. Even if all three of these fossils, *Gleicheniopsis*, *Australopteris* and *Onoclea inquirenda*, could correctly be attributed to "polypodiaceous" families it must be remembered that both Grammitidaceae and Polypodiaceae are most probably of independent origin apart from the main "polypodiaceous" stock (see p. 231) while the affinities of the Onocleoideae remain problematical. Since the Thelypteridaceae appear to represent another independent line of evolution, we are forced to the remarkable conclusion that there is as yet no clear evidence of the main (dennstaedtiaceous) radiation of "polypodiaceous" ferns anywhere in the Mesozoic period.

However, in spite of a paucity of reliable fossils, this situation evidently becomes radically changed early in the Tertiary period. In Palaeocene and Eocene rocks, as well as *Onoclea*, we have acceptable records of examples of all of Adiantaceae, Blechnaceae, Dennstaedtiaceae (*sensu stricto*) and Dryopteridaceae.

The crucial boundary for the appearance of modern angiosperm families lies between the Albian and Cenomanian stages of the Cretaceous period. Cronquist (1974, pp. 20-21) states: "The fossil evidence, especially from the pollen, that the basic diversification of the angiosperms occurred during rather than before the Cretaceous period (Doyle, 1969; Muller, 1970) is now so strong that any attempt to interpret any pre-Cretaceous angiosperm fossil in terms of modern families or genera must be based on very strong evidence in order to be credible." It would seem that the comparable threshold for the appearance of the main radiation of "polypodiaceous" ferns is later, between the Cretaceous and Palaeocene periods. It must surely be true that primitive angiosperms were in existence before the Cretaceous period, and the dennstaedtioid radiation well before the Palaeocene, although neither has yet been certainly detected in the fossil record at an earlier date.

It is, however, quite clear that, as Harris has put it, "In fact the Polypodiaceae came into force rather later than the angiosperms" (Harris, 1973, p. 42). Indeed, it appears that the angiosperm radiation and expansion, with the drastic changes this produced in the structure of the world's vegetation, involving the evolution of ecosystems more complex than any seen previously in the history of the earth, most probably provided the trigger or stimulus for the production of the dennstaedtiaceous radiation, and may well have promoted or accelerated the adiantaceous and polypodiaceous (*s.s.*) radiations. Whether or not such was the case, the appearance of the main "polypodiaceous" radiations does seem to constitute the most recent major innovation in the evolution of the world's flora.

III. DEVELOPMENT OF CYTOLOGICAL STUDIES ON FERNS

It is seldom that the development of an entire field of current research can be traced unequivocally and uniquely to a single publication, but such is the case with respect to fern cytology, which can fairly be said to have its effective origin in the appearance in 1950 of Irene Manton's now classic book "Problems of Cytology and Evolution in the Pteridophyta", a work which was particularly unusual for a publication of seminal importance in the breadth of its scope, the basic cytological facts regarding almost the entire British pteridophyta flora being set forth at one stroke.

Cytological research on ferns had of course been carried out in the earlier decades of this century by a number of workers, but a remarkably high proportion of this work was inaccurate, and some of it was wrong by a quite ludicrously wide margin. Nevertheless, not all of the earlier work was unsound. A valuable survey of this earlier phase is provided by Döpp (1938), whose review, since it itself belongs to the pre-1950 period, has an objectivity not possible today but is nevertheless recent enough to cover all developments prior to the second world war. It can now be seen that a few workers were spectacularly accurate in at least some of their studies, in spite of formidable technical difficulties, and it is no more than their due that attention should be drawn to their work here. Most remarkable perhaps is the achievement of Okabe (1929), who produced correct counts of $n = 52$ in *Psilotum triquetrum*, and $n = 104$ in a polyploid variety of the same species. It required the lapse of appreciable time and effort in the post-war period before this degree of accuracy with respect to this genus was equalled. Correct counts indicating the natural occurrence of polyploidy in ferns were achieved by Friebel (1933) with counts of $n = 38$ for *Anemia rotundifolia*, and $n = 76$ in *A. phyllitidis* and *A. obliqua*.

Credit for the first discovery of polyploidy in ferns should be given, however, to de Litardière (1921), who produced counts which although not quite accurate, were near enough to establish the reality of the phenomenon, both in *Pteris* (*P. cretica, n* = *c*.30 [correct would be 29], and *P. cretica* var. *ouvrardii, n* = *c*.60 [58]) and in *Adiantum* (*A. capillus-veneris, n* = *c*.31 [30] and *A. cuneatum, n* = *c*.60 [60]). Lastly, mention should be made of Döpp himself, who in the course of painstaking studies on European *Dryopteris*, produced (Döpp, 1932, t.II, 37) an elegant count of *n* = 82 in a prothallial cell of *D. dilatata*.

One development of overriding importance differentiated the work of Manton from that of her predecessors. This is the application of the acetocarmine squash method to the study of meiosis in spore mother cells. Previously, workers had available only conventional paraffin section methods, which at best are difficult to apply successfully to ferns. In contrast, the acetocarmine squash is simple, and can yield clear and accurate preparations of unsurpassable quality. So successful was Manton's method for the study of meiosis that it has not been significantly improved in the last 25 years, and it is still widely applied in its original form.

The sole important development in fern cytological technique since 1950 has been the development of a variety of squash methods (Meyer, 1952; Wagner, 1954; Chambers, 1955) for the study of fern chromosomes in somatic tissues. Previously, apart from the occasional amenable tapetal cell, somatic cells had been studied only by sectioning techniques. In the preparation of root-tip (or leaf primordia) squashes, methods chosen for fixation and staining are not materially different from those applied for meiotic studies, but two extra stages are usually included in the preparation schedule: (1) prior to fixation, there is a pre-treatment process designed to shrink the chromosomes and disrupt the spindle; and (2), subsequent to fixation, treatment with a macerating agent renders the tissue soft enough for it to be squashed and its constituent cells dispersed. It is probable that the majority of workers now employ snail stomach cytase extracts for maceration, a technique pioneered by Chambers (op. cit.) on *Blechnum*. This method was subsequently exploited and developed by Roy (Roy and Manton, 1965), who applied it in conjunction with meiotic studies, and in collaboration with several different colleagues produced not only a series of valuable studies on genera of critical phylogenetic importance (*Mohria*, Lovis and Roy, 1964; *Lygodium*, Roy and Manton, 1965; *Cystodium* and *Metaxya*, Roy and Holttum, 1965a, c), but also a study on *Lomariopsis* (Roy and Manton, 1966), discussed in detail below, p. 295, which appears to be of the greatest significance in relation to processes of cytological evolution in ferns, and illustrates very well the great potential of this technique. (In its early years, the snail cytase technique involved some rather gruesome preliminaries, not so much bloodthirsty as mucilaginous, but nowadays snails have their own abbatoir, and serviceable active preparations can be obtained from commercial laboratories.)

Subsequent to her survey of British pteridophytes, and after a brief dress rehearsal on the fern flora of Madeira, Manton undertook a study of the pteridophyte flora of Ceylon in which some 150 taxa of a total flora of approximately 250 species were successfully investigated (Manton and Sledge, 1954). This study is particularly notable as the first concentrated cytological survey in a specific geographical area to be carried out in the tropics for any group of plants. In due course, other surveys, less comprehensive in character but still involving samples of substantial size, were carried out by Manton on the fern floras of Malaya (Manton, 1954b) and West Africa (Manton, 1959).

The publication of the Ceylon study excited considerable interest among cytologists in India, with the consequence that numerous contributions concerning the cytology of Indian ferns have now been published,* by Bir, Ghatak, Loyal, Mehra, Panigrahi, Patnaik, Roy, Sinha, Verma and others, although the most substantial individual publication, involving a sample of 100 species, is a survey carried out in Trivandrum by Abraham *et al.* (1962). This remarkable and elegant paper is notable as the first example of the exploitation on a large scale of root-tip squashes in ferns, preparations being illustrated from 14 genera.

Further afield, the cytology of most of the ferns of New Zealand has been studied by Brownlie, while in the New World studies have been made by Britton in Canada and by Wagner and his associates in the United States (also, more recently, in Mexico). A particularly important survey, the largest single study yet produced, involving some 330 taxa or cytotypes, was executed by Walker (1966a, 1973b) on the extremely rich fern flora of Jamaica. This remains the only published study on a tropical American flora, although the same worker has conducted a similar study on Trinidad ferns. Walker, who has been particularly active in this field of enquiry, has also conducted as yet unfinished surveys in New Guinea and Celebes.

Commencing more recently, there has been substantial activity amongst Japanese workers, principally Kurita and Mitui (e.g. Kurita, 1967; Mitui, 1968, 1975), with the result that chromosome numbers have now been reported for a high proportion of the rich fern flora of Japan. A new and interesting development has been the execution of karyological studies on Japanese ferns by a group established in Hiroshima by Tatuno (cf. pp. 285, 288, 291, 315).

In this brief survey, only the most substantial individual studies have been cited, and elsewhere in this review only contributions of particular immediate relevance will be quoted. References to the general literature can be found in the catalogue of pteridophyte chromosome counts compiled by Chiarugi (1960), in two supplements produced by Fabbri (1963, 1965), and, for more recent items, in the annual indices of plant chromosome numbers sponsored by the IOPB (Ornduff, 1967, 1968; Moore, 1973, 1974) and in Walker (1973a).

In total, the consequence of all these contributions has been that the ferns, of

---

* Recently reviewed at length by Bir (1973).

whose cytology very little was confidently known before 1950, have now become, as far as chromosome numbers are concerned, one of the best-known major groups of plants. The accumulated information has provided much valuable evidence of relevance to problems of phylogeny and classification. This subject will be pursued in section IV below. Cytology has also resolved various minor taxonomic problems, which are not of consequence here, and has indicated the complexity of many species groups. Some of these have now been subjected to cytogenetic investigation, revealing details of evolutionary patterns and processes of speciation at work in these ferns. A discussion of investigations in this field forms another major section (VI) of this review.

IV. CYTOLOGY IN RELATION TO CLASSIFICATION AND PHYLOGENY

A. A REVISED CLASSIFICATION OF LIVING FERNS

The information necessary for evaluation of the contribution that cytology can make towards a true phylogenetic classification of the ferns is presented here in the form of a list showing the basic chromosome number or numbers determined for each genus. This list is included not only because it provides the facts essential for this discussion in a convenient form, but because no similar comprehensive list is at present available elsewhere. It is now nine years since the publication of the last supplement to Chiarugi's catalogue of chromosome numbers. Even with the aid of the annual indices produced under the auspices of the IOPT, it has not been easy to ensure that this list is comprehensive and up-to-date, and it is regrettably most unlikely that it is complete.

Levels of polyploidy known in a genus are also indicated, not merely because these are of interest for their own sake in a separate context, but because this information permits the reader to determine exactly where I have allowed myself the liberty of simple division arithmetic to arrive at a postulated base number. (As will become apparent below, this is a liberty which I believe should be exercised sparingly and with great caution.) Actual chromosome numbers are usually quoted only when those known for a genus present so bizarre a series that their interpretation in terms of base numbers is uncertain.

It has not been possible to adopt the system of classification of Copeland (1947) since, as is now generally recognized (cf. Pichi-Sermolli, 1973), his conception of the limits of both the Pteridaceae and Aspidiaceae (= Dryopteridaceae) were excessively broad and quite unnatural. In the course of preparation of this review I have been forced into production of a system of my own, since for no system subsequent to that of Copeland (other than the newly published scheme of Crabbe et al. (1975), see below) has the distribution of all known genera been indicated by its author, although Holttum (in Willis, 1966, 1973) has performed this task for the scheme of Pichi-Sermolli (1958). However, although my system incorporates elements of Pichi-Sermolli's scheme, in general I find his far too divisive to be acceptable. My conception of family limits is closer to that of Nayar (1970) or Alston (1956), but I differ from Nayar in various respects regarding the relative status or affinity of some groups. Alston made no attempt to distribute the families between orders. For convenience in discussion, and to avoid making what may ultimately prove to be substantial errors in attribution, I would have liked to retain the "polypodiaceous" ferns within a single order rather than disperse them between different orders, but the evidence that these ferns are of polyphyletic origin is so strong that to group them within one order is so unnatural as to be entirely unacceptable.*

What has emerged is unmistakably a cytologist's scheme, in that the primary criterion has been that the distribution of genera should make cytological sense, and where uncertainty exists the cytological evidence has usually been regarded as decisive. An example is *Hemidictyum marginatum*, a monotypic tropical American fern of ostensibly athyrioid morphology. Walker (1973a) reports the chromosome number to be $n = 31$, which is highly discordant for an athyrioid fern, and states that the evidence regarding its true affinity is indecisive. I have listed this genus under the Thelypteridaceae, where at least its chromosome number is consistent. I take some support for this decision from a comment by Copeland (1947, p. 151) that "*Hemidictyum* is more probably an aberrant relative, presumably a derivative of *Athyrium*". My feeling is that until the problem posed by *Hemidictyum* is resolved by further evidence, it is better placed where its cytology fits, rather than left where not only is its cytology aberrant, but to some extent also its morphology. (See Notes Added in Proof, 1, p. 415).

The scheme of Crabbe *et al.* (1975), which provides the first attempt at a generic sequence of pteridophytes since that of Copeland (1947), was published when this paper was in an advanced stage of preparation. Nevertheless, I would have abandoned my scheme and adopted theirs, had it not incorporated certain features which preclude its total acceptance. The onocleoid ferns disappear as a separate entity, being accorded neither family nor subfamily status, a decision which seems both premature and perverse, since in his detailed recent monographic study of this very distinctive group Lloyd is satisfied that these ferns form "a separate offshoot of the indusiate ferns" (Lloyd, 1971, p. 63), and treats them as an independent subfamily of "aspidiaceous" ferns.

Of much greater consequence is the decision to associate the asplenioid ferns, along with athyrioid, tectarioid, dryopteroid, lomariopsidoid and elaphoglossoid ferns as separate subfamilies within one great family which most unfortunately, presumably for purely nomenclatural reasons, Crabbe *et al.* are forced to call Aspleniaceae. Even if such a decision was not controversial, it would be unwelcome, since it places the name Aspleniaceae into much the same state of ambiguity as presently afflicts the name Polypodiaceae in common, as opposed to purely taxonomic, usage. As it is, the relationship of the Aspleniaceae (*sensu stricto*!) to other advanced groups of ferns still remains uncertain, though there

---

* In recent years, amongst pteridologists of international reputation, only Morton (1968, p. 156) has maintained that the "Polypodiaceae" (*sensu lato*) are a natural group of monophyletic origin.

are good cytological grounds for maintaining its independence. In consequence, the scheme of Crabbe *et al.* has not been adopted here, for the sufficient reason that it would increase, rather than decrease, the nomenclatural ambiguities present in the text.

However, in the interests of uniformity the generic sequence of Crabbe *et al.* has been followed inasfar as it is compatible with the concepts of phylogenetic relationships embraced in this review. Their opinion has been accepted regarding the position of such genera as *Gymnocarpium* and *Woodsia*, whose exact placement amongst the "aspidiaceous" ferns is a matter of controversy which study of their cytology does not resolve. On the other hand, a few genera, such as *Taenitis* and *Pleurosoriopsis,** whose placements by Crabbe *et al.* are inconsistent with the cytological evidence, have been removed to more appropriate positions.

Apart from the major disagreement indicated above, the areas of concordance are satisfyingly great, and include the conservative treatment accorded to the Adiantaceae, in which the troublesome subfamily Adiantioideae is preserved as a single unit, notwithstanding that there is a strong argument for granting separate subfamily status to both the cheilanthoid and gymnogrammoid ferns. The Dennstaedtiaceae is a notoriously difficult family to subdivide satisfactorily. Our interpretations of the subfamily Lindsaeoideae are identical (following Kramer, 1957, 1971), but whereas I have preferred to separate the hypolepidoid and dennstaedtioid ferns, Crabbe *et al.* group these genera (with the exception of *Monachosorum*) in a single subfamily Dennstaedtioideae. Again, though the argument for recognizing several families of tree-ferns rather than one is admittedly a strong one, I have preferred, partly on grounds of convenience in usage, to retain Holttum's broad concept of the Cyatheaceae, and to recognize his subdivisions of the family, save only that the three tribes recognized by Holttum (1963) in his subfamily Cyatheoideae are each raised here to subfamily status.

B. BASIC CHROMOSOME NUMBERS OF FERN GENERA

INDENTED NAMES: Crabbe *et al.* (1975) indent the names of a considerable number of genera whose acceptance can be regarded, for one of a variety of reasons (op. cit., p. 143), as optional. These are excluded from the present compilation unless they contain chromosome base numbers different from those

---

Pleurosoriopsis, with $n = 72$ (Mitui, 1970), is placed in Aspleniaceae, following Christensen (1938). Recently, on the basis of gametophyte characteristics, Masuyama (1975) suggested a relationship with Polypodiaceae, where its chromosome number would fit equally well. However, the morphology of the sporophyte appears alien to such an affinity. In a detailed study of *Pleurosoriopsis* (in which, with observations of 36 bivalents with 36 univalents and $2n = c.144$ in different collections of *P. makinoi*, they confirm that $x = 36$ in this genus), Kurita and Ikebe (1976) suggest the establishment of an independent family, Pleurosoriopsidaceae.

Taenitis is placed in Dennstaedtiaceae. Wherever this genus is placed, with $x = 22$ it must be a cytologically rather isolated aneuploid derivative. (See footnote to p. 306).

Synopsis of a Revised Classification of the Living Filicopsida

Class	Subclass	Order	Family	Subfamily
Filicopsida				
	Psilotidae	Psilotales	Psilotaceae	
			Tmesipteridaceae	
	Ophioglossidae	Ophioglossales	Ophioglossaceae	
	Marattidae	Marattiales	Marattiaceae	
	Osmundidae	Osmundales	Osmundaceae	
	Filicidae	Hymenophyllales	Hymenophyllaceae	
		Schizaeales	Schizaeaceae	
			Platyzomataceae	
			* Parkeriaceae	
		Adiantales	* Adiantaceae	* Adiantoideae
				* Vittarioideae
				* Pteridoideae
		Marsileales	Marsileaceae	

Continued

Synopsis of a Revised Classification of the Living Filicopsida–(continued)

Class	Subclass	Order	Family	Subfamily
		Gleicheniales	Stromatopteridaceae	
			Gleicheniaceae	
		Matoniales	Matoniaceae	
		Dipteridales	* Dipteridaceae	
			* Cheiropleuriaceae	
			* Polypodiaceae	* Drynarioideae
				* Platycerioideae
				* Microsorioideae
				* Pleopeltoideae
				* Polypodioideae
			* Grammitidaceae	
		Cyatheales	Cyatheaceae	Thyrsopteridoideae
				Dicksonioideae
				Cibotioideae
				Cyatheoideae
				Lophosorioideae
				Metaxyoideae
			* Thelypteridaceae	
			* Dennstaedtiaceae	* Dennstaedtioideae
				* Hypolepidoideae
				* Lindsaeoideae

* Aspleniaceae
* Dryopteridaceae
  * Onocleoideae
  * Athyrioideae
  * Tectarioideae
  * Dryopteridoideae

* Lomariopsidaceae
* Davalliaceae
  * Davallioideae
  * Oleandroideae

* Blechnaceae

Salviniales
  Salviniaceae
  * Azollaceae

Familia incertae sedis:
  Hymenophyllopsidaceae
  Loxsomaceae
  Plagiogyriaceae

* = Elements included in Polypodiaceae *sensu* Christensen.

known in the genus in which they would otherwise be submerged, or are genera accepted by Copeland (1947), in which case they are shown as synonyms. Elsewhere synonyms are not given except where the name of a genus recognized by Copeland has been changed.

- ● = Names different from those accepted by Copeland (1947), or genera first described since 1947.
- ○ = Name of section at present lacking a valid name at generic rank (Hymenophyllaceae only).
- [ ] = Aneuploid derivative: only thus indicated where the true base number is not in doubt.
- ⟨ ⟩ = Secondary base number derived as a dibasic amphidiploid.
- ( ) = Number requiring confirmation, either on grounds of cytological accuracy or taxonomic identity.
- * = Level of ploidy known to comprise or include apomictic taxa.
- M = Monotypic genus.

HYMENOPHYLLACEAE: Crabbe *et al.* (1975) follow Morton (1968) in arrangement and taxonomy of the Hymenophyllaceae, but accept his sections at generic rank. This treatment is adopted here, with the addition that Morton's subgeneric divisions are also indicated, since these to some degree appear to reflect cytological boundaries within the family. This has the consequence of imposing a system of double indentation on this part of the list.

THELYPTERIDACEAE: the genera recognized by Holttum (1971) in the Old World are adopted, but listed in the order given by Crabbe *et al.* (1975). Records for the New World are listed separately because in the absence of a modern monographic treatment, it is not yet possible to integrate the New World species with the generic framework recognized by Holttum for the Old World representatives.

NUMBER OF SPECIES COUNTED: it must be emphasized that these numbers are only an approximate guide since it was a task beyond my resources to check the generic placement and synonymy of every record. An element of overestimation is therefore inevitable. The principal purpose of these figures is to make it clear where a large number of counted species establishes with confidence the characteristic base number of a genus, in contrast to circumstances where only one or a very few records exists for a particular genus.

List of Genera & Chromosome Numbers (See Note Added in Proof, 2, p. 415).

Base numbers						Number of species	
Order Family	Subfamily	Genus	Subgenus Section	Levels of ploidy	Counted	Known
Psilotales							
Psilotaceae							
52		Psilotum		2,4,8	2	3	
Tmesipteridaceae							
52		Tmesipteris		4,8	6	13	
Ophioglossales							
Ophioglossaceae							
45, [46]		Botrychium		2,4,6	21	40	
94		Helminthostachys n = 94		2?	1	M	
120		Ophioglossum (incl. Rhizoglossum)		2,4,6 <8<10+	15	30<50	
Marattiales							
Marattiaceae							
40		Angiopteris		2,4	2	100?	
—		Archangiopteris		—	—	10	
—		●Protomarattia		—	—	M	
[39], 40		Marattia		2,4	3	60	
—		Macroglossum		—	—	2	
40		Danaea		4	3	30	
40		Christensenia		4	1	M	
Osmundales							
Osmundaceae							
22		Osmunda		2	8	10	
22		Todea		2	1	2	
22		Leptopteris		2	3	7	

Continued

List of Genera & Chromosome Numbers—(continued)

Order	Family	Base numbers (Subfamily)	Genus	Subgenus	Section	Levels of ploidy	Counted	Known
Hymenophyllales								
	Hymenophyllaceae							
			Hymenoglossum			—	—	M
			Serpyllopsis			—	—	M
			Rosenstockia			—	—	M
			Hymenophyllum					
		11,13,18,21,⟨31⟩		*Hymenophyllum*				
					Hymenophyllum			
					n = 11,13,18,21,22,31	2,4	6	25
					Buesia	—	—	5
		21,22,26,28			*Meringium*	2,4	11	75
		—			*Eupectinum*	—	—	3
		21			*Myriodon*	2	1	M
				Sphaerocionium	*Sphaerocionium*			
		36			(incl. *Leptocionium*)	2	9	50
		36			*Apteropteris*	2	1	M
		—			*Craspedophyllum*	—	—	2
		—			*Hemicyatheon*	—	—	M
		21,(27),28,36			*Mecodium* (inc. *Amphiterum*)	2,4	24	100
		36	*Cardiomanes*			2	1	M
			Trichomanes	*Trichomanes*				
		32			*Trichomanes* (non *Trichomanes* sensu Copeland)	4	1	1

○● *Lacosteopsis(=Vandenboschia,* excl. *T.scandens* L.)	2,3*,4	16	60	36
Crepidomanes	2,3*,4	13	30	36
Polyphlebium	2	1	M	36
● *Reediella (=Crepidopteris)*	2	3	10	36
Abrodictyum	2	1	2	36
Pleuromanes	2	1	3	36
Gonocormus	2*,3*,4	5	12~	36
Pachychaetum				
Selenodesmium (incl. *Macroglena*)	2,4	10	25+	33,36
Davalliopsis	2	1	M?	32
Cephalomanes	2,4	5	15	32
Callistopteris	2	1	5	36
Nesopteris	2	1	6	36
Didymoglossum				
Didymoglossum	2,4	3	20	34
Microgonium	2,4	6	20	34
● *Lecanolepis (=Lecanium)*	2	1	M	34
Achomanes (except. Feea et Homoeotes, =*Trichomanes* sensu Copeland)				
● *Achomanes*	4,8	6	30+	32,36
● *Neuromanes*	2	1	3	(c.13)
● *Odontomanes*	—		3	—
● *Lacostea*	2	2	4	32,36
○● *Trigonophyllum*	4,8	2	2	32
● *Homoeotes*	—		M	—
Feea	2	1	4	32
● *Ragatelus*	4	1	M	32
Acarpacrium	4	2	9	32
Microtrichomanes	2,4	3	10	36

Continued

List of Genera & Chromosome Numbers—(continued)

						Base numbers	Levels of ploidy	Number of species	
Order	Family	Subfamily	Genus	Subgenus	Section			Counted	Known
Schizaeales									
	Schizaeaceae								
			Schizaea (sens. strict.) n = 77, c.94,96,103, c.154, c.270			?	?	7	} 30
			● Actinostachys (Schizaea, pro parte) n = 325 ± 30, 350 < 370			?	?	1	
			Lygodium			28,29,30	2,4,6	11	40
			Anemia			38	2,3*,4,6	12	90
			Mohria			38	4	1	3
	Platyzomataceae					38			
			Platyzoma				2	1	M
	Parkeriaceae					39,(40)			
			Ceratopteris n = 39,40,77,78				2,4	3	4
Adiantales									
	Adiantaceae								
		Adiantoideae							
			Actiniopteris			29	2*,3*	2	5
			● Afropteris			—	—	—	2
			Ochropteris			—	—	—	M
			Anopteris			29	4	1	M
			Onychium			29	2,3*,4,6	7	7
			Cryptogramma			30	2,4	3	4
			Llavea			29,30	2	1	M
			Neurosoria			—	—	—	M
			Cheilanthes			29,30	2,3*,4,6	45	125+
					Adiantopsis	30	2	1	20
					Aleuritopteris	29,30	2,3*,4	3	15

Aspidotis	30	2,4	4	5
Cheiloplecton	—	—	—	M
Mildella	29	4	1	M
● *Notholaena*	[27],29,30	2,3*,4	13	<85
● *Sinopteris*	—	—	—	2
● *Negripteris*	—	—	—	M
Pellaea	29,30	2,3*,4*	16	80
Doryopteris	29,30	2,4,8	3	35
Ormopteris	—	—	—	M
Saffordia	29 or 30	2	1	M
Trachypteris	30	2	1	2
Anogramma	[26,27],29	2,4	5	7
Pityrogramma	29,30	2,4,8	8	40
Trismeria	29	4	1	M
Hemionitis (inc. *Gymnopteris*)	30	2,3*,4*	4	12
Bommeria	30	2,3*	5	5
Paraceterach	—	—	—	M
Pterozonium	—	—	—	4
Jamesonia n = 87	29	6	2	17
Eriosorus n = 87,174	29	6,12	6	35
● *Nephopteris*	—	—	—	M
Syngramma (incl. *Craspedodictyum*) n = 58,116	29	4,8	3	20
● *Austrogramme*	—	—	—	2
Coniogramme	30	2,4,6	10	20
Aspleniopsis n = c.58	29?	4	1	3
● *Rheopteris*	—	—	—	2
Cerosora	—	—	—	2
Adiantum	29,30	2*,3*,4, 6*,8,10	55	200
Vittarioideae				
Antrophyum incl. *Polytaenium* and *Scoliosorus*)	30	4,6,8	7	50

Continued

List of Genera & Chromosome Numbers—(continued)

Order	Family	Subfamily	Genus	Subgenus	Section	Base numbers	Levels of ploidy	Number of species Counted	Known
			Anetium (=*Pteridanetium*)			30	4	1	M
			Hecistopteris			—	—	—	M
			Ananthacorus			30	8	1	M
			Vittaria			30	4,8	7	50
			Monogramma			—	—	—	2
			Vaginularia			30	2	1	6
		Pteridoideae	*Pteris* (incl. *Hemipteris* and *Schizostege*)			29	2*,3*,4*,6,8,10?	85	250
			● *Copelandiopteris*			—	—	—	2
			Neurocallis			—	—	—	M
			Acrostichum			30	2	2	3
Marsileales	Marsileaceae		*Marsilea*			20	2	3	60
			Regnellidium			19	2	1	M
			Pilularia			10	2	1	6
Gleicheniales	Stromatopteridaceae		*Stromatopteris*			(c.39)	2	1	M
	Gleicheniaceae		*Gleichenia*			20,22	2	4	10
			Diplopterygium (=*Hicriopteris* sensu Copeland)			56	2?	2	20
			Sticherus (=*Gleichenia* sect. *Mertensia*)			34	2,4	9	100

	Taxon				
39	*Dicranopteris*	2,4	2	10
43,(44)	• *Acropterygium* (=*Dicranopteris*, pro parte)	2	1	M
Matoniales					
	Matoniaceae				
26	*Matonia*	2	1	M
—	*Phanerosorus*	—	—	2
Dipteridales					
	Dipteridaceae				
33	*Dipteris*	2,4	2	8
	Cheiropleuriaceae				
—	*Cheiropleuria*	—	—	M
	Polypodiaceae				
	Drynarioideae				
36,37	*Drynaria*	2	6	20
36	*Photinopteris*	2	1	M
36	*Merinthosorus*	2	1	2
36	*Aglaomorpha* (incl. *Drynariopsis* and *Holostachyum*)	2	1	6
36,37	*Pseudodrynaria*	2	1	M
—	*Thayeria*	—	—	M
	Platycerioideae				
37	*Platycerium*	2	4	17
36,37	*Pyrrosia*	2,4,6	15	100
—	• *Saxiglossum*	—	—	M
—	• *Drymoglossum* (=*Pteropsis*)	—	—	6
	Microsorioideae				
36,37	*Microsorum* (incl. *Phymatodes, Colysis* Dendroconche and Lecanopteris)	2,4,6	39	95
—	• *Podosorus*	—	—	M
—	*Diblemma*	—	—	M

Continued

List of Genera & Chromosome Numbers—(continued)

Base numbers							Number of species	
Order	Family	Subfamily	Genus	Subgenus	Section	Levels of ploidy	Counted	Known
36			Leptochilus (incl. Paraleptochilus and Dendroglossa)			2,4	2	8
—			Christiopteris			—	—	2
—			Pycnoloma			—	—	3
—			Grammatopteridium			—	—	2
—			Oleandropsis			—	—	M
—			Holcosorus			—	—	3
33,35,36			Crypsinus			2,4	10	40
—			Selliguea			—	—	5
36,37			Arthromeris			2,4	4	9
—			● Polypodiopteris (=Polypodiopsis)			—	—	3
[22,23,25,26],35,36,⟨47⟩		Pleopeltoideae	Pleopeltis (incl. Lepisorus and Microgramma)			2,4,6	31	60
—			● Solanopteris			—	—	M
—			Marginariopsis			—	—	M
37			● Neurodium (=Paltonium)			2	1	M
—			● Neolepisorus			—	—	5–6
36			Lemmaphyllum (incl. Weatherbya)			2	2	5
36			Drymotaenium			2	1	2
—			Paragramma			—	—	2
33,35			Belvisia			2,4	3	15
—			● Dicranoglossum (=Eschatogramme)			—	—	6
36			Neocheiropteris			2,4	2	3
37		Polypodioideae	Niphidium (incl. Pessopteris)			4	1	M
37			Campyloneurum			2,4	6	25
—			Dictymia			—	—	4

	Taxon			
37	*Phlebodium*	2,4	1	10
–	*Synammia*	–	1	M
35,36,37	*Polypodium* (incl. *Goniophlebium* and *Thylacopteris*)	2,3*,4,6	71	100
Grammitidaceae				
33,36,37	*Grammitis* (incl. *Xiphopteris* and *Ctenopteris*)	2,4	21	400
–	*Calymmodon*	–	–	25
	Acrosorus	2	3	5
37	*Amphoradenium*	4	1	6
36 or 37	*Prosaptia*			20
	Glyphotaenium			4
	Oreogrammitis			M
	Nematopteris			M
	Scleroglossum			6
	Cochlidium			7
	● *Hyalotricha*			M
35,36	*Loxogramme*	2,4	6	40
37	*Anarthropteris*	2	1	M
Cyatheales				
Cyatheaceae				
	Thyrsopteridoideae			
–	*Thyrsopteris*		–	M
66–68	*Culcita* sect. *Culcita*	2	1	M
58	,, sect. *Calochlaena*	2	1	4
	Dicksonioideae			
65	*Dicksonia*	2	5	30
56	*Cystodium*	2	1	M
	Cibotioideae			
68	*Cibotium*	2	2	10

Continued

List of Genera & Chromosome Numbers—(continued)

Order	Family	Base numbers	Subfamily	Genus	Subgenus	Section	Levels of ploidy	Counted	Known
			Cyatheoideae						
		69 (70)		Cyathea (incl. Trichopteris, Cnemidaria, Gymnosphaera and Schizocaena)			2	36	600
			Lophosorioideae						
		65		Lophosoria			2	1	M
			Metaxyoideae						
?		95 or 96		● Metaxya (=Amphidesmium)		$n = 95\text{-}96$, $2n = 190\text{-}192$	?	1	M
	Thelypteridaceae								

Old World: Genera of Holttum (1971), arrangement of Crabbe *et al.* (1975)

Base numbers	Genus	Levels of ploidy	Counted	Known
35	Thelypteris	2	1	4
29	Amauropelta	6	1	6
30	Phegopteris	2,3*,4	3	3
31	Pseudophegopteris	2,4,6	7	20
?	Cyclogramma $n = c.136$	8?	1	7 < 8
27,31,(c.36)	Parathelypteris[†]	2,4	5	10
32	Coryphopteris	2,4	2	30
31	Macrothelypteris	2,4,6	3	9
34	Oreopteris	2	3	3
(31),35,(36)	Metathelypteris[†]	2,4	4	12
36	Cyclosorus	2,4	3	3
36	Trigonospora	2,4	1	8
36	Pronephrium	2,4	8	60
36	Mesophlebion	4	2	18
36	Plesioneuron	2	1	39
36	Glaphyropteridopsis	2,4	1	4
36	Chingia	2	1	12+

Haplodictyum	—		—	2
Nannothelypteris	—		—	4
Stegnogramma	36	2,4,8	3	15
Sphaerostephanos	36	2,4	9	120+
Ampelopteris	36	2	1	M
Menisorus*	—		—	M
Pneumatopteris	35,36	2,4	10	76
Christella	36	2,4	11	50
Amphineuron	36	4	2	12+
New World: Genera of Copeland (1947)				
Lastrea	27,(28),29,30,31,32,34,35,36		30	
incl. Amauropelta	29	2,4,8	18	175
Glaphyropteris‡	36	2,4	1	3
Phegopteris	30	2,3*	2	2
Steiropteris	36	4	1	12
Cyclosorus	36	2,4	14	17
Goniopteris	36	2,3*?,4	13	70
Meniscium	36	2,4	1	12
Hemidictyum	31	2	1	M
Dennstaedtiaceae				
Dennstaedtioideae				
Dennstaedtia	30,31,32,33,34,46,47	2,4,6?	14	70
Microlepia (incl. Oenotrichia) $n = 43,86,88?,129,$	40?,42,43,c.44	2,4,6	11	45
$2n = 84,160$				

* See Holttum (1974, p. 154).   ‡ See Smith (1974, p. 89).

† Certain records, i.e. $n = 62$ in *Thelypteris uraensis* Ching (Hirabayashi, 1969) and $n = 72$ in *Lastrea laxa* (Fr. *et* Sav.) Copeland (Kurita, 1963) (both = *Metathelypteris*), $n = c.72$ in *L. glanduligera* (Mitui, 1968) (=*Parathelypteris*), suggest that these genera are cytologically more complex than otherwise appears to be the case. It is clearly desirable that the accuracy and/or taxonomic placement of these records be confirmed. Heterogeneity of base number within *Parathelypteris* has now been established by counts of $n = c.27$ (Mitui, 1975) and $n = 54$, $2n = 108$ (Kurita, 1976), all recorded for *Thelypteris* (*Lastrea*) *cystopteroides* (Eat.) Ching.

Continued

List of Genera & Chromosome Numbers—(continued)

Base numbers (Order, Family, Subfamily)	Genus, Subgenus, Section	Levels of ploidy	Counted	Known
c.47	Leptolepia	2	1	2
44†	Saccoloma	2	1	M
47?	Orthiopteris $n = 188$	8?	1	9
Hypolepidoideae				
26?,29,(39),49?,52?	Hypolepis $n = 29,(39),52,98,104$	2,4? ($n = 52,98$), 8?($n = 104$)	6	45
26	Paesia	2,8	3	12
26	Preridium	2,4,8	4	7
? 24 or 48	Histiopteris (incl. Lepidocaulon) $n = 48,96$	2,4 or 4,8	1	7
50	● Lonchitis (=Anisosorus)*	2,4	2	2
38	● Blotiella (=Lonchitis auct.)*	2,4	3	15
22,?	● Taenitis (incl. Holttumiella and Schizoloma sensu Copeland) $n = c.44,108,110,114,$ $2n = c.88,c.108-114$	4,10	4	15
27	● Idiopteris	2,4	1	M
? 28 or 56	Monachosorum $n = 56,112,168$	2,4,6 or 4,8,12	4	5
Lindsaeoideae				
34,c.40,(41),42, 44,47,c.50	Lindsaea (incl. Isoloma)	2,4,6,8 or 10? ($n = c.220$)	27	200
—	Ormoloma	—	–	2
38,(39),c.41,c.44 47,48,c.50	Sphenomeris	2,4,6? ($n = 145-7$), 8? ($n = 162-164$)	6	18

c.48	*Odontosoria* n = c.96	4	2	11
?	*Tapeinidium* n = c.120	?	1	14
	●*Xyropteris*	—	—	M
Aspleniaceae				
36,[40]	*Asplenium*	2,3*,4,5*, 6,8*,10*, 12,16	143	650
36	*Camptosorus*	2	2	2
36	●*Phyllitis*	2,4	3	8
36	*Ceterach*	2,4,6,8	3	5
36	●*Ceterachopsis*	2,4,8	2	3
36	*Pleurosorus*	2,4	2	3
36	*Pleurosoriopsis*	4	1	M
?35 or 70	*Loxoscaphe*	2 or 4	1	4
36	*Diellia*	2	2	5
—	*Holodictyum*	—	—	2
—	*Antigramma*	—	1	2
36	●*Diplora*	4	1	4
—	*Schaffneria*	—	—	M
? 38 or 76	●*Boniniella*	2 or 4	1	M
Dryopteridaceae				
Onocleoideae				
[39],40	*Matteuccia*	2,4	3	3
37	*Onoclea*	2	1	M
40	*Onocleopsis*	2	1	M
Athyrioideae				
40	*Athyrium* (excl. *Diplazium*)	2,4,6,8	83	180
40,41	●*Cornopteris* (=exindusiate derivates of *Athyrium* and *Diplazium*?)	2,4	6	12

† *fide* Mickel (1973, p. 139)
*See Tryon, R. (1962, p. 93 *et seq.*)

Continued

List of Genera & Chromosome Numbers—(continued)

Base numbers					Levels of ploidy	Number of species	
Order	Family	Subfamily	Genus — Subgenus — Section			Counted	Known
	—		• Rhachidosorus		—	—	
	41		Diplazium (incl. Diplaziopsis and Callipteris)		2*,3*,4*, 5*,6,8*	68	400
	40		Anisocampium		4	1	2
	40		Gymnocarpium		2,4	5	5
	42		Cystopteris		2,4,6,8	12	20
	[33,38,39],41		Woodsia		2,4	11	40
	40		• Lunathyrium		2,3*,4,6	13	?
	—		Adenoderris		—	—	2
	—		Cheilanthopsis		—	—	1
	40,41		Hypodematium		2,4	2	3
	—		• Kuniwatsukia		—	—	
		Tectarioideae					
	—		• Trichoneuron		—	—	1
	41		Ctenitis		2,4	21	150
	41		• Ctenitopsis		2,4	1	20
	41		• Lastreopsis		2,4	12	30
	41		Psomiocarpa		2	1	M
	—		Atalopteris		—	—	3
	41		• Pleocnemia (ex Tectaria)		2	2	15
	41		• Arcypteris (ex Tectaria)		2	2	4
	41		Pteridrys		2	2	8
	—		Dryopolystichum		—	—	M
	40		Tectaria (incl. Luerssenia)		2,4,6,8	28	200
	—		• Pseudotectaria		—	—	2
	40		Hemigramma		2,4	2	6

40	*Quercifilix*	2,4	1	M
40	*Cionidium*	2	1	2
—	*Tectaridium*	3*	—	2
40	*Fadyenia*	—	1	M
—	*Pleuroderris*	—	—	M
40	*Hypoderris*	2	1	2
—	*Amphiblestra*	—	—	M
—	*Dictyoxiphium*	—	1	M
40	*Camptodium*	2	1	M
40	*Stenosemia*	2	1	2
40	*Heterogonium*	2,4	2	12
41	*Cyclopeltis*	2	1	6
41	*Didymochlaena*	2	1	M
	Dryopteridoideae			
41	*Peranema*	2,4	1	2
41	*Diacalpe*	2,4	1	M
41	*Polystichum*	2,3*,4,6,8	72	135
c.42	*Cyclodium*	2	1	2
41	● *Cyrtomium* (=*Phanerophlebia*)	2,3*	8	25
41	● *Arachniodes* (=*Rumohra* Ching, *pro parte max.*, non Raddi)	2,4	16	30
41	● *Polystichopsis*	2	1	4
—	*Lithostegia*	—	—	M
41	*Polybotrya*	2	2	25
41	*Maxonia*	2	1	M
41	*Dryopteris*	2*,3*,4*,6	108	150
41†	● *Nothoperanema*	2	2	5

† Reported as *Dryopteris hendersoni* C.Chr. and *D. shikokiana* C.Chr. (Mitui, 1968; Hirabayashi, 1970, 1974, p. 14); = *Nothoperanema hendersonii* (Beddome) Ching & *N. shikokianum* (Makino) Ching (1966).

Continued

List of Genera & Chromosome Numbers—(continued)

Order	Family	Base numbers	Subfamily	Genus	Subgenus	Section	Levels of ploidy	Number of species	
								Counted	Known
		41		Stigmatopteris			2	1	26
		41		Acrophorus			2	1	2
		–		Stenolepia			–	–	M
	Lomariopsidaceae								
		41		Bolbitis (incl. Egenolfia)			2,4	17	95
		–		Thysanosoria			–	–	M
		–		Arthrobotrya			–	2	3
		41		Teratophyllum			2	2	9
		41		Lomagramma			2	2	15
		[16,31,39],41		Lomariopsis			2,4	5	40
		41		Elaphoglossum			2,4,6	30	400
		–		● Peltapteris (=Rhipidopteris and incl. Microstaphyla)			–	–	6
	Davalliaceae								
		40	Davallioideae	Humata			2,3*	3	50
		–		Trogostolon			–	–	2
		–		Scyphularia			–	–	8
		–		Parasorus			–	–	M
		40		Davallia			2,6	7	40
		–		Davallodes			–	–	8
		40†		● Paradavallodes			2	1	4
		40		Araiostegia			2	1	4
		41		Leucostegia			2	1	2
		–		● Gymnogrammitis			–	–	M
		41		● Rumohra Raddi, sensu stricto			2	1	M
		40,41	Oleandroideae	Oleandra			2,4	5	40

281

41	*Arthropteris*	2	4	20
–	*Psammiosorus*	–	–	M
41	*Nephrolepis*	2,4	13	30
Blechnaceae				
28,29,31,32,33, 34,36,c.37	*Blechnum*	2,4,6	26	220
40	*Salpichlaena*	2	1	M
32	*Doodia*	2,4,6	4	11
33	*Brainea*	2	1	M
33	*Sadleria*	2	1	6
34,35	*Woodwardia* (incl. *Lorinseria*)	2,4	8	12
–	●*Pteridoblechnum*	–	–	M
37	*Stenochlaena*	4	2	5
Salviniales				
Salviniaceae				
9	*Salvinia*	2	3	10
Azollaceae				
22	*Azolla*	2	1	6
Familia Incertae Sedis				
Hymenophyllopsidaceae				
–	*Hymenophyllopsis*	–	–	2
Loxsomaceae				
50	*Loxsoma*	2	1	M
–	*Loxsomopsis*	–	–	3
Plagiogyriaceae				
25?,33?	*Plagiogyria* n = 66,c.75,c.100,c.125 & c.132	4,6,8,10	8	36

† Reported as *Davallodes membranulosum* (Wall.) Copeland (Verma in Mehra, 1961, p. 149), = *Paradavallodes membranulosum* (Wall.) Ching (1966).

C. FUNDAMENTAL CYTOLOGICAL DIVISIONS AMONGST MODERN FERNS

With the breakthrough in cytological technique heralded by the appearance of Manton's book there was considerable optimism with respect to the advances that increasing knowledge of chromosome numbers would bring with regard to resolution of the many intractable problems associated with fern phylogeny and classification. This optimism was quickly justified, since the results obtained by Manton and Sledge (op. cit.) on the Ceylon flora clearly indicated the solution of two major problems.

Two systems of classification commanded attention at that time, those of Copeland (1947) and Holttum (1947). Fundamental differences between the two systems resided in Copeland's very broad conception of the natural limits of his Pteridaceae and his Aspidiaceae.

With regard to the Aspidiaceae ($\equiv$ Dryopteridaceae), Holttum had contended that *Thelypteris* and its relatives were entirely distinct from the true dryopteridaceous ferns, and were of a quite independent origin, being related to *Cyathea* (Fig. 1). The cytological data emphatically supported Holttum's belief in the separateness of the thelypteridaceous genera, a marked discontinuity existing between the chromosome numbers in thelypteridaceous and dryopteridaceous ferns, the base numbers found being 31, 35, 36 and 40, 41, 42 respectively, with 36 and 41 predominating.

Subsequent research has more than adequately confirmed the cytological distinction between the Thelypteridaceae and the Dryopteridaceae. The Thelypteridaceae is a large and taxonomically difficult family, whose classification is in an only partially resolved state. Holttum (1971) has made an extensive study of the Old World representatives, and recognized 23 genera. The New World representatives of the family are in substantial measure different from the Old World forms, but have received no comprehensive recent taxonomic revision, with the consequence that it is quite uncertain to what proportion of the New World species Holttum's treatment can be extended. The New and Old World forms are therefore listed separately here. In the Old World, base numbers range from 30 to 36, and in the New World, from 27 to 36, though in both hemispheres, a majority of genera have $x = 36$. Combined, the two hemispheres show a complete sequence of numbers from 27 to 36, with ɪne sole exception of 33.

In contrast, the Dryopteridaceae as interpreted here is a large and morphologically diverse but cytologically remarkably uniform family. Apart from *Cystopteris*, with $x = 42$ (and two maverick genera, *Woodsia* and *Lomariopsis*, which display a range of numbers, but which are nevertheless clearly based on 41), the only base numbers encountered in a long list of genera are 40 and 41. One further apparent exception, *Hemidictyum marginatum*, has already been mentioned. Although possessing athyrioid sori, its chromosome number ($n = 31$) is discordant here, and this fern is suspected of being an aberrant thelypterid.

With regard to Copeland's Pteridaceae, the difference between his treatment and that of Holttum was substantial, Holttum having distributed the elements of Copeland's Pteridaceae between the Adiantaceae and no fewer than three separate subfamilies of his Dennstaedtiaceae (Dennstaedtioideae, Lindsaeoideae and Pteridoideae). The results obtained from the Ceylon flora indicated that neither treatment was natural. A cytological discontinuity divides Copeland's family into two parts and also cuts across Holttum's subfamily Pteridoideae. All of the ferns investigated from Holttum's Adiantaceae (which includes the vittarioid ferns), together with *Pteris* and a few related genera, were found to be cytologically uniform, all possessing base numbers of 29 or 30, whereas the first part of Copeland's Pteridaceae possessed very diverse chromosome numbers.

All subsequent research has confirmed the existence of a great natural group of genera, the Adiantaceae as interpreted in this paper. Thirty-two of the genera attributed to this family, approximately two-thirds of the whole, have now been cytologically investigated, and all have base numbers of 29 or 30. Only one, *Anogramma*, shows any departure from this rule, containing elements with 26 (Kurita, 1971) and 27 (Mickel *et al.*, 1966) but since this genus also contains 29 (Brownlie, 1958; Mehra & Verma, 1960; Mickel *et al.*, 1966; Gastony and Baroutsis, 1975), 26 and 27 may clearly be interpreted as aneuploid derivatives of 29. No fewer than eight genera in the Adiantoideae are reported to contain some species with both 29 and 30, an indication that single aneuploid changes are of widespread occurrence in this subfamily.

Thus by 1954 cytological evidence already existed confirming the independence of the thelypteridaceous ferns, and establishing the existence of a large natural group, composed of pteroid, cheilanthoid and adiantoid ferns, all based on 29 and 30. It has to be admitted that in comparison with these early successes, progress in the cytological solution of phylogenetic problems during the last 20 years has been disappointingly slow, and to a considerable degree extension of knowledge has served to confirm the intransigence of problems rather than resolve them. Many problems remain highly controversial. Thus cytology can render no assistance with respect to the question of the existence of any relationship between grammitoid and polypodioid (*sensu stricto*) ferns. Further discussion of problems of phylogenetic relationship will be postponed until after the cytology of the ancient ferns has been reviewed.

D. CYTOLOGY OF THE ANCIENT FERNS

1. Psilotales

There are substantial differences between the two genera of Psilotales in their foliar appendages, shoot architecture and synangia, sufficient to support their placement into two separate families. They also differ in their ecological range and behaviour. Surprisingly, in view of its archaic character, *Psilotum* is a plant of weedy tendencies, and is sometimes an effective colonizer of open situations such as recent roadside cuttings, volcanic scoria, or even the edges of pavement asphalt (in Australasia). In contrast, *Tmesipteris* is characteristically an epiphyte in rain forest, though two species,* *T. oblanceolata* and *T. vieillardii*, are known to adopt a terrestrial habit.

Nevertheless, *Psilotum* and *Tmesipteris* are cytologically indistinguishable, thus confirming that, notwithstanding their morphological and ecological differences, they are closely related. Both possess the same base numbers, $x = 52$, and share a characteristic laxity of spiralization of their chromosomes at the first metaphase of meiosis.

2. Ophioglossales

Ophioglossum is now widely renowned for its enormous chromosome numbers, which show that the level of tolerance of cytological mechanisms for the multiplication of chromosome number can be truly remarkable, and extraordinarily high numbers are possible before a fatal level of meiotic and mitotic constipation is reached. Chiarugi (1960), Ninan (1956b) and Kurita and Nishida (1965) all diagnose the base number of both *Ophioglossum* and *Botrychium* as $x = 15$, on the basis that this is the highest common factor of their lowest known numbers ($n = 120$ and $n = 45$ respectively). This leads to such exotic inferences as that the lowest ploidy level known in *Ophioglossum* is 16-ploid, and the highest 84-ploid! There is in my opinion no justification for this sort of arithmetical exercise. Clearly 120 is not a primitive number, and must have evolved from some lower number, but there is at present no means of determining just how it arose. In fact, chromosome numbers in *Ophioglossum* proceed in multiples of 120 until $n = 360$ is reached. Inconsistent numbers above this level ($n = 370/380$, 410/420, 436, 451, 564/572, 630) can reasonably be interpreted as aneuploid, because at this level of chromosome number it cannot always be of great consequence if some chromosomes are lost or added to the complement. However, the very fact that there is little evidence for even relatively minor aneuploid variations until the number exceeds the $n = 360$ level suggests that the effective base number today is $x = 120$, however surprising this may be. Those who would prefer to interpret this entire sequence of chromosome numbers from a base of 15 must explain why, although what are on this basis 48-, 50-, 58-, 60-, 64- and 68-ploids are known, the only lower levels yet discovered are 16- and 32- ploids. There is some evidence for aneuploidy even at the $x = 120$ level. Thus Verma (1956) claims and illustrates $n = 116$ in *O. polyphyllum* as well as 120 and 122 (Verma, 1957) and states that the error in the case of $n = 116$ cannot be more than +1. In her original report Manton (1950, pp. 263-265) was unable to give an exact number for *O. lusitanicum*, but states "not less than 125." Even with very clear preparations,

---

* *Tmesipteris* is now known to contain many more distinct species than was earlier realized (Wakefield, 1943; Barber, 1954; Braithwaite, 1973; Chinnock, 1975).

exact determination of chromosome number at even the lowest level encountered in *Ophioglossum* is difficult.

An interesting example of the practical problems that can arise is the case of *Botrychium virginianum*, apparently unique in the genus in being aneuploid with $n = 92$ (Britton, 1953; Wagner, 1955; Niizeki *et al.*, 1963). The claim by Gopal-Iyengar (1957) that this species has in reality only $n = 90$, and that reports of higher numbers are due to precocious disjunction of some of the bivalents at metaphase I of meiosis, has been refuted by the mitotic count of $2n = 184$ by Löve and Löve (in Löve and Solbrig, 1964).

3. Marattiales

The other order of eusporangiate ferns, the Marattiales, are only partially known cytologically but the four genera so far investigated, *Angiopteris, Christensenia,** *Danaea* and *Marattia*, are all based on 40. Counts of $n = 39$ (Brownlie, 1961) and $n = 78$ (Ninan, 1956c) known in *Marattia* are evidently simple aneuploid derivatives from $x = 40$. Indeed, $n = 40$ and $n = 78$ are reported from ostensibly the same species of *Marattia, M. fraxinea* (Walker, in Manton, 1959; Ninan, op. cit).

4. Osmundales

The cytology of the Osmundales is distinctive and uniform. All three genera, and all investigated species, have $n = 22$. The chromosomes are large, and are convenient material for karyotype analysis, which has been undertaken for five species of *Osmunda* by Tatuno and Yoshida (1966, 1967), who claim on the basis of their analysis that the true base number is $x = 11$.

5. Hymenophyllales

The Hymenophyllaceae is one of the most cytologically distinctive of all fern families, to an extent that suggests, considered in conjunction with the equally characteristic morphology of the family, that its relationship to any other living family can only be very remote.

Chromosome size is very variable in the Hymenophyllaceae. The great majority of genera share with the Osmundaceae the possession of very large chromosomes, which makes these two families excellent subjects for cytological study. However, chromosome size is relatively small in species of *Microgonium*, e.g. *M. motleyii* (Manton and Sledge, op. cit., fig. 35) and *M. hookerii* (Walker, 1966a, fig. 19), and certainly very much below the norm for the Hymenophyllaceae. In contrast to the uniformity of chromosome numbers in the Osmundaceae, those in the Hymenophyllaceae are very diverse, ranging from the lowest chromosome numbers known in homosporous ferns, $n = 11$ in *Hymenophyllum peltatum* (Brownlie, 1958) and $n = 13$ in *H. tunbrigense* (Manton, 1950), to a whole range of genera based on $x = 36$.

---

* *Christensenia aesculifolia*, $n = 80$ (A. F. Braithwaite, unpublished).

Prior to a major revision of the family by Copeland (1938), all of the some 650 species in the family were customarily distributed between only two genera, *Hymenophyllum* and *Trichomanes*. In his treatment, which is based in some measure on a very much older work by Presl (1843), Copeland dispersed the family around some 30 genera, a revision so radical as to have remained controversial. In consequence, there has been considerable interest with respect to the extent that cytology would prove to endorse Copeland's treatment. Walker (1966a) and Braithwaite (1969, 1975) have both produced excellent analyses of this question.

Morton (1968) produced a taxonomic revision of the Hymenophyllaceae in which, apart from four distinctive monotypic genera, he retained only *Hymenophyllum* and *Trichomanes* but divided these two great genera into a total of nine subgenera and 36 sections, the latter for the most part corresponding to Copeland's genera.

Walker has recently pointed out (Walker, 1973a, p. 93) that although about 100 species of Hymenophyllaceae have now been investigated, this may be a less reliable sample than is apparent at first sight because, though counts are available for 24 of Copeland's genera, the distribution of chromosome counts about these genera is very uneven. More than half of the counts refer to only four genera. Those of Copeland's genera in which most species have been counted (*Hymenophyllum, Mecodium, Meringium, Trichomanes* and *Vandenboschia*) have all proved to be cytologically heterogeneous. Two other genera (*Macroglena* and *Selenodesmium*) are already known to be heterogeneous, and it is evident that more may follow when more species have been counted. This point is acutely demonstrated by *Trichomanes sensu* Copeland. Walker (1966a) reported on the cytology of eight species, approximately one-third of the genus. All had numbers based directly on $x = 32$ ($n = 32$, 64, 128), providing a strong and reasonable presumption that this genus was cytologically uniform. Subsequently, Pignataro (1971) recorded $n = 36$ in *T.* (*Lacostea*) *akersii*, and recently Tryon *et al*. (1975) report $n = 72$ in *T.* (*Achomanes*) *cristatum*, while Bierhorst (1975) reports $n = 13 \pm 1$ in *T.* (*Neuromanes*) cf. *pinnatum*.*

Although not all genera which appear to be cytologically heterogeneous will necessarily prove to be unnatural genera, since this may depend on the course of cytological evolution within individual genera, nevertheless it is likely that pairs

---

* Bierhorst (1975), in describing an apomictic life-cycle in *Trichomanes pinnatum* Hedw., gave $2n = 26$ with "the possible error . . . no more than one pair" (pp. 448/449), but made no comment on the fact that such a low number is otherwise unknown in *Trichomanes sensu lato*. The count was made on serial-sectioned material, and the illustrations given are not adequate to provide more than a rough approximation of the number. It is clearly desirable that the chromosome number of this taxon be determined accurately by the squash method. (*T. pinnatum* Hedw., though included in *Trichomanes sensu stricto* by Copeland, was placed in *Trichomanes* subgen. *Achomanes* sect. *Neurophyllum* by Morton (1968, pp. 194/5), not in *Trichomanes* sect. *Trichomanes*, as stated by Bierhorst (op. cit., p. 448)).

of genera like *Trichomanes sensu* Copeland and *Vandenboschia*, both with $n = 32$ and 36, and *Macroglena* and *Selenodesmium*, both with $n = 33$ and 36, will prove to have been unnaturally defined. The presence of $n = 21$ and 22 in both *Hymenophyllum* and *Meringium*, two genera which are not always clearly distinguishable, again suggests unnatural boundaries.

In certain respects, Morton's treatment accords rather better with the cytological evidence. Morton (1968, p. 175 *et seq.*) argues that *Trichomanes* is correctly typified by *T. scandens* L. (a species included by Copeland in *Vandenboschia*), which he separates in a monotypic section, leaving the rest of *Vandenboschia* (= sect. *Lacosteopsis*) as a cytologically uniform group. He unites *Macroglena* and *Selenodesmium*, which at least has the merit of creating one probably unnatural group where formerly there were two. To some extent Morton's subgenera reflect apparent cytological boundaries within the family. Thus in *Hymenophyllum sens. lat.*, subgen. *Hymenophyllum* includes a range of lower numbers, while subgen. *Sphaerocionium* is uniformly $x = 36$. In *Trichomanes s.l.*, subgen. *Trichomanes* is almost exclusively based on 36, subgen. *Achomanes* has predominately $x = 32$, while all species with $x = 34$ fall in subgen. *Didymoglossum*. Other subgenera, e.g. *Mecodium* and *Pachychaetium*, cannot be so simply categorized.

The wide range of low chromosome numbers encountered in the Hymenophyllaceae has encouraged speculation regarding the fundamental base numbers involved and the course of cytological evolution in the family. Vessey and Barlow (1963) and Walker (1966a) have produced hypothetical schemes of cytological evolution for the family, but these are too detailed to be discussed at length here. Vessey and Barlow emphasise two ancestral base numbers, $x = 7$ and $x = 11$, on which their scheme is constructed. Walker is critical of their refusal to countenance the possibility of $x = 9$ and certainly, in view of the evidence of $n = 18$, their disbelief does appear to be rather perverse. Walker admits the possibility of a whole series of low base numbers participating in the evolution of the family, from $x = 6$ to 9, 11 and 13, but emphasizes that our knowledge is still too slight to permit firm conclusions to be drawn. The real pattern may well be very complex, but the series of hypothetical changes by which Vessey and Barlow derive existing numbers from $n = 18$ is too intricate to be a likely approximation of the truth.

Nevertheless the recent discovery by I. Manton *et al.* (unpublished) of a widely dibasic polyploid species of *Hymenophyllum*, with $n = 31$, derived from *H. tunbrigense*, $n = 13$, and *H. wilsonii*, $n = 18$, suggests that equally remarkable events may have had a part in the earlier cytological evolution of the family, thus giving credence to Vessey and Barlow's contention that $n = 18$ arose in the family by a very similar event, i.e. as a dibasic amphidiploid compounded from two species with $n = 7$ and $n = 11$ respectively.

It is also noteworthy that the karyotype of *Hymenophyllum tunbrigense* is strongly heteromorphic (Manton and Vida, 1968, pl. 45, fig. 6) suggesting that this number may have evolved by relatively drastic aneuploid reduction (cf. p. 295 *et seq.*).

Tatuno and Takei (1969a) have published karyotype studies on one species of *Hymenophyllum* and three species of *Mecodium*. *Mecodium* is one of the genera which is most clearly cytologically heterogeneous. A whole group of species has the effective base number $x = 36$, while the discovery of $n = 21$ in two species (*M. exsertum* and *M. serrulatum*) and $n = 28$ in a related species (*M. polyanthos*) induced Manton & Sledge (op. cit., p. 144) to describe this "as the first clear proof of a monoploid number as low as 7 in an existing fern genus". The three *Mecodium* species investigated by Tatuno and Takei all belong to this latter cytological group, with chromosome numbers $2n = 42$, $2n = 56$ and $2n = 84$. It is therefore disconcerting to find that Tatuno and Takei conclude that the true base number is $x = 11$ and the chromosome numbers studied are aneuploid derivatives, so that in *M. oligosorum* $2n = 42 = (44 - 2)$ and in *M. polyanthos* $2n = 56 = (66 - 10)$. I do not find the evidence presented by Tatuno and Takei in support of their conclusion to be convincing, since the published karyotype analyses could with just as much reason be rearranged to provide a quite different interpretation. Though the degree of heterogeneity present in these karyotypes may not be sufficient to permit an unequivocal analysis, it is clearly desirable that the studies pioneered by Tatuno and Takei be repeated and extended by other independent workers.

However, the value of karyotype analysis in Hymenophyllaceae is not likely to lie so much in attempts to determine ancestral base numbers directly from analysis of a few species as in detailed comparison of a wide range of karyotypes, which can confidently be expected to yield information regarding affinities and paths of evolutionary development. Ultimately a better understanding of the pattern of cytological evolution within the family and recognition of more natural taxonomic groupings may result. No other family of ferns offers more favourable material for such a study.

6. Gleicheniales

Stromatopteris was formerly (e.g. by Bower and Copeland) classified in the Gleicheniaceae. The morphology of this remarkable fern has been reinvestigated by Bierhorst (1968b, 1969), who has stressed the many features it has in common with *Psilotum* and *Tmesipteris*. However, the synangia of the Psilotales are entirely different from the sporangium of *Stromatopteris*, which is of typically gleicheniaceous character. *Stromatopteris* is therefore given here the status of a separate monotypic family within the Gleicheniales. Its chromosome number has not yet been determined with accuracy. Bierhorst states (1968b, p. 261): "The chromosome number of *Stromatopteris* may be reported from the study of sporangium smears as $2n = c.$ 39 pairs (probably exactly 39)." Bierhorst is an histologist of undoubted ability, who customarily illustrates his publications with photomicrographs on an admirably lavish scale. The paper in question contains 143 figures, of which 100-odd are photomicrographs. In these circumstances, it is particularly regrettable that a photograph of the chromosome count was not included, no matter how imperfect the cell, in order that the degree of approximation involved in the count should be open to some measure of independent assessment.

The Gleicheniaceae, deprived of *Stromatopteris* and *Platyzoma* (see below, p. 291) constitute a very natural family of very characteristic morphology. Some 30 taxa have now been cytologically examined, and all are consistent with a very distinctive cytological pattern. The chromosomal evidence corresponds with the subgenera recognized by Holttum (1957, 1959a), a particular distinctive basic chromosome number characterizing each subgenus, i.e. 20 and 22, 34, 39, 43, (and 44?), 56 (Walker, 1966a, 1973a; Sorsa, 1968). As Walker writes (1973a, p. 94): "All of these are very distinctive numbers, forming neither a simple aneuploid series nor a polyploid one." The relationship between these numbers is not obvious and cannot be simple. It is clear that the cytology indicates "that the taxonomic divisions are fundamental and of very ancient origin" (Walker, 1966a, p. 188). There is evidently strong cytological support for the recognition of the five taxonomic and cytological groups as distinct genera, a decision readily implemented since it is in accord with Copeland's treatment, save only that the genus *Dicranopteris* requires to be divided.

Mehra (1961) proposes that the fundamental base numbers in Gleicheniaceae are 7, 10, 13 and 17, corresponding to the numbers 56, 20, 39 and 34 known today. There really is no justification for the implied assumption that cytological evolution in the group has proceeded simply by polyploid multiplication, save only for the origin of a new secondary base, 43. The Gleicheniaceae are a very ancient family, and it is much more likely that the effective base numbers we find today are of more devious and complex origin. The Gleicheniaceae are also a very homogenous and natural family, which can reasonably be assumed to be of ultimately monophyletic origin from some unknown very ancient source. It follows that to postulate a series of base numbers 7, 10, 13 (17 = 10 + 7) implies the origin of this series by aneuploid changes, e.g. $7 \rightarrow 10 \rightarrow 13$, or $7 \leftarrow 10 \rightarrow 13$. Even if 13 originated as 14 ($= 7 \times 2$) $- 1$, the appearance of 10 would require aneuploid changes. It is difficult to understand why the proponents of such low base number series are not prepared to accept the equal likelihood of aneuploid changes occurring at higher levels of chromosome number.

Sorsa (1968) has shown how the existing base numbers in Gleicheniaceae could have been derived from two original bases of 11 and 17, by a scheme involving two dibasic polyploids (11 and 17, 22 and 17), and two subsequent aneuploid changes ($22 \rightarrow 20$ and $44 \rightarrow 43$). Though it does not explain how the two postulated distinct fundamental base numbers came into existence, this hypothesis is ingenious, but to my mind not very acceptable. Cytological evolution is not obliged to follow the arithmetically most direct or least complex route.

7. Matoniales

There is reason to believe that the Matoniaceae has some affinity both with the Gleicheniaceae and the Dipteridaceae. Very few species of Matoniaceae remain extant. The only one to have been cytologically investigated is *Matonia pectinata* (Manton, 1954a), which has $n = 26$, a number different from any found in the Gleicheniaceae, and different from $n = 33$, found in *Dipteris* (Manton, 1954b; Patnaik and Panigrahi, 1963). It would be imprudent to assume that the arithmetical relationship between $n = 26$ in *Matonia* and $n = 39$ in *Dicranopteris* is more than coincidence.

8. Schizaeales

The Schizaeaceae is a cytologically and morphologically heterogeneous assemblage and there is a good case (cf. Bierhorst, 1971) for separation of its elements into separate families, but for convenience in discussion a more traditional approach is followed here.

The genus *Actinostachys* is recognized here as distinct from *Schizaea*, following the discovery by Bierhorst (1968a) that these two genera possess fundamentally different gametophytes. As far as I can tell, only one species of *Actinostachys, A. digitata*, has been cytologically examined (Lovis, in Holttum, 1959b; Abraham *et al.*, 1962). These authors encountered great technical difficulties and an extremely high chromosome number, which could only be reported very approximately as of the order of $n = 325$ and 360 respectively, with a wide margin of error.

The series of chromosome numbers known in *Schizaea* itself is bizarre indeed, namely $n = 77$, $c.94$, 96, 103, $c.154$ and $c.270$ (Lovis, 1958; Brownlie, 1965; Wagner, 1963; Tryon *et al.*, 1975) and defies any attempt at arithmetical analysis. It is noteworthy that prior to the higher numbers becoming known the base number of *Schizaea* was given by Chiarugi (1960) as $x = 11$, on the basis of the first recorded count, $n = 77$.

Chromosome numbers elsewhere in the Schizaeaceae are less exotic. The significance of $x = 28$, 29, 30 in *Lygodium* as providing support for the derivation of the Adiantaceae from a schizaeaceous source will be discussed below (p. 309).

Roy and Manton (1965, p. 290) argue that in *Lygodium* "species with $n = 28$ and $n = 30$ are likely to be polyploid derivatives from precursors with $n = 14$ and $n = 15$ respectively ... [with $n = 29$] ... a relatively late innovation". It is not clear why they should suggest that $n = c.70$ in *L. articulatum* (Brownlie, 1961) will, if this number proves to be correct, constitute evidence for a lower base of 7, since $n = 70$ is five times 14, and does not require the postulation that $x = 7$.

In contrast, Abraham *et al.* (1962) proposed a base of $x = 10$, in which case $n = 30$ is the primary number in *Lygodium*, with 29 (and 28) as aneuploid derivatives, and $x = 38$ in *Anemia* and *Mohria* is derived from 20, i.e. $(20 - 1 = 19) \times 2$. The strength of this hypothesis depends on the reality of a relationship between Schizaeaceae and Marsileaceae, where the actual numbers known are $n = 20$ in *Marsilea* (Abraham *et al.*, op. cit.), $n = 19$ in *Regnellidium* (Abraham *et al.*, op. cit.; Loyal, 1962) and $n = 10$ in *Pilularia* (Chopra, 1960). Relationship between these two families was suggested long ago by Campbell (1918) and Bower (1926), who considered the Marsileaceae to be specialized derivatives of the Schizaeaceae, modified for a semiaquatic habitat. If this relationship is real then it constitutes a powerful argument in favour of $x = 10$ in Schizaeaceae, but the morphological differences between the two families are great.

In view of the general acceptance (see p. 309 below) of the hypothesis that the Adiantaceae arose from the Schizaeaceae, the karyotype studies of Kawakami (1971) on three species of *Pteris* ($2n = 58$, 116) are of interest here, since he interprets the *Pteris* karyotype as derived from $x = 10$. However, with the exception of the two pairs of chromosomes with negatively heteropycnotic regions, and three subcentric pairs, all the rest of the chromosomes in the complement are acrocentrics of remarkably uniform size. In consequence, the karyotype evidence for $x = 10$ is not strong.

Whether or not the aberrant genus *Platyzoma*, a monotypic Queensland endemic, is of particular relevance with respect to the question of affinity between Schizaeales and Marsileales is an interesting point. The same number, 38, characteristic of *Anemia*, has now been reported (as $2n = 76$) for *Platyzoma* by Tryon and Vida (1967). Prior to the determination of its chromosome number, Tryon (1961) had already concurred with Holttum (1956) that *Platyzoma* was incorrectly placed in Gleicheniaceae (cf. Bower, 1926; Copeland, 1947) and noted a morphological affinity with the Schizaeaceae. The sporangia of *Platyzoma* are, however, more "polypodiaceous" than schizaeaceous in character, and Tryon (1964) proposed that *Platyzoma* be treated as a separate subfamily in Polypodiaceae *sensu* Christensen, next to the Gymnogrammoideae. Following the disruption of the old "Polypodiaceae", it now seems appropriate to treat *Platyzoma* as a monotypic family (as has already been proposed by Nakai, 1950), probably of schizaeaceous origin and therefore included in the Schizaeales.

Platyzoma is now known to possess one particularly individual characteristic. It has long been known (Thompson, 1917) that it produces spores of two sizes. Some sporangia produced 32 small spores, others 16 large spores. Tryon (1964) was able to germinate spores and establish gametophytes, and discovered that whereas small spores produced filamentous gametophytes without rhizoids and bearing only antheridia, the large spores produced gametophytes of a spathulate form, with rhizoids and archegonia. The prothalli are thus effectively dioecious, although older female gametophytes may later produce antheridia, so the condition should perhaps be described as hemidioecious. The twin features of incipient heterospory and dioecy are unique to *Platyzoma* among terrestrial ferns, strict heterospory being known only in the aquatic fern families

Marsileaceae and Salviniaceae, both of uncertain affinity. Clearly, since an association between Marsileaceae and Schizaeaceae has been suggested on other grounds, the presence of incipient heterospory in *Platyzoma* may not be purely coincidental, but instead have some significance with respect to this postulated relationship.

Two small groups of uncertain affinity can conveniently be mentioned here, although they cannot with certainty be described as ancient ferns, since neither is known fossil, and their affinities are controversial.

The Loxsomaceae is an isolated and distinctive family including two clearly related genera *Loxsoma* and *Loxsomopsis*. Comparative morphological studies on gametophytes have shown correspondence between Loxsomaceae and Cyatheaceae *sensu stricto* (= Cyatheoideae *sensu* Holttum), with regard to several characters, which collectively "make a common ancestry seem highly probable" (Stokey and Atkinson, 1956, p. 260). More recently van Cotthem (1970, 1973) has shown that the Loxsomaceae possess paracytic stomata, otherwise known in the ferns only in Cyatheaceae (*sensu lato*), Dipteridaceae and *Cheiropleuria*. The cytology of *Loxsomopsis* remains unknown, but Brownsey (1975) has now established beyond doubt that the chromosome number of *Loxsoma cunninghamii* is $n = 50$. Though this result can neither confirm nor disprove the affinity between Cyatheaceae and Loxsomaceae suggested by micromorphological evidence, it does emphasize that this relationship is probably remote.

Plagiogyriaceae contains a single genus *Plagiogyria*. The systematic position of this genus has always been controversial. It occupies a central position in Bower's scheme, as a supposed intermediary between the Osmundaceae and gymnogrammoid (incl. cheilanthoid and adiantoid) ferns, a suggested line of evolution which can now be seen to have been one of Bower's less fortunate ideas. However, the finding of $n = 66$ in both *P. semicordata* (Walker, 1966a) and *P. tuberculata*, with $n = c.132$ in *P. glauca* (Walker, 1973a), might well be accepted as strong evidence in evidence in favour of the first part of Bower's proposed lineage, i.e. the origin of *Plagiogyria* from Osmundaceae, were it not for the report by Kurita (1963) of the numbers $n = c.75$, $c.100$ and $c.125$ in three Japanese species, respectively *P. matsumureana*, *P. japonica* and *P. euphlebia*. That two workers should discover two so distinct series in the one genus is an extraordinary event. At present, the relationship between these two polyploid series is beyond interpretation. It is obviously desirable that the numbers in the Japanese species be determined with greater certainty, but the photograph provided by Kurita for *P. matsumureana* is of a cell of sufficient quality to indicate that the margin of error in the existing count cannot be large. (See Note Added in Proof, 3, p. 415.)

The example of *Plagiogyria* is a salutary warning against any tendency that might exist for complacent assumption that chromosome numbers yet unknown will not provide any unexpected surprises.

E. CYTOLOGY OF THE TREE-FERNS

The tree-ferns not only have a dominant importance in the phyletic scheme of Bower (1926, p. 333; 1928, p. 2), but are also of central significance in the contemporary schemes of Nayar (1970) and Holttum (1973a), yet there is still no general agreement as to whether or not those with marginal or submarginal sori (the dicksonioids and thyrsopteroids) are related to those with superficial sori (the cyatheoids). These two groups are combined within one family (Cyatheaceae) in earlier treatments by Mettenius (1856) and Diels (1899), but were widely separated by Bower, to whom the distinction between a primitively marginal and primitively superficial origin of the sorus was of fundamental importance. His division of the tree-ferns into two groups of quite distinct origin was subsequently accepted by Christensen (1938), Copeland (1947), Holttum (1947) and also by Pichi-Sermolli (1958) for whom they characterize two different orders, Dicksoniales and Cyatheales.

More recently, on the basis of morphological and anatomical studies (Holttum and Sen, 1961), Holttum has changed his opinion. Bower, influenced by his conviction that the cyatheoid ferns were derived from the exindusiate Gleicheniaceae, had considered that the cyatheoids were also primitively exindusiate, and that the indusium had arisen *de novo* in the development of the indusiate members of the family. The re-examination of details of soral development in the tree-ferns by Holttum and Sen led them to the decision that Bower was mistaken in his interpretation of the sorus of *Cyathea*, whose ontology in their opinion shows its origin to be effectively marginal and that its indusium (p. 411) "originates in the same way as the indusium in *Dicksonia* not ... as postulated by Bower".* They concluded, in accord with Goebel (1915/18, pp. 1148/1149) that the indusium of *Dicksonia* was indeed strictly homologous with that of *Cyathea*. In consequence they reunited the tree-ferns within one family, Cyatheaceae. This approach constitutes the exact opposite of that of Pichi-Sermolli (1970), who has now recognized a whole series of families of tree-ferns, e.g. Culcitaceae, Lophosoriaceae, Metaxyaceae etc.

Chromosome numbers in the tree-ferns are now tolerably well known, apart from the important exception of *Thyrsopteris*. Thanks to the efforts of Roy and Holttum, and of Walker, our knowledge now includes the monotypic and possibly phylogenetically important genera *Cystodium* (Roy and Holttum, 1965a), *Lophosoria* (Walker, 1966a) and *Metaxya* (Roy and Holttum, 1965c). The latter two genera constitute Bower's Protocyatheaceae.

The situation in the family has been well summarized by Walker (1973a, p. 95): "The majority of the genera possess chromosome numbers of 65

---

* A recent study by Tryon and Feldman (1975) of the development of the indusium in *Cyathea fulva* and the range of diversity of soral and indusial structure in the Cyatheoideae is also relevant here.

(*Lophosoria* and *Dicksonia*), 68 (*Culcita* subgenus *Culcita* and *Cibotium*), or 69 (*Cyathea* and *Cnemidaria*). This represents a series of numbers which are close enough together not to suggest major discontinuities, whilst at the same time individually characterizing genera." The coincidence of $n = 65$ in *Lophosoria* and *Dicksonia* is worthy of particular note, since *Lophosoria* combines morphological features of *Dicksonia* (hairs instead of scales) and *Cyathea* (unequivocally superficial, not submarginal sori) (cf. Walker, 1966a, p. 207). Somewhat lower numbers are found in *Cystodium* ($n = 56$), which is morphologically nearly related to *Dicksonia*, and in *Culcita* subgenus *Calochlaena* ($n = 58$). (Though the morphological difference is not great, there is evident cytological and geographical justification for the separation at generic level of the Austro-Malaysian *Calochlaena* from the Macaronesian and New World *Culcita*.) The high chromosome number of *Metaxya*, $n = 94$-96, $2n = 190$-192, is of problematical significance. The constancy of the curious number of 69 throughout the large genus *Cyathea* is very remarkable. R. Tryon (1970), who does not accept the inclusion by Holttum and Sen (op. cit.) of the dicksonioids within the Cyatheaceae, proposes a revision of the cyatheoid ferns wherein the constituents of *Cyathea sensu lato* (incl. *Cnemidaria*) are distributed into six separate genera. Walker (1973a, p. 96) reports that he has counted representatives of all of Tryon's genera and all have $n = 69$.

Overall, the cytological evidence favours the unification of the tree-ferns into a single family. As a group, they show distinctive and curious cytological features, a virtual absence of polyploidy being combined with very high and very characteristic base numbers, which are in general higher than those of any other ferns, excepting only certain very ancient genera, e.g. *Ophioglossum* and *Schizaea*. The fact that each of the numbers known in the Cyatheaceae is peculiar and distinct, and the ubiquitous nature of 69 in the enormous genus *Cyathea* suggest that these high numbers are fundamental to, and ancient in, the family, a point which, as will become apparent in later discussion, may be very significant.

The possession of 65 by both *Dicksonia* and *Lophosoria* may indicate the direction of the main trend of cytological evolution in the Cyatheaceae. Thus the fundamental base within the family might be 65 or 66, with *Culcita* and *Cibotium* (68) and *Cyathea* and *Cnemidaria* (69) representing two independent lines of accretion by aneuploidy, in which case *Cystodium* and *Calochlaena* (56, 58) represent the opposite trend of aneuploid reduction. On the basis of this hypothesis the position of *Metaxya* (94-96) would remain an anomaly unless the true ancestral base of the family was $x = 33$. This last suggestion would, however, require the rather uncomfortable assumption that all extant tree-ferns are anciently tetraploid from this base, yet none show any further polyploid development, (see Note Added in Proof, 4, p. 415) almost as if the tree-fern habit in itself in some quite mysterious way imposes a brake on higher levels of polyploidy; in other words that $2n = 130$-138 is approximately the maximum chromosome number their tissue organization can tolerate (Note: *Metaxya* is not, in habit, a tree-fern.)

In view of the prevailing emphasis (to which only Mehra (1961) and Mickel (1973) constitute exceptions) on the dicksonioid, thyrsopteroid and cyatheoid lineages as the origin of a high proportion of modern groups of ferns, the high chromosome numbers found in the Cyatheaceae present an evident difficulty, unless the $x = c.33$ base number hypothesis is correct. Further consideration of this problem is postponed until the frequency of major aneuploid changes and their possible significance in fern evolution has been discussed.

F. SIGNIFICANCE OF ANEUPLOIDY IN MODERN FERNS

Two quite unexpected developments in the cytological study of the genera *Lomariopsis* and *Lepisorus** are in my opinion of very great significance in relation to our understanding of cytological evolution in ferns.

Lomariopsis is the type genus of the Lomariopsidaceae, a very distinctive family of controversial affinity. *Lomariopsis cochinchinensis*, an Asiatic species, was found to have the base number $x = 41$, in common with all other genera so far investigated within the family, namely *Bolbitis*, *Egenolfia*, *Elaphoglossum*, *Lomagramma* and *Teratophyllum* (cf. Roy and Manton, 1966, p. 344). In contrast, four Ghanaian species of *Lomariopsis* investigated by Roy and Manton (1966) had very different chromosome numbers, namely $n = 39$ and $2n = 78$ in *L. guineensis*, $2n = 78$ in *L. palustris*, $2n = 62$ in *L. rossii* and $2n = 32$ in *L. hederacea*. The base number 39 found in *L. guineensis* and *L. palustris* presents no real difficulty in interpretation, since it can readily be derived from 41 by means of two aneuploid changes. However, a number as low as 16 could be derived from 41 only by the most drastic reorganization of the chromosome complement. Nevertheless, there is no plausible alternative to the hypothesis that the extraordinary number $n = 16$ in *L. hederacea* constitutes one end of a reduction series, in which $n = 31$ in *L. rossii* represents an approximately halfway stage, and $n = 39$ an initial phase, in the sequence. The constancy of 41 in other genera of the Lomariopsidaceae dictates that the sequence be read in this direction.

The chromosome complements of *Lomariopsis rossii* and *L. hederacea* both show a great range of variation in chromosome size (in other words, an extremely heterogeneous chromosome complement) to an extent unequalled elsewhere in the ferns, though a very much less extreme degree of this phenomenon is known in some low-numbered Hymenophyllaceae, e.g. *Hymenophyllum tunbrigense* (Manton, 1950; Manton and Vida, 1968). This diversity in

---

* Current taxonomic practice (Holttum, in Willis, 1973; Crabbe *et al.*, 1975) submerges *Lepisorus* within *Pleopeltis*. The name *Lepisorus* is retained here purely as a matter of convenience, since it is used in all the principal contributions discussed here.

chromosome size can materially assist explanation of these aberrant chromosome numbers.

Reduction of chromosome numbers can occur by three principal means. Firstly, simply by loss of entire chromosomes from the complement by mitotic misadventure. It is obvious that such an event, though it may be the cause of some examples of small aneuploid changes, cannot possibly be responsible for reduction in number as massive as from 41 to 16, or even from 41 to 32, since the loss of genetic material involved must certainly be unacceptably great. Secondly, by fusion of centromeres. A sequence of such events can transform a complement of chromosomes with acrocentric* centromeres into one containing fewer chromosomes, some of which are longer than the norm, and possess centric or subcentric centromeres. This process is widespread and very well studied in *Drosophila* (Stone, 1962) and in several groups of Orthoptera (White, 1973) but is less common in plants (Stebbins, 1971, p. 92), though well documented examples are known, e.g. in *Fritillaria* (Darlington, 1963, p. 110). Roy and Manton have not presented karyotype diagrams for any of the species of *Lomariopsis* they studied, but the quite excellent photographs presented (op. cit., pl. 2, figs 4 and 5) of the chromosome complements of *L. rossii* and *L. hederacea* are clear enough to make it quite evident that only very few chromosomes in either species have subcentric or centric centromeres. It is possible however for metacentric or subcentric chromosomes to revert to an acrocentric state by means of subsequent pericentric inversions.

The third means of reduction of chromosome number is by translocation (either by one event, or in a succession of stages) of the main body of one chromosome away from its centromere to another chromosome, leaving only a small centric fragment which, if it contains no essential genetic material, will ultimately become lost by mitotic error. This mechanism, which is known as the dislocation hypothesis (Navashin, 1932; Stebbins, 1971, p. 86), has been clearly demonstrated to have operated several times in *Crepis* in reduction of species with $x = 5$ to $x = 4$, and from $x = 4$ to $x = 3$ (Babcock and Jenkins, 1943; Tobgy, 1943; Stebbins, 1950, pp. 446-450). To my mind it is clear from the evidence of the extraordinary karyotypes possessed by *Lomariopsis rossii* and *L. hederacea*, that it is this process which has been mainly responsible for the drastic reduction in numbers they have undergone. Thus somatic cells of *L. hederacea* contain 12 excessively large chromosomes and eight very small ones, of which four are small enough to be described as centric fragments. In a similar fashion, *L. rossii* contains about 14 very small chromosomes, with the rest of the chromosomes displaying a range in size, though none appear to be as large as the largest chromosomes of *L. hederacea*, some of which, if this explanation is correct, must have become the recipients of the essential genetic material of more than

---

* The cytological terminology employed here is that of White (1945, 1973). See Stebbins (1971, p. 13 *et. seq.*) for a brief and clear account.

one other chromosome, since almost all of the chromosomal substance of the *L. hederacea* genome is concentrated in only 12 chromosomes, rather than 41.

I have allowed myself to be more dogmatic in my interpretation of this phenomenon than Roy and Manton themselves felt to be justified on the basis of the existing evidence, but I can visualize no real alternative to interpreting this wholly remarkable series as a reduction sequence, in which we have knowledge of the stages $41 \rightarrow 39 \rightarrow 32 \rightarrow 16$, and in which the great bulk of the changes involved are produced by chromosomal fusions. In other words, the chromosome complements of *Lomariopsis rossii* and *L. hederacea* have been not so much reduced as reorganised.* I will permit myself a further speculation. The abundance of small chromosomes in *L. hederacea* and *L. rossii* suggests to me that these dramatic cytological changes are of relatively recent age, and that there has not yet been time for all of the resulting centric fragments to become lost. There are also grounds for belief, though this is a difficult and mostly hypothetical field, that a drastic chromosome revolution of this character may be followed by a further redistribution of chromosomal material, an equalization in fact, restoring greater homogeneity to the karyotype by means of further interchanges or translocations restoring longer arms to fragments which have survived because though small, they nevertheless carry essential information. Such a process constitutes a form of karyotypic orthoselection (White, 1973). It follows that the product or products of a similar sequence of chromosomal reduction of substantially greater age may not display the striking internal evidence shown in *Lomariopsis*.

It must further be borne in mind that cytological revolution and reorganization of this general type can logically be expected to provide foci for fresh bursts of evolutionary radiation. In other words, a reconstituted genome of 16 chromosomes, if a successful innovation, may itself be the parent of a great many variants possessing the same basic karyotype, so that what first appeared as an aberrant number may in the course of time become the norm for a group of species or genera. It is necessary, in the course of analysis of the cytological data, to take care that we do not interpret such a secondary base number as one of primary origin, which we may be particularly likely to do if, as may happen in the course of evolutionary time, the group from which our seminal $n = 16$ plant was derived has become extinct.

Be that as it may, the great interest of these remarkable cytological facts discovered in *Lomariopsis* by Roy and Manton cannot be denied, and it is to be regretted that no worker has yet had opportunity either to extend their

---

* This remarkable series in *Lomariopsis* suggests an alternative possible explanation for the unexpected chromosome number seen in *Hemidictyum*, mentioned above (p. 259), and included here in the Thelypteridaceae. Just as in *Lomariopsis* $41 \rightarrow (39) \rightarrow 32 \rightarrow (16)$, so perhaps Hemidictyum may have evolved by 40 or $41 \rightarrow 31$. As Manton (1973, p. 261) points out, if this is the case, then examination of the karyotype may produce residual evidence of the chromosome changes involved.

observations to other African species of *Lomariopsis* or to examine other populations of the species already sampled.

There are good grounds for regarding the remarkable situation in the genus *Lepisorus* as representing a stage comparable in some respects to one in the hypothetical sequence of events just visualized, although fortunately the events in this genus have been sufficiently recent to permit enough evidence to have survived to prevent it from being fundamentally misinterpreted.

Lepisorus belongs to the Polypodiaceae (*sensu stricto*) in which every genus yet studied possesses at least one of the base numbers 35, 36 and 37. The commonest numbers are 36 and 37; only one genus, *Belvisia*, does not possess either of these. Apart from $x = 35$, *Belvisia* possesses, in common with *Crypsinus*, the number $x = 33$, the only base number other than 35, 36 or 37 encountered in any of the Polypodiaceae, excepting the unique series found in *Lepisorus* (Patnaik and Panigrahi, 1963; Mitui, 1971).

Six taxa in *Lepisorus* have the base 35, three possess 36 and one has $n = 74$. There is therefore no question that 35 or 36 (or perhaps 37) is the primary base number for the genus. Nevertheless, a second series of much lower base numbers exists in *Lepisorus*; two taxa with $n = 22$, one with $n = 23$, three with $n = 25$ (also two with $n = 50$), one with $n = 26$ (and another with $n = 52$).* Two further numbers are known. These are $n = 51$, which presents no problem in interpretation because it occurs in the species *L. thunbergianus* wherein all the numbers $n = 25$, 26, 50 and 52 are also known, and $n = 47$, which can reasonably be interpreted as most probably a dibasic polyploid derived (as 22 + 25) from the low-numbered series.

There can be little doubt that this remarkable sequence of low numbers, 22 to 26, had its origin in some initial drastic cytological reorganization comparable in extent to that seen in *Lomariopsis rossii*. It is of great interest that Mitui (op. cit.) finds that the two distinct cytological series (22-26 and 35-37) in *Lepisorus* coincide with the separation by Tagawa (1959) of the Japanese representatives of the genus into two distinct morphological groups. Thus the species based on $x = 35$ or 36 have transparent scales and irregular paraphyses, while the second series, based on $x = 22$-26, have scales with opaque centres and circular paraphyses. He is also of the opinion that the former group retains more primitive characteristics and the latter group is younger genetically. The morphological distinctions between the two cytological groups suggests to me that the "hole" in the cytological sequence between 26 and 35 is of some antiquity, and that the range of numbers now seen between 22 and 26 may be a secondary aneuploid radiation of more recent date, perhaps originating from some particularly successful product persisting from the initial chromosomal

---

* Another low-numbered taxon is *Lepisorus pseudonudus*, $2n = 39$ with an irregular meiosis ($11 < 17$ bivalents and $17 > 5$ univalents) (Patnaik and Panigrahi, 1963, pl. 5, figs 7 and 8; Panigrahi and Patnaik, 1964a, pl. 1 fig. 3), which is too bizarre to be interpreted with confidence, but is presumably an aneuploid derivative of some kind.

reorganization. Unfortunately there is at present very little evidence regarding karyotypes in *Lepisorus*. One published mitotic figure from a low-numbered taxon, *L. thunbergianus*, $2n = 50$ (Mitui, op. cit., fig. 1F), shows some indication of size differences, but nothing comparable with those seen in *Lomariopsis rossii* and *L. hederacea*. Nevertheless, a careful study of karyotypes in *Lepisorus* would be of great interest.

Lastly, it is indeed fortunate that both the unique series seen in *Lepisorus*, and the equally remarkable group of species of *Lomariopsis* occur in families whose cytological behaviour is otherwise so uniform that the aberrant behaviour of these two genera can be interpreted unequivocally as sequences involving a massive reduction in the chromosome number. Against a more diverse background, such an interpretation would not be so self-evident.

What appears to be the remains of another aneuploid series can be seen elsewhere in the Dryopteridaceae in the genus *Woodsia*. Kurita (1965) provides a valuable compilation of the cytological facts relating to this genus, though I am not able to support his analysis of the evidence. At least four species,* *W. elongata*, *W. ilvensis*, *W. macrochlaena* and *W. polystichoides*, have $n = 41$ or 82, while *W. glabella* has $n = 39$ (Meyer, 1959; Britton, 1964), *W. obtusa*, *W. oregana* and *W. scopulina* all have $n = 38$ or 76 (Wagner, 1963; Brown, 1964; Taylor and Brockman, 1966), and *W. manchurensis* has $n = 33$ (Mitui, 1965; Kurita, op. cit.). Kurita also presents a plate showing a mitotic count ($2n = 66$) for this last species, which though not of sufficient quality to establish details of the karyotype nevertheless suggests that appreciable differences in chromosomes size are present.

A curious and significant feature is that the low numbered species with $x = 33$, 38 and 39 are located variously in all three different sections of the genus, while each of these sections also contains species based on 41. This circumstance can only mean that either the recognized sections are not natural groups or else that aneuploid changes have occurred in three separate lines within the genus *Woodsia*, in which case the direction of these changes must have been reductional. Kurita (1965) appears not to consider the possibility that the low numbers in *Woodsia* are reduced states. His own analysis is not entirely sound. Thus he writes (p. 361), "the author considers that 33 and 39 in *Woodsia* suggest affinity with $n = 34$ and 39 in Gleicheniaceae." However, 34 and 39 belong to different genera in the Gleicheniaceae. It follows that though it

---

* *Woodsia mexicana* is also reported as $n = 82$ (Knobloch and Correll, 1962) but Brown (1964) states that examination of a single cell suggested the number $2n = 152$. *Woodsia alpina* is recorded as $n = c.82$ by Manton (1950) and Vida (1964). Prior to the demonstration of $n = 39$ in *W. glabella*, Manton (op. cit.) reported "of the order of 40" pairs and univalents in *W. ilvensis* x *alpina*, and suggested (p. 151) that *alpina* might be derived by allopolyploidy from *ilvensis* x *glabella*. Brown (op. cit., p. 20/21) believes this hypothesis is correct, but points out that in this case the true chromosome number of *W. alpina* should be $n = 80$. (The situation is further confused by a report by Sorsa (1963) of $n = c.82$ in *W. glabella*. See also Löve and Löve (1976) Taxon, **25**, 486-487.)

is remotely conceivable that there is some real relation between 33 in *Woodsia* and 34 in Gleicheniaceae, the coincidence of 39 and 39 makes a relationship no more likely, because *Woodsia* could not be derived from both *Sticherus* (34) and *Dicranopteris* (39). Species evolve along lines of descent, not in parallel on a broad front.

Wagner (1973b, pp. 245/246), taking a lead from a biophysicist (Platt, 1964), has recently reminded pteridologists that they should devise alternative hypotheses and test them against the data as sympathetically as those they personally support. The possibility that 33, 38, 39, 41 is a progressive series, and that the lower numbers really are relict numbers indicating the path by which $x = 41$ has arisen in the Dryopteridaceae must therefore be examined. We shall see below (p. 312) that the number 33 has a considerable potential significance. One is surprised to find that Ching (1940) actually separated *Woodsia manchurensis* and called it *Protowoodsia*. However, the characteristic feature of this new genus, tetrahedral (= trilete, and therefore primitive) spores, was a misinterpretation on Ching's part. The spores of this species, like other species of *Woodsia*, are bilateral (Copeland, 1947; Brown, op. cit.). Nevertheless, Copeland (op. cit., p. 105) writes that the section (*Physematium*) containing *W. manchurensis* "may be presumed to be the more primitive element in the genus". Furthermore, Holttum (1949, p. 291) writes that *Woodsia* and its relatives "show most primitive characters" amongst the Dryopteridoideae, endorsing the earlier opinion of Bower (1928), who also regarded *Woodsia* as a primitive dryopteroid fern though, since he believed that the genus illustrated the formation of a cup-like indusium by fusion of encircling hairs, he would hardly have regarded *W. manchurensis*, with its conspicuous complete indusium, as one of the most primitive of living species of the genus. Bower was confident that the origin of *Woodsia* and related genera was from the cyatheoid ferns, and indeed he goes so far as to write (op. cit., p. 112) of *Woodsia*, "This genus may be regarded as comprising small arctic and alpine representatives of the type of *Cyathea* itself." However, with this aspect of Bower's opinion Holttum is not in agreement, believing "that an origin from *Dennstaedtia* or an allied genus is more likely." (Holttum, 1947, p. 150).

Irrespective of the question of their point of origin, it is clear that a plausible case can be made for the primitive nature of *Woodsia* amongst dryopteridaceous ferns. It follows that the possibility that the numbers 33, 38 and 39 in *Woodsia* are relict numbers showing steps by which the characteristic dryopteridaceous number 41 has evolved cannot be dismissed out-of-hand, even though, as we have seen, it would necessarily involve the consequence that the taxonomic subdivision of the genus established by Hooker and Ching was entirely unsound.

In his monograph of *Woodsia*, Brown (1964) does not retain the sectional divisions of any earlier author. However, he presents a chart of the relationships he suggests exist within the genus, in which he accepts aneuploid reduction from a base of 41. Unfortunately, no cytological information is available for any of the four species, *W. andersonii*, *W. cinnamonea*, *W. cycloloba* and *W. lanosa*, which Brown believes to be the most primitive living species. It is clearly very desirable that more species in this interesting genus be studied cytologically and that, if possible, karyotypes be accurately analysed throughout the whole genus.

An undisputable and extensive aneuploid series within the confines of a single genus is present in *Blechnum*, in which an almost unbroken sequence from 28 to 36 is known, i.e. 28, 29, 31, 32, 33, 34, 36 (and *c*.37). Walker (1973a, p. 99) writes "There is a suspicion that possibly the primitive number in the genus is 33, this being found in *Brainea* . . . and also in *Sadleria*." If true, this would indicate that both additive and subtractive aneuploid changes have taken place in *Blechnum*. At present, the evidence for 33 being primitive for the family Blechnaceae is not strong. Copeland (1947) regards both *Brainea* and *Sadleria* as being derivative forms, while Holttum (1949, p. 289) writes, "The most primitive frond form . . . is seen in *Woodwardia* . . . and this genus may well show the most primitive sorus form also." *Woodwardia* possesses both $x = 34$ and $x = 35$. It may be significant, though equally possibly misleading, than an indubitable example of fossil *Woodwardia* (see p. 251 above) is known from Palaeocene rocks.

Blechnum is one genus wherein karyotype studies can be expected to be highly informative, and it is surprising, since *Blechnum* constitutes relatively complaisant material for study of mitosis and the existence of an aneuploid series has been apparent for some years (cf. Manton and Sledge, 1954), that no study of this kind has yet been made.

Aneuploidy has undoubtedly played an important role in the differentiation of genera within the Thelypteridaceae. In the Old World, although a majority of genera are based on $x = 36$, other individual genera are characterized by bases of 30, 31, 32, 34 or 35. In the Old World, where generic limits are as yet more broadly defined, the large genus *Lastrea* contains a whole series of numbers from 27 to 36. It is noteworthy that most of the genera which conserve one or more characters believed by Holttum to be primitive for the group have numbers intermediate within the range of the family. Thus *Coryphopteris* has $x = 32$, while *Parathelypteris*, *Pseudosphegopteris* and *Macrothelypteris* have $x = 31$, which suggests that these genera might also be conservative in chromosome number, and that a radiation based on 36 is a relatively late development in the evolution of the family. On the other hand, the only genus which retains a trilete spore, *Trigonospora*, has $x = 36$.*

One group which has given much aggravation to cytologists, taxonomists and phylogenists alike is that containing the dennstaedtioid, lindsaeoid and hypolepidoid ferns. In a recent valuable review, Mickel (1973, p. 136) treats the Dennstaedtiaceae as comprising these three groups. This convenient approach is

---

* Both Loyal (1963) and Smith (1971, p. 46) have presented hypothetical schemes representing aneuploid evolution in the family from an ancestral base number of 36.

followed here, in spite of misgivings that a fundamental natural discontinuity might exist between the Hypolepidoideae and the rest of the Dennstaedtiaceae. However, if such a boundary exists, at present I would not know exactly where to draw it.

It is necessary to have entered into this brief taxonomic preamble because of the differing concepts various authors have of the limits of the Dennstaedtiaceae. Thus for Holttum (1947) the family included the majority of the modern ("polypodiaceous") ferns, as opposed to Ching (1940), who included therein only eight genera, the dennstaedtioid ferns, *sensu stricto.*

For cytologists the aggravation comes in part from the technical difficulties involved in investigating some of these ferns, more specifically the lindsaeoids, which are frugal in their production of meioting sporangia, and for the most part refuse to persist, let alone flourish, in cultivation. In consequence, a substantial number of the chromosome numbers recorded for *Lindsaea, Schizoloma* and *Odontoloma* are only approximate counts (cf. Walker, 1973a, p. 98).

For taxonomists there are problems in that many of these genera are difficult to define and difficult to separate satisfactorily from one another.

More general aggravation is occasioned by the perplexing range of chromosome numbers which have been found in the Dennstaedtiaccae, which is disconcerting in view of the critical position these ferns occupy in most modern schemes of fern phylogeny, being of central importance in both of the schemes of Holttum (1949, 1973a) and that of Mickel (1973). A typical example is the key genus *Dennstaedtia*, which includes numbers based on all of 30, 31, 32, 33, 34, 46 and 47. The gradual accumulation of knowledge of chromosome numbers has, however, made it possible for cytologists to discern some signs of pattern, which was singularly lacking from the known data when fewer species had been examined (cf. Manton, 1958a). Walker (1973a) draws attention to the bimodal character of the series of base numbers now known in *Dennstaedtia* itself, and proposes a scheme dependent on the co-existence of an ancestral series of much lower numbers, 15, 16 and 17, with the extant series 30-34. The series 30 to 34 is itself considered to have originated as a series of in part dibasic polyploids derived from combinations of 15, 16 and 17, the higher numbers 46 and 47 having their origin also as dibasic polyploids, e.g. $17 + 30$ or $16 + 31 = 47$, $16 + 30$ or $15 + 31 = 46$. This scheme may be unnecessarily complicated, since there is no need to assume the existence of more than one ancestral base number (say 16, for argument's sake) or any need to postulate more than one dibasic polyploid step, since aneuploid changes in both the 30s and the 40s groups of numbers such as seen to be common, at least at a modest level, in many fern groups, can account for all the other numbers known, e.g. $16 \times 2 \rightarrow 32$, and $32 - 1 \rightarrow 31$, $31 + 16 = 47$, $47 - 1 = 46$. Walker points out (op. cit, p. 96) that his "scheme is advanced purely to indicate how it is *possible* such numbers have been derived. ... It cannot be too strongly emphasized that there is no direct evidence that the postulated ancestral numbers 15, 16 and 17 did exist". I would prefer not to postulate such low ancestral numbers at all, unless there is no alternative. As it is, I believe an alternative explanation is possible. Mention has already been made of the taxonomic difficulties associated with generic delimitation in these ferns. There is general agreement between Copeland, Holttum and Mickel that though some species of *Microlepia* are distinct enough from some species of *Dennstaedtia*, another element in *Dennstaedtia* is, as Copeland writes (op. cit., p. 51), "so near to *Microlepia* that the difference is not always certain". It is therefore interesting that the base numbers recorded in *Microlepia*, x = 40?, 42, 43 and 44?, go some way towards filling the gap in the sequence known in *Dennstaedtia*. I am induced to wonder whether or not there might be a long aneuploid series underlying the entire subfamily Dennstaedtioideae, more fundamental and more ancient than the present generic divisions.

Having gone so far into the realm of speculation, I am inclined to go further. If one compares the range of base numbers known in the Dennstaedtioideae with that known in the Lindsaeoideae (Table II) one cannot but be struck by the degree of coincidence. Can it be that there is here the influence of one long great aneuploid explosion more ancient than the differentiation of these two subfamilies? The length of this suggested aneuploid series, from 30 to 48 (or c.50), is in relative terms no greater than that known today in *Lepisorus*, 22-36 (37?) (disregarding the secondary dibasic polyploid *L. subconfluens*, with n = 47). If such an aneuploid series did exist, it would imply that the generic and subfamily boundaries currently drawn in the Dennstaedtiaceae do not all accurately delimit natural evolutionary groups, and that there are more lines of evolutionary development than we can yet recognize. As already pointed out, this is a group in which taxonomic confidence with regard to the definition of genera is not high, though there is general acceptance of the distinction between the dennstaedtioids and the lindsaeoids, to the extent that they are now often placed in independent families (Pichi-Sermolli, 1958; Nayar, 1970). Nevertheless, the pattern of chromosome numbers shared by these two groups is unique to them amongst ferns, and it would be curious indeed if they shared only a single common ancestral base number, and all existing similarities in their cytological architecture were due to coincidences of parallel cytological development.

An advantage of this hypothesis is that it provides a convenient range of base numbers from which other groups could be held to have evolved, should the origin of such families as the Davalliaceae and the Dryopteridaceae be sought from within the Dennstaedtiaceae. We shall return to this aspect of the problem later.

The question of the nature of any relationship between Dennstaedtioideae and Hypolepidoideae is an extremely difficult one. The genus *Hypolepis* itself presents very decided problems. Morphologists are agreed on the existence of a strong affinity between *Dennstaedia* and *Hypolepis*. Thus Mickel (op. cit., p. 137) refers to the difficulty of distinguishing these two genera in the Old

TABLE II

Chromosome base numbers in the three subfamilies of the Dennstaedtiaceae.

Subfamily	22	23	24	25	26	27	28	29	30	31	32	33	34	35	36	37	38	39	40	41	42	43	44	45	46	47	48	49	50
HYPOLEPIDOIDEAE	X		?°		X	X	?*	X										?									?°		
DENNSTAEDTIOIDEAE									X	X	X	X	X						c.		X	X	X		X	X			
LINDSAEOIDEAE													X				X	X	c.	X	X	X	X			X	X		c.

* *Monachosorum n* = 56, 112, 168
° *Histiopteris n* = 48,96

World. Holttum (1949, p. 285) writes, "I am by no means sure that the line of distinction between *Dennstaedtia* and *Hypolepis* is sharp . . ."; while Copeland states (op. cit., pp. 57/58), "The more primitive element in the genus [*Hypolepis*] . . . is hardly distinguishable from a similar element in *Dennstaedtia*." Yet, although *Hypolepis* shows a rather peculiar range of chromosome numbers (n = 29, 39, 52, 98, 104) which might suggest the genus was not a natural one, there is no clear affinity with any number known in the Dennstaedtioideae.

One particular incongruent element in the list of chromosome numbers in *Hypolepis* requires further investigation. Sorsa (in Fabbri, 1965), gives n = 39 for *H. repens* from Costa Rica, whereas Wagner and Chen record n = 104 in the same species from Florida. It is hardly conceivable that one and the same species can have two such different numbers; it is possible to hypothesize a relationship as hexaploid and 16-ploid derivates from a base of x = 13 (even though the combination of two such ploidy levels would be decidedly unusual), but this is most probably a false trail. A real base number as low as x – 13 in *Hypolepis* is in my opinion highly unlikely. It is clearly desirable that the identity of the Costa Rica plant be confirmed.

Another distinctive number, n = 29, is not in doubt, having been established for *Hypolepsis nigrescens* independently by Mickel *et al.* (1966) in Mexico, and by Walker (1966a) in Jamaica. Both Mickel (1973, p. 139/140) and Walker (1973a, p. 97) draw attention to the determination by Brownlie (1957) of n = 26 in *Paesia scaberula*. In *Pteridium* the great majority of counts are n = 52, or $2n$ = 104, but Löve and Kjellqvist (1972) record $2n$ = 52 in *P. herediae* from Spain and in *P. aquilinum* var. *gintlii* from Yugoslavia.* The existence of x = 26 in both *Paesia* and *Pteridium* provides strong support for the former existence of 26 in *Hypolepis*, as a base from which the existing numbers n = 52 and n = 104 are derived. Both Mickel and Walker point out that once the former existence of n = 26 in *Hypolepis* is accepted, then it is reasonable to regard 26 and 29 as parts of an ancient aneuploid series of which n = 29 still survives.

Whether aneuploidy from a base of 26 truly exists in Hypolepidoideae other than in *Hypolepis* is uncertain (the origin of n = 48, 96 in *Histiopteris* is obscure), but I have placed here the genus *Idiopteris* described by Walker (1957) as clearly distinct from *Pteris* not only in its chromosome number of n = 27, but also in its submarginal sorus. I have also included in the Hypolepidoideae, though not without substantial misgivings, the troublesome genus *Monachosorum*. The chromosome number of *M.* (*Ptilopteris*) *maximowiczii* is confirmed as n = 56 (Mitui, 1968; Hirabayashi, 1968). Hirabayashi interprets some other members of the genus as constituting a polyploid series based on x = 56, but the

* Löve & Kjellqvist (op. cit.) interpret $2n$ = 52 in *Pteridium* as tetraploid on a base of 13, rather than diploid with x = 26, but they do not explain why they assume this lower base number.

numbers recorded, $n = 112$ in *M. subdigitatum* and $n = 168$ in *M. flagellare* (cf. Mitui, op. cit., $n = c.155$) are not clearly established and require confirmation. If $n = 56$ in *M. maximowiczii* is itself derived from $x = 28$, then this count is not excessively discordant placed amongst the Hypolepidoideae, particularly if this subfamily really contains an ancient aneuploid series between 26 and 29. *Monachosorum* has clearly perplexed taxonomists. Christensen (1938, p. 544) regarded it as an offshoot of *Thelypteris* (*sensu lato*). Copeland (op. cit.) placed it amongst dennstaedtioid ferns, although he states that he formerly regarded as related to *Acrophorus* in the Aspidiaceae (= Dryopteridaceae). Mickel (1973, pp. 141/142) believes it still fits better in this latter family, where its chromosome number will not allow it to go, unless it be a cytological aberration unique in the family. Most recently, Crabbe *et al.* (1975) have treated *Monachosorum* as a monogeneric subfamily, Monachosoroideae, within the Dennstaedtiaceae.

Some may consider that I have treated the Hypolepidoideae as a "dust-bin" group, a repository for genera which will not fit comfortably elsewhere, and such a charge is clearly to some extent well founded. But even if *Idiopteris, Taenitis,** and *Monachosorum* are removed elsewhere, the essential problem presented by the Hypolepidoideae remains unresolved.

Observing the principle of examining alternative hypotheses, the possibility remains that the fundamental number in the Hypolepidoideae is really 29 (or 30) and that this group is after all related to the Adiantaceae and more specifically to the Pteridoideae as an early divergent line which soon lost

---

* Holttum (1968) redefined the taxonomic boundary of *Syngramma* and *Taenitis*, transferring some species from the former genus to the latter. This revision has had the effect of clarifying the cytological distinction between these two genera. *Syngramma*, with a polyploid series based on $x = 29$, is cytologically a conventional member of the Adiantaceae. In contrast, *Taenitis* is cytologically unique. *T. blechnoides* includes both $n = 44$ (Manton, 1958a) and $n = 110$ (Walker, 1968) indicating that $x = 22$ in this species. It is just possible that all of the cytological records known for *Taenitis* are derived from a base of $x = 22$, but this would require that the material studied of *T. diversifolia* with $2n = ca.108-114$ (Vida in Holttum, 1975) is pentaploid, and therefore either an apomict or a sterile hybrid. Unfortunately, no information about meiosis is available. The high numbers $n = 108$ in *T. pinnata* (Vida in Holttum, 1975) and $n = 114$ in *T. hookeri* (Walker, 1968, as *Syngramma lanceolata*) could be aneuploid variants at the decaploid level. However, the true cytological structure of the genus may be more complex.

Recently, Holttum (1975) has discussed the systematic position of *Syngramma* and *Taenitis*, which he believes are related. By means of transverse sections he demonstrates clearly that the sorus of *Taenitis cordata* (= *Schizoloma cordatum* Gaud.) does not possess a true indusium, and is accordingly very different in structure from that of *Lindsaea ensifolium*, thus demolishing the principal basis for the association by Copeland (1947) of *Taenitis* with the lindsaeoid ferns. He concludes (op. cit., p. 341) that "I feel sure that *Syngramma* and *Taenitis* both belong to the Gymnogrammeoid ferns, not to the Dennstaedtiaceae."

If Holttum is right, then *Taenitis* (in Adiantaceae, with $x = 22, 27?, ?$) would constitute another example, like *Lepisorus, Lomariopsis* and *Woodsia*, of drastic aneuploid reduction in just one genus in an otherwise cytologically stable family, but here complicated and disguised by subsequent high levels of polyploidy and, apparently, by extinction of the original base number.

(almost) the characteristic pteroid chromosome number. However, in view of the close morphological relationship between *Hypolepis* and *Dennstaedtia*, this hypothesis is most unlikely to be correct.

The last hypothesis leads to consideration of another extremely uncomfortable concept. This is in effect the converse hypothesis, the idea that the Adiantaceae is not a natural group inasmuch as some part of it is really of hypolepidoid origin, having evolved the number 29 by aneuploidy from 26. We may recall with some alarm that $n = 29$ is already present in *Hypolepis*. In this respect a questioning finger has been pointed specifically at *Pteris*, which Holttum (1973a, p. 5) gives alternative positions in his latest phyletic scheme, either in Adiantaceae or Dennstaedtiaceae, and furthermore writes (op. cit., p. 8): "*Pteris* is usually associated with the 'gymnogrammoid' ferns, but it seems to me to show little definite relationship to any of them, and I still think it is probably related to *Pteridium*, and so to *Dennstaedtia*." Doubt is also expressed by Mickel (1973, p. 142) on account of some species of *Pteris* possessing dennstaedtiaceous anatomical characters, although he does conclude that on the balance of evidence he still places *Pteris* with the Adiantaceae. Certainly, the existence of $n = 29$ in *Hypolepis* raises a spectre which study of chromosome numbers cannot dismiss, and which must instead be resolved on other evidence.

I return to the question of the degree of relationship between the Hypolepidoideae and the rest of the Dennstaediaceae. If this relationship is indeed close, then their very different chromosome numbers have to be explained. Can it be that the long ancestral aneuploid series I have proposed for Dennstaedtioideae and Lindsaeoideae is even longer and more fundamental than I suggested, and extends back further from 30 to 26? In other words, the possibility exists that the Dennstaedtiaceae is based on one very ancient long aneuploid series. This explanation may not appeal, but I am not able to conceive of another cytological explanation compatible with the existence of a genuine relationship between *Dennstaedtia* and *Hypolepis*.

G. CONTRIBUTION OF CYTOLOGY TO PHYLOGENY

It is necessary at the outset to remind ourselves that there are severe intrinsic limits to what cytology alone can tell us with any certainty regarding phylogeny. In particular, we are in uncharted territory when we endeavour to interpret what we find in the ancient groups. The evidence of cytology consists of facts obtained from living plants and therefore belongs to the present day. The more remote in time are the events we are trying to elucidate or recapitulate the more improbable is it that contemporary evidence can represent or reflect these events with any fidelity. We must not allow ourselves to forget, when endeavouring to reconstruct the early stages in the cytological evolution of the older fern families, that hundreds of millions of years separate our modern cytological evidence and past reality.

I have already criticized the low basic numbers hypothesized for such groups as Ophioglossaceae and Gleicheniaceae, the justification of which even in the former case is inadequate, while in the latter case it is tenuous to an extreme. Abraham *et al.* (1962) have produced a phyletic chart, a "tentative scheme showing cytological evolution of the ferns" in which all existing families are traced back to a series of base numbers 9, 10, 11, 13 and 15 (Fig. 4a). To my mind this scheme, which is undoubtedly elegant and interesting, is nevertheless built on shifting sand.

I am more impressed by the great differences between the ancient families in their cytology, and by the irrationality of the different numbers they each contain. This is particularly true of the Gleicheniaceae and Schizaeaceae and suggests the intervention of some dramatic cytological changes at times in their long evolutionary history, rather than a quiet progression up the polyploid levels, punctuated only by an occasional small aneuploid change. I am also impressed by the contrast between the cytological diversity of the Gleicheniaceae and Schizaeaceae on the one hand, and the majority of modern families on the other, most of which, although they encompass an astonishing range of morphological diversity, are extraordinarily uniform in their cytology. Dryopteridaceae and Aspleniaceae are excellent and perhaps extreme examples of this phenomenon, to which only the Thelypteridaceae, Blechnaceae and Dennstaedtiaceae (though this last family, as we shall see, may be a unique case) constitute substantial exceptions. This phenomenon is understandable, now that it is confirmed how very modern the modern fern families are, probably younger than their flowering plant equivalents (see p. 253 above). The great diversity of morphological types in the Aspleniaceae and the Dryopteridaceae is, in geological terms, of recent date, and has been accomplished, apart from a few interesting exceptions, without unlocking and drastic revision of the karyotype. Indeed, our modern ferns may not, in genetic terms, be so different from one another as their morphological diversity might induce us to believe they are. There is certainly evidence to that effect in the Aspleniaceae, where the most extreme phenotypes can be combined in a viable hybrid plant (Lovis and Vida, 1969; Lovis, 1973).

It will also be evident, from comments made above in relation to *Lepisorus* and *Lomariopsis*, that I believe that in general the possibility of radical reduction in the chromosome number has been unappreciated, and the possible evolutionary significance of radical rearrangement of the karyotype underestimated, by fern cytologists. Regarding this latter point I would reiterate that in such karyotype explosions may well lie the opportunity for a new phase of evolutionary radiation and development, based on a particular new and successful karyotype combination produced in the turmoil of cytogenetic revolution. The possible significance of this potentiality for drastic reduction in chromosome number in relation to schemes of cytological evolution will not need further emphasis, but it should be noted that some authors appear not to consider the possibility of reduction in chromosome number at all, save for the loss of an odd chromosome. Thus the scheme of phyletic shrubs drawn by Mehra (1961) is to my mind ruled out of serious consideration on this ground alone, only upward progression being countenanced to the extent that both the Dicksoniaceae and Cyatheaceae appear at the top of their particular trees.

There is no question that the modern "polypodiaceous" ferns are of polyphyletic evolution, though there is no agreement concerning how many lines of derivation are involved. Origins for different modern groups from all of the Schizaeaceae, Dipteridaceae (or Gleicheniaceae) and Cyatheaceae (*sensu lato*) are variously suggested. The cytological evidence relevant to these hypotheses will be reviewed in turn.

1. Schizaeaceous derivatives

Chromosomal evidence has indicated very clearly the existence of a large cytologically homogeneous group based on $x = 29$ and 30, and comprising the pteroid, cheilanthoid, gymnogrammoid, adiantoid and vittarioid ferns, all included here within the family Adiantaceae, which is quite distinct from the dennstaedtioid, lindsaeoid and hypolepidoid ferns. There does now seem to be general agreement (cf. Nayar, 1970; Holttum, 1973a; Mickel, 1973) that the origin of the adiantoid/vittarioid radiation is to be found in the Schizaeaceae. This is so, notwithstanding the fact that the base numbers 29 and 30 have been preserved in the Schizaeaceae only in the morphologically highly specialized genus *Lygodium*, and not in *Mohria* which represents a morphologically far more satisfactory parental type, providing a surprisingly good imitation of a cheilanthoid fern. The Vittarioideae, which are uniformly based on $x = 30$, are considered by some authors (Alston, 1956; Pichi-Sermolli, 1958; Nayar, 1970) to be morphologically sufficiently distinct to justify separation as a distinct family. It is an open question whether the vittarioids are an early and distinctive branch of the adiantoid radiation or are of independent origin from schizaeaceous stock. The fundamental base number in Adiantaceae is almost certainly 30, the aneuploid change to 29 having occurred commonly in the Adiantaceae and coincidentally also in *Lygodium*. However, the Pteridoideae might have originated directly by independent origin from a source with $x = 29$ in the Schizaeaceae. Thus $n = 30$ in *Acrostichum* may well be a secondary reversion, since it is apparently now universally accepted by taxonomists that this unquestionably specialized genus is a relative of *Pteris*. It is equally possible that the adiantoid, cheilanthoid and gymnogrammoid groups are each of independent origin from the schizaeaceous stock, in which case the separate family status often granted to these groups would be justified, but this problem remains an open question which cytology cannot resolve. In contrast, *Ceratopteris*, here included in the Schizaeales as the only genus of the Parkeriaceae, an aquatic or subaquatic genus usually considered to be related to cheilanthoid or gymnogrammoid ferns (cf. Bower, 1926; Copeland, 1947), has $x = 39$, 40 (the actual numbers recorded are $n = 39$, 77, 78, 40, 80). These base numbers would be discordant with the Adiantaceae as interpreted here but are entirely consistent with an independent origin from a schizaeaceous source, $x = 39$ being readily derived from $x = 38$, as known in *Anemia* and *Mohria*.

2. Dipteridaceous derivatives

The sole living survivor of the Dipteridaceae, so abundant in the Jurassic, is the genus *Dipteris*. The relation between *Dipteris* and Mesozoic Dipteridaceae is not seriously in question (cf. p. 237 above), although *Dipteris* does differ from Mesozoic Dipteridaceae in its monolete, not trilete, spore. *Dipteris* differs from both Polypodiaceae and Grammitidaceae with regard to a number of characters, including the complexity of its venation, the relatively simple (solenostelic cf. dictyostelic) anatomy of its rhizome, which bears hairs, not scales, the relatively massive stalk to the sporangium (four rows of cells, not two or one), the complete annulus and undifferentiated stomium. The gametophyte is also distinctive and arguably more primitive (Stokey, 1945), but none of these characteristics is so distinctive either alone or regarded as a whole as to prevent the Dipteridaceae being considered as the potential source of either or both of the Polypodiaceae and Grammitidaceae.

The position is somewhat complicated by the existence of *Cheiropleuria bicuspis*, a monotypic Malaysian genus, which shares with *Dipteris* its rhizome hairs, massive sporangial stalk and complete annulus, but possesses a well-defined stomium. In contrast, its rhizome structure is even more primitive (protostelic), the spores are trilete, and the output of its sporangium (*c.* 128 spores) is twice that of a typical leptosporangiate sporangium. Both of these last characteristics reflect those of Mesozoic Dipteridaceae (cf. Harris, 1961, p. 119), and *Cheiropleuria* is best regarded as a relic of an early and independent departure from dipteridaceous stock. It is therefore segregated in recent treatments (Pichi-Sermolli, 1958; Nayar, 1970) into a separate monotypic family, the Cheiropleuriaceae. (See Note Added in Proof, 5, p. 415.)

In *Dipteris* the base number $x = 33$ is established by counts of $n = c.33$ in *Dipteris conjugata* (Manton, 1954b) and $n = 66$ in *D. wallichii* (Patnaik and Panigrahi, 1963). The number 33 is known in both Grammitidaceae and Polypodiaceae, but only in association with other numbers in the same genus. Thus Walker (1966a) found $n = 33$ in *Grammitis hartii* (*Grammitis* otherwise has $x = 36$ or 37), and in the Polypodiaceae *Crypsinus enervis* has $n = 33$ (Manton, 1954b), as does also a species of *Belvisia* (Walker, 1973a). Other species of *Belvisia* have $n = 35$, while both $n = 35$ and $n = 36$ are known in *Crypsinus*. Genera of both Grammitidaceae and Polypodiaceae mostly possess base numbers of 36 or 37. It is probable that these isolated and apparently random occurrences of species with $n = 33$ are accidental rather than of genuinely relict character,[*] 33 here being of secondary origin due to aneuploid reduction.

---

[*] Panigrahi and Patnaik (1964b, p. 9) suggested that $n = 33$ in *Crypsinus, n = 22* in *Lepisorus*, and $n = 33$ in *Dipteris* represents a common ancestry "characterized by $x = 11$" for Polypodiaceae and Dipteridaceae.

The question of the existence of a relationship between Polypodiaceae and Grammitidaceae is a vexed one, though there does appear today, in contradiction to Copeland's treatment, to be general agreement that these two groups of ferns ought at least to be segregated into distinct families. Nayar (1970) believes the two families to be totally unrelated. (He associates Grammitidaceae with Thelypteridaceae and Aspleniaceae.) Holttum (1949) also believes the two families to be entirely unrelated, but more recently (1973a) he is undecided whether the Polypodiaceae are derived from a davallioid stock or, like the Grammitiaceae, originate (though independently) from a dipteridaceous line of evolution.

The Grammitidaceae differ from the Polypodiaceae in their trilete, not monolete, spores, in their unarticulated fronds, in their simpler rhizome and stipe anatomy, in their peculiar hairs and in the structure of the gametophyte. There appears to me to be nothing in these differences of sufficient consequence to exclude Holttum's second alternative, independent origin from an ancient dipteridaceous source. The exact correspondence of the cytology of Grammitidaceae and Polypodiaceae may in part be coincidental, but certainly cytology can give no support to their wide separation.

It seems to me that, as a general rule in the present state of our knowledge, in a case where the cytology gives no indication of discontinuity, and the morphological evidence is equivocal, in part suggesting affinity, and in part not, it is more prudent to retain families together in the same sector of the phylogenetic system, rather than impose a separation which is not decisively indicated. It is therefore appropriate to associate this group of families all together within a distinct order, the Polypodiales. (It would be less confusing, in view of the different ways in which all names beginning "Polypod-" have been used, to call this order Dipteridales, but this is not permissible.)

The question of a direct relationship between Dipteridaceae and Gleicheniaceae is complicated by the fact that it is possible to propose an origin for the Grammitidaceae (or the Polypodiaceae) from a gleicheniaceous source rather than via the Dipteridaceae. This case is argued by de la Sota (1973) in a valuable review of the Polypodiales. However, his views on basic origins and relationships are clearly influenced by an uncritical acceptance (op. cit., p. 235) of a record of *Polypodites* from the Triassic of Russia (reference not given) as a genuine example of the Polypodiaceae (*sensu stricto*), which has induced him to make the surprising claim that the "the Polypodiaceae would be *at least* as old as the Dipteridaceae" (op. cit., p. 242: my italics). Notwithstanding the recent discovery of a convincing *Drynaria*-like fossil, *Australopteris coloradica*, from Cretaceous rocks (cf. p. 247 above), there really is no justification for this statement. This question of a direct relationship between the Gleicheniaceae and either or both of the Dipteridaceae and Grammitidaceae is one not likely to be resolved by cytological evidence, since it could be only coincidence that the genera of Gleicheniaceae which appear to have the closest morphological affinity to the Polypodiales, namely *Sticherus* and *Dicranopteris*, have $x = 34$ and $x = 39$

(and 43) respectively, although these base numbers are clearly compatible with base numbers of 33, 36 and 37 in the sense that not very many aneuploid changes would be necessary to derive one group from the other.

3. Cyatheaceous derivatives

Ever since the publication of Bower's views on fern phylogeny, an important position has been postulated for the tree-ferns in relation to the origin of modern fern families. The question as to whether the dicksonioid and thyrsopteroid ferns are closely related to or entirely distinct in origin from the cyatheoid ferns is therefore fundamental to any further discussion. As we have seen (p. 293 above), majority opinion now once again regards the two main groups of tree ferns as closely related, and they are usually combined in a single family, Cyatheaceae. Moreover, the cytological evidence is consistent with such a treatment.

As far as derivation of modern groups from the Cyatheaceae is concerned, there is very strong morphological evidence in favour of an origin for *Dennstaedtia* and its relatives from a dicksonioid or thyrsopteroid source, and for the origin of the Thelypteridaceae from a cyatheoid stock. In each case the postulated lineage presents us with essentially the same difficulty, the necessity to explain the derivation of plants with $x = 30 < 47$ from a source represented today by species with $x = 56 < 68$, and the derivation of a family with $x = 27 < 36$ from a source now possessing $x = 65$ or 69.

One is struck by the circumstance that both of the modern groups, the Demmstaedtioideae (and, for that matter, the related Lindsaeoideae) and the Thelypteridaceae, are characterized by a range of chromosome numbers which must reflect a fundamentally volatile karyotype. We have noted that this is a decidedly unusual feature in modern fern families. Can it be more than coincidence? It is surely possible that the origins of the Dennstaedtiaceae and the Thelypteridaceae were both associated with violent revolution in the karyotype, with an explosive and drastic aneuploid reduction of chromosome number, such as we see clearly in *Lomariopsis*, and which must also have occurred in *Lepisorus*.

The logical alternative is that both of these groups had their origin in a more tranquil fashion, differentiating from ancestors in the Cyatheaceae at a time when these ferns possessed lower chromosome numbers in the range $n = 32 < 34$, their inception requiring no radical revision of the karyotype, though changes in karyotype of more modest scale must have taken place during their subsequent development and radiation. Of course, if the hypothesis proposed by Walker (1973a; cf. p. 302 above) regarding the origin of the chromosome numbers in *Dennstaedtia* is correct, then at the time this genus first evolved, an even lower level of ploidy, $n = c.16$, must have been co-existent with numbers in the low 30s.

If evolution did take this course, we have to come to terms with the curious fact that no low chromosome numbers are known today in Cyatheaceae ($n = 56$, 58 in *Cystodium* and *Calochlaena* do not materially affect the issue), yet the family appears to be quite devoid of any recent polyploids. This is however a rather weak objection, since base numbers in the sixties cannot be primitive, and must have evolved from lower numbers. The isolated genus *Metaxya*, with $n = 95$-96, could be hexaploid on a base of $c.32$, but it is always dangerous and unwise to draw firm conclusions on the strength of a single undoubtedly isolated and aberrant type.

My personal preference is for a more complicated explanation, only partially because I tend to look for complexity in evolutionary patterns (I believe there is little virtue in the persistent application of William of Occam's Razor to problems of cytological evolution), but also because it will fit the chronological evidence better. I envisage the Dennstaedtiaceae as evolving by means of a karyotype revolution involving drastic aneuploid reduction, but the Thelypteridaceae as evolving in a more orthodox fashion (Fig. 3).

We now have reason to believe (cf. p. 239 *et seq.*) that, compared with other "polypodiaceous" ferns, the Thelypteridaceae are a relatively ancient group, with good evidence of their existence in Jurassic time. There is no such clear evidence for the Dennstaedtiaceae, though it must be admitted that it will be very difficult to distinguish with certainty an incipient dennstaedtiaceous fern from some of the thyrsopteroid and dicksonioid ferns known to be abundant in the Mesozoic. At present, the fossil record is consistent with an earlier origin for the Thelypteridaceae, at a time before the Cyatheaceae had reached the level of cytological evolution at which we find them today.

There is at present no evidence to refute the conclusion (cf. p. 253 *et seq.*) that the bulk of modern families (with the certain exception of only Thelypteridaceae and Polypodiaceae, whose origins are independent and have already been discussed) are of remarkably recent origin, their main radiation materializing in the fossil record even later (i.e. at the beginning of the Tertiary period) than the appearance of modern flowering plant families. There is little enough evidence from which to proceed, but let us presume, as in the phyletic schemes of Holttum (1949) and Mickel (1973), that such groups as the Davalliaceae, Dryopteridaceae, Aspleniaceae and Blechnaceae all have their origins within the basic dennstaedtiaceous radiation, even though their points of origin cannot yet be defined. We have to explain the sudden appearance of this very diverse radiation, containing the great bulk of modern ferns, at a relatively recent geological date. I have suggested above (p. 255) that it was the diversification of the angiosperms and the new complexity they brought to the ecosystems of the world which acted as the trigger for the new burst of evolutionary activity on the part of the ferns. It is my belief that this astonishing late burst of evolutionary activity on the part of the ferns, which had, after all, already been in existence for a very long time, is correlated with a spectacular cytological event providing a drastic reorganisation of their genetic material. The

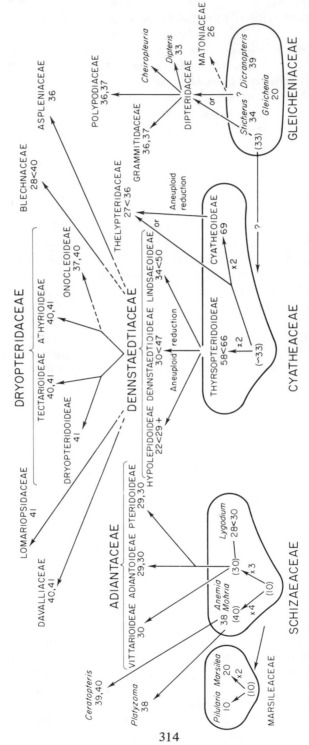

Fig. 3. A phyletic scheme for the modern ferns. The distinctive feature of this scheme is the derivation of the long series of different chromosome numbers known in Dennstaedtiaceae (cf. Table III) by a process of drastic aneuploid reduction from a source in the Cyatheaceae.

diversity of chromosome numbers in the Dennstaedtiaceae represent the remains of this cytological explosion, the consequential fragments continuing to evolve away from the epicentre of the upheaval, much as the Crab Nebula represents the aftermath of a supernova explosion.* I envisage that families like Aspleniaceae and Dryopteridaceae represent the exploitation of particularly felicitous karyotypes thrown up in this period of evolutionary turmoil in the initial dennstaedtioid radiation, karyotypes of so advantageous a constitution that there has been strong selection pressure for their maintenance without drastic alteration.

It is now necessary to return to earth after this astronomical simile. The work done in karyotype analysis by the Hiroshima school on Osmundaceae, Hymenophyllaceae and *Pteris* has already been mentioned. Apart from these studies, Tatuno and Okada (1970) have produced analyses for two species of *Diplazium* (Athyrioideae) and for one species of *Matteuccia* (Onocleoideae) on the basis of which they conclude that the primary base numbers for these three species is 11, so that $2n = 82$ in *Diplazium esculentum* and *D. wichuriae* $= 8x = 88 - 6$, and $2n = 80$ in *Matteuccia orientalis* $= 8x = 88 - 8$. Tatuno and Kawakami (1969) and Kawakami (1970) have provided karyotypes for a total of five species of Aspleniaceae and propose a primary base number of 12. On the basis of study of *Lemmaphyllum microphyllum* and two species of *Pyrrosia*, all with $2n = 72$, Takei (1969) proposes the same base number (12) for Polypodiaceae (*s.s.*). He furthermore compares the karyotypes of these species with those found in species of *Hymenophyllum* and *Mecodium* by Tatuno and Takei (1969a) (cf. p. 288 above) and suggests that in these latter genera "$b = 11$ may have derived from $b = 12$, as found in . . . Polypodiaceae" (op. cit., p. 482). Such a relationship is, however, morphologically incredible. I can only repeat my opinion that the degree of heterogeneity encountered in all these karyotypes is not so great as to exclude interpretations different from those advanced by these authors. Though I am sceptical of their conclusions, I do not underestimate the potential value of the studies they have undertaken. I consider their real value for phylogeny will prove, once a sufficient number of different species have been studied, to lie in comparative studies between different genera and families, rather than in attempts to determine directly ancestral base numbers. Nevertheless, in a discussion of this character, prejudice must be avoided, and it must be admitted that a base number of $x = 11$ for the Dryopteridaceae can be correlated with the phylogenetic relationships under discussion (Fig. 4b). However, as will be obvious from the earlier discussion, I do not believe it represents the truth. The facility with which alternative speculative schemes of this kind can be devised is shown by comparison with that of Abraham *et al.* (1962, pp. 410/411), in which the modern families are derived from bases of 9 and 10 (Fig. 4a).

* 58 in *Calochlaena* may represent a relict trace in the Cyatheaceae of the inception of this active phase of aneuploid activity.

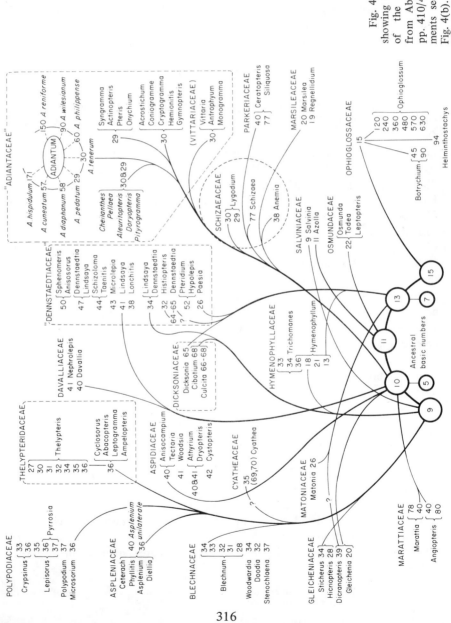

Fig. 4(a). Tentative scheme showing cytological evolution of the ferns. (Reproduced from Abraham *et al.*, 1962, pp. 410/411, insert.) For comments see pp. 308, 315; cf. Fig. 4(b).

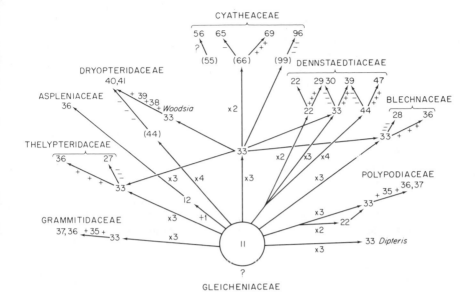

Fig. 4(b). Diagram to show how a hypothetical scheme can be devised displaying the origin of the basic chromosome numbers of all the modern fern families, other than those of schizaeaceous affinity, from an ancestral genome of 11, principally by simple polyploid multiplication. This scheme incorporates hypotheses proposed by the Hiroshima school of karyotype studies, e.g. Kawakami (1970), Tatuno and Okada (1970) and by Panigrahi and Patnaik (1964b). For the sake of simplicity neither Davalliaceae nor Lomariopsidaceae are shown, but since these families have the same base numbers as in Dryopteridaceae, their origin presents no arithmetical problem. It is argued in the present article that this type of scheme does not reflect the probable course of cytological evolution.

H. CONCLUSION

There are severe limitations on what cytology can tell us about relationships between fern families. This is well illustrated by consideration of the Thelypteridaceae. The cytological evidence informs us quite unequivocally that Thelypteridaceae are a distinct group from the Dryopteridaceae, but it cannot specify the degree of their mutual unrelatedness. Holttum (1949, 1973b) regards the Thelypteridaceae as remote from all other families of modern ferns and unique in their derivation from a cyatheoid ancestor. In contrast, Mickel (1973) derived them from a hypothetical stock ancestral not only to the Cyatheaceae and to Dennstaedtiaceae, but also the direct source of the main radiation of modern ferns, of which Thelypteridaceae are one branch, along with Aspleniaceae, Blechnaceae and Dryopteridaceae etc. Our cytological knowledge cannot settle this controversy, which must be resolved on the basis of other evidence.

Cytology can help us to define the limits of natural families amongst the modern ferns, but gives little information about their interrelationships. There is a depressing lack of variety about the chromosome numbers of modern ferns. If the Dennstaedtiaceae are put on one side as a special case then nearly all the rest fall into three arithmetical groups: (1) 29 or 30 = Adiantaceae; (2) 36 (or 37) = Aspleniaceae, Grammitidaceae, Polypodiaceae (and Thelypteridaceae— majority of genera and species); (3) 40 or 41 = Davalliaceae, Dryopteridaceae and Lomariopsidaceae. All that is left is the Onocleoideae, which with 37 and 40 has feet in two camps, and Blechnaceae, which is a law unto itself. In these circumstances, there is clearly substantial difficulty involved in sorting out coincidence from consanguinity, and one can with benefit quote here: "a word of warning about the limitations of the cytological method. It is obviously true that similar numbers can be synthesized independently from widely different sources, and therefore cytological resemblance is only a valid index of relationship if other morphological evidence points in the same direction" (Manton & Sledge, 1954, p. 174). Elsewhere, Manton (1953, p. 177) has elaborated on the different ways in which 36 and 40 might arise.

No one has yet felt the evidence of chromosome numbers to be so reliably diagnostic as to advocate associating together as putatively related all the families based on 36, and in a separate assemblage, all the families based on 40 or 41, though Nayar has come very close to doing this in his phylogenetic tree (Nayar, 1970, p. 231), in which he does indeed associate all the $x = 40/41$ families, and comes very near to it with respect to the $x = 36$ families, bringing together Aspleniaceae, Thelypteridaceae and Grammitidaceae, excluding only the Polypodiaceae.

The magic of chromosome numbers is not felt too strongly by Holttum, who in his latest scheme (Holttum, 1973a, Fig. 2) goes so far as to imply that the Dryopteridaceae (as defined here) are unnatural, the Tectarioideae together with the Lomariopsidaceae being derived from what one can call a "dicksonioid line", and the Dryopteridoideae and Athyrioideae being associated with a quite different line, a "thyrsopteroid line". Cytology can do nothing as yet to resolve this problem.

I feel bound to end this discussion on a pessimistic note. Knowledge of chromosome numbers very soon established the reality of the Adiantaceae as an independent and unified group, confirmed the plausibility of their origin from Schizaeaceae, and established the reality of the basic distinction between Thelypteridaceae and Dryopteridaceae, but it has since been able to contribute very little of substance to fundamental phylogenetic problems, save only to corroborate, albeit rather tentatively, the natural unity of the Cyatheaceae. To me, as a cytologist, this is clearly a disappointment. We are still profoundly ignorant about important aspects of the phylogenetic structure of the modern ferns. For instance, we really have no clear idea from whence came either the Aspleniaceae or the Blechnaceae.

It should perhaps be pointed out that it is not only the cytological evidence which proves to be frustratingly difficult to interpret. Much parallel morphological evolution has taken place in the development of the "polypodiaceous" ferns. Striking characteristics like the perispore and the monolete as opposed to trilete spore have clearly originated quite independently in different groups.

If anyone feels this agnosticism is being overdrawn, I ask him to compare the two phyletic schemes produced by Holttum, one in 1949, the other in 1973. (Figs 1, 2.) They are not only very different, which is a tribute to the openness of mind that Holttum has maintained in maturity; the later scheme is much more tentative, with more points of admitted uncertainty, than the earlier scheme. These come from one whose knowledge and experience of ferns is unrivalled amongst living pteridologists, and who moreover has not only been aware of the importance of the work of cytologists, but has been one of the most active and enthusiastic of collaborators in chromosome studies.

Future advances in cytological knowledge can be expected to resolve some of the outstanding problems of fern phylogeny, but it is not going to solve them all. A synthesis with other, newer, approaches and techniques will be required.

V. POLYPLOIDY IN FERNS

The different types of polyploidy found in ferns (autopolyploidy, allopolyploidy, segmental alloploidy and autoallopolyploidy) (Figs 5, 6) and their occurrence in different species complexes are discussed below (section VIA, B). Here the systematic and geographical distribution of polyploids, irrespective of their genetic nature, are reviewed in turn.

A. SYSTEMATIC DISTRIBUTION OF POLYPLOIDS

The interpretation of chromosome numbers in terms of grades of ploidy is clearly dependent upon decisions regarding base numbers. In this review the principle established by Manton and Vida has been followed, i.e. "In each case the number used as a monoploid is selected as the lowest gametic number for which there is direct evidence in any group" (Manton and Vida, 1968, p. 365). These authors were aware that "where this is not a prime number further discoveries may lower it in future with the consequent need to upgrade the derivatives . . ." but explain that ". . . In spite of such risks . . . the use of imaginary monoploids based solely on arithmetic has been avoided as more likely to obscure than to clarify the situation". I am entirely in accord with this attitude, since, for example, the true base number in Psilotales may reach back from $x = 52$ to 26 or even 13, but there is no valid criterion which we can apply to tell when to stop this reductional process in order to arrive at, but not overshoot, the truth. Furthermore, much earlier, Manton (1953, p. 177) had pointed out very clearly, using the frequently encountered numbers 36 and 40 as examples, that both of these numbers could have been derived in several different ways and, moreover, not all of these involve just simple multiplication of a lower base.

It is noteworthy that even once this principle is accepted and applied, several

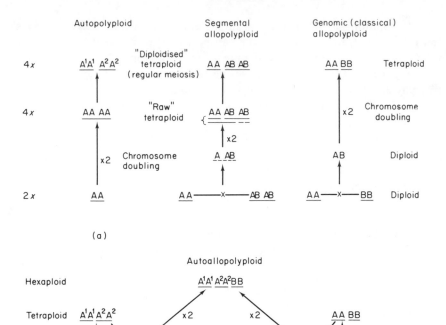

(a)

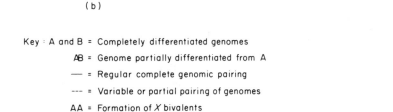

(b)

Key : A and B = Completely differentiated genomes

AB = Genome partially differentiated from A

— = Regular complete genomic pairing

--- = Variable or partial pairing of genomes

AA = Formation of X bivalents

AAAA = Formation of X tetravalents

Fig. 5. The nature and origin of the four principal categories of polyploids. (a) Distinctions between autopolyploid, segmental allopolyploid and genomic allopolyploid. (b) Alternative modes of origin of an autoallohexaploid.

of the most archaic fern families are seen to show active polyploidy today, with polyploid series sometimes built upon very high base numbers. Thus tetraploid and octoploid levels from a base of 52 are known in both *Psilotum* and *Tmesipteris* (Okabe, 1929; Manton, 1950; Ninan, 1956a; Barber, 1957; Brownlie, 1965). *Ophioglossum*, in spite of the handicap of a base number as large as 120, includes levels as high as hexaploid and octoploid, with a further

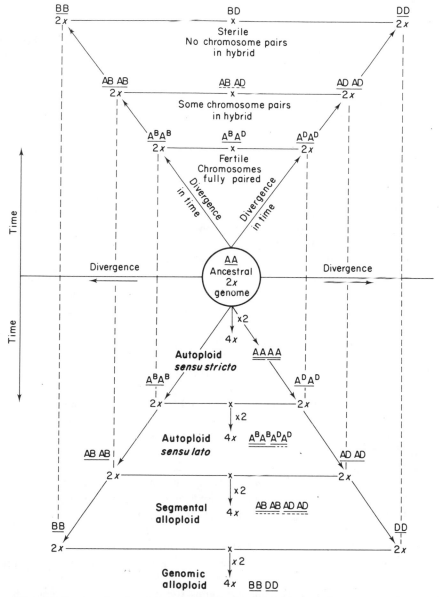

Fig. 6. Diagram showing how segmental allopolyploidy constitutes a category encompassing a continuous range of conditions intermediate between autopolyploidy and genomic allopolyploidy. It envisages what results if at any stage in the course of differentiation of two non-homologous genomes from a common ancestral diploid the two diverging taxa come into contact, hybridize, and generate a tetraploid derivative. The top and bottom halves of the diagram are equivalent in time. The top half represents the cytology of the diploid hybrid formed at different stages in divergence, while the bottom half displays the nature of the respective tetraploid derivatives. It is evident that in nature two diverging taxa might meet and give rise to a new polyploid at any stage in differentiation between A^BA^B/A^DA^D and BB/DD. Thus all conditions intermediate between autopolyploidy and genomic allopolyploidy may be encountered.

irregular level (n = 630) beyond decaploid. Polyploidy, at the tetraploid level, is known in Marattiaceae in three of the four genera as far investigated, and its existence can confidently be inferred in the fourth, *Christensenia*.

High levels of polyploidy are frequent in Schizaeaceae. Thus hexaploids are known both in *Anemia* and *Lygodium*. Polyploidy must also occur in *Actinostachys* and *Schizaea*, but it is not possible to diagnose base numbers in either of these genera, and therefore the levels of polyploidy present are unknown. In contrast, polyploidy is uncommon in Gleicheniaceae, wherein only a few tetraploid species have been detected, and it is unknown in the remarkably stable Osmundaceae. Polyploidy is also unknown in Cyatheaceae, in spite of the high chromosome numbers present (see Note Added in Proof, 6, p. 415). A single possible exception is *Metaxya*, but it is quite unclear how the aberrant chromosome number of this monotypic genus should be interpreted in terms of ploidy.

Polyploidy is frequent or abundant in all of the leptosporangiate ("polypodiaceous") families. Even the Polypodiaceae, regarded by Panigrahi and Patnaik (1963, 1964a, b) as exceptional in their low frequency of polyploids, includes tetraploids in more than half of the genera for which numbers are yet known, while hexaploids are known in *Pleopeltis, Polypodium* and *Pyrrosia*.

However, polyploidy is of decidedly uneven frequency within individual families, though the fact that the higher levels of polyploidy tend to be concentrated within large genera in which a substantial number of species have been investigated may only reflect the larger size of the samples available from these genera. This feature is seen very clearly in the very large family Dryopteridaceae, in which the tetraploid level is known to be exceeded only in *Dryopteris* ($6x$), and in *Athyrium, Cystopteris, Diplazium, Elaphoglossum, Polystichum* and *Tectaria* (all to $8x$ level). The same tendency is seen in the Adiantaceae, where the large genera *Adiantum* and *Pteris* both rise to the decaploid level, and *Pellaea* to the octoploid level. However, in this family the much smaller genera *Doryopteris* and *Pityrogramma* also include octoploids, while *Eriosorus* is unique in that all but one of the species examined are hexaploid (A. F. Tryon, 1970) and the one exception, *E. cheilanthoides*, is 12-ploid, with n = 174 (Manton and Vida, 1968), a level of polyploidy otherwise encountered in the ferns only in the vast genus *Asplenium*.* High levels of polyploidy are also frequent in the Dennstaedtiaceae. Hexaploidy is known in *Microlepia* and can be presumed in *Dennstaedtia*. Octoploids exist in *Hypolepis, Sphenomeris* and *Paesia*. (In this last genus the only other species counted is diploid.) *Orthiopteris dominguensis* with n = 188 (Walker, 1966a) may also be interpreted as octoploid. A count of n = c.220 in *Lindsaea* is probably decaploid.

---

* On the basis of the principle stated earlier, I am not prepared to accept x = 8 and x = 9 as base numbers in *Trichomanes*, in which the lowest gametic numbers known in the genus (irrespective of whether it is interpreted *sensu lato* or *sensu stricto*) are 32 (33), and 36 (but see footnote, p. 286). I therefore interpret *T. crispum* with n = 128 (Walker, 1966a) as octoploid, not 32-ploid.

The Vittarioideae are unique in that all except one of the counts (a total of 18 species) so far recorded in this subfamily are polyploid (4x, 6x or 8x). The sole exception is the report by Manton (1954b) of $n = c.30$ in *Monogramma trichoidea*.* As Walker has commented, "It is evident that polyploidy has played a very large part in evolution of the Vittariaceae as a whole." (Walker, 1966a, p. 228).

B. GEOGRAPHICAL DISTRIBUTION OF POLYPLOIDS

Hagerup (1932) was the first to propose that polyploids are more tolerant of extreme ecological conditions than their diploid relatives, and he furthermore maintained that this was true both for arid conditions in the tropics and extreme cold in arctic latitudes. The actual examples he produced as evidence in support of this concept are not immune from criticism. This is also true, as Stebbins (1950, pp. 342/5) has made clear, of the data assembled by Tischler (1935), whose comparison of four different European floras indicated a markedly lower level of polyploidy (31%) in the most southerly flora, that of Sicily, than in that of Schleswig-Holstein (44%), with still higher percentages in the floras of the Faröe Islands and Iceland. However, Löve and Löve (1943) collected more substantial data for northern European flowering plant floras, which showed very clearly a cline of increasing percentage of polyploids in relation to increase in latitude, from 48.7% in Schleswig-Holstein (54/55°N) to 77.4% in Spitzbergen (77/81°N). In consequence, Löve and Löve concluded that in these floras polyploidy was an adaptation to cold conditions, correlated with greater hardiness on the part of the polyploids. More recently, further data for various northern hemisphere floras has been collated by Löve and Löve (1971).

Manton's initial study on the pteridophytes in the British flora indicated a higher proportion of polyploidy (53%), though without figures for any other pteridophyte floras available for comparison it was then not possible to tell whether or not this high level of polyploidy was characteristic of pteridophyte floras in general, or to decide to what extent this figure showed an influence related to higher latitudes parallel to that detected in higher plants. Not until much later was it clear that Manton's figures for the British flora were also characteristic of the central European continental flora, when Vida (1966, unpublished thesis, quoted in Manton and Vida, 1968) reported a virtually identical proportion of polyploids (54%) in the fern flora of Hungary.

The only fern flora available in 1950 for comparison with that of the British Isles was that of Madeira, which since it has a lower percentage of polyploids (42%), appears at first sight to conform in a straightforward manner with the correlation between latitude and proportion of polyploids established by

---

* Walker (1966a, p. 227) points out that this species is stated by Copeland (1947, p. 226) to be the type of *Vaginularia*, and consequently this count is classified under *Vaginularia* here.

Tischler and by Löve & Löve. The Madeiran flora contains however two very distinct phytogeographical elements, one related to oceanic Europe, the other of African affinity. The European fraction of the Madeiran flora does indeed include a much lower proportion of polyploids (35%) than is characteristic of either Britain or continental Europe. On the other hand, though the proportion of polyploids (50%) present in the smaller African element is not notably distinctive, the levels of polyploids encountered were unprecedently high. Thus even in the small fern flora (only *c.*50 species) of Madeira, the three genera *Polystichum, Adiantum* and *Asplenium* each contain one high polyploid, namely *P. falcinellum, A. reniforme* and *A. aethiopicum*, on octoploid, decaploid and 12-ploid levels respectively (Manton, 1950; Manton and Vida, 1968). Further-more, in contrast to the lower polyploids found in Britain, many of which are still in contact with their related diploids, these very high Madeiran polyploids are "undoubtedly ancient and isolated types totally devoid of local relatives and . . . must therefore be ancient species with a long past history" (Manton, 1950, p. 283). Manton clearly regarded these unexpected discoveries in Madeira as highly significant, and continues thus: "My personal conclusion from this is that polyploidy as such is not in itself either ancient or modern or an adaptation to cold or any other single ecological factor but that it is correlated rather with climatic or geographical upheavals however caused" (Manton, 1950, p. 283).

This appreciation of the small sample of high polyploids found in Madeira soon proved to be highly prescient, being emphatically confirmed by the cytological survey of the Ceylon fern flora. In the first instance, the Ceylon flora produced a proportion of polyploids which is actually higher, not lower, than that found in northern temperate latitudes, a fact which, as Manton pointed out, "disposes at once of any contention which may still be upheld that polyploidy as such is an adaptation to cold and to be correlated primarily with the Ice Age" (Manton, 1953, p. 181). Manton further concluded that any stimulus to species formation provided by the Pleistocene glaciations would be associated with the recession of glacial conditions providing unusual opportunities for colonisation of open habitats by new forms, rather than with glacial conditions as such.

Of greater significance than the mere percentage of polyploid forms in Ceylon was the finding that, as in Madeira, genera common to Europe and Ceylon mostly show higher grades of polyploidy in Ceylon. This tendency is more emphatic in Ceylon than in Madeira inasmuch as a greater number of species are involved. Examples compared in detail by Manton were *Asplenium, Athyrium* and *Polystichum.* The data she presented can be quantified as figures for mean ploidy level, which are as follows (Europe first : Ceylon second): *Asplenium* 3.0 : 5.7, *Athyrium* 2.0 : 4.4 and *Polystichum* 2.7 : 3.7.

Considered in their entirety, the results obtained from the survey of Ceylon ferns (the first cytological survey of any substance on any group of tropical plants) led Manton to the important conclusion that "evolution is proceeding (and has in general always proceeded) faster in the tropics than in temperate latitudes" (Manton, 1953, p. 185).

The cytological facts established by Manton for Ceylon were in due course provided with an extraordinarily exact parallel in the New World by the extensive survey of the pteridophytes of Jamaica conducted by Walker (1966a), who found a virtually identical proportion of polyploids (c.60%), and moreover established with regard to the grades of polyploidy present that once allowance was made for some differences in detail "the overall statistics for the distribution of polyploidy in the two floras are strikingly similar" (Walker, 1966a, p. 230).

Evidence that the data for Ceylon and Jamaica are not unduly influenced by the increased rate of speciation that may prevail in island floras is provided by an analysis of a substantial sample of the tropical African fern flora, which also produced a figure of 60% polyploids, with high individual levels present (Manton, 1959; Manton and Vida, 1968, p. 374).

Substantial confusion exists regarding the importance of polyploidy in tropical angiosperm floras. Stebbins (1971, p. 182) writes that "present data suggest that the lowest percentages of polyploidy are found in floras of warm temperate and subtropical regions, and that percentages increase as we go from these regions either toward the cooler or the tropical areas", but Löve and Löve (1971, p. 469) state that "les polyploïdes sont plus nombreux sous les Tropiques, bien qu'ils ne représentent en fait qu'un très faible pourcentage des flores énormes des régions tropicales". This direct contradiction arises because whereas Löve and Löve quote the figure of 26% polyploids presented by Morton (1966) for an assemblage of counts on some 700 species in the West African flowering plant flora, Stebbins cites the work of Mangenot and Mangenot (1962), who find of the order of 85-90% polyploids among a sample of 510 species in the Ivory Coast rain-forest flora. However, Mangenot and Mangenot's figure comprises, "en grande majorité" (p. 427), palaeopolyploids, that is, species ultimately of polyploid origin, but whose diploid ancestors are extinct. In effect, only numbers close to or equalling 6, 7 or 8 are treated as true base numbers, numbers like $n = 12$, 13, 14 etc. being regarded as having originated principally by ancient polyploidy. Since more than half of the species counted in this sample have chromosome numbers in the range $n = 10 < 14$, the influence of this attitude on the calculated level of polyploid frequency is very considerable.

The approach adopted by Mangenot and Mangenot introduces an unnecessary and controversial complication, which is best avoided in the present state of our knowledge of the evolution of base numbers in flowering plants. We can therefore accept, for flowering plants, the statement by Morton (1966, p. 73) that "the overall incidence of polyploidy in West Africa is appreciably lower than that in non-tropical regions".

In detail, the pattern is more complicated, both geographically and taxonomically. Thus the flora of Cameroons Mountain has a higher proportion of polyploids (49% in a sample of 156 species: Morton, 1966) than the flora as a whole, from which it can be inferred that the percentage of polyploids in the lowland vegetation is appreciably lower than that of the West African tropical region in general and must be close to 20%.

The occurrence of polyploidy also varies widely between different angiosperm families. Morton (1966) studied several individual families in the West African flora, and reported a low percentage of polyploidy (29%) in Compositae, in contrast to high levels in the Commelinaceae (64%)* and the Labiatae (71%; details in Morton, 1962). Low levels were also found in Liliaceae and Amaryllidaceae (32% and 22% respectively) but the sample sizes here were smaller. The Commelinaceae are a predominately tropical and subtropical family, so no comparison with temperate representatives is possible, but Morton (1961) has given figures of 27% and 71%† polyploids respectively for Compositae and Labiatae in the British flora. The frequency of polyploidy found for each of these families is thus remarkably similar in two very different floras. The same feature is true for the ferns. The level of polyploidy known for the West African tropical fern flora is 55%,‡ while that found in British ferns is 53%.

Furthermore, not only may levels of polyploidy be widely different for various plant groups within a flora, and for mountain and lowland floras within the same region, but it is clear from the evidence of the ferns that the proportion of polyploids within a particular major taxonomic group is not necessarily consistent throughout the tropics, since a sample of more than 100 taxa from the strictly equatorial flora of Malaya produced only 39% polyploids (Manton, 1954b, 1969).

Why the Malayan fern flora should be so different in this respect from other tropical fern floras is unknown, but it has become apparent that the concept of tropical rain forest as a uniformly stable environment, which has been effectively insulated from the drastic climatic changes of the Quaternary period by its latitudinal and altitudinal position, is less than the full truth. Thus evidence has now been adduced from studies on the distribution and speciation patterns of birds (Haffer, 1969), lizards (Vanzolini and Williams, 1970), and heliconian butterflies (Brown et al., 1974), as well as from plants (Prance, 1973), all

---

* See Morton (1967) for details of counts. Cytological studies on Indian representatives of Commelinaceae have been reviewed by Rao et al. (1968). Of counts actually made on Indian material, about 50% are polyploids. See Jones and Jopling (1972) for a survey of the literature, and for another extensive series of counts, principally on American taxa.

† See also Morton, J. K., (1973) Watsonia 9, 239-246 and Morton, J. K., (1977) Watsonia 11, 211-223 for details of counts on British Labiatae and Compositae giving revised figures of 69.8% and 44.6% polyploids respectively.

‡ Adapted from data of Manton (1959), Manton and Vida (1968), by exclusion of counts on East African collections.

suggesting that the great forests of the Amazonian basin have at some time in the Quaternary been dissected and reduced to several small relict areas. Regrettably, none of the fern floras of the South American continent have yet been cytologically studied. To what extent other rain-forest floras have suffered similar drastic contractions in the past is less certain, but it is probable that similar events have occurred in parts of tropical Africa.

In view of these considerations it is to be expected that global cytogenetic patterns are likely to prove complicated, and unlikely to conform to any simple principle. However, one region that does appear at first sight to correspond to Stebbins's generalization is the Indian subcontinent, where there is a crude cline showing a decrease in the proportion of polyploids present in fern floras as one proceeds northwards. Thus studies in South India (Abraham *et al.*, 1962; Bir, 1965) show about 54% polyploids, as compared to 60% in Ceylon, whereas according to calculations made by Manton (1969), the proportion of polyploids found in the Himalayan flora is approximately 40%. She points out that it is not possible to explain this low figure by treating the Himalayan flora as temperate in the sense of being comparable to the flora of Europe, as is suggested by Mehra and Bir (1960), since the floristic affinity of Himalayan ferns is overwhelmingly tropical in character. The low frequency of polyploids in the Himalayan fern flora may therefore be related in some way to the equally low percentage (39%) in the strictly equatorial flora of Malaya (Manton, 1954b, 1969). Mehra (1961, p. 7) gave figures of 23.8% polyploids for the ferns of the western Himalayas and 36.2% for the eastern Himalayas.* Bir (1973, pp. 108/109) points out that though the climate and flora of the eastern Himalaya is predominantly subtropical, that of the western Himalaya is temperate with "great resemblances with European flora". However, 23.8% is very different from the 54% polyploids recorded for Europe, and emphasizes that the Himalayas do contain an exceptionally rich reservoir of diploid taxa.

In this context, an explanation proposed for the data available from the Canary Islands is of particular interest. Larsen (1960) gives a figure as low as 23% of polyploids for this flora. The figure for the Canary ferns is similar. Thus Page (1973) reports that "more than 70% of the Canary Islands fern species are diploids". Furthermore, the very high polyploid levels ($10x$, $12x$) encountered in Madeira are apparently not present in the Canaries, though the reasons for these striking differences between two Macaronesian floras are far from obvious. Page regards the very low of polyploidy in the Canary Islands as directly related to the origin of the flora. He concludes that "the Canarian fern flora is very largely composed of ancient species which existed in Tertiary Europe, which have survived little changed as relics in the mountains of the Canaries" (Page, 1973, p. 86). It is noteworthy that the Mediterranean region is relatively rich in diploids, some of which are also present in Madeira and in the Canaries. Perhaps

* Including only species growing below 7000 ft.

all of these regions of low polyploid frequency, e.g. the Canary Islands, the Mediterranean, the Himalayas and Malaya, represent or contain substantial elements from relatively ancient and little altered mid-Tertiary floras (cf. Manton, 1969, p. 220).

There is some indication that substantial differences exist between the levels of ploidy found either side of the equator, though at present this evidence is only fragmentary, owing to the total lack of information from South America. The fern flora of this continent remains cytologically unexplored.

The most direct evidence comes from the genus *Asplenium*. A recent analysis (Lovis, 1973) of the frequency and levels of polyploidy known in the vast genus *Asplenium* throughout the world demonstrates that in this genus mean levels of ploidy are much lower in the northern temperate zone (2.9-3.8) than in the southern temperate zone where levels (4.7-5.5) are approximately similar to those determined in tropical floras (3.75-5.5). This is associated with an irregular but unmistakeable cline in the frequency of diploids extending along a NW/SE axis across the world, from Europe (53%) through the Himalayas (21%) and Ceylon (13%) to New Zealand (0%).* The meaning of this phenomenon is at present quite obscure, and is likely to remain so while we are so ignorant regarding the likely geographical origins of the modern families of ferns. It may be that, other influences being equal, a genus tends to evolve its highest grades of polyploidy where it has existed longest. On the contrary, it may be that the production of polyploids occurs most frequently in response to pressures which a group meets in the course of its migration in space and time, with the majority of diploids persisting as relicts close to the original centre of diversification. A major radiation at the diploid level is probably a necessary prerequisite to the extensive production of polyploid derivatives and the appearance of high polyploid levels. Such bursts of diploid diversification may well occur at places other than the point of primary origin of the genus. I prefer to regard the extraordinary diversity of diploid species of *Asplenium* in Europe as representing a secondary more recent radiation of this kind, rather than the remains of the original primary diversification of the genus, which I believe more probably took place in the Southern Hemisphere, though I would be hard put to justify this opinion, in view of the lack of factual evidence.

Some important information has been obtained from study of oceanic island floras. Manton and Vida (1968) presented an extremely interesting report on the fern flora of the island of Tristan da Cunha, remotely situated in temperate latitudes in the southern Atlantic Ocean. The sample of species is by its very nature small, only 20 taxa, but shows a high proportion of polyploids (65%),

---

* It might be supposed that this effect is at least in part an artifact produced by the more intensive sampling of northern temperate floras, since more exhaustive studies did result in a higher proportion of diploid taxa becoming known in European Aspleniaceae. However, the reverse result has recently obtained in New Zealand, where reinvestigation by Brownsey (in press), has uncovered more high polyploid taxa, but no diploids.

with examples of high grades of ploidy, namely two hexaploids, *Elaphoglossum hybridum* and *Thelypteris tomentosa*, one octoploid, *Adiantum poirettii*, and a 12-ploid, *Eriosorus cheilanthoides*. The flora of Tristan da Cunha must be primarily of immigrant origin, since the Tristan group are all volcanic in origin, being situated near the mid-ocean ridge, and all the existing islands are of recent age. Tristan da Cunha itself is dated at only 1×10^6 years, while the oldest island in the group, Nightingale, is but 18×10^6 years old (Wace and Dickson, 1965). Owing to the prevailing wind direction, the predominant influence on the Tristan fern flora must be from South America. It is therefore quite uncertain to what extent the high level of polyploidy, and the high grades encountered, are attributable simply to the character of the source material, and to what extent to subsequent evolutionary changes on the Tristan group.

Another sample of a remote island flora, the Galapagos group (Jarrett *et al.*, 1968) is of interest here, because it also comes from an island group whose flora is principally of South American origin. The sample yet investigated is only small (17 cytotypes) but again shows a high proportion of polyploids (72%).

Furthermore, recent studies on the flora of Madeira (I. Manton, J. D. Lovis and G. Vida, unpublished) have established the existence of another category of polyploids distinct in origin from the three high polyploids of relict status mentioned above. Thus a widely dibasic allotetraploid species derived from *Hymenophyllum tunbrigense* and *H. wilsonii* and a new incipient hexaploid species in the *Asplenium trichomanes* complex have been discovered in the island. The circumstances of their discovery leave no room for doubt that both of these new polyploids have originated in Madeira. Their existence in Madeira but, as far as is known, not elsewhere, suggests that the relaxed conditions of competition in the relatively simple ecosystems formulated by the limited flora of this oceanic island have permitted the appearance and persistence of polyploid forms not viable in the more complex structure of continental communities.

In *Ceterach* the existence has been detected of a hexaploid form in Madeira (I. Manton *et al.*, unpublished) and of both tetraploid and octoploid forms in the Canaries (T. Reichstein and G. Vida, unpublished). None of these three forms is known outside Macaronesia. Although this complex undoubtedly requires further study, the most likely interpretation of this situation is that these three taxa constitute elements persisting as relicts of a complex which evolved within Macaronesia.

These recent studies in Madeira have thus produced good evidence indicating that oceanic islands may not only provide a refugium for polyploid (and diploid) forms which are no longer capable of survival elsewhere, but also an environment suitable for the creation of new polyploid taxa.

All the available evidence now indicates that the environmental factors which influence the creation and the survival of polyploids are both numerous and complex. It is remarkable that in spite of the numerous studies carried out in the last 25 years, our understanding of the operation of these processes cannot be briefly summarized now any more effectively than by Manton in 1950: "Under stable conditions, the natural spread of species is probably accompanied by some, though perhaps infrequent, polyploidy as new species come into contact with old ones and hybridize with them. In a relatively undisturbed flora the incidence of polyploidy might therefore be expected to be low. Under changing climates or topography, on the other hand, the opportunities for hybridization and therefore for allopolyploidy can hardly fail to be increased. The nature of the significant climatic or topographical changes may be very varied and include perhaps cold, heat, drought, inundation, mountain building, volcanic action, changes in the distribution of land and sea or any other vicissitude which may affect the whole earth or portions of it. All of these may be expected to leave their mark on the evolution of vegetation" (Manton, 1950, p. 283).

VI. BIOSYSTEMATICS OF FERNS

A. GENOME ANALYSIS

The analysis of genome interrelationships, principally by study of chromosome pairing behaviour at meiosis in interspecific hybrids, has already proved to be of great value in detecting relationships between species and in resolving patterns of micro-evolution in ferns.

Chromosome numbers alone, without knowledge of genome homologies, can in certain circumstances prove to be of diagnostic value. The mere knowledge of the levels of ploidy, simply to be able to sort the various taxa into diploids and tetraploids, may prove sufficient additional information to enable a skilful morphologist to resolve correctly the relationships of a complex. A good example is the Appalachian spleenwort complex, which was correctly analysed by Wagner (1954) on the basis of morphology and chromosome numbers, with the aid of only a minimal amount of chromosome pairing evidence. Previously, the complex had appeared to be an inextricable morphological tangle. Wagner's interpretation has been amply confirmed, first by gradually accumulating cytogenetic evidence (Wagner, 1956; Wagner and Darling, 1957; Wagner and Boydston, 1958; Smith, Bryant and Tate, 1961; Wagner, 1963; Wagner and Wagner, 1969), and subsequently by chromatographic (Smith and Levin, 1963) and phytochemical analysis (Harborne *et al.*, 1973) (Figs 7, 8). However, this complex is unusual in that the three basic parental diploid species are morphologically exceedingly distinct.

Fig. 7. Relationships in the Appalachian spleenwort complex, showing interpretation of chromosome pairing patterns observed in wild and synthetic hybrids (references in text, above). The complex is based on three morphologically very distinct diploids (inner triangle). All three possible amphidiploid (= allotetraploid) combinations based on these three diploid species are known (outer triangle). In part redrawn and modified after Wagner (1954, p. 104, fig. 1).

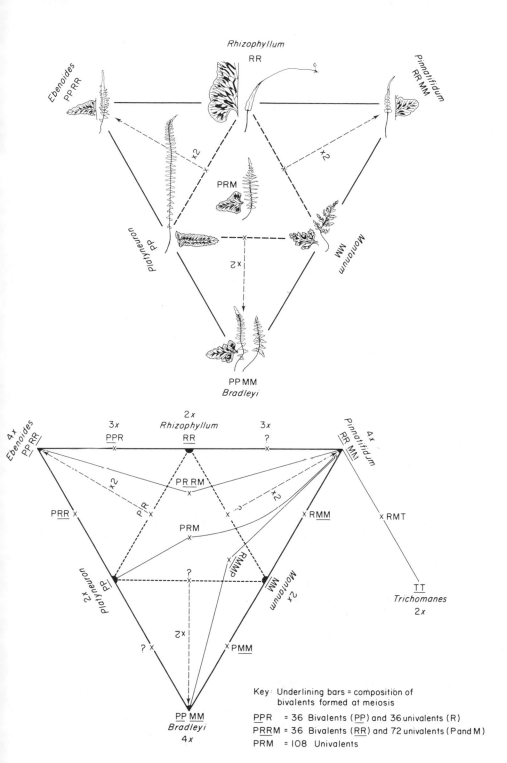

Key: Underlining bars = composition of
bivalents formed at meiosis

$\underline{PP}R$ = 36 Bivalents ($\underline{PP}$) and 36 univalents (R)

$PR\underline{R}M$ = 36 Bivalents ($\underline{RR}$) and 72 univalents (P and M)

PRM = 108 Univalents

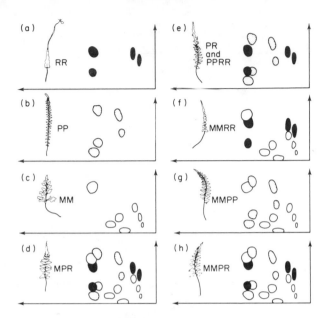

Fig. 8. Chromatographic analysis of the Appalachian spleenwort complex. (a), (b) and (c) = parental diploid species; (e), (f) and (g) = allotetraploid derivatives; (d) and (h) = trigenomic hybrids. (a) *Asplenium rhizophyllum*, (b) *A. platyneuron*, (c) *A. montanum*, (d) *A.* x *kentuckiense* = *A. pinnatifidum* x *platyneuron*, (e) *A. ebenoides* (and 2x hybrid *A. platyneuron* x *rhizophyllum*), (f) *A. pinnatifidum*, (g) *A. bradleyi*, (h) *A.* x *gravesii* = *A. bradleyi* x *pinnatifidum*. The distribution of flavonoid compounds is additive in the allotetraploids and in the two hybrids, *A.* x *kentuckiense* and *A.* x *gravesii*, in which all three genomes are combined. (Redrawn after Smith and Levin, 1963.)

In more complex situations, where a greater number of diploid species exist, any one of which could conceivably have contributed to the polyploid complex, interpretations based on ploidy levels and morphology alone may prove to be unsound. Thus it was first suggested that the parents of the tetraploid species *Asplenium majoricum* were *A. obovatum* and *A. trichomanes* (Meyer, 1967). Analysis of genome relationships demonstrated that neither of these two species were involved in the origin of *A. majoricum*, the true parentage of which is *A. fontanum* x *petrarcheae* ssp. *bivalens* (Sleep, 1967; Lovis and Reichstein, 1969).

The technique of genome analysis, or at least the degree of reliance placed upon the evidence it provides, has been criticized in recent years by Solbrig (1968, p. 90), and with particular force by de Wet and Harlan (1972). These authors emphasize the wide variety of genetic factors that can influence chromosome pairing, a topic reviewed by Riley and Law (1965). There can indeed be little doubt that the so-called VB system in hexaploid wheat, discovered simultaneously by Sears and Okamoto (1959) and Riley and his colleagues (Riley and Chapman, 1958; Riley *et al.*, 1959), in which a localized genetic element is responsible for restricting meiotic chromosome pairing to truly homologous partners, pairing between homoeologues* being excluded in the presence of the element, is exceptional only in the virtuosity of the cytogenetic engineering by which the controlling element was determined to be located on a specific chromosome arm. (Riley, 1960; Riley *et al.*, 1960; Riley and Chapman, 1964). However, the complexities and difficulties encountered in certain grass genera (de Wet and Harlan, 1970) have no relevance to this present discussion since the circumstances encountered in different groups of plants must be evaluated independently each on its own merits. If the method can be shown to work consistently in ferns, then it is manifestly a valid technique for use on ferns, irrespective of the difficulties encountered elsewhere. Experience has indeed shown that in the best-studied examples amongst ferns, e.g. European Aspleniaceae, the *Dryopteris spinulosa* complex in North America, and the *Adiantum caudatum* complex in the Old World tropics (see below for discussions and references), the results provided by analysis of chromosome pairing are consistent and convincing.

However, in the same paper as he summarizes details of the complexities of interrelationships in Appalachian *Dryopteris* revealed by cytogenetic analysis, Wagner shows signs of reacting to this present climate of scepticism with some diminution of confidence in the reliability of the technique. He writes: "In the

---

* Outside of the narrow circle of specialists in cytogenetics, the distinction between the very similar terms **homologue** and **homoeologue** undoubtedly causes considerable confusion. However, the term homoeologue is now in general use, and has no exact equivalent. In brief, corresponding chromosomes in the genomes of genetically differentiated diploid species are homoeologues. The point can only be made clearer with an example. For this purpose it is better to select the well-known artificial tetraploid species *Primula* x *kewensis* (Newton and Pellew, 1929; Upcott, 1939), rather than an example from the ferns, in order to avoid the complications inherent in the question of the degree of ancient polyploidy built into the basic chromosome number of such a genus as *Asplenium*. *Primula* x *kewensis* originated from a diploid hybrid between two distinct species, *P. floribunda* and *P. verticillata*, each with $n = 9$. These two species must themselves be differentiated from a hypothetical ancestral *Primula*, with a genome containing nine different chromosomes, which can be designated $P_1, P_2, P_3, \ldots P_9$. In the course of differentiation of the various species in the genus, these ancestral chromosomes have evolved genetically in different directions in different species, but the basic cytological structure of the genome has persisted. Thus the genome of *P. floribunda* can be symbolized $F_1, F_2, F_3, \ldots F_9$, and that of *P. verticillata* $V_1, V_2, V_3, \ldots V_9$. F_1 and V_1, being both derived from the same ancestral chromosome (P_1), are said to be homoeologous chromosomes. Somatic (sporophytic) tissues of *P. verticillata* will of course contain $V_1V_1, V_2V_2, \ldots V_9V_9$, wherein V_1 and V_1 are truly homologous chromosomes, and meiotic pairing proceeds between homologues thus: $\underline{V_1V_1}$, $\underline{V_2V_2}$ etc. In the diploid hybrid *P. floribunda* x *verticillata* up to nine bivalents are formed. In this case, pairing is between homoeologues: $\overline{F_1V_1}, \overline{F_2V_2}$ etc. In the tetraploid derivative of this hybrid (i.e. in *P.* x *kewensis*) pairing is usually between true homologues only, thus: $\underline{F_1F_1}, \underline{V_1V_1}, \underline{F_2F_2}, \underline{V_2V_2}$, but a few quadrivalents are formed in which both homologues and their homoeologues participate, thus: $\underline{F_1F_1V_1V_1}$.

genus *Dryopteris* we should entertain the possibility that what we observe in chromosome pairing does not necessarily have to be a direct clue to overall evolutionary differentiation between the taxa. Thus the scheme of relationships shown . . . may be subject to potential revision of a profound nature" (Wagner, 1971, p. 186). To my mind, in view of the essentially consistent character of what is a substantial body of cytological evidence, the variations in pairing behaviour observed in certain hybrids do not justify this loss of confidence in the method. Some variability in chromosome pairing is to be expected in certain hybrid combinations. Where in course of time minor differentiation has occurred between two originally identical genomes, it is to be expected that not every spore mother cell in a hybrid involving these genomes will necessarily show the maximum possible chromosome pairing between them, because the chromosomes of the ancestral genome will not all have differentiated to the same degree in the evolution of the two descendant genomes, and therefore in the hybrid individual homoeologous pairs will differ in the ease and frequency with which they synapse at meiosis. In such circumstances, the maximum number of bivalents observed is more significant than the minimum number observed. In a hybrid combining two, three or even four well-differentiated genomes, it is not surprising if in the absence of truly homologous partners, residual homologies are expressed by the formation of some pairs between homoeologues, and moreover, equally not surprising if the extent of this homoeologous pairing is decidedly variable. Thus the hybrids *D. clintoniana* x *intermedia* and *D. clintoniana* x *marginalis*, each of which is believed to contain four non-homologous genomes, showed a range of $1 < 19$ pairs (in 11 cells analysed) and $0 < 11$ pairs (in six cells) respectively (Wagner, 1971). In brief, pairing between true homologues can be expected to be consistent, but pairing between homoeologues will be variable in extent.

It is clear, however, that great care must be taken in interpreting cytogenetic evidence to ensure that the conclusions drawn from the available facts do not exceed what the evidence will strictly permit. The case of triploid hybrids formed between a tetraploid and a putative diploid parent is of particular importance, since the presence of an equal number of bivalents and univalents ("*n* pairs and *n* singles") at meiosis in a hybrid combination is, considered in isolation, uninterpretable (Fig. 9). Before the *n* bivalents present in such a hybrid can be interpreted as being formed by pairing between the genome of the diploid and one of those contributed by the tetraploid, and therefore as direct evidence of the participation of the diploid in the origin of the tetraploid as one of its parents, it is necessary to obtain an independent demonstration that the tetraploid is indeed an allotetraploid. A simple example where this principle was violated is the report by Panigrahi (1965) on a synthetic triploid hybrid obtained between diploid and tetraploid cytotypes of *Dryopteris villarii*, the pairs observed being interpreted as being formed between chromosomes contributed

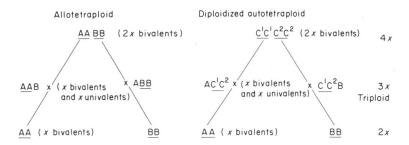

Fig. 9. Interpretation of chromosome pairing in triploid hybrids, showing that without further evidence the pattern of chromosome pairing obtained in the triploid hybrids formed between an allotetraploid and its parent diploids is indistinguishable from that obtained between an autotetraploid and two unrelated diploid species. (Underlining bars indicate composition of bivalents formed in meiosis. A, B and C = unrelated genomes.)

respectively by the diploid and tetraploid parents, i.e. as being of allosyndetic* origin, on the assumption that the tetraploid is of allotetraploid nature, presumably, though this is not specifically stated, because of the regular meiosis of the tetraploid which shows no sign of multivalents. However, with the only partial exception of *Asplenium ruta-muraria* (Vida, 1970; Lovis and Reichstein, unpublished), those few wild tetraploid ferns which have been demonstrated unequivocally to be of autopolyploid constitution possess diploidised meiosis, forming only bivalents (Lovis, 1964b; Lovis and Vida, 1969; Manton *et al.*, 1970; Bouharmont, 1972a, b; Callé *et al.*, 1975). In ferns, as in flowering plants, the absence of multivalents is therefore not valid evidence of allopolyploidy. In view of the close morphological similarity of diploid and tetraploid *D. villarii*, the assumption made by Panigrahi was particularly hazardous, but would still have been unjustified had the two parents of the triploid been morphologically very distinct.

The case just quoted is not an isolated example. There are many instances in the literature where unjustified conclusions have been drawn from the observation of equal numbers of bivalents and univalents in a triploid hybrid. Indeed, most authors involved in the earlier stages of the post-1950 development of fern cytogenetics, myself included (Lovis, 1955b), have transgressed in this way at some time or another.

A more recent example (Smith, 1971) will serve to demonstrate the

---

* **Allosyndesis** = pairing of chromosomes derived from different parental gametes. The opposing condition, pairing of chromosomes contributed by one and the same parental gamete, is known as **autosyndesis**. Stebbins (1947, pp. 412/413) points out that these terms have been used by a number of authors with exactly the opposite meanings. However, their present usage (which follows the original definitions of Ljungdahl (1924)) now appears to be generally accepted and there seems no longer to be any likelihood of confusion arising through their use.

difficulties and frustration that can attend the analysis of chromosome pairing in hybrids. In the course of an extensive taxonomic and biosystematic study of *Thelypteris* sect. *Cyclosorus* in the New World, Smith proposed the scheme of interrelationship of five species (Fig. 10a) wherein *T. normalis* and *T. augescens* are both allotetraploid species which share one parent. Forced to rely solely on wild interspecific hybrids, Smith had available for study only three hybrids involving combinations of species involved in the putative complex, namely *T. normalis* x *ovata* (*n* bivalents and *n* univalents), *T. augescens* x *normalis* (*n* bivalents and 2*n* univalents) and *T. augescens* x *ovata* (*n* bivalents and *n* univalents). The chromosome pairing patterns found are entirely consistent with Smith's hypothesis, but unfortunately, in the absence of other material evidence, equally compatible with a second interpretation (Fig. 10b) wherein one of the two tetraploids is of autopolyploid constitution. Furthermore, a variant of this second interpretation, wherein *T. normalis*, not *T. augescens*, is the autoploid, is equally possible. Thus all that the available cytogenetic evidence can tell us with any certainty is that *T. ovata* is one parent of one or both of *T. normalis* and *T. augescens*. Of course, it may well prove that the hypothesis presented by Smith

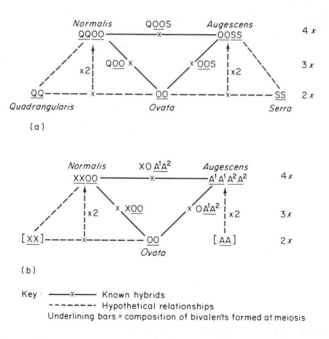

Fig. 10. Relationships in the *Thelypteris* (sect. *Cyclosorus*) *ovata* group. (a) Interpretation of Smith (1971). (b) Alternative interpretation of the available cytogenetic evidence (X could be *quadrangularis*, Q). If the cytogenetic evidence is considered in isolation, (a) and (b) are equally likely. In the absence of further cytogenetic information a decision between (a) and (b) must be based on morphological criteria alone.

regarding the interrelationship of these species is entirely correct, but the strength of his interpretation rests almost entirely on grounds of comparative morphology and chromosome numbers.

Wild and synthetic interspecific hybrids have been variously studied in cytogenetic investigations on ferns. The most intricate studies, those on North American *Dryopteris* and *Asplenium* in Europe, have been based on both, though synthesized hybrids have been utilized to a somewhat lesser extent in the former studies than in those on European *Asplenium*. The use of wild hybrids has two disadvantages. One is that there may occasionally be serious doubt regarding the exact parentage of a particular hybrid. The second disadvantage is one of availability. The specific combination necessary to resolve a particular problem may simply never occur in a state of nature. The use of synthetic hybrids is therefore, although initially possibly slower, in two respects at least to some degree ultimately more certain, even though some desirable combinations may be impossible to obtain through being totally genetically incompatible. It is therefore surprising that, apart from a few isolated hybrids made in the USA (Wagner, 1956; Wagner and Darling, 1957; Wagner and Boydston, 1958, 1961; Hickok and Klekowski, 1974), and with the notable exception of the activities of Gabor Vida in Budapest, the production and cytogenetic study of synthetic interspecific hybrids has remained virtually the exclusive province of the school established in Leeds by Irene Manton. Experimental hybridization studies have been carried out in Leeds on various complexes in *Adiantum, Asplenium, Dryopteris, Phyllitis, Polypodium, Polystichum* and *Pteris*, all of which will be to some extent discussed below. Other studies, of less conclusive scope, which will therefore not be mentioned further here, have been executed on complexes in *Aleuritopteris* (Panigrahi, 1962; Manton *et al.*, 1966) and *Christella* (= *Cyclosorus s.l.*) (Panigrahi and Manton, 1958; Ghatak and Manton, 1971).

B. CYTOGENETICS AND SPECIES INTERRELATIONSHIPS: TEMPERATE FERNS

1. Dryopteris filix-mas complex

Dryopteris filix-mas, the familiar Male Fern, has a strong claim to be the first allopolyploid fern to be recognized as such. The history of its investigation is of special interest for two reasons. Firstly, as a clear demonstration of how, in the absence of further experimental hybridization studies, and in spite of the development meantime of new analytical techniques, a partly resolved problem can remain incompletely resolved for decades. Secondly, as a salutary warning as to how even the most familiar species may contain unsuspected complexities and difficulties.

Manton (1950) obtained a hybrid between the tetraploid *Dryopteris filix-mas* and a morphologically closely related montane diploid species, *D. abbreviata*. This triploid hybrid showed the now familiar pattern of equal numbers of bivalents and univalents. Subsequently, Manton and Walker (1954) secured a diploid (i.e. polyhaploid) sporophyte of *D. filix-mas*, obtained by successful induction of apogamy in a potentially normal gametophyte, fertilization being prevented by withholding free surface water. Meiosis in this sporophyte showed a very slight degree of chromosome pairing, from two to five bivalents being observed. The production by experiment of a diploid sporophyte of *D. filix-mas* was also described by Döpp (1961). This plant showed only univalents.

Considered in combination, these observations constitute strong evidence that *Dryopteris filix-mas* is an allotetraploid species, and that *D. abbreviata* is one of its parents (Fig. 11a).

The second parent of *Dryopteris filix-mas* has still not been certainly identified, though no fewer than three candidates have been proposed in recent years, namely *D. villarii* ssp. *villarii* (Widén *et al.*, 1971), *D. villarii* ssp. *pallida* (Vida, 1972) and *D. caucasica* (Fraser-Jenkins and Corley, 1972). The first two of these are well-known members of the *villarii* complex, whereas *D. caucasica*, although not new to science, had been overlooked for very many years until its recent rediscovery in Asian Turkey and the Caucasus by Fraser-Jenkins. In its spore characters and general frond morphology, this species provides an accurate simulant for the characteristics to be expected in the second parent of *D. filix-mas*. Fraser-Jenkins and Corley (1972, p. 230, footnote) report that an approximately equal number of bivalents and univalents are present at meiosis in a wild triploid hybrid, *D. caucasica* x *filix-mas*. At first sight (Fig. 11b), in view of what is already known of the cytogenetics of *D. filix-mas*, this might seem to be conclusive evidence establishing *D. caucasica* as the second parent of *D. filix-mas*, but regrettably this is not the case since *D. abbreviata* and *D. caucasica* are morphologically sufficiently close not to preclude totally the possibility that these two species are morphological variants of the same ancestral genome (Fig. 11c). Clearly, study of the diploid hybrid between these two species is required.

In recent years, the phloroglucides present in the rhizomes of different taxa of *Dryopteris* have been extensively studied by Widén with a variety of collaborators (Table III). One study (Widén *et al.*, 1971) demonstrated that two compounds, para-aspidin and desaspidin, present in *D. filix-mas*, are absent from *D. abbreviata*. It is therefore to be expected that they will be present in the other parent of *D. filix-mas*, and it was shown in the same study that they are indeed present in *D. villarii* ssp. *villarii* and in *D. villarii* ssp. *pallida*. However, it has subsequently been found (Widén *et al.*, 1973) that these compounds are also present in *D. caucasica*! The chemical evidence is therefore compatible with the hypothesis that either *D. caucasica* or one of the subspecies of *D. villarii* is the second still unknown ancestor of the allotetraploid *D. filix-mas*, but does not indicate which taxon is in fact the second parent. It is significant that Widén and his co-authors (who on this occasion included T. Reichstein, an organic chemist of outstanding eminence) comment that their results "also demonstrate the limitations of the chemical method which is quicker but less informative than the classical hybridisation procedures" (Widén *et al.*, 1973, p. 831). It is indeed

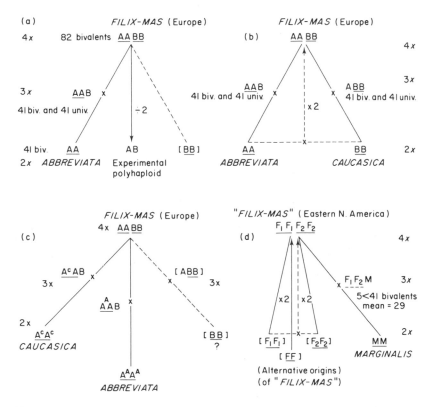

Fig. 11. Cytogenetic relationships and origin of *Dryopteris filix-mas*. (a) Evidence of Manton (1950) and Manton and Walker (1954) demonstrating that European *D. filix-mas* is of allopolyploid origin, with *D. abbreviata* representing one of its parental diploids. (b) Probable interpretation of origin of European *D. filix-mas* incorporating evidence of Fraser-Jenkins and Corley (1972). (c) Relationships if *D. abbreviata* and *D. caucasica* are only variants of the same (A) genome. Study of the hybrid *D. abbreviata* x *caucasica* should resolve choice between hypotheses (b) and (c). (d) Interpretation of *D. 'filix-mas'* in eastern North America as a segmental allopolyploid or autopolyploid on basis of chromosome pairing in *D. 'filix-mas'* x *marginalis* hybrid (data of Wagner, 1971).

true that this problem will only be ultimately solved by a comprehensive programme of experimental hybridization, culminating in the resynthesis of *D. filix-mas* but, owing to the slow maturation of the sporophytes of this group of *Dryopteris*, this will take many years to accomplish.

The phytochemical evidence indicates that the distinctions between *Dryopteris abbreviata* and *D. caucasica* extend beyond morphological character-

TABLE III

Semiquantitative analyses of phloroglucinol derivatives in rhizomes of some European and North American species of Dryopteris, greatly condensed and simplified from data of Britton and Widen (1974), Widen and Britton (1971a, b, c), Widen et al. (1970, 1971, 1973, 1975).

Species \ Compound	3	4/4a	5	6	7	9/9a	10	11	12	13	14	16	20	21
Filix-mas Europe	+++	-	+	-	+	-	-	-	-	-	-	+++	-	-
Filix-mas W. America	+++	-	+	-	++	-	-	-	-	-	-	+++	-	(+)
Filix-mas E. America	+++	-	+++	-	+/++	-	-	-	-	-	-	+	-	(+)
Abbreviata	+++	-	-	-	-	-	-	-	-	-	-	++	-	-
Caucasica	+	-	+++	-	++	-	-	-	-	-	-	+++	-	+
Villarii ssp. *Villarii* 2x	++	-	+++	-	+	+	-	-	-	-	-	+++	-	-
Villarii ssp. *Pallida*	++	-	+++	-	+	++	-	-	-	-	-	-	++	+
Marginalis	+	-	+/++	++	+++	-/(+)	+	++	+++	+	-	-	-	-
Assimilis Europe	+/++	+++	-/+++	-	-/+++	(+)/+	-	-	-	-	+++	-	-	-
Assimilis W. America	+	+++	+++	-	-/(+)	(+)	-	-	-	-	+++	-	-	-
Assimilis E. America	+	(+)++	+++	-	-/(+)	(+)	-	-	-	-	+++	-	-	-
Intermedia	+/++	+++	-	-	-	(+)	-	-	-	-	-	-	-	-
Azorica	(+)/+	+++	-	-	-	(+)	-	-	-	-	-	-	-	-
Maderensis	+	+++	++	-	-	(+)/++	-	-	-	-	-	-	-	-
Dilatata sensu stricto	+	+++	+++	-	-/(+)	(+)	-	-	-	-	(+)	-	-	-
Campyloptera	+/++	(+)/++	+++	-	-/(+)	+	-	-	-	-	+++	-	-	-
Spinulosa	+/++	+++	+++	-	++	++	-	-	-	-	-	-	-	+
Cristata	+++	-	-/+++	-	++	++	-	-	-	-	-	-	-	(+)
Clintoniana	+++	-	(+)	-	++	++/+++	-	-	-	-	-	-	-	(+)
Celsa	+++	-	++	-	-	++	-	-	-	-	-	-	-	-
Goldiana	+++	-	+	-	-	++	-	-	-	-	-	-	-	-
Ludoviciana	+++	-	+++	-	-	++	-	-	-	-	-	-	-	-

KEY to compounds (numbered according to Widen et al., 1971, 1973, 1975.)

3	= flavaspidic acid	10	= phloraspin
4	= aspidin BB	11	= phloraspidinol
4a	= aspidin AB	12	= margaspidin
5	= para-aspidin	13	= methylene-bis-desaspidinol
6	= methylene-bis-aspidinol	14	= phloropyrone
7	= desaspidin	16	= filixic acid
9	= albaspidin 1	20	= trispara-aspidin
9a	= albaspidin 2	21	= trisdesaspidin

KEY to symbols		
	−	nil/absent
	(+)	< 5%/trace amount
	+	5 < 10%
	++	10 < 20%
	+++	> 25%

istics to qualitative differences in their phloroglucide content, and makes it most unlikely that these two species are very closely related. It also suggests that *D. caucasica* may itself be related to the *villarii* complex, but this at present is a subject of some controversy. Taken in combination, the available phytochemical and cytogenetic evidence does indicate very strongly that the second parent is to be found within this group of species, and it is reasonable to conclude that the most likely candidate is indeed *D. caucasica*.*

A recent communication, (Fraser-Jenkins *et al*., 1975) has added a further complication. Attention is drawn to an overlooked and misunderstood mediterranean taxon, *D. tyrrhena*. It is suggested, principally but not exclusively on morphological grounds, that this plant, which is tetraploid, may be the amphidiploid derivative of *D. abbreviata* x *villarii* ssp. *pallida*. If this hypothesis is correct, it eliminates this combination as a possible origin for *D. filix-mas*. However, as Fraser-Jenkins *et al*. recognize, their hypothesis requires confirmation by synthesis and cytogenetic investigation of appropriate hybrid combinations.

Dryopteris filix-mas occurs in the New World as well as the Old World, where, like many other species, it has a disjunct distribution, with separate ranges on the western and eastern sides of North America. Both Wagner (1971) and Widén and Britton (1971c) have reported on the cytogenetics of wild triploid hybrids found in the east of North America between *D. filix-mas* and *D. marginalis*, a very distinct diploid species. In view of what was already known regarding *D. filix-mas* in Europe, their findings were unexpected. Both reported that a variable number of bivalents, from $5 < 41$ (mean 29) and $15 < 30$ respectively, were present in the hybrids they studied, and both concluded that these bivalents were most probably of autosyndetic origin, which would indicate that the *filix-mas* parent is of either autopolyploid or segmental alloploid origin (Fig. 11d). Wagner comes to the conclusion that the eastern North American form of *D. filix-mas* is of autopolyploid constitution principally because the morphology of *D. marginalis* is "so sharply distinct", but Widén and Britton also present phytochemical evidence showing that the phloroglucinol content of *D. marginalis* is extremely different to that of *D. filix-mas*, no less than five compounds present in *D. marginalis* not being found in *D. filix-mas* (Table III).

There is other evidence favouring the hypothesis that *Dryopteris marginalis* and the eastern North American *D. filix-mas* are unrelated. The very variable chromosome pairing, which has been studied in a surprisingly large number of different hybrid individuals, and only exceptionally approaches *n* pairs, is quite

* An example of the hybrid *Dryopteris abbreviata* x *caucasica* (= *D.* x *initialis*) found in the Caucasus by Fraser Jenkins and cytologically investigated by himself and Mary Gibby has shown 5-14(mean)-25 bivalents (*Fern Gaz.* (1976) **11**: 263-267). The logical interpretation of this finding is that *D. abbreviata* and *D. caucasica* are indeed the parents of *D. filix-mas*, which must be a segmental allopolyploid and not, as has long been presumed, a genomic alloploid. *Dryopteris filix-mas* thus provides another example where the full potential for homoeologous pairing between two partially differentiated genomes is not realized in a polyhaploid plant raised by induced apogamy (cf. p. 364, footnote).

uncharacteristic for a hybrid between an allotetraploid and one of its parents. Furthermore, Widén and Britton were also able to establish minor differences in the proportions of phloroglucides present in eastern *D. filix-mas* and *D. filix-mas* from the west side of North America, the phloroglucide composition of the latter being indistinguishable from that of *D. filix-mas* from Europe. There is also some indication of possible morphological differences between the western *D. filix-mas* and the plant found in the east of America, but this aspect clearly needs more detailed study. The sum total of the available evidence certainly suggests that whereas *D. filix-mas* as found in western North America may well be identical with the European plant, *D. filix-mas* in the east of America, though morphologically very close to true *D. filix-mas*, is in fact a separate taxon of quite different origin. At present its origin remains unknown. Widén and Britton state: "The eastern *D. filix-mas* could well be an autotetraploid of the European diploid, *D. abbreviata*. The morphological and cytological evidence would favour this hypothesis" (Widén and Britton, 1971c, p. 1599). However, it is probable that this opinion was written before the phloroglucide composition of *D. abbreviata* had been studied (Widén *et al.*, 1971), since the two compounds absent in *D. abbreviata* but present in European and western American *D. filix-mas* are also present in the eastern American *D. filix-mas*. Indeed, the phloroglucide content of the eastern plant is somewhat less similar to that of *D. abbreviata* than is that of true *D. filix-mas*.

More recently, Britton and Jermy (1974), in presenting a SEM study of spore ornamentation in the *Dryopteris filix-mas* complex, favour the conclusion that both the North American forms of *D. filix-mas* have "a similar origin to that of European *D. filix-mas*". Regarding the contrary cytological evidence, they simply comment, "It would appear that the chromosomal evidence from the pairing in the hybrid *D. filix-mas* x *marginalis* must be misleading" (Britton and Jermy, 1974, p. 1926).

In this confused situation, the only point with regard to which there can be no argument is that just as a programme of experimental hybridization is required to resolve the outstanding problems presented by the *Dryopteris filix-mas* group in Europe, so equally is a similar study required in relation to the American members of this complex before their relationships and origins can be known. The example of *D. filix-mas* also serves as an indication that it is unsafe to assume that results obtained with respect to a wide-ranging taxon in one region necessarily hold good in some other part of its range.

2. Dryopteris spinulosa complex

The *Dryopteris clintoniana/spinulosa/dilatata** complex falls naturally into two main parts: (1) the *clintoniana* group, confined (with the exception of *D.*

---

* It appears that more correct names for *Dryopteris spinulosa* and *D. dilatata* are respectively *D. carthusiana* and *D. austriaca* (cf. Gibby *et al., Bot. J. Linn. Soc.*, in press), but since the former names are used universally in the biosystematic literature discussed here, they have been retained in this review.

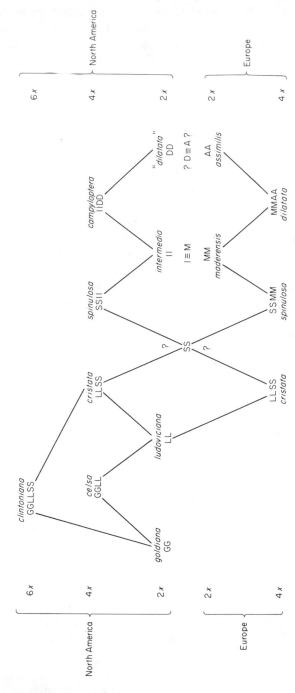

Fig. 12. Relationships in the *Dryopteris clintoniana/spinulosa/dilatata* complex on basis of data and interpretations of Wagner, S. Walker and Widen *et al.* (see text for references). Genome symbols for American taxa according to Wagner (1971). For the purpose of this figure Madeira is regarded as part of Europe; *D. azorica* is omitted because although there is strong morphological and phytochemical evidence for its close affinity to *D. intermedia* and *D. maderensis* (see p. 348) there is at present no cytogenetic evidence regarding the relationships of this taxon.

cristata) to the eastern side of North America, and (2) the *dilatata* group, with a circumboreal distribution, the two being linked together by the allotetraploid species *D. spinulosa*. Present knowledge of relationships in this complex is represented in Fig. 12.

In contrast to the *dilatata* complex, which presents an intricate and most imperfectly understood pattern of relationships, the structure and origin of the *clintoniana* group is in very large measure now resolved. The interrelationships of this taxonomically confusing group were first resolved by Stanley Walker (1962a, 1969), principally but not exclusively by study of wild hybrids provided by American correspondents. As Wagner has written, "The cytomorphological analysis by Stanley Walker of this overwhelmingly complicated taxonomic situation stands as a classic example of the value of cytogenetic studies" (Wagner, 1971, p. 175). Subsequently, Wagner and Wagner (1965), Wagner *et al.*, (1969) and Wagner (1971), studying a more extensive and remarkably wide range of wild hybrids, confirmed Walker's analysis.* The keys to the *Dryopteris clintoniana* complex are the two diploids, *D. goldiana* and *D. ludoviciana*. The latter is of southern subtropical affinity. These two species are no longer sympatric, but at some stage in the past have given rise to an allotetraploid derivative, *D. celsa. Dryopteris ludoviciana* is also involved in the parentage of the widespread tetraploid *D. cristata*, as well as the endemic allohexaploid *D. clintoniana*, whose parents, *D. cristata* and *D. goldiana*, both swamp species, still grow together today. *Dryopteris cristata* and *D. clintoniana* also include another genome, here designated as S, which is also present in *D. spinulosa*, but which is not yet known at the diploid level. The absence of this ancestral diploid, which Wagner has christened "semicristata", is unfortunate. Since the two tetraploids of which it is believed to be one parent are both of widespread distribution and presumably of some age, this missing species is not necessarily to be sought in eastern North America, but might survive elsewhere in the northern hemisphere, though of course, it may be that "semicristata" is in fact extinct.†

* The cytology of a total of 22 hybrids is summarized in this last paper (Wagner, 1971). The range of genome combinations examined is impressive. Excluding species, these are: MD, MI, MG and IG; ILG, ILS, MDI, MSL, MLG, FFM, DII, SII, LLG and LGG; MSLG, ISLG, ISSL, SLLG and SLGG; ISSLG and SSLLG. Fifteen of these hybrids are directly relevant to the *clintoniana* group. The allopolyploidy of the polyploid taxa is amply demonstrated by study of hybrids between each of *Dryopteris clintoniana, D. celsa* and *D. cristata* with unrelated diploids, in every case both with *D. intermedia* and *D. marginalis*.

The complex has also been studied independently by Widén and Britton (1971b), who have reported on the phloroglucinol contents of the species involved in the complex and of six hybrids. The phloroglucinols were found to be consistent with the interpretation of the complex obtained by Walker from cytogenetic analysis. Study of the meiotic behaviour of the hybrids confirmed the analyses found in these combinations by either Walker or Wagner.

† Hickok and Klekowski (1975) have just proposed a radically different interpretation of relationships within this complex which, though broadly consistent with the known cytogenetic facts, does not require the postulation of an S genome. How radical this reinterpretation is can be judged from the genome attributions: *Dryopteris celsa* LLM'M', *D. cristata* G'G'LL, *D. spinulosa* IILL and *D. clintoniana* GGG'G'LL (= autoallohexaploid).

Initial studies on the *Dryopteris dilatata* group were carried out on European material by Manton (1950) and extended to North America taxa by Walker (1955, 1961), who made and analysed a substantial number of synthetic hybrids, though these were not sufficient to resolve all the problems presented by this intricate complex.

Dryopteris campyloptera constitutes an element in America comparable to *D. dilatata* sensu stricto in Europe. Both are tetraploid, and apart from *D. spinulosa** itself, which is partly related to both, (and apart from two newly distinguished endemic taxa, *D. crispifolia* and *D. guanchica*, restricted to the Azores and the Canaries respectively (Gibby *et al.*, unpublished, *fide* Widén *et al.*, 1975)), they are the only polyploids in this part of the *D. spinulosa* complex. In each case, there is evidence that they are of allopolyploid origin. Thus Manton and Walker (1954) found failure of chromosome pairing (82 univalents) in an experimental polyhaploid (diploid) plant of *D. dilatata*, and Wagner (1971) reported almost complete failure of pairing (0 < 3 bivalents) in a wild example of *D. campyloptera* x *marginalis*.† However, *D. campyloptera* and *D. dilatata* are

---

There is not space here to evaluate this hypothesis at length, but it can be briefly noted that it does require acceptance of: (1) one drastic weakness in terms of cytogenetic logic in that very substantial differentiation of the M genome must have occurred either in *D. marginalis* itself or in *D. celsa* subsequent to its formation (*D. celsa* x *marginalis* forms only 2 < 17 bivalents!); (2) the concept that the phloroglucides peculiar to *D. marginalis* are entirely recessive in *D. celsa* (see Widén and Britton, 1971b, c; Widén, Britton *et al.*, (1975)); and (3) the implication that those who have been engaged in the field, herbarium and laboratory study of these plants have made, continuously over a considerable period of years, an astonishing number of misjudgements regarding the morphological relationships of these taxa.

Even if a choice between the two rival hypotheses cannot already be made by more thorough analysis of the available facts, the controversy is capable of resolution by further cytogenetic analyses of certain critical hybrids, though these would need to be obtained by synthesis, since an essential feature of Hickok and Klekowski's argument is that the wild hybrids attributed to *Dryopteris celsa* x *goldiana* by Wagner and others are all incorrectly named.

Furthermore, Hickok and Klekowski suggest that the problematical triploid *Dryopteris 'leedsii'* studied by Walker (1962b) "seemingly represents the predicted hybrid *D. celsa* x *goldiana*" (Hickok and Klekowski, 1975, p. 568). They have apparently overlooked the statements by Wagner and Wagner (1966, p. 133) and Wagner (1971, p. 173) indicating that reinvestigation of the locality from which the plants studied by Walker originated has confirmed that *D. goldiana* is absent there, and the triploid plants found in this locality are *D. celsa* x *marginalis*. This point may be of crucial significance.

* Hickok and Klekowski (1975, p. 565) state that "at present there is no substantial experimental evidence to indicate that *Dryopteris spinulosa* is an allotetraploid". However, Wagner (1971, p. 154) recorded 0-(4)-8 bivalents in the wild hybrid combination *D. marginalis* x *spinulosa*.

† It is not clear why Widén and Britton (1971a, pp. 256/7) when reporting the presence of more than 41 bivalents in a wild triploid hybrid *Dryopteris campyloptera* x *'dilatata'* should interpret this as evidence in favour of *D. campyloptera* being an autotetraploid. They state that six clear plates (regrettably, none are illustrated) showed the formation of 46 < 49 bivalents. The aberrant 5-8 bivalents can, however, only have been found from within the third genome, in other words from pairing of chromosomes that were not only not homologues, but not even strictly homoeologues. Such events can arise by pairing of homologous segments duplicated in the genome as a consequence of independent

morphologically distinct. Although it is apparent that they share one genome in common, the exact extent of their interrelationship is not known.

The diploid taxa recognized in this group are rather numerous, and the nature of their interrelationships are only most imperfectly known. The experimental studies carried out by Walker (1955, 1961) indicated that *Dryopteris assimilis* of Europe, *D. maderensis* of Madeira and *D. intermedia* of North America are all representatives of the same ancestral genome. Apart from *D. intermedia*, two other diploid taxa are also present in North America. It is generally accepted that the plant found on the western side of the continent should be interpreted as a form of *D. assimilis*, but the status of the form of *D. 'dilatata' s.l.* found in the Lake Superior region is controversial. It is clearly not only morphologically but cytogenetically distinct from the sympatric *D. intermedia*, since Wagner and Hagenah (1962) found almost complete failure of pairing ($0 < 4$ bivalents) in a wild hybrid *D. 'dilatata' x intermedia* discovered in Michigan.

The phloroglucinol chemistry of the European and North American representatives of this group has now been studied very intensively (Wieffering *et al.*, 1965; Widén and Sorsa, 1966, 1969; Sorsa and Widén, 1968; Widén, 1969; Widén and Britton, 1969, 1971a; Widén *et al.*, 1970, 1975; Britton and Widén, 1974; Widén, Britton *et al.*, 1975; Widén, Lounasmaa *et al.*, 1975), and has produced a body of data which has contributed very significantly to resolution of relationships within this complex, even though some of the taxa (e.g. *Dryopteris assimilis*) have proved to be as variable in their phloroglucinol chemistry as in their morphology (Table III).

An initial finding by Widén and Britton (1969) that the Lake Superior *Dryopteris dilatata* was quite distinct in its phloroglucinol content and therefore "may not be conspecific" with the western diploid (Widén and Britton, 1969, p. 1338) was negated by further studies (Widén and Britton, 1971a; Britton and Widén, 1974), which have shown this taxon to be highly variable in its phloroglucinol content,* the outcome being that now these "authors consider

---

assortment after the occurrence of interchanges between non-homologous chromosomes, or by recognition of residual homologies persisting either from ancient aneuploid duplication, or from an equally ancient element of polyploidy involved in the origin of the base number of the genus, but they are certainly very rare indeed in ferns. In any case, these aberrant bivalents are not evidence for autopolyploidy from a base of 41.

On the contrary, the available cytogenetic evidence (all from wild hybrids, i.e. *Dryopteris 'dilatata' x intermedia* = $0 < 4$ bivalents, rest univalents (Wagner and Hagenah, 1962), *D. campyloptera x marginalis* = $0 < 3$ bivalents, rest univalents (Wagner, 1971), *D. campyloptera x intermedia* = c. 41 (41 $<$ 45) bivalents and 41 univalents (Wagner 1963b, 1971), and now *D. campyloptera x 'dilatata'* = $46 < 49$ bivalents, rest univalents (Widen and Britton, 1971a)), is essentially consistent with the original hypothesis of Wagner and Hagenah (1962, p. 98) that *D. campyloptera* is the allotetraploid derivative of *D. 'dilatata' x intermedia*.

* The collections originally studied by Widén and Britton (1969), and some of those investigated later (Widén and Britton, 1971a) proved to contain either no phloroglucinols, or only traces of these components. This phenomenon, whereby individual plants are deficient in phloroglucinols, is now known in another taxon, *Dryopteris campyloptera* (Britton and Widén, 1974, p. 634) but its explanation is unknown.

that the diploid taxon found in northern Quebec and the Lake Superior basin, and in western North America should be called *D. assimilis*" (Britton and Widén, 1974, p. 631).

The total accumulated chemotaxonomic evidence (Widén, *et al*., 1970; Britton and Widén, 1974; Widén, Lounasmaa *et al*., 1975) shows (Table III) that a consistent pattern of phloroglucinol distribution is shared by all of *Dryopteris intermedia, D. maderensis* and *D. azorica* (endemic to the Azores) which is quite different from that found in collections of *D. assimilis* (including the Lake Superior taxon). Britton and Widén (1974, p. 637) state: "The chromatographic results . . . indicate that *D. assimilis* and *D. intermedia* have evolved separate and distinctive phloroglucinol patterns."

These two patterns are complementary to a degree which has induced Widén, Lounasmaa *et al*. (1975, p. 881) to write, "Die chemischen Resultate wären ausgezeichnet mit der Annahme verträglich, dass *D. dilatata* eine allotetraploide Sippe ist, die einmal durch Chromosomenverdoppelung aus einer Hybride von *D. assimilis* und *D. intermedia* entstanden ist (*D. azorica* und *D. maderensis* sind conspecifisch mit *D. intermedia*." This same hypothesis had been proposed earlier by Widén and Sorsa (1969), who further suggested that different races of *D. assimilis*, together with *D. intermedia*, were also the parents of *D. spinulosa*, a proposal which though broadly compatible with most of the known cytogenetic evidence (though not with the synthetic putative *D. assimilis* x *maderensis* hybrids (Walker, 1961, p. 607)), nevertheless seems decidedly improbable, because it requires that a quite unexpected degree of cytogenetic differentiation has subsequently taken place between both the *D. assimilis* and *D. intermedia* genomes, as incorporated into *D. dilatata s.s.* and *D. spinulosa* respectively, since the *D. dilatata* x *spinulosa* hybrid shows only 33 or 34 bivalents of the possible maximum of 82 (Walker, 1955, p. 206).

Leaving on one side the question of the parentage of *Dryopteris spinulosa*, the hypothesis that *D. assimilis* and *D. intermedia* are the parents of *D. dilatata s.s.*, a known allopolyploid, is of course in direct conflict with Walker's conclusion that *D. assimilis, D. intermedia* and *D. maderensis* share the same basic genome. So also, once the Lake Superior taxon is equated with *D. assimilis*, is the finding by Wagner and Hagenah (1962) that *D. 'dilatata'* (Lake Superior) x *intermedia* shows *c*.82 univalents.

Fortunately, resolution of this difficulty appears to be nearly at hand, for Widén *et al*. (1975, p. 898) quote a personal communication from Gibby and Walker to the effect that new cytological results from wild hybrids show that the concept that *Dryopteris assimilis* and *D. intermedia* are conspecific is incorrect, and corroboration is currently being sought by means of a programme of experimental hybridizations.*

---

* Even so, it will be necessary in due course to provide some alternative explanation of the results presented by Walker (1955, 1961) which appear in themselves to be convincing enough. The two critical synthetic *Dryopteris assimilis* x *maderensis* hybrid individuals

In circumstances such as obtain here, where in recent years opinions have changed rather more commonly than is usual, it would be rash to assume that this problem is now near to solution. The conclusion that *Dryopteris assimilis* and *D. intermedia*, are not, after all, closely related, does not mean that all of the diploid taxa extant in the *dilatata* complex will necessarily prove to polarize into two discrete cytogenetically unrelated groups. Little information is yet available about the relationship of other diploid taxa in the complex, e.g. *D. amurensis* and *D. 'austriaca'*, both from Japan (Hirabayashii, 1974). There is as yet no cytogenetic evidence linking the European and Macaronesian diploid species *D. aemula* with the complex.

Furthermore, it should be noted that if we disregard Walker's "*D. assimilis* x *maderensis*" no information is yet available about the cytological behaviour in hybrids between different diploid taxa, apart from Wagner and Hagenah's wild *D. 'dilatata'* (Lake Superior) x *intermedia* hybrid, which may prove to be exceptional in the complete cytogenetic isolation of its parents, since it is well established that sympatry can result in selection for increased isolation between related species. With so many diploid taxa known, it would be surprising if evidence of a state of partial differentiation of genomes is not found at some point in this complex. In other words, not all the possible diploid hybrid combinations may display either virtually complete bivalent formation or failure

---

showed ". . . normal meiotic pairing, but approx. 40% abortive spores which suggests gene imbalance. . . . More hybrids are necessary to establish this" (Walker, 1961, p. 607). These plants may well have been abnormal selfs of one or other intended parent in a subfertile condition on account of some physiological cause, since it is not always easy to discriminate hybrids when the parents are of the same level of ploidy, and are as morphologically close as are these two species. With the hybridization technique used, a proportion of selfed progeny has to be tolerated (Lovis, 1968a).

Concerning the synthetic *Dryopteris assimilis* x *spinulosa* hybrids, the undetected contamination of cultures by air-borne spores is always a possibility with pot-grown prothalli, particularly when cultures of similar taxa of the same ploidy are being raised simultaneously for hybridization experiments, though in practice such accidents are fortunately rare. This hazard is greater where a possible contaminant is much more fertile in the attempted cross than the intended parent. In this respect, it may be significant that whereas Britton (1965, p. 7) stated that in Ontario "*intermedia* x *spinulosa* is by far the most common hybrid", elsewhere (Britton and Widén, 1974, p. 636) he comments on his "inability to find hybrids of *D. assimilis*". Furthermore the *D. assimilis* x *spinulosa* hybrid has as yet been found only once in Finland (Sorsa and Widén, 1968), where both species are common. (See also Gibby and Walker, *Fern Gaz.* in press).

The situation is complicated by the report by Sorsa and Widén (1968) on meiosis in their wild example of *Dryopteris assimilis* x *spinulosa*, in which they record the presence of 15-20 bivalents, together with 5-8 multivalents, a total of 20-28 associations. (Walker (1955, p. 202) found 38 ± 1 pairs in his corresponding synthetic hybrids.) However, the claim that multivalents are present is best discounted, because the quality of the illustrated cell (Sorsa and Widén, 1968, fig. 17) is not adequate to demonstrate their presence. Its appearance suggests an indifferent quality of fixation (often attributable to an unbalanced genotype, rather than technique) with "sticky" chromosomes. My candid opinion is that the level of true chromosome pairing present in this cell is probably appreciably lower than is indicated by Sorsa and Widén. Nevertheless, regardless of their exact number, the origin of the pairs present in this hybrid remains an unsolved problem.

of pairing. What is clear is that the *D. dilatata* complex appears to provide an unrivalled opportunity for investigation of this aspect of fern evolution, and the results of the synthesis and investigation of different diploid combinations will be awaited with considerable interest.

3. Polypodium vulgare complex

One of the most dramatic results of Manton's initial cytological survey of European ferns (Manton, 1950) was the discovery that the familiar Common Polypody included three distinct taxa, corresponding to different levels of ploidy, diploid, tetraploid and hexaploid, which differed in their morphology, phenology and distribution. Investigation of chromosome pairing in wild triploid and pentaploid hybrids indicated clearly that not only was the hexaploid plant an allopolyploid derived from the sympatric diploid and tetraploid, but also that the European diploid and tetraploid forms were unrelated, their triploid hybrid showing complete failure of pairing. In due course the three fertile cytotypes were recognized as distinct species, with these names: *Polypodium australe* Fée (diploid), *P. vulgare* L. sensu stricto (tetraploid) and *P. interjectum* Shivas (hexaploid) (Shivas, 1961b).

A study of collections of *Polypodium* from North America showed that at least four distinct diploid species, including *P. glycyrrhiza* and *P. virginianum*, occurred there (Manton, 1951). The latter species proved to include both diploid and tetraploid cytotypes (Manton and Shivas, 1953) which were incorporated, together with *P. glycyrrhiza*, in a programme of experimental hybridizations involving all the three British taxa (Shivas, 1961a). Study of meiosis in the resultant synthetic hybrids proved highly informative, not only confirming the relationships between the British taxa indicated by Manton (1950), but also providing a clear demonstration that one of the genomes of *P. vulgare* was also present in the diploid form of *P. virginianum*. Another hybrid, *P. glycyrrhiza* x *vulgare*, showed that *P. vulgare* also shared a common genome with *P. glycyrrhiza*. The data obviously permitted presentation of the hypothesis that the parentage of *P. vulgare* is in fact *P. glycyrrhiza* x *virginianum*, but Shivas refrained from advocating this interpretation without considerable reservation, not so much from reluctance to accept the rather bizarre concept that the origin of an exclusively Old World tetraploid species was by hybridization between two North American diploids, but because of the absence of any demonstration whether *glycyrrhiza* and *virginianum* ". . . represent distinct genomes or only geographically differentiated variants of one. This cannot be known with certainty until additional hybrids, preferably the direct cross between diploid *P. virginianum* and *P. glycyrrhiza*, can be set up and the chromosome pairing analysed. The degree of morphological difference between these species makes it somewhat unlikely that such a hybrid would show regular chromosome pairing. . . . As long as the possibility is open, however, it would be wrong to claim that the full parentage of *P. vulgare* L. *sens. strict.* has been found" (Shivas, 1961a,

p. 22). Shivas thus very correctly refuses to extend interpretation of her data beyond what they in fact demonstrated. No further experimental studies have yet been carried out, and this question remains unresolved, although increased knowledge of the morphology of the North American taxa has perhaps made it more probable that this interpretation of the parentage of *P. vulgare* is correct.

Further cytological and morphological studies (Lloyd and Lang, 1964) on American representatives of the *Polypodium vulgare* complex have revealed the existence of more taxa, in particular the existence of both diploid and tetraploid cytotypes in both *P. californicum* and *P. hesperium* (*s.l.*). The diploid taxon of the latter has now been separated as *P. montense* (Lang, 1969). However, progress in elucidation of the interrelationships of these American taxa has been impeded by a lack of the experimental hybridization studies necessary to resolve what has every appearance of being a complex situation.

Lloyd and Lang (1964) quote a previously unpublished hypothesis of Evans that the tetraploid cytotype of *Polypodium virginianum* is derived from *P. virginianum* (2x) x *montense*. This is a plausible hypothesis, but lacks any experimental confirmation. It should be pointed out that the implied conclusion of Shivas (1961a, p. 23, fig. 6) that the tetraploid *P. virginianum* is of alloploid constitution is not very securely based, being in effect dependent on the absence of trivalents in its hybrids with diploid *P. virginianum* and *P. interjectum*. Though taxa of autotetraploid constitution frequently do produce a conspicuous number of trivalents in triploid hybrids with their diploid progenitor (e.g. *Asplenium ruta-muraria*, Vida, 1970; Bouharmont, 1972a), this is not always the case, as for instance in *Asplenium trichomanes* (Lovis, 1955a, and unpublished). It is certainly true that the two cytotypes of *P. virginianum* are morphologically very close. It is equally true that the morphological differences between diploid *P. virginianum* and *P. montense* are slight. Indeed there is little doubt that in practice taxonomic separation of these two plants would be substantially more difficult were it not that their distributions are allopatric, each being restricted to a different side of North America. Clearly this problem can only be resolved by experimental study.

Lloyd and Lang (1964) also suggested that the tetraploid form of *Polypodium californicum* may be an allotetraploid, the product of hybridization between the diploid cytotype of *P. californicum* and *P. glycyrrhiza*. This hypothesis also lacks supporting cytogenetic evidence. The *P. californicum* complex is clearly a difficult one. At present the possibility cannot be excluded that the tetraploid cytotype of *P. californicum* may include plants of both autopolyploid and allopolyploid origins. It appears to be generally assumed that the morphology of *P. scoulerii* is so distinctive that it cannot be involved in the origin of any of the tetraploid taxa, but an admittedly superficial examination of some plants of tetraploid *P. californicum* has persuaded me that this assumption may be unjustified.

Lang (1971) has produced an elegant and detailed cytogenetic and comparative morphological study of *Polypodium glycyrrhiza, P. hesperium* and *P. montense*, as a result of which he proposes what he recognizes to be "... a tentative hypothesis ... concerning the origin of the tetraploid *P. hesperium*. It seems likely that *P. hesperium* is an allotetraploid that arose from the diploids *P. glycyrrhiza* and *P. montense* or their progenitors" (Lang, 1971, p. 252). The observation of *n* bivalents and *n* univalents in each of two wild putative backcross hybrids, *P. glycyrrhiza* x *hesperium* and *P. hesperium* x *montense*, is suggestive evidence, but is of course not a reliable indication of relationship without an independent demonstration of the alloploid nature of *P. hesperium*. However, the morphological evidence produced by Lang in support of his hypothesis is highly persuasive.

The *Polypodium vulgare* complex remains a source of great interest with many unresolved problems. The apparent origin of *P. vulgare* L., *sensu stricto*, demonstrates that the phytogeographical history of the complex is surely complicated. Nothing less than an ambitious programme of experimental hybridization, which should include Asian as well as European and North American taxa, will serve to elucidate its interrelationships. Any such programme ought not to neglect the synthesis of diploid x diploid hybrids, since we are particularly ignorant of the nature of affinities at the diploid level in this group. Unfortunately, these plants are slow-growing, and such a programme, even if commenced now, would take many years to bring to fruition. The scheme of possible interrelationships presented in Fig. 13 is highly speculative but may act as an incitement to further experimental work.

Detection of three morphologically distinguishable cytotypes amongst the European *Polypodium vulgare* complex inspired several detailed taxomonic studies (Benoit, 1966; Roberts, 1966, 1970), which though not directly relevant here, are of interest as a clear demonstration of the manner in which quantitative characters first regarded as clearly diagnostic can become weakened by intensive study of more and larger samples, which inevitably extend the absolute range known within each cytotype of the characters concerned. Unfortunately, the specific status of the three fertile taxa is not universally accepted, in spite of the lucid exposition of the argument by Manton (1958b). They are still treated as subspecies in some important floristic publications (e.g. Perring, 1968), even though this involves the thoroughly unnatural association within one species, *P. vulgare* L., of a diploid and a tetraploid which are not related, while another diploid to which the tetraploid is related remains separated in another Linnean species, *P. virginianum* L. This insular approach to taxonomy, whereby specific limits are in effect determined by geographical boundaries, is not only poor taxonomy, but positively harmful in that it disguises relationships and discourages the inception of the broadly based investigations which are needed if the evolutionary patterns of genera are to be effectively elucidated. It is interesting to speculate what might be the taxonomic treatment afforded to the diploid cytotype of *P. californicum* and *P. australe* if they occurred in the same flora instead of separated on opposite sides of the world.

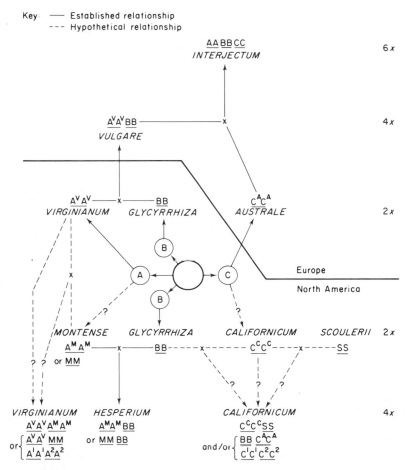

Fig. 13. Hypothetical scheme of relationships in the European and North American *Polypodium vulgare* complex. Solid lines joining taxa indicate relationships established by Shivas (1961a) and Lang (1971). Dotted lines indicate purely speculative relationships.

4. Polystichum

The genus *Polystichum* is unique amongst those genera of ferns which have been investigated inasmuch as it seems that homology is rarely if ever totally lost even between the genomes of morphologically very distinct species.

In her pioneer investigation of the three British species Manton (1950) suggested that the tetraploid *Polystichum aculeatum* was derived from the other two species, *P. setiferum* and *P. lonchitis*, which are both diploids. Subsequent investigations of various wild hybrids involving these species and the only other European species, *P. braunii*, provided cytogenetic confirmation of this hypothesis, and also demonstrated that *P. braunii* is not directly related to any of the other three species (Manton and Reichstein, 1961; Sleep and Reichstein,

1967). Nevertheless, none of the three hybrids *P. braunii* x *aculeatum, P. braunii* x *setiferum*, and *P. braunii* x *lonchitis* is devoid of bivalents, even though these are somewhat imperfect pairs which tend to commence dissociation prematurely in metaphase. In spite of the fact that one of these three hybrids is tetraploid, and the other two only triploid, the numbers of bivalents encountered were very similar (10-(mean 13)-18, 12-(14)-17, and 6-(11)-24 respectively), in each hybrid. The simplest explanation of this finding would be to attribute all of the pairs formed to partial pairing between the two genomes, both still unidentified, of *P. braunii*, which must then be a segmental alloploid, but other evidence shows that this hypothesis possesses an important weakness.

Polystichum lonchitis and *P. setiferum* have very distinct ecological preferences, and only very rarely grow together. However, the wild hybrid between these two species has very recently been discovered in Eire (Sleep *et al.*, in press). Pairing at meiosis is very similar to that seen in synthesized examples of *P. lonchitis* x *setiferum* produced by Sleep. Both show a variable but appreciable number of bivalents (6-(15)-25 in the synthetic examples, 8-(15)-24 in the wild hybrid), indicating that *P. aculeatum* is not strictly of true or genomic allotetraploid origin, but instead arose as a segmental alloploid.* The evidence of this hybrid combination shows that the L and S genomes are capable of forming an average of 12-15 bivalents. If indeed the genomes of *P. braunii* (X and Y) are alone responsible for the pairing seen in the *P. braunii* x *lonchitis* and *P. braunii* x *setiferum* hybrids (mean values 11 and 14 respectively), then it would be logical to expect the expression of both of these potentials (L and S, X and Y) in

---

* The question of the distinction between a **segmental allopolyploid** and a genomic allopolyploid and the manner in which the latter can develop from the former may cause confusion. The terminology followed here is that of Stebbins (1947, 1950), who defines a segmental allopolyploid as "a polyploid containing two pairs of genomes which possess in common a considerable number of homologous chromosomal segments or even whole chromosomes, but differ from each other in respect to a sufficiently large number of genes or chromosome segments, so that the different genomes produce sterility when present together at the diploid level" (Stebbins, 1950, p. 318). This is clear enough, but difficulty arises when later (op. cit., pp. 321/2) Stebbins states that "the most important genetic difference between segmental allopolyploids and typical or genomic ones is the ability of the former to segregate in respect to some of the characteristics by which their ancestral species differ from each other". Species like *Polystichum aculeatum* and *Cystopteris fragilis* (tetraploid cytotype, see p. 358), with their regular diploidized meiosis, do not display the ability to form multivalents, but their hybrids show that they have nevertheless not lost that ability, which is only suppressed. The segmental alloploid condition is an unstable one which can be expected, in examples like those we are considering, to evolve by selection for increased fertility in the direction of cytological allopolyploidy. Segmental alloploids, just like autopolyploids, can acquire genetic mechanisms stabilizing their meiosis by inhibiting multivalent formation. The point made here, the distinction between genomic allopolyploids and allopolyploids which have differentiated from a segmental alloploid origin, is not an academic one, since should the regulating mechanisms in the segmental alloploid become upset, as might be induced purely temporarily by some environmental extreme, then multivalent formation and consequent segregation with regard to parental differences will occur. The genetical outcome of such occasional cytological atavism might well be of significance in the subsequent evolution of a polyploid species.

the *P. braunii* x *aculeatum* combination, since the constitution of this hybrid is LSXY. Why this expectation is not in fact realized is by no means evident.

A substantial number of synthesized *Polystichum* hybrids have been produced and studied by Sleep (1966, and unpublished). It appears from her investigation that it is universal in the genus that whenever two genomes are present, each lacking truly homologous partners, then residual homologies are recognized, permitting an appreciable degree of pairing to occur between homoeologues. The alternative explanation that interchanges have occurred on a substantial scale within the genomes of *Polystichum* seems increasingly less likely in view of the generality of the phenomenon. A series of four diploid and five triploid hybrids, all between unrelated species, have been studied (Sleep, in press). These show a remarkable consistency in the level of pairing seen, the ranges observed being 6-(13-25)-28 bivalents in the diploid hybrids and 6-(11-20)-28 in the triploid hybrids. In the course of this study, one diploid, *P. acrostichoides*, has been hybridized with three other diploid species, namely *P. lonchitis*, *P. munitum* and *P. setiferum*. It forms bivalents with all of these, the overall range being 9-25 bivalents. (See Note Added in Proof, 7, p. 415.)

The universal presence and remarkably constant level of homoeologous pairing between differentiated genomes in *Polystichum* suggests an explanation in the form of some intrinsic property of the ancestral *Polystichum* genome. It may be that some chromosomes have been in part "conserved" against extensive change, presumably because certain complex linkages have some peculiar advantage for the genus as a whole. In other words, some segments of the genome may have been protected and maintained by selection, and these allow mutual recognition and pairing of certain homoeologues throughout the genus.

Wagner has recently reported in some detail on some remarkable situations among species of *Polystichum* on the west coast of the USA (Wagner, 1973) (Fig. 14). Three tetraploid species are interpreted as being allopolyploids. There is at present no cytogenetic evidence available concerning one of these, *P.*

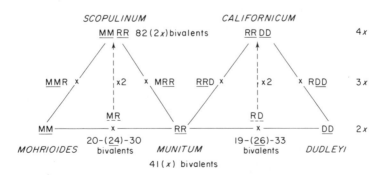

Fig. 14. Segmental allopolyploidy in *Polystichum* in the western United States, based on data of Wagner (1973). Symbols as in Fig. 5.

kruckebergii (which it is suggested has been formed from *P. lonchitis* x *mohrioides*), and in consequence it will not be discussed further here. The other two tetraploid species, *P. scopulinum* (ex *P. mohrioides* x *munitum*) and *P. californicum* (ex *P. dudleyi* x *munitum*), are of especial interest because in each case it has been possible to find localities where the parent diploid species, their sterile diploid hybrid, its tetraploid derivative, and triploid backcrosses, can all be found growing together. For neither of these tetraploids is a formal cytogenetic demonstration of the nature of their polyploidy available, (such has been obtained for *P. aculeatum* and *P. braunii*), but to insist on this would be excessively pedantic, since in view of the existence of the diploid hybrids the origin of the tetraploids is obvious from the evidence of morphology and chromosome numbers alone. Once more, and again in spite of considerable morphological differences between the parent species, in both cases the sterile diploid hybrid shows a substantial degree of chromosome pairing, the range found in *P. mohrioides* x *munitum* being 20-(24)-30 bivalents, and in *P. dudleyi* x *munitum* 19-(26)-33 bivalents. It is evident that though *P. scopulinum* and *P. californicum* are now stabilised, as Wagner has described them, as allotetraploid species, they are nevertheless of segmental alloploid origin.

Wagner comments on the remarkable fact that in spite of the considerable degree of chromosome pairing in the sterile diploid hybrids, their derivative tetraploids show regular pairing without, as might well have been expected, the formation of multivalents. This paradox can however be explained. In view of the very substantial and diverse morphological distinctions between the parent diploid species, there can be no doubt that the pairing seen in the diploid hybrids consists of bivalents formed between homoeologues, and not between truly homologous chromosomes. In such circumstances, preferential pairing can readily become established in the tetraploids by a genetic mechanism inhibiting the association of homoeologues, since this mechanism will be favoured by selection because of its beneficial effect on fertility. Though the potentiality for formation of new tetraploid individuals by chromosome doubling clearly exists today, this does not mean that the tetraploid individuals encountered in these localities are necessarily of recent polyploid origin. Indeed, apart from a very few isolated individuals too recent in origin to have yet been eliminated (selection has an element of chance in its operation), the opposite will be the case, because newly formed tetraploids displaying any degree of meiotic instability will lose out against the pressure of competition from existing tetraploids which have the advantage of the stabilising effect of generations of selection.

5. *Cystopteris fragilis complex*

Cystopteris is dominated by the widespread and polymorphic *C. fragilis* complex, which constitutes perhaps the most formidable biosystematic problem in the ferns. The genus was recently monographed by Blasdell (1963), who provided what is best described as an inadequate treatment of this intricate complex. Readers who have already tackled section II of this paper will appreciate that I am unsympathetic towards an argument which seeks to explain "a disjunct distribution of the primitive element ... by postulating that the genus *Cystopteris* evolved during the Carboniferous period" (Blasdell, 1963, p. 30). There is not space here to criticize this monograph in detail, but one point of technique deserves attention in case it should otherwise become more widespread. The use of spore size as a means of predicting the ploidy level of a taxon whose chromosome number is unknown is not in itself a wholly reprehensible procedure, even though, as one of its practitioners has recognized (Smith, 1971, p. 28), it must be applied with considerable caution if false predictions are not to result. It is, however, essential that such predictions are always clearly distinguished from real chromosome counts. The practice of inserting such "pseudocounts" into species descriptions in a manner totally indistinguishable from genuine counts needs to be deprecated in the strongest possible terms, since not every reader will realize that he needs to consult a table in an earlier section of the monograph to discover whether the chromosome numbers given have or have not been actually counted. Thus the chromosome number of *C. tenuisecta* is given as $n = 42$, although the only recorded count (apart from a triploid hybrid) is $n = 84$ (Bir, 1971). Even more confusing is *C. montana*, which is given as $n = 42, 84$. In fact, tetraploid counts have been recorded from different areas by several authors (Manton, 1950; Brögger, 1960; Britton, 1964), but $n = 42$ is only a "pseudocount." In the circumstances it is debatable whether on balance it is a good or bad thing that no information is given concerning these "pseudocounts" with regard to the actual measurements made, the number of plants "counted", or the locality of origin of the specimens concerned.

Blasdell's treatment differs from previous taxonomic practice in recognizing *Cystopteris diaphana* as an aggregate species, parallel to the *C. fragilis* complex, but quite separate, and indeed placed in a different section of the genus. The widespread occurrence of intermediate individuals or populations far outside the main range of *C. diaphana* is regarded as probably attributable to the long range dispersal and establishment of hybrid individuals from zones of introgression. This extraordinary hypothesis is not necessary to explain the observed facts, which can be more simply interpreted by accepting that although certain features of vein termination, epidermal and indusium cell shape are consistent in and characteristic of a particular Central and South American taxon, these characteristics also occur as a normal but variable feature of some other populations of the *C. fragilis* complex, of which this taxon is really just one element.

Modern European taxonomic practice (e.g. Crabbe, in Tutin *et al.*, 1964) accepts *Cystopteris dickieana*, which possesses distinctive morphological features other than its highly characteristic spores, as a valid species related to but distinct from *C. fragilis*. In contrast, Blasdell dismissed *C. dickieana* "as a form not worthy of taxonomic designation" (Blasdell, 1963, pp. 4/5), but this opinion has now been refuted by study of a synthetic hybrid between tetraploid races of the two species, which is sterile and shows meiotic disturbance, only 54 of a possible maximum of 84 bivalents being formed (Vida, 1974).

Cytogenetic investigation of the *Cystopteris fragilis* complex is severely handicapped by the high levels of ploidy encountered, and the scarcity of extant diploids, which indeed appear to be absent from Europe in all species of the genus. The initial study by Manton (1950) showed tetraploidy in *C. dickieana* and *C. montana*, both tetraploid and hexaploid levels in *C. fragilis*, and hexaploidy in *C. regia* (= *alpina*). Subsequently this picture has been complicated further by the discovery of a hexaploid cytotype of *C. dickieana* (Manton and Reichstein, 1965), and an octoploid form of *C. fragilis* (Reichstein and Vida, in Vida, 1974), both in the Alps. Diploid members of the *C. fragilis* complex are at present only recorded in North America, as *C. protrusa* (Wagner and Hagenah, 1956), and from Chile, as *C. diaphana* (Blasdell, 1963). Because of the high level of polyploidy involved and the substantial amount of chromosome pairing encountered, wild pentaploid hybrids within *C. fragilis* itself and between the tetraploid cytotype of *C. fragilis* and *C. regia* are, considered in isolation, not very informative about the interrelationships of these taxa, on account of the plurality of interpretations possible.

In a skilful and courageous experimental investigation Vida (1974, and unpublished) has made use of the North American diploid *Cystopteris protrusa* to obtain useful information regarding the nature of polyploidy in the *C. fragilis* complex. He has synthesized and analysed hybrids between *C. protrusa* and (1) the tetraploid cytotype of *C. dickieana*, and (2) both tetraploid and hexaploid cytotypes of *C. fragilis*. Vida has also obtained, through inducing apogamy in an otherwise normal gametophyte, a polyhaploid (diploid) sporophyte of *C. fragilis*, which has provided indispensable evidence. The cytogenetic information obtained is summarized below ($x = 42$):

$2x$ polyhaploid ex *Cystopteris fragilis* $(4x)$ = $3 < \underline{12}$ bivalents, rest unpaired
$3x$ hybrid : *C. fragilis* $(4x)$ x *protrusa* $(2x)$ = $13 < \underline{15}$ bivalents, rest unpaired
$3x$ hybrid : *C. dickieana* $(4x)$ x *protrusa* $(2x)$ = $\underline{18}$ bivalents, rest unpaired
$4x$ hybrid : *C. dickieana* $(4x)$ x *fragilis* $(4x)$ = $\overline{54}(= 42 + \underline{12}?)$ bivalents, rest unpaired
$4x$ hybrid : *C. fragilis* $(6x)$ x *protrusa* $(2x)$ = $40(<)$ bivalents, rest unpaired
$5x$ hybrid : *C. fragilis* $(6x)$ x *fragilis* $(4x)$ = $c.84$ bivalents, rest unpaired

These cytogenetic analyses present a complicated pattern, whose interpretation must at present remain subject to some degree of uncertainty, but which undoubtedly indicates a most interesting web of relationships (Fig. 15). Some conclusions are not open to dispute. The low number of bivalents in the hybrids between *Cystopteris protrusa* and the tetraploid cytotypes of both *C. dickieana* and *C. fragilis* indicates that this diploid taxon is not one parent of either

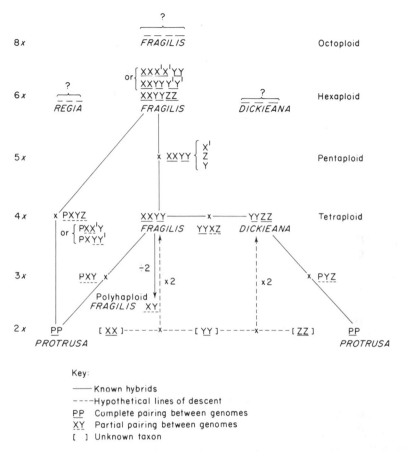

Fig. 15. Interpretation of cytogenetic relationships in the *Cystopteris fragilis* complex in Europe, based on data of Vida (1974).

tetraploid, and furthermore indicates that neither tetraploid is of autopolyploid origin, a conclusion confirmed in the case of *C. fragilis* by the polyhaploid plant. Possession of this information makes it possible to interpret the hybrid between these two tetraploids, whose high degree of pairing ($54 (= 42 + 12)$ bivalents) clearly indicates not only that its parents share a common genome, but also indicates that "the other two genomes present . . . are also closely related" (Vida, 1974, p. 188). It must be significant that partial pairing of two genomes is observed also in the polyhaploid and in the two triploid hybrids. If, to clarify discussion of the analysis, we accept the genome code presented by Vida (1974, p. 188), it is seen that the origin of tetraploid *C. fragilis* and tetraploid *C. dickieana* requires the postulation of three participant diploids, all not known extant today, possessing X, Y and Z. These genomes are not completely differentiated, however, there being direct evidence that X can in part pair with Y, and likewise X with Z. The same must also be true of the third combination, Y and Z (the constituents of tetraploid *C. dickieana*) unless there is partial morphology between P and Z, as well as between Z and X, X and Y, but not between Y and Z. This alternative is not impossible, but is less likely. Vida interprets both tetraploid *C. fragilis* and tetraploid *C. dickieana* as allopolyploids, which is now effectively true, but the evidence indicates that the former is clearly of segmental alloploid origin, and it is more likely that the same is true of tetraploid *C. dickieana.*

In the existence of substantial partial or residual homology between genomes, the available evidence in *Cystopteris* provides an interesting parallel to the situation now well known in *Polystichum.* In *Polystichum* it is clear that loss of cytogenetic homology does not keep pace with morphological differentiation, since diploid hybrids between extremely distinct species show the phenomenon of residual homology. The same is not necessarily true in *Cystopteris*, where the taxa involved in the *C. fragilis* complex are morphologically very close, and it seems reasonable to interpret the evidence of partial homologies as truly representing the state of only partial differentiation of the diploid genomes involved at the base of the *C. fragilis* complex.

Vida (1974, p. 182) comments that "we are dealing with a rather old polyploid complex", an opinion which is amply supported by the scarcity of diploids, the high polyploid levels involved, and the worldwide distribution of the complex.

6. European Aspleniaceae

Thanks to the efforts of a number of investigators, the interrelationships of the representatives of the family Aspleniaceae in Europe are now much better known than those of any other group of ferns of comparable size in any flora. Of some 38 taxa which can be given specific rank* the great majority belong to *Asplenium* (31), the rest to *Phyllitis* (3), *Ceterach* (2) and *Pleurosorus* (1), smaller genera which are sometimes merged into *Asplenium*, on account of their ability to hybridize with that genus (Lovis, 1973). The entire group is remarkable for its extreme morphological diversity, and for its capacity for hybridization, hybrids being possible between species of very different morphology.

In relation to its size, Europe contains the richest concentration of diploids known in the family (Lovis, 1973, p. 216), half of the species* being diploids. Excluding sterile hybrids, the polyploid taxa are all tetraploids, apart from one

---

* For the sake of simplicity, "species" here is used in the sense of biological species, inasmuch as autotetraploid derivatives are counted as being separate species from their diploid progenitors. Certain authors (e.g. Rothmaler, 1963, p. 5; Löve, 1970, p. 64) do indeed treat these forms taxonomically as distinct species, but this is a minority opinion, the more usual practice being to recognize such plants as subspecies.

triploid apomict (*Asplenium monanthes*) known in the European region only from the Azores. The nature and origin of all of the 18 fertile tetraploid taxa is now known, with the sole exception of one species of evidently American affinity, also known only in the Azores.†

(a) Allotetraploids

Eleven of these tetraploids are of allopolyploid constitution. In every case their diploid parents are still extant in Europe. Their parentages are as follows:

Asplenium aegaeum x *ruta-muraria*	(Lovis *et. al.*, 1966; Vida,
ssp. *dolomiticum* = $\begin{cases} A. \ haussknechtii \\ \quad lepidum \end{cases}$	1970; Reichstein *et. al.*, 1973; Brownsey, 1973, 1976a,b)
A. aegaeum x *viride* = *creticum*	(Brownsey 1973, 1976a; Reichstein *et al.*, 1973)
A. cuneifolium x *onopteris* = *adiantum-nigrum*	(Meyer, 1968, Shivas, 1969; Lovis and Vida, 1969)
A. fontanum x *obovatum* = $\begin{cases} foresiense \\ macedonicum \end{cases}$	(Sleep, 1966; Meyer, 1968)
A. fontanum x *petrarcheae* ssp. *bivalens* = *majoricum*	(Sleep, 1967; Lovis and Reichstein, 1969)
A. obovatum x *onopteris* = *balearicum*	(Shivas, 1969; Lovis *et. al.*, 1972)
A. ruta-muraria ssp. *dolomiticum* x *seelosii* = *eberlei*	(Meyer, 1967)
A. trichomanes ssp. *trichomanes* x *viride* = *adulterinum*	(Lovis, 1955b, 1968b)
Ceterach javorkeanum x *Phyllitis sagittatum* = *P. hybrida*	(Vida, 1963, 1972, and unpublished; Emmott, 1964)

Remarkably, as far as is known, with one exception all of these are strictly genomic allopolyploids. The exception is *Asplenium adiantum-nigrum*, but even here, since the genomes of *A. cuneifolium* and *A. onopteris* form only $0 < 6$ bivalents (Lovis and Vida, 1969; Shivas, 1969), it would be academic rather than realistic to describe this plant as a segmental allopolyploid.

The distributional ranges of these various allotetraploids are very diverse (Jalas and Suominen, 1972). Most are confined to Europe, but *Asplenium adiantum-nigrum* is evidently an old species, since although its parent diploids are virtually confined to Europe (*A. cuneifolium* is a serpentine endemic of Central and East Europe; *A. onopteris* is of oceanic preference, found predominantly in the south-west of the region), *A. adiantum-nigrum* is known outside Europe not only in North America, but also in the mountains of Central and Southern Africa. In this, *A. adiantum-nigrum* is quite exceptional; the majority of these allotetraploids have very restricted distributions and correspond well to the concept of apoendemics (Favarger and Contandriopoulos,

---

† Judged solely within the context of the Aspleniaceae the case for inclusion of the Azores within the European region (Tutin *et al.*, 1964, p. xvi) is rather weak.

1961), being polyploids of much more restricted range than their parental diploids.

The most extreme example of an apoendemic is *Asplenium eberlei*, which is, as far as is known, confined to a single rock ledge in the Dolomites. It may be a new species discovered on the site of its inception, but is more likely to represent a recent re-creation of a marginally viable combination capable of only ephemeral persistence.

More solidly established apoendemics are *Phyllitis hybrida* (confined to the Quarnero islands in the N.E. Adriatic, and remarkable as being an example of a bigeneric allopolyploid), and *Asplenium majoricum*, which is restricted to Majorca, and most probably originated there, though it is no longer in contact with either of its parents. This circumstance illustrates the possibility that an allotetraploid, once established, may under conditions of severe competition oust one or both of its parents completely from a particular locality. *Asplenium majoricum* is one of the only two allopolyploid fern species to have yet been resynthesized from its diploid parents (J. D. Lovis, unpublished).

In the same category, though on a rather larger scale, is *Asplenium adulterinum*, which is the most completely studied allotetraploid fern, having been resynthesized from its diploid parents and the analysis completed by crossing the synthetic allotetraploid with the natural species (Lovis, 1968b) (Fig. 16). It is widely dispersed but disjunct in North and Central Europe, being confined to serpentine and other ultrabasic rocks, whereas both of its parents are

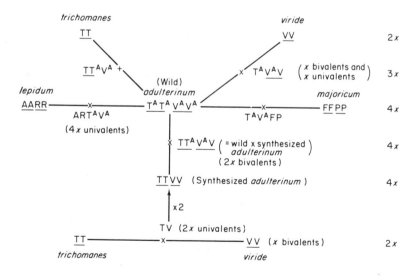

Fig. 16. Cytogenetic analysis and resynthesis of *Asplenium adulterinum*, from data of Lovis (1968b).

of pan-Boreal distribution and do not share its edaphic restriction. Thus *A. trichomanes* ssp. *trichomanes* has a catholic preference, with the sole exception that it avoids limestone, whereas *A. viride* grows on a variety of basic rock types. The restriction of *A. adulterinum* to ultrabasic rocks must be related to the peculiar edaphic environment derived from these rocks, which contains high levels of normally toxic metallic ions. Growth in this environment requires special inherent physiological adaptations, which have been shown by competition experiments (Kruckeberg, 1954) to place a serpentine-adapted form at a selective disadvantage on normal soil. Equally strong selection pressures must operate in the comparable but less complex metal toxicity situations that occur on copper, lead and zinc mine workings, where the distribution of tolerant individuals corresponds very exactly to the distribution of abnormal levels of the toxic ion (Jain and Bradshaw, 1966; McNeilly, 1968; Bradshaw, 1971). The restriction of *A. adulterinum* to serpentine must therefore be correlated with its formation on serpentine from serpentine-adapted races of its parent diploids, which thus donated to their offspring a genetic prison from which it has been unable to escape.

(b) Autotetraploids

The remaining six tetraploids are all believed to be of autopolyploid constitution, though in every case their meiosis has become diploidized and is invariably regular, the potential for formation of multivalents being somehow inhibited. A small minority of populations of *Asplenium ruta-muraria* (about 10% in the study by Vida) constitute an exception to this rule, and do show multivalent formation and irregular segregation (Vida, 1970; J. D. Lovis and T. Reichstein, unpublished). These autotetraploids and their related diploids are:

4x	2x	
Asplenium billotii	?*Asplenium obovatum*	(Sleep, 1966; Lovis and Vida, 1969; Callé *et al.*, 1975)
A. petrarcheae ssp. *petrarcheae*	*A. petrarcheae* ssp. *bivalens*	(Sleep, 1966; Lovis *et. al.*, 1969)
A. ruta-muraria ssp. *ruta-muraria*	*A. ruta-muraria* ssp. *dolomiticum*	(Lovis, 1964b; Lovis and Reichstein, 1964; Vida, 1970; Bouharmont, 1972a)
A. septentrionale (4x cytotype)	*A. septentrionale* (2x cytotype)	(Lovis, 1964b; Bouharmont, 1972b; Callé *et al.*, 1975)
A. trichomanes ssp. *quadrivalens*	*A. trichomanes* s.l. (2x cytotype)	(Lovis, 1955a; Bouharmont, 1972c)
Ceterach officinarum	?*Ceterach javorkeanum*	(Vida, 1963)

The inclusion of *Ceterach officinarum* in this list is subject to qualification, since in this case the cytogenetic evidence of autopolyploidy is not conclusive. In the case of *Asplenium billotii*, its autoploid constitution is not in doubt, but the exact nature of the relationship between it and the diploid *A. obovatum* requires clarification.

The best studied autoploids in this list are *Asplenium ruta-muraria* and *A. septentrionale* (*A. trichomanes* presents special problems and will be considered separately). In both of these species the tetraploid subspecies predominates and is presumably of some age, since it is very widespread, being distributed all round the Northern Hemisphere. In contrast, their related diploids are patroendemics (Favarger and Contandriopoulos, 1961), being confined to relict localities. Thus *A. ruta-muraria* ssp. *dolomiticum* is known only from limestone localities in Europe recognized as being centres of relict survival (e.g. the Dolomites and also the Gorge du Verdon, the only known locality of the diploid endemic *A. jahandiezii*), while the diploid cytotype of *A. septentrionale* has not been found in Europe and is known only from the Caucasus.

In the case of *Asplenium septentrionale*, which is an extremely isolated morphological type, the mere existence of both diploid and tetraploid forms is a strong indication of autopolyploidy. However, the investigation of this species does provide a good illustration of how autopolyploid construction can be demonstrated cytologically in spite of the presence of a mechanism inhibiting multivalent formation in the polyploid. Hybrids with the tetraploid cytotype of *A. septentrionale* as one parent are now known (either as wild or synthetic examples), and have been cytologically investigated at meiosis, involving respectively all of the following taxa: *A. adiantum-nigrum*, *A. billotii* and *A. foresiense* (Callé *et al.*, 1975); *A. haussknechtii* and *A. ruta-muraria* ssp. *dolomiticum* (P. J. Brownsey, unpublished); *A. ruta-muraria* ssp. *ruta-muraria* (Lovis, 1963, 1964b; Vida, 1970); *A. trichomanes* ssp. *trichomanes* (Manton, 1950; Lovis and Shivas, 1954; Bouharmont, 1966); *A. trichomanes* ssp. *quadrivalens* (J. D. Lovis and T. Reichstein, unpublished; G. Vida, unpublished). In all of these hybrids not less than 36 bivalents is the maximum number observed, owing to the ability of the reduced (diploid) genome of the tetraploid cytotype of *A. septentrionale* to form autosyndetic pairs.* Further corrobor-

---

* It is interesting to note that Bouharmont (1972b) recorded variable chromosome pairing in a polyhaploid (diploid) sporophyte obtained by induction of apogamy in an otherwise normal gametophyte of the tetraploid cytotype of *Asplenium septentrionale*. This plant showed a range of only 5-(13)-21 bivalents at metaphase. He also noted that a higher proportion of pairs were evident in early diakinesis, but many dissociated before metaphase. More extensive data are available for a polyhaploid obtained by similar means from *A. trichomanes*, though here the shortfall in pairing behaviour was much less pronounced, the analyses obtained being 30-(34)-36 bivalents at diakinesis and 20-(30)-36 bivalents at metaphase (Bouharmont, 1972c). Similar phenomena may also occur in *Cystopteris*, wherein a range of only 3-12 bivalents was recorded in a polyhaploid sporophyte obtained from the tetraploid cytotype of *C. fragilis*, whereas 13-15 bivalents "could most frequently be observed" at metaphase in *C. fragilis* (4*x*) x *protrusa* (Vida, 1974, p. 186).

These examples are instructive, because they do seem to indicate that the polyhaploid, generally regarded as the ultimate *desideratum* in investigations of polyploids, is in fact a less reliable indicator of maximum pairing than a series of "wide" hybrids. This difference may be related to the comparative physiological weakness of the polyhaploid in contrast to

ation is provided by synthetic triploid hybrids between the two cytotypes of *A. septentrionale*, which show a substantial number of trivalents (J. D. Lovis, unpublished).

(c) Delayed Allopolyploidy

Of particular interest in the above list of hybrids are those like *Asplenium septentrionale* (4x) x *ruta-muraria* ssp. *ruta-muraria, A. septentrionale* (4x) x *trichomanes* ssp. *quadrivalens* and *A. septentrionale* (4x) x *billotii*, wherein both parents are of autotetraploid constitution, and the possibility exists that meiosis may be completely regular, chromosome pairing being entirely autosyndetic in character.

This possibility is realized in some wild examples of *Asplenium septentrionale* x *ruta-muraria* (otherwise known as *A.* x *murbeckii*), which display a regular meiosis indistinguishable from that of a tetraploid species. This situation is of particular interest, because this hybrid is in effect an allotetraploid. However, its mode of origin is different from that of a conventional allotetraploid inasmuch as instead of chromosome doubling occurring after hybridization, it precedes it (Fig. 17). Because of the very substantial delay which would most usually intervene, between the inception of the autotetraploids from their diploid parents and their meeting and hybridizing together, this type of allopolyploid

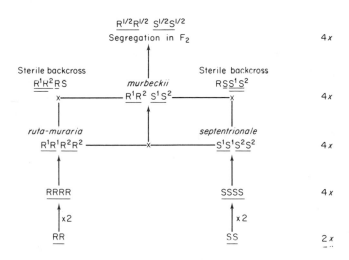

Fig. 17. Delayed allopolyploidy in *Asplenium* x *murbeckii*, based on unpublished results of J. D. Lovis and T. Reichstein. Symbols as in Fig. 5.

the hybrid vigour of many interspecific hybrids. Whether or not this finding may ultimately prove to have relevance in *Dryopteris*, where evidence from two polyhaploid plants is presently of critical importance in the cytogenetic analysis (see pp. 338, 346), only time and further study can tell. (See footnote added in proof, p. 342.)

origin might well be termed **delayed allopolyploidy**. This distinction is not purely academic, because their different modes of origin are reflected in an important genetic difference between the structure of a conventional genomic allopolyploid and a delayed allopolyploid. The former, having just originated by chromosome duplication, initially must be homozygous, and can acquire variability only by mutation. In contrast, if one or both of the autotetraploids participating in a delayed allopolyploid is of any substantial age, its chromosome sets will have become partially differentiated (e.g. $\underline{SS}\ \underline{SS} \to \underline{S^1 S^1}\ \underline{S^2 S^2}$), and chromosome pairing in the F_1 delayed allopolyploid will be between chromosomes that are no longer strictly homologous ($\underline{S^1 S^2}$), with the consequence that the F_2 generation will show substantial segregation. This initial high variability should be an asset to the new species in increasing its chances of evolving a genotype capable of competing successfully. A new delayed genomic allopolyploid will also be isolated from its parents, since backcross hybrids will be sterile for cytogenetic reasons.* From first principles one may then very well expect delayed genomic allopolyploidy to be an uncommon but very effective means of production of a polyploid. Obviously, for some unexplained reason this does not hold true in the case of *A. ruta-muraria* x *septentrionale*, otherwise *A.* x *murbeckii* would have long since become an established and widespread species rather than a very rare hybrid.

Nevertheless, the discovery near Kassel in Germany of three localities containing most unusual numbers of *Asplenium* x *murbeckii* was a matter of considerable interest. These plants showed very regular meiosis. Backcross hybrids were, however, not detected.† The predicted segregation of the F_2 generation and the sterility of backcross hybrids were confirmed in the experimental garden (Lovis, Nieschalk and Reichstein, unpublished). Unfortunately, once the existence of these localities became known, the populations of *A.* x *murbeckii* have declined, no doubt due to the interest and attention of botanists who did not appreciate that they were unwittingly interfering with a natural evolutionary experiment.

---

* An interesting comparison can be made with an example described in flowering plants (Fagerlind, 1937; Clausen *et al.*, 1945, p. 144). Both diploid and tetraploid cytotypes occur in *Galium mollugo* and *G. verum*, though the two diploids are of more restricted range and are now allopatric. Hybrids occur between the two tetraploid cytotypes, known as *G.* x *pomeranicum* (= *G.* x *ochroleucum*). This example of a delayed allopolyploid is different from *Asplenium* x *murbeckii* in one important respect is that it is evidently a segmental alloploid, not a genomic allopolyploid, since multivalents have been observed in *G.* x *pomeranicum*, showing that the genomes of *G. mollugo* and *G. verum* can pair to a variable extent. Because of this facility, backcross hybrids between *G.* x *pomeranicum* and its parents may be partially fertile (in contrast to the sterility of backcross hybrids formed by *A.* x *murbeckii*) thus allowing the possibility of gene exchange by introgression at the tetraploid level between two species which are now spatially (and apparently also reproductively) isolated at the diploid level.

† A single example of the backcross hybrid *A.* x *murbeckii* x *septentrionale* has been detected in Hungary (Vida, unpublished).

An exactly similar fate has befallen a comparable but smaller unique population of *Asplenium* x *clermontiae* (= *ruta-muraria* ssp. *ruta-muraria* x *trichomanes* ssp. *quadrivalens*) in Austria. These examples of a delayed genomic allopolyploid did not show regular meiosis quite so consistently as the *A.* x *murbeckii* populations, but nevertheless produced a segregating F_2 generation in culture (Lovis, Melzer and Reichstein, unpublished). The very existence of this hybrid presents a problem, for it is normally intensely rare, in spite of the fact that the two parents grow together in mixed populations of hundreds or thousands of individuals in innumerable localities across Europe. It would not be surprising if the hybrid did not occur at all, for the two parents are morphologically extremely different, but that it should occur, but only so extremely rarely, is difficult to understand.

(d) Intraspecific Differentiation

It will be noted that so far the evolutionary patterns and processes described in European Aspleniaceae all involve either new combinations or duplications of already highly differentiated genomes. However, there is evidence of earlier stages of differentiation and divergence in progress in this group.

Relatively recent diversification is seen in *Asplenium seelosii* and *A. celtibericum*, two diploid species virtually confined to the Dolomites and the Pyrenees respectively. These two plants are clearly morphologically very closely related, and were until recently treated as subspecies of *A. seelosii* (Rothmaler, in Cadevall and Font Quer, 1936; Becherer, 1962; Rivas-Martinez, 1967). Experimental studies have shown regular meiosis in a synthetic F_1 hybrid, but substantial segregation and weakness in the F_2 generation (J. D. Lovis and T. Reichstein, unpublished).

At the tetraploid level the allopolyploids derived from *Asplenium fontanum* x *obovatum* and *A. aegaeum* x *ruta-muraria* ssp. *dolomiticum* both show diversity. Thus the *A. foresiense*/*macedonicum* complex shows a crude and interrupted west–east cline in pinna dissection, with the disjunct relict *A. macedonicum* being sufficiently distinct to merit specific rank (Sleep, 1966; T. Reichstein *et al.*, unpublished).

The morphological pattern in the *Asplenium haussknechtii*/*lepidum* complex, recently investigated by Brownsey (1973, 1976b), is much more intricate. The ranges of *A. lepidum* and *A. haussknechtii* are contiguous, with *A. lepidum* confined to Central and South-east Europe, and *A. haussknechtii* centred in Asia Minor, extending eastwards to Afghanistan (as *A. samarcandense*). Their distributions, which are highly discontinuous, on account of their very narrow ecological tolerance, adjoin today only in the island of Crete, where *A. lepidum* occurs in the western mountains, and *A. haussknechtii* in the east, with a population of intermediate character present in the central mountain massif. These two taxa, which are distinct in glandulosity, annulus morphology and sporangial behaviour, are nevertheless best treated as subspecies of one polymorphic species, since they show a bewildering and overlapping range of variation in frond form and dissection, in pinna number and shape, and moreover readily hybridise in culture. The great range of inter-population variation encountered in the complex may be influenced not only by the spatial isolation associated with its very discontinuous distribution, but also by the small size of many populations. To what extent the diversity encountered in both the *A. haussknechtii/lepidum* complex and the *foresiense/macedonicum* complex is due to polytopic origin of the allotetraploids from different strains of their parent diploids, as opposed to post-polyploidization differentiation at the tetraploid level, is quite uncertain.

7. *Asplenium trichomanes complex*

The most evident signs of evolutionary vigour are seen in the *Asplenium trichomanes* complex, a widely distributed, extremely intricate and relatively ancient group, whose evolutionary history cannot possibly now be deciphered exactly. This complex has been studied by myself over a period of more than 20 years. The summary which follows is principally based on unpublished work.

Two diploid subspecies of *Asplenium trichomanes* are recognized in Europe (Lovis, 1964a). One of these, ssp. *trichomanes*, is a morphologically relatively invariable taxon of pan-Boreal distribution found only on non-calcareous rock. Related but not identical diploid forms are known from Tibet, Malaysia and Australia. The second European diploid subspecies is ssp. *inexpectans*, known only from C. and S.E. Europe. In direct contrast to ssp. *trichomanes*, it is confined to limestone rock, and in spite of its much more limited distribution, is a more variable taxon. The two subspecies are completely interfertile in culture. Some hybrid combinations yield plants with morphology close to that of tetraploid forms.

Also relevant are a related vicarious pair of diploid species, *Asplenium anceps* of the Atlantic Islands, and *A. tripteropus* of China and Japan. Synthetic hybrids between these two species and *A. trichomanes* ssp. *trichomanes* show approximately half of the chromosomes paired. Some of these hybrids also simulate the morphology of certain tetraploid forms.

Tetraploid races of *Asplenium trichomanes*, at present collectively known as ssp. *quadrivalens*, are extremely widespread, appearing in all of the major mountain ranges of the world except those of Mexico and the Andes. The tetraploid cytotype displays a wide range of variation, but it is difficult to characterize distinct taxa. Some European forms "mimic" very closely races of ssp. *inexpectans*, but are not sympatric with them, though ssp. *inexpectans* usually grows together with some form of the tetraploid cytotype.

Evidence that the tetraploid cytotype of *Asplenium trichomanes* is, as far as is yet known, always of essentially autopolyploid constitution is shown by the levels of autosyndetic chromosome pairing seen in (1) "wide" hybrids involving as second parent variously *A. foresiense*, the tetraploid cytotypes of *A. ruta-muraria* and *A. septentrionale* (J. D. Lovis and T. Reichstein, unpublished),

A. lepidum (Lovis *et al.*, 1966; Vida, 1970) and *A. majoricum* (Lovis and Reichstein, 1969); and (2) an experimental polyhaploid sporophyte (Bouharmont, 1972c).

The origin of the tetraploid forms of *Asplenium trichomanes* is, however, without doubt complex. Several possibilities exist. These are: (1) direct autopolyploidy from diploid forms, (2) hybridization between diploid subspecies, followed by chromosome doubling, producing intersubspecific autopolyploidy; and (3) hybridization between *A. anceps* (in Pliocene time) or *A. tripteropus* and a diploid form of *A. trichomanes*, again followed by chromosome doubling, producing segmental allopolyploidy. All three of these alternatives may have occurred at different times and in different places. The history of the complex may well have been complicated by hybridization and subsequent recombination between tetraploids of distinct origin, which individually may have arisen by any of processes (1), (2), and (3) (Fig. 18). The almost ubiquitous range of the aggregate species, which is rivalled amongst ferns only by *Cystopteris fragilis, s.l.*, *Pteridium aquilinum, s.l.*, and *Anogramma leptophylla, s.l.*, demonstrates its relatively great age, indicating a history long enough for separate cycles of differentiation and hybridization, comparable to those postulated in the evolution of European *Achillea* and *Galium* complexes by Ehrendorfer (1959), to have occurred more than once, and in different parts of the great range of the complex. Present indications that the centre of diversity of the complex at both diploid and tetraploid levels probably lies within Europe may be an artifact, produced by the more intensive study made of populations occurring in this region.

Further complications, in the form of hexaploid plants, are known in Madeira, Australia and New Zealand, though only in New Zealand is a hexaploid form the prevalent race.

Our techniques are quite inadequate to resolve such a potentially complex evolutionary history. It is apparent, however, that the *Asplenium trichomanes* aggregate is in the evolutionary sense an extremely fit complex, with ample resources and flexibility to enable it to respond to changing conditions. In many respects it parallels the extraordinary fit and successful grass complex, *Dactylis glomerata, s.l.*, which it resembles in its wide range of diploid and tetraploid forms, with the diploids more clearly differentiated into individual taxa than the tetraploids, a close resemblance between particular diploid and tetraploid forms, and particularly distinctive taxa occurring in the Atlantic islands (Stebbins and Zohary, 1959; Borrill, 1961).

Finally, mention must be made of an exceedingly interesting situation which has been encountered in one locality in the Swiss Jura. Forms of *Asplenium trichomanes* with more or less prostrate fronds are often associated with rock overhangs, though probably always in the near proximity of more normal forms with erect fronds growing on the open rock face. In this particular station, a distinctive prostrate race is genetically partially isolated from the normal plant,

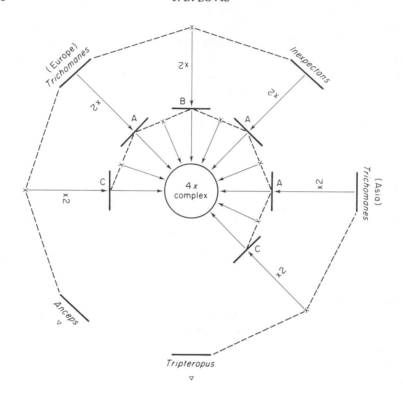

Fig. 18. Hypothetical origin of tetraploids in the Euroasian sector of the *Asplenium trichomanes* complex. Outer circle = parental diploid taxa. Middle circle = hypothetical basic derivative tetraploid taxa (A = autopolyploids, *s.s.*, B = intersubspecific auto-polyploids, C = segmental allopolyploids): W → NE = polyploidization events occurring in Europe, E & SE = events occurring in Eastern Asia. Innermost circle = existing tetraploid complex, believed to be derived in part directly from all or some of the basic tetraploid derivative lines, and in part by subsequent hybridization between tetraploids of different origins. For the sake of clarity, not all such possible events are shown, e.g. [4x *ex* (*anceps* x *trichomanes* ssp. *trichomanes*)] x [4x *ex* (*trichomanes* ssp. *trichomanes* x ssp. *inexpectans*)] is not indicated.

although both are tetraploids (J. D. Lovis and T. Reichstein, unpublished). Wild hybrids between the two forms could be detected by their positive heterosis and subsequently proved to show substantial meiotic disturbance. These findings of heterosis and irregular chromosome pairing were later confirmed in artificial hybrids produced between the two forms. Whether or not this is an example of incipient sympatric speciation is a particularly interesting question. It may well be so simply because it is difficult to visualize circumstances in which differentiation of the prostrate overhang form could occur in isolation from the normal form, since, by definition, an overhang is inevitably associated with an adjacent exposed vertical rock face.

The existence of the prostrate forms can be explained by postulating that the prostrate habit is advantageous in that it prevents the development of competing individuals of the same or of other species close to the crown of the plant. It can be demonstrated in culture that this form is clearly disadvantageous where surface water is available, because water tends to remain trapped around the crown, with subsequent rotting of the lower pinnae and, ultimately, of the stem apex. Under overhangs in the wild, it is clear from their cobwebby condition that, in spite of the influence of wind, the prostrate forms do not receive any significant amount of rain.

C. CYTOGENETICS AND SPECIES INTERRELATIONSHIPS: TROPICAL FERNS

1. Adiantum caudatum complex

Very few experimental biosystematic studies have yet been carried out on tropical ferns. Of those which have, by far the most impressive and informative has been the investigation of the *Adiantum caudatum* complex carried out by Manton and her associates over a considerable number of years (Manton *et al.*, 1967, 1970; Sinha and Manton, 1970). The distribution of this complex extends through Africa and the Indo-Malaysian region. Both African and Asian representatives of the group were incorporated into this study, which can well serve as a model of analytical experimental cytogenetic technique, and can be presented as the best answer to the criticisms raised regarding the validity of this approach in recent years (see p. 332), since it demonstrates how a sufficiently extensive and thorough investigation can unequivocally discriminate between and identify three different types of polyploid origin.

The experimental details of this investigation can be set out more economically than is done here (Figs 19, 20), but this mode of presentation has

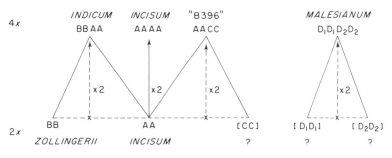

[ ] = Postulated ancestral taxa, not yet known alive

Fig. 19. Cytogenetic relationships between sexual taxa in the *Adiantum caudatum* complex, based on data of Manton *et al.* (1967, 1970) and Sinha and Manton (1970), displayed on Fig. 20.

(a)

$4x$ KILIMANJARO
$\underline{AA}BC$
B396
$\underline{AA}CC$

$\underset{ABD_1D_2}{\times}$

MALESIANUM
$\underline{D_1}D_2D_2$ 10<20 biv. and 80<100 univ.

$\underset{\times}{INCISUM}$
$\underline{AA}\underline{BB}$

30 biv. and
60 univ.

$4x$ $\underset{AAAB}{\times}$ INCISUM

30 biv./triv. and
60 univ. (<7 triv.)

$\underline{AA}B \times$ 30 biv and
30 univ.

$3x$ **30 biv and
30 univ.**

$\underline{A}BB$

ZOLLINGERII
$\underline{BB}$

$2x$ INCISUM
$\underline{AA}$

(b)

$4x$ MALESIANUM
$\underline{D_1}D_2D_2$ 10<20 biv. and
80<100 univ.

$\underset{ACD_1D_2}{\times}$

KILIMANJARO
B396
$\underline{AA}CC$

$\underset{CAAB}{\times}$

INDICUM
$\underline{AA}\underline{BB}$

30 biv. and
60 univ.

$4x$ $\underset{AAAC}{\times}$ INCISUM

$\underline{AAAA}$ 30 biv/triv
and 60 univ.
(<3 triv.)

$\underline{AA}C \times$ 30 biv. and
30 univ.

$3x$ **90 univ.** $\times$ CAB

INCISUM
$\underline{AA}$

$2x$ ZOLLINGERII
$\underline{BB}$

(c)

$4x$ INCISUM
$\underline{AA}\underline{AA}$

$\underset{AAD_1D_2}{\times}$ 45<50 = (30 + 15<20) bivalents

$\underline{ABD_1D_2}$

$4x$ KILIMANJARO
B396
$\underline{AA}CC$

$\underset{ACD_1D_2}{\times}$

MALESIANUM
$\underline{D_1}D_2D_2$

10<20
bivalents

$\underset{\times}{INDICUM}$
$\underline{AA}\underline{BB}$

10<20
bivalents

$\underline{AD_1}D_2 \times$ 10<20
bivalents

$3x$ $\underline{BD_1}D_2$

ZOLLINGERII
$\underline{BB}$

$2x$ INCISUM
$\underline{AA}$

(d)

$4x$ MALESIANUM
$\underline{D_1}D_2D_2$ $\times$ 45<50 biv. and 20<30 univ.

$\underline{AAD_1D_2}$

$\underset{INCISUM}{\times}$
$\underline{AAAA}$

$\underline{AAAB}$

$4x$ KILIMANJARO
B396
$\underline{AA}CC$

$\underset{INCISUM}{\times}$
$\underline{AAAA}$

30 biv/triv.
and 60 univ.
(<3 triv.)

$\underline{AA}A \times$ 30 biv/tri.
and 30 univ.
(<3 triv.)

$3x$ **30 biv. and
30 univ.**

$\underset{\times}{INDICUM}$
$\underline{AA}\underline{BB}$

30 biv/triv
and 60 univ
(<7 triv.)

$\underline{A}AB$

ZOLLINGERII
$\underline{BB}$

$2x$ INCISUM
$\underline{AA}$

been preferred because the form of the analysis can thus much more readily be appreciated. To date, six cytogenetically distinct taxa have been included in the investigation. A seventh taxon, *Adiantum caudatum* L., *sensu stricto*, is a triploid apomict. Two of the tetraploids are genomic allopolyploids. The two diploids *A. incisum, sensu lato* (AA) and *A. zollingerii* (BB) are the parents of *A. indicum* Ghatak (AABB). The diploid cytotype of *A. incisum* is also one parent of the second allotetraploid (AACC), which though related to *A. incisum* Forsk., *s.s.*, is not identical with that species, and has at present no name, being therefore known by its collection number B396, or by its only known locality Kilimanjaro. The second parent (CC) of this tetraploid is not known. Another African tetraploid, *A. incisum* Forskkål, *s.s.* (AAAA), is of autotetraploid constitution. A curious feature is that although the A genome of the diploid cytotype of *A. incisum, s.l.* is parental to both *A. incisum* Forsk. and "B396", this diploid taxon has not yet been found in Africa. The fourth tetraploid, *A. malesianum*, a segmental alloploid $(D^1 D^1 D^2 D^2)$, is not closely related to any of the other known members of the complex. The analysis of the complex as a whole requires the participation of two genomes, C and D, which are not known today in the diploid state. This does not by any means necessarily indicate that these ancestral diploids are extinct, since the tropical regions are still very largely underexplored.

Apart from its intrinsic interest, this investigation has demonstrated that the same diversity of processes of polyploid formation and differentiation take place in tropical regions as in temperate latitudes.

2. *Asplenium aethiopicum complex*
Another African complex which has proved to have unique interest is the *Asplenium aethiopicum* group. This complex has outliers in the Caribbean, the Indo-Malaysian region and Australia, but only the African representatives have yet received detailed investigation. The group presents severe technical problems combined with great taxonomic difficulty, both principally associated with the high levels of polyploidy involved. Apart from octoploid and decaploid

---

Fig. 20. Cytogenetic analyses of synthetic hybrids obtained in the *Adiantum caudatum* complex establishing the relationships shown in Fig. 19. (a) Hybrids of *A. indicum*, an allotetraploid of which both parents, *A. incisum s.l.* and *A. zollingerii*, are known. Note the low level of chromosome pairing in *A. indicum* x *malesianum*. (b) Hybrids of "B396", an allotetraploid of which only one parent, *A. incisum s.l.*, is known. The only obvious difference in the cytogenetics of the hybrids of this taxon and those of *A. indicum* is seen in their respective hybrids with *A. zollingerii*. Total failure of chromosome pairing in "*B396*" x *zollingerii* shows that these two taxa are not related. (c) Hybrids of *A. malesianum*, a segmental allopolyploid. None of the hybrids of this taxon shows less than $10 < 20$ bivalents, reflecting partial pairing between the genomes (D_1 and D_2) of *A. malesianum*. (d) Hybrids of *A. incisum s.s.*, an autotetraploid. Here it is seen that none of the hybrids displays less than $30 (= x)$ bivalents, representing the capacity of the genomes of *A. incisum s.s.* for complete autosyndesis. Symbols as in Fig. 5.

apomicts, tetraploid, octoploid and duodecaploid sexual forms are known, but no diploids have been discovered.

The complex has been the subject of a detailed experimental investigation by Braithwaite (1964b). A report on the apomictic forms has been published (Braithwaite, 1964a; see p. 390 *et seq.* below), together with analyses of two subsidiary complexes, including the *Asplenium splendens* group of Southern Africa (Braithwaite, 1972a, b).

Though morphologically very close to *Asplenium aethiopicum*, the distinct nature of *A. splendens* is shown by a synthetic hybrid between these two species, which shows complete failure of pairing. *Asplenium splendens*, a tetraploid species, includes two subspecies with distinct ecological preferences: (1) ssp. *splendens*, a forest taxon extending to 5000 ft, and (2) ssp. *drakensbergense*, a high altitude rock crevice plant occurring above 5000 ft. Braithwaite (1972b) was able to demonstrate by cytogenetic analysis that the octoploid species *A. multiforme*, of intermediate morphology, is in fact derived from these two subspecies, which are interfertile. In view of the very close relationship of its two parents, Braithwaite considers, with good reason, that *A. multiforme* is best regarded as an intervarietal (or intersubspecific) autopolyploid rather than a segmental alloploid. The main *A. aethiopicum* complex also probably contains a substantial autoploid element in its construction.*

3. *Ceratopteris*

Ceratopteris is a small, taxonomically very isolated, genus of aquatic or semiaquatic annual ferns of circumtropical distribution which includes perhaps only four rather ill-defined and polymorphic species (Lloyd, 1974b).

Hickok and Klekowski (1974) have investigated the degree of genetic relationship existing between two diploid species, *Ceratopteris richardii* and *C. pteridoides*. Synthetic interspecific F_1 hybrids frequently showed regular meiosis with $n = 39$ bivalents, though a range of 34-39 bivalents was observed. Occasionally, odd multivalents were present, with two trivalents and one quadrivalent seen in one cell. The most probable cause of these aberrant multivalents is translocation or interchange heterozygosity, due to structural differences existing between the genomes of *C. richardii* and *C. pteridoides*. Sixty per cent of the spores of the F_1 hybrid are abortive, principally no doubt because of genic unbalance created by independent assortment of the chromosomes of the parental genomes. An F_2 generation of homozygous sporophytes raised by self-fertilization of isolated gametophytes represented a virtually complete range of morphological intermediates between the two parent species.

---

* Panigrahi (1963), on the basis of pairing observed in synthetic hexaploid and decaploid hybrids, suggests that this complex is entirely alloploid, with no less than six distinct genomes involved in the duodecaploid. In view of the very restricted morphological range of the entire complex, this seems extremely unlikely.

The F_2 generation also displayed a range of variation in meiotic behaviour and spore fertility. Some individuals showed normal meiosis, whilst variable degrees of meiotic abnormality were observed in others. Spore abortion ranged from 30–100%.

It is thus clear that the degree of genetic isolation existing between *Ceratopteris richardii* and *C. pteridoides* is slight, with some segregates in the F_2 generation showing normal meiosis and relatively high spore viability.

Of potential interest in relation to elucidation of the course of evolution within the genus is the observation that "many of the F_2 segregates . . . fall into the morphological limits of [*Ceratopteris thalictroides*]" (Hickok and Klekowski, 1974, p. 445).

Klekowski and Hickok (1974) have described the spontaneous apogamous production of polyhaploid sporophytes by a male-sterile mutant strain of a Malayan race (C230) of *Ceratopteris thalictroides*. The parent sporophyte being tetraploid, these polyhaploids are diploids. Meiotic pairing in the polyhaploids is variable, ranging from 9-33 bivalents. The most straightforward interpretation of this evidence is that this race of *C. thalictroides* is a segmental alloploid, though comparison with the study by Bouharmont (1972b) of a polyhaploid obtained from *Asplenium septentrionale*, which produced only 5-21 bivalents (of a possible maximum of 36) in a taxon known from its behaviour in wild hybrids (see p. 364, footnote) to be capable of complete autosyndesis, suggests that (tetraploid) *C. thalictroides* lies towards the autoploid end of the segmental alloploid range, and could be of essentially autoploid character.

A report by Hickok and Klekowski (1973) of a study on a hybrid *Ceratopteris* plant of uncertain identity and unknown origin is nevertheless worthy of consideration here because it may shed further light on the nature of polyploidy in this genus in a manner not perhaps fully appreciated by Hickok and Klekowski.

This plant, designated the SG hybrid by Hickok and Klekowski, was discovered in a college greenhouse. Its cytology (mean count = 115 chromosomes) clearly indicates that it is triploid though an exact count was not possible because of meiotic irregularity. Hickok and Klekowski do not speculate on its parentage and origin beyond stating that "it probably originated through the hybridization of a diploid ($n = 39$) and tetraploid taxa" (op. cit., p. 1011).

Hickok and Klekowski state that "the average of several diakineses gave 46 bivalents, 20 univalents and an occasional multivalent" (p. 1011), but make no comment on this extraordinary result, which is in itself very surprising, since there is abundant evidence both within this paper and elsewhere (Klekowski, 1973b; Hickok and Klekowski, 1974; Klekowski and Hickok, 1974) that they were actively seeking evidence of homoeologous pairing, which, if taken at its face value, this cytological analysis appears to provide in a very direct form. Since $x = 39$ in *Ceratopteris*, 46 bivalents could only be encountered in a triploid if either (1) two-thirds of the strictly non-homologous chromosomes within the third genome somehow paired together (which is effectively impossible, unless one is prepared to presuppose structural alterations on a quite grandiose and totally unrealistic scale), or (2) if sufficient traces of homology remain from the polyploid origin of the present base number to permit anciently homoeologous chromosomes to pair. Put in other words, it would be necessary to conclude that $x = 39$ is not the true base number of the genus, but had been derived from 13, 10 or 20 sufficiently recently to permit chromosomes representing the same member of the ancestral genome to be able still to recognise one another. That Hickok and Klekowski are aware of this possibility, at least in another context, is clear from their reference to inherent polyploidy in their discussion of the diploid hybrid *C. richardii* x *pteridoides* (Hickok and Klekowski, 1974, p. 443).

However, another explanation is possible. The question arises as to whether the analyses reported are correct. What is beyond dispute is that the analysis of multivalent associations in ferns, outside of cytologically amenable families like Hymenophyllaceae and Osmundaceae, is always difficult. In circumstances such as hold here with subhexaploid F_2 progeny obtained by Hickok and Klekowski from the SG hybrid, where the total number of chromosomes present in the nucleus is uncertain, the degree of technical difficulty is such as to cause any cytologist who has experienced it to feel considerable sympathy with anyone attempting such an analysis. It is easy in such circumstances to adopt an unduly conservative attitude to the identification of multivalents, and it appears to me that such an approach has resulted here in an underestimation of the member of multivalents present in the cell illustrated from an F_3 sporophyte, A2 (Hickok and Klekowski, 1973, Fig. 20), which surely shows more than two multivalents.

It is also an intrinsic truth that in a triploid where n bivalents and n univalents might be expected, the larger the number of trivalents formed the greater is the difficulty experienced in obtaining a cell showing a clear spread of bivalents and univalents yielding an unequivocal analysis. Regrettably, no interpretative diagram is presented to accompany Fig. 5, a cell in diakinesis from the SG hybrid, said to display $c.46$ bivalents, 20 univalents, 1 multivalent. It is apparent from this illustration that the number of multivalents present may well have been seriously underestimated. To me it seems more probable that the combined number of bivalents and trivalents present in the SG hybrid does not significantly exceed the base number, 39; and that an appreciable proportion of trivalents are present.

This interpretation would be entirely consistent with an origin by hybridization between a segmental allotetraploid or a near autotetraploid with a related diploid species. Attention can be drawn to the fact that the one tetraploid race of *Ceratopteris* so far investigated (C230, see above) has precisely such a constitution, and that a high degree of homology exists between the genomes of the only two diploid species so far known.

All of the available evidence in *Ceratopteris* shows a situation unique in ferns in that all of the strains so far investigated are imperfectly differentiated one from another genetically, and in consequence it appears likely that a high degree of homology still exists between all of the genomes extant in the genus.

Whether this apparently unique (amongst ferns) biosystematic structure of *Ceratopteris* is related to its undoubtedly unique mode of life (the only other genus of ferns with an annual sporophyte is *Anogramma*, and there are no other truly aquatic ferns other than the heterosporous Salviniaceae and Marsileaceae), or is only coincidental, remains to be determined.

D. INTROGRESSION

Hybrid swarms, populations of fertile segregating hybrid individuals, appear to be of decidedly rare occurrence in ferns. It is remarkable that the example occurring in the *Pteris quadriaurita* complex in Ceylon (Sri Lanka), which was the subject of an elegant and complete investigation by T. G. Walker, described in a now classic paper (Walker, 1958), remains virtually the only recorded instance.

The 1951 Leeds expedition to Ceylon found itself faced with a bewildering array of different forms of *Pteris*, many of which could not be named. Plants resembling a form which had been named *P. otaria* (Beddome, 1873) occupied an approximately central position in this assemblage. This form was found to be diploid and to reproduce sexually.

Subsequently, progenies were raised by Walker from different examples of *Pteris otaria*. All showed a similar very wide range of variation, the two extremes being forms with simply pinnate or regularly bipinnate fronds, with the intermediates showing different degrees of abortion of the pinnules. The simply pinnate extreme segregate corresponded with another named species, *P. multiaurita* Ag., which had also been found to be diploid and sexual. On a visit to Ceylon in 1954 Walker was able to find pure populations corresponding to the other extreme (bipinnate) segregate. He later established that this plant was also a sexual diploid, and that it corresponded to *P. quadriaurita* Retz, *s.s.*

Synthetic hybrids were readily raised between *Pteris multiaurita* and *P. quadriaurita*. These hybrids proved to be fully fertile, with perfect meiosis and viable spores. Though the F_1 hybrids were closer to *P. quadriaurita* than to *P. multiaurita*, the F_2 generation showed a series of forms ranging from one parental extreme to the other, including forms characteristic of *P. otaria*, and indeed duplicating the populations previously raised by self-fertilization of wild examples of *P. otaria*.

This investigation conclusively demonstrated that *Pteris otaria* is only a hybrid state, representing a particular segregate of the hybrid swarm arising from hybridization between *P. multiaurita* and *P. quadriaurita*, *s.s.*

Today, in Ceylon, *Pteris quadriaurita*, *s.s.*, only grows in parts of the virgin forests on the fringe of the Dry Zone, though it is abundant in some of its localities. *P. multiaurita* is more widely distributed, but tends to occur in small populations on the borders of forests. The two species thus meet at the edge of

Dry Zone forests, resulting in the formation of hybrid swarms capable of invading disturbed ground. Such hybrid swarms are now to be found growing on roadside banks, forest paths and similar examples of disturbed ground over all those parts of Ceylon suitable for the growth of ferns. As is to be expected, backcrossing occurs wherever hybrid swarms adjoin the forest populations of *P. quadriaurita*, with hybrid individuals close to *P. quadriaurita* predominating on the fringe of the forest.

Walker is surely right when he suggests that the economic development of the interior of the island during the British administration, with the establishment of coffee and tea estates involving the construction of new roads and the destruction and dissection of the rain forests, created conditions ideal for the wider dissemination and increase of the hybrid swarms.

Smith (1971, pp. 37-39) has suggested that introgression may occur in the West Indies between *Thelypteris* (*Cyclosorus*) *grandis* and *T.(C.) serra*. Unfortunately, the evidence at present available is confined to that which can be obtained from herbarium material. Smith's study provides a strong prima facie case that hybrids intermediate between these two species do exist in nature, and there is some indication that backcross hybrids with *T. grandis* may also occur. The hybrids clearly have a much reduced spore fertility. There is no reasonable ground to doubt that these two species do hybridise in nature, but the frequency with which this occurs, and whether or not this is such as can be considered to constitute introgression, can only be determined by field investigations and experimental studies.

E. BREEDING SYSTEMS IN FERNS

Klekowski and Baker (1966) drew attention to the striking difference that exists between the levels of chromosome number found in heterosporous and homosporous pteridophytes respectively. It is scarcely possible to make a comparison of this kind restricted to ferns, because only two small isolated families of aquatic ferns, Marsileaceae and Salviniaceae, are heterosporous. More substantial data are obtained by extending the comparison to the pteridophytes as a whole, since both the Selaginellaceae and the Isoetaceae are heterosporous.

The facts are remarkable. Ninety per cent of heterosporous pteridophytes but only 4% of homosporous pteridophytes have chromosome numbers below $n = 28$. The mean values given for the two groups are $n = 13.6$ and 57.0. These figures clearly indicate that for some reason polyploidy is more common in homosporous than in heterosporous pteridophytes.

In this same paper, Klekowski and Baker suggested that "fern populations are characterized by greater homozygosity than that previously suspected" (p. 306). They pointed out that a sporophyte resulting from self-fertilization by a single gametophyte will be completely homozygous. Less convincingly, they argued that some fern gametophytes are adapted to self-fertilization. They proposed that polyploidy was prevalent in homosporous ferns because it counterbalanced this innate tendency towards homozygosity, since the duplication of loci in itself materially assists in the maintenance of heterozygosity.

In recent years, there has been an awakening of interest in the genetic system of homosporous ferns. The basic technique involved has capitalised on the independent existence of the fern gametophyte by determining the proportion of isolated but watered gametophytes which prove incapable of producing viable sporophytes in order to obtain a measure of the recessive lethal genes concealed in the parent sporophyte and thus provide, in turn, a measure of the heterozygosity it contains. The initial publications (Klekowski and Lloyd, 1968; Klekowski, 1969, 1970a, b) are open to criticism on the grounds that the source material was for the most part of garden origin, the samples investigated were very small, and the experiments entirely devoid of replication. More recent and more substantial studies have indicated that extensive levels of heterozygosity exist in wild populations of *Osmunda regalis* (Klekowski, 1970c, 1973a), *Thelypteris palustris* and *Onoclea sensibilis* (Ganders, 1972). These and other studies have been reviewed by Lloyd (1974a). The existence of considerable levels of heterozygosity may seem paradoxical in relation to the ready propensity in the fern life-cycle for the formation of completely homozygous sporophytes. However, the experimental data obtained for these three species show overall that not only are a high proportion of gametophytes incapable of producing a sporophyte in isolation, but also that those isolated gametophytes which do produce sporophytes do so with much diminished freedom, and that moreover a substantial proportion of these sporophytes are demonstrably abnormal or weak. It is therefore logical, as Ganders suggests, "that natural selection against complete homozygotes may be ... a major mechanism for retaining heterozygosity" (Ganders, 1972, p. 220).

1. Incompatibility

The now widely observed phenomenon that self-inseminated gametophytes are in general less fertile, in terms of frequency or ease of production of viable sporophytes, than are crosses between sister gametophytes was first demonstrated in three populations of *Pteridium aquilinum* by Wilkie (1956), who interpreted his data as demonstrating the existence of a single locus multiple allele incompatibility system, though he recognized that the "incompatibility alleles appear to be rather weak in their action, since selfing was 8, 16 and 17% respectively in the M, B and K populations" (op. cit., p. 255).

More recently Klekowski (1972) has shown that such a system is certainly not universal in the *Pteridium aquilinum* complex. He tested the self and intrapopulational compatibility of gametophytes raised from 18 collections, belonging to four different intraspecific taxa from origins as diverse as the Galapagos Islands, Machu Picchu (Peru), Costa Rica, Victoria (Australia), California and Massachusetts by culture of series of random pairs and series of isolated gametophytes. The most distinctive results were obtained from the four collections from Galapagos, in which the range of fertility of the isolate cultures was almost as high (91-100%) as in the pairs experiments (95-100%). In contrast, one trial on a collection from California produced a result (pairs 24%, isolates 0%) almost identical with the theoretical expectation for an incompatibility system involving two unlinked loci (25%, 0%), but in the isolate trials for five other collections from California of the same taxon no less than 30-58% of gametophytes produced sporophytes. Fertility was relatively high (32-85%) in the isolate trials of all the other American collections tested. On the other hand, the two collections from Victoria gave figures (Aus 1—pairs 45%, isolates 12%; Aus 2—pairs 76%, isolates 18%) comparable to those obtained by Wilkie, and are thus equally consistent with a one locus self-incompatibility system where a small amount of intragametophytic selfing is permitted.*

Wilkie performed two types of intrapopulation experiments (crosses between populations were all compatible, indicating that if a self-incompatibility system existed, it must be a multiple-allelic system), both simple random pairs experiments, like those performed by Klekowski, and replicated trials of individual gametophyte combinations, made possible by establishing clones from individual prothalli. Since Wilkie presented his data *in extenso*, but did not completely summate or analyse his figures, it may be of service to examine them further here. The results of the random pairs trials were: B, 19 of 27 pairs compatible (70%); K, 9/18 (50%); M, 22/34 (65%).† The results of the replicate trials were remarkably consistent and convincing. Of 39 combinations tested, only two gave an equivocal result. All the others gave figures for the number of replicates producing sporophytes which either exceeded 70% (= "compatible"), or less than 20% (= "incompatible"). (The number of replicates tested varied from 9-(15.7)-26.) For each population, the proportions of combinations tested which were compatible are as follows (excluding the two equivocal results, and of course excluding also the intraclonal tests, which would seriously unbalance the data): B, 7 compatible/3 incompatible; K, 7/5; M, 5/10. If the results of both series of experiments are combined, the figures are: B, 24 compatible/11

---

* Klekowski, while conceding that his data from Aus 1 are "possibly consistent with the hypothesis of genetic self-incompatibility", states that "... data from another spore collection [Aus 2] indicate that this hypothesis is untenable in this variety" (op. cit., p. 68). In this he is mistaken. In testing the hypothesis of single locus incompatibility, adjustment must be made for the observed level of failure of incompatibility in isolate tests (18%). Thus the true expectation is not that 25 of 50 random pairs should prove compatible, but 25 + [(18/100) × 25 × 2] in 50, = 25 + 9 = 34 "positives" in 50 trials, because according to expectation 25 pairs of prothalli will not produce sporophytes by cross-fertilization, and are therefore available to produce selfed progeny. A 2 × 2 chi-square test for the Aus 2 data gives a χ^2 figure of 0.32 ($P = 0.7 > 0.5$). In other words, if a single locus incompatibility system was present in Aus 2, a random deviation as great as that actually obtained could be expected in slightly more than half of the trials made.

† A few pairs which produced a sporophyte on only one prothallus are classed as compatible. If these are excluded as being equivocal results, the figures are: B, 17 of 25; K, 9/9; M, 19/31.

"incompatible" (= 71% compatible); K, 26/17 (= 63%); M, 14/19 (= 42%). Overall figures for intragametophytic selfing for each population were: B, 23 in 188 (13%); K, 33 in 218 (15%); M, 22 in 297 (7%).

The data obtained by Wilkie for the three Scottish populations and by Klekowski for two Victorian collections are thus remarkably similar. This does not mean that Klekowski may not be correct in preferring to present an interpretation based on a genetic load model, where failure of intraclonal or isolated gametophyte self-insemination is attributed to the presence of a recessive sporophytic lethal gene, as an explanation of this data. The issue remains unresolved.

What is clear is that the processes involved in sporophyte production are most imperfectly understood. In particular, the phenomenon described by Klekowski as "leaky lethality" and demonstrated by him in *Pteridium aquilinum* (Klekowski, 1972) as well as in *Osmunda regalis* (Klekowski, 1970, 1973a), whereby an isolated gametophyte may ultimately produce a viable sporophyte after persistent self-insemination, although previous fertilizations have aborted at an early embryonic stage, lacks any satisfactory explanation.

One development of considerable potential importance is the discovery by Löve and Kjellqvist (1972) that forms of *Pteridium* with $2n = 52$ exist in Spain (*P. herediae*) and in Yugoslavia (*P. aquilinum* var. *gintlii*). All previous counts for *Pteridium*, including counts made by Wilkie on his material, have been $n = 52$ or $2n = 104$. It follows that since these taxa with $2n = 52$ are presumably diploid, then Wilkie's material was tetraploid. (The ploidy of the collections studied by Klekowski was not determined.) There are many instances known where incompatibility systems originally established and efficient at the diploid level are less effective in polyploid derivatives (Lewis, 1943, 1949, pp. 480 *et seq.*; Ehrendorfer, 1959). The discovery that British *P. aquilinum* is tetraploid and not diploid clearly creates the possibility that herein lies the explanation of the weak (7-15%) intraclonal fertility observed by Wilkie, and thus increases the likelihood that his interpretation of his data was right; in other words, a weak incompatibility system may exist in Scottish (and Australian) *Pteridium*.

Clearly, it is highly desirable that experiments similar to those executed by Wilkie be carried out on the newly discovered diploid forms of *Pteridium*.

2. Homoeologous pairing

Klekowski, still impressed by the possible significance of the implications of the potential in homosporous ferns for formation of completely homozygous individuals, has proposed ". . . an hypothesis which describes a novel means of genetic recombination in homosporous ferns. Meiosis in these plants is envisaged as allowing chromosomes to pair within homoeologous sets while restricting pairing so that only bivalents are formed. The altered meiosis results in a genetic system which allows the release of genetic variability in the meiotic products of completely homozygous sporophytes or sporophytes of apomictic origin"

(Klekowski, 1973b, p. 535). An explanation as to why this system cannot operate in Döpp-Manton apomicts is given below (p. 395-6). Here we are concerned only with the evidence for, and the extent of, the operation of this system in sexually reproducing ferns.

Klekowski (1973b, p. 540) recognizes that the frequency of homoeologous pairing can be expected to vary from taxon to taxon, from a situation where pairing is truly at random between homologues and their homoeologues, to circumstances where pairing between homoeologues occurs only occasionally. This reflects the fact than an entire range of different polyploid types are theoretically possible, covering all gradations from a raw autotetraploid, in which chromosomes pairing is indeed at random amongst sets of four (though in this case all the chromosomes are true homologues) to a strictly genomic allopolyploid, in which homoeologous pairing will only occur as a very rare and exceptional accident.

In essence, the point at issue is thus a question of degree. Klekowski states that in his paper the "discussions are based upon a radical departure with regard to the nature of meiosis in homosporous ferns" (Klekowski, 1973b, p. 543). The extent to which this in fact represents natural truth is clearly open to question, and concrete evidence regarding the extent of homoeologous pairing in different taxa is evidently highly desirable. Klekowski points out that two approaches are possible. Firstly, "the degree of homoeologous pairing may be estimated indirectly by studying the segregation of marker genes in homozygous sporophytes" (op. cit., p. 543). Secondly, homoeologous pairing may be studied directly by cytological observation, utilizing chromosome aberrations such as translocations and paracentric* inversions. Crossing-over within the inversion (reverse pairing) loop in a bivalent heterozygous for a paracentric inversion will produce a chromosome bridge and an acentric fragment at anaphase I. The method proposed is to search for these phenomena in the meiosis of homozygous sporophytes, wherein bivalents heterozygous for a paracentric inversion could only arise from pairing between homoeologues (but of course still only possible where homoeologous genomes in a polyploid do happen to differ with respect to the presence of such an inversion.)

This second approach appears at first sight to be very promising, but is regrettably subject to a very important weakness since, as has been cogently argued by Lewis and John (1966), following earlier observations by Matsuura (1950) and Haga (1953), the dicentric bridge and acentric fragment phenomenon can also be produced by spontaneous chromosome breakage and exchange in meiosis, i.e. by errors in chiasma formation. Specifically, this can occur by iso-locus breakage of non-sister chromatids with inverted reunion (Fig. 21b) or by iso-locus sister-chromatid union beyond a chiasma (Fig. 21c). (Other patterns of exchange can yield true or false dicentric half-chromatid bridges without fragments.)

---

* Paracentric inversion = one not including the centromere.

Pachytene   Anaphase I

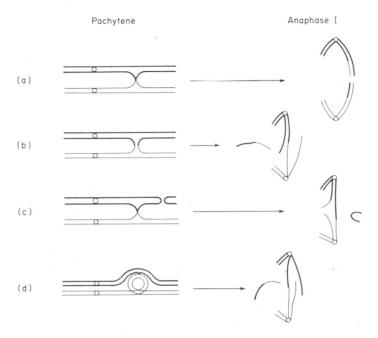

Fig. 21. Consequences of meiotic breakage. (a) Normal crossing-over and chiasma formation. (b) and (c) Two different modes of chromatid breakage and reunion leading to a dicentric chromatid bridge and acentric fragment. See text (p. 382) for further details. (After Lewis and John, 1966, p. 297, fig. 32.) (d). Crossing-over in reverse pairing loop in inversion heterozygote leading to dicentric chromatid bridge and acentric fragment.

Lewis and John consider that "the occurrence of bridges and fragments does not constitute a satisfactory basis for inferring inversion hybridity" (Lewis and John, 1966, p. 303), and state that "there seems little doubt that the extent to which inversions may have been involved in karyotype changes has been grossly exaggerated" (p. 300). They point out that "following crossing-over within a heterozygous paracentric inversion the fragment produced should be constant in size" (p. 299), whereas if meiotic breakage is involved the size of the fragment will be variable, since it is dependent on the site of breakage, and will also occur in different chromosomes in different cells, not being confined to specific chromosomes. They also present cytological evidence from their own experience in support of their argument, the most impressive element in which is their statement that although the anaphase bridge and fragment phenomenon can be observed in grasshopper meiosis, in which "Pachytene is an exceptionally clear stage to analyse ... we have examined many hundreds of pachytene nuclei without finding any trace of reverse loop pairing" (p. 301).

Although aware of Lewis and John's arguments, Hickok and Klekowski appear to discount them since, when discussing the causation of anaphase bridge and fragment figures seen in the synthetic hybrid *Ceratopteris richardii* x *pteridoides*, they consider that these are attributable to inversion heterozygosity, commenting that "the extreme differences in fragment size suggest that each of these figures represents the result of crossing-over within different paracentric inversions" (Hickok and Klekowski, 1974, p. 443), even though Lewis and John (1966, pp. 299/300) state that in their opinion Swanson (1940) was not justified in presenting what is exactly the same argument and conclusion with regard to comparable phenomena he found in *Tradescantia canaliculata* x *humilis*. The ultimate development of this line of reasoning was apparently that of Geitler (1937), who postulated the presence of 21 distinct inversions in *Paris quadrifolia*, in order to encompass the range of fragment sizes encountered.

Furthermore, (Klekowski and Hickok, 1974), in presentation of experiments involving evidence of bridge and fragment formation with regard to which the presence of paracentric inversions on a rather liberal scale is claimed, make no reference to the study by Lewis and John. Nevertheless, in view of the arguments set forth by Lewis and John, the first part of the evidence presented by Klekowski and Hickok (1974) to demonstrate the existence of non-homologous chromosome pairs in *Ceratopteris* must be regarded as insecurely based and is best disregarded, because it is concerned with cytological aberrations observed in homozygous sporophytes derived from spores that had been subjected to a substantial dosage of ionising radiation, and which therefore can reasonably be considered to be abnormally liable to the display of breakage phenomena.

The second line of evidence presented in this paper concerns plants of rather complicated geneology (Fig. 22), associated with the appearance of a male-sterile (gametophytic) mutant in the progeny of sporophyte C230H7, which is itself a homozygous descendant of C230, a strain of *Ceratopteris thalictroides* originating from a wild collection in Malaya. A sporophyte (C230X) was established from this male-sterile gametophyte by inseminating it with spermatozoids from one of its sisters. The progeny of this sporophyte segregated with regard to the male sterility character, in a ratio very close to 2 sterile : 1 fertile, (though subsequent experimentation showed that no simple genetical model would serve to explain its inheritance). The family of sporophytes raised from the fertile gametophytes amongst this progeny proved to be heterogeneous both for spore viability and the presence of meiotic abnormalities, though the incidence of low spore viability and frequency of meiotic abnormality were not correlated (unless negatively!). Of the eight sporophytes examined, spore viabilities were: 37% (1 individual), 50% (2), and $90 < 95\%$ (5). The incidence of meiotic abnormalities was: 0% (4), $12 < 36\%$ cells showing acentric fragments (4).

In this context, the origin of the meiotic abnormalities is fortunately of only secondary moment, the important point being the fact that the progeny of a nominally homozygous sporophyte show variation with respect to their liability to meiotic abnormalities (and in spore viability). In other words, whether the individual plants differ with respect to the presence of paracentric inversions in

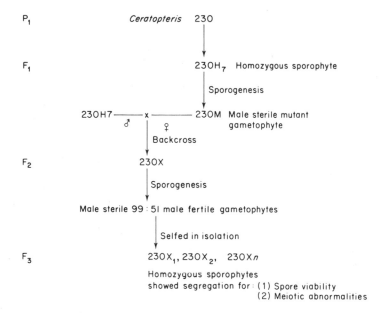

Fig. 22. Geneology of progeny of Ceratopteris 230H7. (Adapted and simplified from Klekowski and Hickok, 1974, p. 427, fig. 20.)

heterozygous combination or in their susceptibility to errors in chiasma formation is in this case irrelevant to the main argument.

The interpretation of this experiment would be more secure had it been possible to culture the sporophytes under completely controlled conditions, thus excluding the possibility of C230X possessing a facultatively unstable genotype susceptible to its expression being influenced by minor variations in environmental conditions. However, the simplest explanation of the independent variation observed in this progeny with respect to spore viability and meiotic disturbances is undoubtedly segregation due to pairing of homoeologues.

Principally because the frequency of bridge and fragment phenomena seen is substantially less than expectation if chromosome pairing was random with respect to homologues and homoeologues when heterozygosity for a paracentric inversion was present (as they maintain is the case), Klekowski and Hickok conclude that "homologous pairing predominates, but homoeologous pairing does occur" in this strain of *Ceratopteris thalictroides*, and "occasional homoeologous pairing released the genetic variability stored homoeologously into the meiospores" (Klekowski and Hickok, 1974, p. 431).

In this same paper Klekowski and Hickok (1974) reported important information regarding the nature of polyploidy in the parent strain of *Ceratopteris thalictroides*, obtained by study of two polyhaploid sporophytes which arose spontaneously by apogamy on the male sterile gametophytes. Study of their meiosis indicated the nature of polyploidy in the original parent, C230. Pairing is very variable, ranging from 9 to 33 bivalents (of a possible maximum of 39 bivalents). This evidence provides a clear indication that the parent race of *C. thalictroides* is of segmental alloploid constitution, though as has already been discussed (p. 364, footnote), comparison with a study by Bouharmont (1972b) shows that chromosome pairing of even rather less than this order of magnitude can occur in a polyhaploid derived from a tetraploid known from other evidence to be of essentially autopolyploid character.

This discovery regarding the nature of tetraploidy in the C230 strain of *Ceratopteris thalictroides* does render it likely that, irrespective of the correctness of the details of their interpretation of the evidence, Klekowski and Hickok's conclusion regarding the frequency of homoeologous pairing in this plant is indeed correct.

It does, however, equally make it evident that it would be hazardous to relate this conclusion obtained from a study of *Ceratopteris* to homosporous ferns in general. The situation in segmental alloploids or autoploids is clearly very different from that prevailing in genomic alloploids, which are probably the commonest type of polyploid in ferns. (Even when other things are equal, there must always be an exceptional element of risk involved in extrapolating from experience in the thoroughly atypical genus *Ceratopteris*, with its unique mode of life (see p. 377), to ferns in general.)

Klekowski (1973b) is surely unwise to disregard the evidence of wide hybrids formed between genomic allopolyploids and unrelated species, in which complete failure of chromosome pairing results (unless the second parent is of autoploid or segmental alloploid constitution). This phenomenon is too general in occurrence to be dismissed as "anomolous meiotic behaviour" (Klekowski, 1973b, p. 536). The totality of the evidence known to me relating to European Aspleniaceae is that no less than 32 of different pairs of genomes tested fail to pair together at all. If homoeologues fail to pair when there are no apparent restrictions to prevent them doing so, it is extremely difficult to see why they should ever do so when faced with competition from true homologues.

It is also hard to see how the occurrence of occasional pairing of homoeologues in old autopolyploids and in segmental alloploids, coupled with the complete absence of any evidence that homoeologous pairing occurs in true genomic alloploids (except as a very rare accident), can be considered to be compatible with the suggestion by Klekowski (1973b) of the existence of "a *radical* departure with regard to the nature of meiosis" (p. 543) (my italics), which he envisaged as having arisen as a special response to the innate propensity of these plants towards production of homozygous offspring.

In fact the level of homoeologous pairing so far claimed to have been demonstrated in *Ceratopteris* by experiment is no more extreme than that already envisaged as occurring in angiosperm polyploids, which have no such special problem to negotiate. Thus Stebbins (1950, p. 329) stated that "another property of many allopolyploids which results from the presence of some duplicated genetic material is the ability of their chromosomes to undergo occasional heterogeneous association, and so to segregate with respect to some of the characteristics which differentiated their parental species". (In this quotation "heterogeneous association" is synonymous with "homoeologous pairing".)

F. APOMIXIS IN FERNS

1. Reproductive apomixis

It is customary to describe ferns showing obligate apomixis as apogamous ferns, but this is a terminological inexactitude which I wish to avoid. Regular apomixis requires the alternation of two quite separate phenomena, the avoidance of meiotic reduction, and the spontaneous development of a new sporophyte without fertilization. Since strictly speaking the term apogamy refers solely to the production of a sporophyte from a gametophyte without involvement of sexual reproduction, which can be induced or even occur spontaneously in gametophytes belonging to species which normally reproduce sexually with a regular reduction/fertilization cycle and are not apomictic, it is clear that the use of the term "apogamous" in a sense synonymous to "apomictic" is an imprecision which could lead to confusion. It has indeed led to infelicitous and illogical terminology, such as "meiotic apogamy" (Wagner, 1968, p. 125; Evans, 1969a, p. 205). A clear example of the degree to which the term "apogamy" has been misused is seen in the otherwise quite excellent paper of Braithwaite (1964a) which is entitled "A new type of apogamy in ferns", although the process described is in fact a mode of meiotic non-reduction which was indeed previously unknown in ferns. Imprecise use of the term "apogamy" in this fashion is almost universal in the literature of this subject, though a valuable earlier review by Walker (1966b) constitutes an important exception.

Gustaffson (1946, 1947a, b) produced a now celebrated monograph on apomixis in flowering plants which included an ordered terminology and classification for the variety of phenomena encountered in flowering plants which can perfectly well be extended to the ferns. Indeed Gustaffson himself points out: "It can be transferred without difficulty to apomictic pterido-phytes" (Gustaffson, 1946, p. 7). According to his system, all obligately apomictic ferns (with the exception of purely vegetatively apomictic forms) show diplospory* (since the unreduced spores are all produced from spore mother cells) and apogametry (because the new sporophyte embryo develops from vegetative cells of the gametophyte and not from an unfertilised egg).

---

* In commenting that "diplospory which is so widespread in apomictic angiosperms has so far not been observed as a functional mechanism in apomictic ferns", Mehra (1961, p. 11) is evidently using the term diplospory in a manner different from the usage of Gustaffson (1946, p. 7).

An excellent review of the early development of this subject is provided by Manton (1950, ch. 10), who states that though certain aspects of the cytology of the production of unreduced spores were correctly observed by Allen (1914) and Steil (1919) the only previous account "which can be accepted as adequate" is that of Döpp (1932) on *Dryopteris remota*. Manton herself independently elucidated the detail of the process, extending her observations to a variety of genera. The account she presented is detailed and has in no respect been improved in the last quarter-century. There is room here to extract only the essence of the process she observed in all of *Dryopteris borrerii* (= *pseudomas* = *paleacea*) *D. remota, D. atrata, Pteris cretica, Cyrtomium* (*Phanerophlebia*) *falcatum* and *Phegopteris polypodioides*, with more limited observations on certain species of *Asplenium* and *Pellaea* (Fig. 23a).

In leptosporangiate ferns, usually four successive mitotic divisions take place in the archesporium, producing a total of 16 spore mother cells, which subsequently divide meiotically to produce a total of 64 spores with reduced chromosome number. In apomictic sporangia, the last mitotic division is imperfect, becoming arrested in metaphase, and subsequently returns to a resting stage without an anaphase separation having occurred, though at some time before the onset of meiosis the centromeres divide. The number of cells in the archesporium remains unaltered at eight. These are now the spore mother cells,

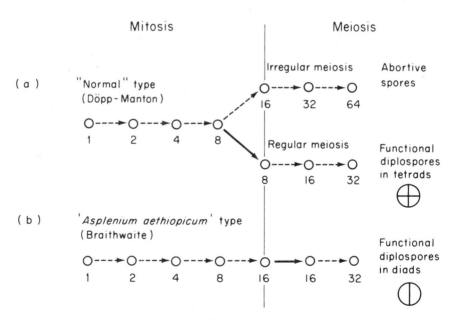

Fig. 23. The course of sporangial development in apomictic ferns. (a) The "normal" (Döpp-Manton) type. (b) The "*Asplenium aethiopicum*" (Braithwaite) type. The numbers below each circle indicate the number of cells per sporangium. Broken lines indicate normal nuclear divisions and thick unbroken lines "compensating" divisions. Further explanation in text, p. 388 *et seq*. (Redrawn after Walker, 1966, p. 153, fig. 8.)

but the restituted nuclei they contain are of doubled chromosome number. Thus, to take *Cyrtomium falcatum* as an example, ordinary sporophytic tissues are triploid, but apomictic spore mother cells are hexaploid. Meiosis proceeds in due course, and is entirely regular, reducing the chromosome number to the normal (triploid) sporophytic level. The effect of the abortive last pre-meiotic mitosis is thus to compensate for the subsequent meiotic reduction.

Most remarkably, this process, referred to as the "normal" type by Walker (1966b) and christened the Döpp-Manton scheme by Klekowski (1973b), is identical in all of the species studied by Manton, belonging to no less than seven different genera. As Manton wrote, "A more striking example of parallel evolution would be hard to find" (Manton, 1950, p. 195).

The gametophytes produced by the apomictic spores will possess the same chromosome number as the sporophyte. The new embryonic sporophytes develop as an outgrowth of the central cushion, which remains thin. Functional antheridia are usually present and can participate in hybrids with sexual species. Manton (1950, p. 58) states that archegonia are absent in *Dryopteris borrerii*, but it seems that either rudimentary, abortive or apparently normal archegonia are present on the gametophytes of some other apomictic ferns (e.g. *Anemia tomentosa* var. *anthriscifolia* (Mickel, 1962)), though whether these archegonia are ever capable of function is unknown. Bierhorst has recently reported on a remarkable condition in *Trichomanes pinnatum*. This species possesses a filamentous prothallus bearing club-shaped archegoniophores. Not only are archegonia present, but "the apogamous embryo originates from the archegonial jacket and contiguous tissue" (Bierhorst, 1975, p. 448).

Usually not all sporangia proceed through the apomictic sequence. Some sporangia, even in the same sorus as apomictic sporangia, will proceed through the normal pre-meiotic mitotic sequence to the 16-spore mother cell stage.* Meiosis in these sporangia is irregular. The reason for this is that, just as in flowering plants, there is a strong correlation between the incidence of apomixis and both hybridity and sterility. In short, the great majority, if not all, obligately apomictic ferns are basically sexually sterile interspecific hybrids. The nature of the meiotic abnormalities observed in the "16-cell sporangia" will depend on the degree of homology existing between the parent genomes combined in the hybrid.

It is therefore not surprising that in a census of apomictic ferns (Manton, 1961) of the 24 taxa of apomictic ferns then known, a high proportion, namely 15, were triploids. Two were diploids. Further interesting statistics on the relation of the level of ploidy to apomixis were provided by Walker (1962), who

---

* Walker comments on the evident existence of a "switch mechanism" of unknown character determining whether or not an individual sporangium shall follow the reductional (16-cell) or non-reductional (8-cell) pathway, pointing out that "it is difficult to see why a 'switch mechanism' is necessary at all (instead of the plants producing only the functional type of sporangium)" (Walker, 1966b, p. 155). It is indeed surprising that the reductional (non-functional) pathway has not been eliminated by natural selection at least in the more ancient apomictic species (e.g. *Asplenium monanthes, Dryopteris borrerii*).

found, in the course of an extensive biosystematic survey of the predominantly tropical genus *Pteris*, that 28 of the 82 species investigated were apomictic, about one-third in all. Of these 28 apomicts, 10 were diploid, 13 triploid and five tetraploid. Regarding the incidence of apomixis in ferns generally, Walker later wrote, "A survey of the literature . . . and of my own unpublished results indicates that out of a total of approximately 1055 species of ferns for which the breeding system is known or can be confidently deduced about 106, or almost exactly 10%, are apomictic" (Walker, 1966b, p. 155). Apomixis has now been detected in many genera other than those already mentioned here, including *Actiniopteris, Adiantum, Cheilanthes, Diplazium, Hemionitis* and *Polystichum*, but is clearly very unevenly distributed within different genera.

It should be noted that either cytological evidence or observations on gametophytes in culture are necessary to demonstrate the presence of apomixis. Vida *et al.* (1970) demonstrated the production of both 16-spore (rare) and 32-spore sporangia (frequent) in a diploid race of *Cheilanthes catanensis, s.l.*, known to reproduce sexually. This discovery undermines a survey of American species of *Cheilanthes* by Knobloch (1966) in which apomictic reproduction was inferred from the production of 32-spore sporangia. Hickok and Klekowski report that *Ceratopteris richardii* and *C. pteridoides* are characterized by the formation of 16 large and 32 smaller spores per sporangium respectively, although both reproduce sexually, and they comment that "these spore numbers are commonly interpreted as being evidence for an apogamous life-cycle" (Hickok and Klekowski, 1974, p. 441).

Though it is now abundantly clear that the Döpp-Manton mechanism is by a wide margin the prevalent mode of sporangial apomixis in ferns, it is not the only one. An entirely different sequence of cytological events by which unreduced (apomictic) spores are produced was described by Braithwaite (1964a) in members of the *Asplenium aethiopicum* complex (Fig. 23b). In this process the full complement of 16 spore mother cells are produced in the usual way, but in the ensuing meiosis only univalents are formed, and the first division aborts in metaphase, restitution nuclei being formed. Because this is a meiotic division, no division of the centromeres occurs until after the restitution phase. Subsequently, ". . . the restitution nuclei undergo what at first sight could be mistaken for a simple mitotic division. Except for the presence of only one spindle the chromosome behaviour is however that of a typical second meiotic division" (Braithwaite, 1964a, p. 301). Ultimately the 32 spores are produced in diads, not in tetrads as is the case in normal (reductional) spore production, or in the more familiar Döpp-Manton apomictic process. The presence of complete asynapsis in the first meiotic division is noteworthy, because the plant on which Braithwaite made his main observations is a high polyploid and a member of a complex believed to have a substantial autoploid content. It is therefore most unlikely that the failure of pairing in the apomict is due to lack of homology, and much more probable that asynapsis is genetically imposed.

Complete asynapsis has been observed in three species of Hymenophyllaceae. Manton and Sledge (1954) recorded this phenomenon in two separate collections of *Hymenophyllum javanicum* fixed in cultivation. Mehra and Singh (1957) described the formation of diads of apparently good spores by an abbreviated meiosis involving complete asynapsis in a form (*β*) of *Trichomanes insigne*, but concluded only that this plant was an asynaptic triploid. Bell (1960) observed asynaptic diakinesis and sporangia containing 32 apparently good spores in *T. proliferum*, and suggested that both this species and *T. insigne* forma *β* were obligately apogamous. Thus, though in none of these examples is the evidence complete, there is a strong inference that apomixis of the type described by Braithwaite occurs also in the Hymenophyllaceae.*

More recently, this type of apomixis has been detected in a near hexaploid form ("*n*" = *c*.212 univalents) of *Asplenium flabellifolium* (Lovis, 1973). Only incomplete cytological observations have been made on spore formation in this species, but it is now known from observation of prothalli in culture that the young sporophytes are produced apogamously.

Evans (1964, 1969a) in describing an apomictic life-cycle in *Polypodium dispersum*, states that the process of spore-formation is ameiotic, i.e. purely mitotic. In each sporangium, 16 spore mother cells are present, and 32 spores are formed in diads, as is the case in *Asplenium aethiopicum*. Comparing his findings in *P. dispersum* with those of Braithwaite, Evans writes that there "would appear to be very similar processes in these two different genera" (Evans, 1969a, p. 206). Nevertheless, and in spite of describing one of his figures of *P. dispersum* as showing "111 univalent chromosomes at metaphase in the spore mother cells" (Evans, 1969a, pp. 270, 284, fig. 13E), Evans states that meiosis is absent. Writing before the appearance of Evans's illustrated second paper, Walker commented that "Evans (1964) described a different process in the American *Polypodium dispersum* Evans, but from his description and my own observations on this species it is possible that it really belongs to the pattern of development here termed the 'Asplenium aethiopicum' type" (Walker, 1966b, p. 152, footnote). However, in a recent review Klekowski (1973b) has accepted that the processes of spore formation in *A. aethiopicum* and *P. dispersum* are distinct, and refers to them as the Mehra-Singh† scheme and the Evans scheme respectively.

---

* Braithwaite (1975, p. 180) also reports an asynaptic diakinesis and the regular production of diads of diplospores in *Gonocormus prolifer* (= *Trichomanes proliferum*) in the New Hebrides.

† The appropriateness of this name is questionable, since whereas the entire course of events in *Asplenium aethiopicum* is documented in some detail, it is still not established that *Trichomanes insigne* forma *β* is actually apomictic. Indeed, the possibility that the cause of the observed asynapsis was associated with apomixis was specifically "ruled out" by Mehra and Singh (1957, p. 338). Moreover Mehra (1961, p. 3) stated that abortive spores are formed and this "taxon multiplies only by vegetative means". This process will therefore be referred to here as the Braithwaite scheme.

Nevertheless, the question is still open as to whether or not the pattern of events in these two plants is really different, in view of the technical difficulties involved in accurate cytological study of these processes. Although in some respects (e.g. chromosome number) more difficult to analyse, serial section preparations allow greater confidence in interpretation, particularly in identification of stages of the sequence. There can be no doubt that the illustrative cytological evidence presented by Braithwaite, a combination of section and squash preparations, is of a different order of quality to that provided by Evans. However, the preparations illustrated by Evans (1969a) are adequate to raise doubts concerning the correctness of the proposed interpretation. The one high power figure presented (fig. 13E) showing metaphase in a spore mother cell, does indeed show what I would accept without question as a meiotic figure showing dispersed univalents. The chromosome appearance seen could only be simulated by a mitotic figure showing drastic chromosome shrinkage and spindle inhibition, such as can be induced by a c-mitotic (spindle-disrupting) drug, which for obvious reasons would not have been applied in this particular study. Evans states (1969a, p. 206) that "squashes of metaphase figures in the eight-celled stage of the archesporium and in the 16 spore mother cells show that the chromosome configuration is essentially identical, both stages showing 111 unpaired chromosomes (fig. 13, A, D)". The dividing cells in figs 13A and 13D are indeed essentially identical. Why should both of these stages, if truly mitotic, simulate asynaptic meiosis? The solution to this difficulty could well be that fig. 13A shows the partial contents of a sporangium in meiosis, not mitosis in the eight-celled archesporium. It may be significant that only six dividing cells can be distinguished in the figure. In view of these criticisms and Walker's independent observations on this species (see above) the process of spore formation in *Polypodium dispersum* clearly requires further investigation, and for the time being the concept of a purely mitotic apomictic life-cycle in ferns should be regarded as unproven.

A regular apomictic life-cycle depends upon two independent conditions, the avoidance of reduction, and the vegetative (apogamous) production of sporophytes. (The different factors required for regular gametophytic apomixis in ferns are analysed in greater depth by Walker (1966b).) The inception of a successful apomict is thus dependent upon a fortuitous combination in one individual of these two potentialities. No artificial hybrid yet synthesized between two sexual species has proved to be an apomict.

Some diploid *Asplenium* interspecific hybrids show complete asynapsis on account of lack of homology between their parents. In these hybrids meiosis may abort at metaphase in some mother cells with the consequent formation of (unreduced) diplospores. This occurs very commonly in *A. trichomanes* ssp. *trichomanes* and ssp. *inexpectans* x *viride* hybrids, up 10 or 11 large spores being present in some sporangia (Lovis and Reichstein, 1968a, b). In *A. fontanum* x *viride* the number of well-formed spores (<26) may exceed the number of spore mother cells, which suggests that some restituted mother cells may subsequently divide mitotically, although there is as yet no cytological evidence to support this interpretation (Lovis, 1970). In all of these examples the unreduced gametophytes reproduce sexually. Therefore the meiotic phenomena, rather than constituting the first phase in a new apomictic life-cycle, merely make possible the ready inception of a new fertile allotetraploid from the otherwise sterile diploid hybrid. However, some authors (Morzenti, 1962, 1967; Evans, 1969b)* have adduced evidence suggesting not only that diplospores may be formed in some essentially sterile hybrids, but that the prothalli produced by these unreduced spores may form apogamous sporophytic outgrowths, at least in culture. It is thus possible that in nature ageing unreduced prothalli may occasionally successfully produce an apogamous sporophyte, and that therefore such sterile hybrids may, rarely, reproduce apomictically.†

Walker (1962) has drawn attention to the absence in the ferns of large agamic complexes such as that known in *Rubus* amongst the flowering plants, and has pointed out that this is largely attributable to the apparent absence in ferns of facultative apomixis, all known fern apomicts being obligate apomicts. Only one possible exception to this rule is known. Lloyd (1973) has reported facultative apomixis in *Matteuccia orientalis*. However, the evidence in this case seems less than conclusive at present, and in particular lacks cytological documentation. The level of ploidy of the strain used as sporophyte parent in the experiments reported had not been determined.

Small-scale agamic complexes can occur in ferns. The best known example involves *Dryopteris borrerii* (= *pseudomas*) and was investigated in some detail by Manton (1950) and Döpp (1941, 1949, 1955). This apomictic fern occurs on two levels of ploidy, diploid and triploid. It very frequently grows with the sexual tetraploid species *D. filix-mas*, and can hybridize with it through the medium of the antheridia borne on its apogamous prothalli. The hybrid

---

* I have not seen the widely quoted unpublished thesis of DeBenedictis (née Morzenti) (1969, Univ. of Michigan) which records observations on large putatively unreduced spores in a wide range of fern hybrids (*teste* Hickok and Klekowski, 1973).

† Hickok and Klekowski (1973) have reported the production of diads and triads of spores (as well as tetrads of abortive spores) by a triploid ($2n = c.117$) *Ceratopteris* hybrid of unknown provenance. They suggest that either one or the other of the two meiotic divisions fails in the origin of diads, and one of the first division products fails to divide again in the formation of spore triads. Approximately half of these spores proved to be capable of germination, and a proportion in due course produced sporophytes, which were divisible into two categories: (1) subhexaploid sporophytes with $2n = 210$ or 178, which had clearly been formed sexually, and subsequently proved capable of giving rise to a further sporophyte generation; and (2) circa-triploid sporophytes with $2n = 81$, 104 or 119 which they presumed had arisen by apogamy, and which were completely sterile. In view of the wildly varying chromosome numbers found in the progeny, it seems extremely doubtful whether these spores could function in nature, either as a viable source of a new polyploid, or as a means of propagation of the triploid parent, except by means of the very occasional fortunate individual which happened by chance to have obtained a balanced chromosome complement.

sporophytes (*D.* x *tavelii*) so produced are tetraploid and pentaploid respectively, and can be distinguished morphologically from their parents by their robust growth and intermediate characteristics. A significant feature of these hybrids is that the proportion of 16-mother cell sporangia is very substantially increased (in comparison with *D. borrerii*), with a consequent diminution of the proportion of apparently good spores produced. It appears that the apomictic faculty is diluted by the addition of the genomes of the sexual parent. It is at present uncertain whether these two hybrids can reproduce themselves successfully by apomictic means in nature though clearly they possess the potential to do so. The tetraploid hybrid in particular can be frequent in some populations, but this may only reflect the relative ease with which this hybrid is formed *de novo* in nature.

Other examples of natural apomict hybrid combinations involving both apomict and sexual parents are *Asplenium heteroresiliens*, formed from the tetraploid race of the sexual *A. heterochroum* and the triploid apomict *A. resiliens* (Morzenti, 1966; Wagner, 1966), and an apomictic tetraploid hybrid formed from the diploid sexual species *Phegopteris hexagonoptera* and the triploid apomict *P. polypodioides* (Mulligan *et al.*, 1972).

Since the apogamous prothalli produced by an apomict carry the unreduced sporophytic chromosome number, these gametophytes can only give rise to a hybrid sporophyte of a higher level of ploidy than their own parent sporophyte. The only exception to this rule can be where a non-apomictic spore proves to be functionally viable, which is likely to be a decidedly rare event.

In *Pteris*, Walker (1962) has compared the spore production of a series of synthetic apomictic hybrids produced between the diploid apomict *P. confusa* (or in one case *P. argyraea*) and different sexual diploids, with a series of wild apomicts of the same (triploid) ploidy. The percentage of good spores produced by the natural apomicts (57-85%) was generally much greater than that produced by the synthetic apomictic hybrids (19-55%). As Walker indicates, these figures suggest that selection is able to act upon newly formed apomictic species so as to improve their spore fertility, presumably by influencing the proportions of the two different types of sporangia present.*

The possibility that natural selection can be effective on apomicts, at least on such characters as spore fertility, raises the question of the source of their

---

* Walker overextends this argument when he applies it to the vigour of the spermatozoids, which was seen to be greater in the natural apomicts (op. cit., p. 35). Natural selection could not possibly act directly with respect to the vigour of spermatozoids produced by an apomict, since these have no relevance whatsoever to its own survival. It is true that an increase in vigour of spermatozoids would increase the possibility of the production of hybrid progeny, but its own hybrid progeny are irrelevant to the fitness of an obligate apomict, except possibly in a purely negative fashion if they compete with it too successfully. If the observed difference in spermatozoid vigour between natural and synthetic apomicts is attributable to the past action of natural selection on the former, it can only be as a correlated response.

variability. That even obligate apomicts are not quite the evolutionarily inert material they were previously generally assumed to be has become increasingly recognized with respect to angiosperm apomicts, particularly since the pioneer studies of Turesson (1943) on *Alchemilla*. As far as is known, the apomictic members of this genus, which provide wretched cytological material, are all obligate apomicts, yet it has been possible to demonstrate that some species (e.g. *A. vestita*) have evolved dwarf upland ecotypes comparable to those formed by sexual species (Bradshaw, 1963).

Walker draws attention to the *Pteris biaurita* apomictic complex, in which "there are many true-breeding populations of characteristic morphology and definable geographical and ecological distributions which might potentially be given specific recognition if they were less numerous and the differences distinguishing them more marked. . . . In any one geographical area one often finds numerous populations resembling one another closely but differing from corresponding populations from another geographical area" (Walker, 1962, p. 33).

Similar morphological complexities have long been recognized (von Tavel, 1937) in *Dryopteris borrerii* though in this plant the situation is complicated by the existence of two levels of ploidy, as well as its extremely wide geographical range. Thus, outside Europe, a diploid form is known in Madeira (Manton, 1950) and a triploid in Ceylon (Manton and Sledge, 1954). A closely related species, *D. parallelogramma*, exists both in Gough Island in the far South Atlantic and in North America, while a number of different taxa which may properly belong to this complex are recognized in Japan (Hirabayashi, 1974).

With the aid of undergraduate students I have made a close study, involving monitoring the cytology of many wild individuals, of mixed populations of *Dryopteris* near Brathay, in the English Lake District. It is now quite clear that in this area the diploid form of *D. borrerii* is consistently more extreme in all its characters than the triploid, in the sense of being more distinct from the sexual *D. filix-mas*, and it is evident that a hybrid between this diploid apomict and the sexual diploid species *D. abbreviata* would be very difficult to distinguish from the sympatric triploid race of *D. borrerii*. It is, however, most unlikely that all triploid forms of this apomict have the same origin. Both levels of ploidy in *D. borrerii* are variable. The question of the source of variability, which it would seem gene mutation alone is not adequate to provide, is still unanswered.

Klekowski (1973b) has recently suggested that homoeologeous pairing at meiosis, disguised by an accompanying restriction of chromosome pairing to the formation of bivalents, is an important source of variability in homosporous ferns. Whether or not this system exists in sexually reproducing ferns is a quite separate question, but the cytological facts clearly indicate that such a system cannot possibly operate, as he proposes, in the case of the Döpp-Manton apomictic process.

In the last pre-meiotic mitosis of the Döpp-Manton process the division aborts in metaphase so that the newly formed daughter chromosomes are not separated by the anaphase mechanism, but instead retain their relative positions in the restitution nucleus until the onset of meiosis, when each chromosome pairs with its adjacent sister. The exact regularity of this process is remarkable, extending both to aberrant cells where cleavage has divided the restitution nucleus without affecting the regularity of the subsequent meiotic pairing, and to other aberrant cells where both the penultimate and ultimate pre-meiotic mitoses fail, producing a quadrupling of the chromosome number, but nevertheless again resulting in a regular meiosis, without multivalent formation, in which sister pairs with sister, and never with sister-once-removed, as it were (Manton, 1950, pp. 166/168, fig. 169, 170d). The possibility of the pairing of homoeologues is thus precluded.*

The question as to whether or not recombination can occur at meiosis in the Braithwaite apomictic process is still unresolved. In *Asplenium aethiopicum* the enormous chromosome number and small chromosome size prevent satisfactory observation of early prophase stages. Chromosome pairing (bivalents and trivalents) has been observed in *Trichomanes insigne* forma β (Mehra and Singh, 1957, p. 387, pl. 17, fig. 17), although the associations fall apart after pachytene, and it is quite uncertain whether crossing-over may sometimes occur. Moreover, it has yet to be established that this taxon really is a diplosporous, rather than merely vegetative, apomict. (See p. 391, footnote).

It is my own belief that the question of the origin of variability in apomicts, other than that provided by mutation, can be answered by mitotic recombi-

---

* Klekowski (1973b) clearly believes that mechanisms for the restriction of chromosome pairing to bivalent formation are generally present in ferns. Much of the supporting evidence he cites come from autotetraploids like *Asplenium billotii* and *A. ruta-muraria* wherein the selective value of a multivalent inhibition mechanism is evident and its existence no longer in doubt. However, he also claims that "data presented by Manton (1950) on the cytology of apogamy contains important evidence on the physiological control of bivalent formation in ferns" (Klekowski, 1973b, p. 537) and indeed quotes the formation of bivalents in 4-cell sporangia (referred to above) as supporting evidence. In direct contradiction, I believe that specific mechanisms for the inhibition of multivalent formation are generally absent from Döpp-Manton apomicts. The regular pairing in 8-cell sporangia is simply attributable to the spatial relationships of the chromosomes in the nucleus prior to the onset of meiosis. In contrast, 16-cell sporangia "display the true pairing homologies of the chromosomes" (Manton, 1950, p. 163). In the 16-cell sporangia multivalents are obvious in all three cytotypes of *Pteris cretica* studied by Manton, while trivalents are formed in *Cyrtomium fortunei* (Manton, 1950) and in *Asplenium monanthes* (J. D. Lovis, unpublished) It is scarcely possible to conceive of a control mechanism which is effective in one type of sporangium but inoperative in the other!

Very recently Bierhorst (1975) has claimed to have obtained "confirmation of Klekowski's prediction on homoeologous pairing" in the course of a study on a filmy fern, *Trichomanes pinnatum*, which he found to follow the Döpp-Manton apomictic cycle. The critical observation is the presence in a single cell of an anaphase chromosome bridge, accompanied by an acentric fragment, such as can result from crossing-over in a bivalent heterozygous for a paracentric inversion. (The observation in some other meiotic cells of chromosome bridges without fragments is not of particular significance, since a similar bridge was also observed in a pre-meiotic mitosis.) Essential to the argument is that the

nation. I am not aware that this suggestion has been made in print before, even though the concept of mitotic recombination is by no means a new one. As Pontecorvo has written, "That segregation may occur regularly though rarely in heterozygous somatic cells of *Drosophila* is unquestionable since the classic work of Stern (1936)" (Pontecorvo, 1959, p. 101). Nevertheless, it does appear that in spite of its undoubted elegance, Stern's discovery was largely ignored (as has been the initial fate of so many other perfectly good discoveries that seem to conflict uncomfortably with orthodox theory) until the growth of an entire new field of research, based on a technique for genetic analysis of mitotic recombination, was developed in another quite different organism, the fungus *Aspergillus nidulans* (Pontecorvo and Kafer, 1958). It is true that mitotic crossing-over has not been convincingly demonstrated cytologically (and is not likely to be, in view of the rarity of its occurrence), but the genetic evidence that it does occur in these two organisms is irrefutable. If this process exists in two such widely different organisms, why not in others? Logic demands acceptance of the concept that the same process can occur in ferns and in flowering plants, though there is no likelihood whatsoever that it can be proved cytologically to exist in ferns. However, the fact that our techniques are not adequate to demonstrate a phenomenon is not a conclusive argument against its existence. For any individual nucleus, mitotic crossing-over must be a rare and exceptional

---

observation of bridge and fragment is made on a spore mother cell with doubled chromosome number and therefore containing strictly homozygous homologous chromosomes. It is asserted that in such a cell the bridge and fragment phenomenon could only be produced by crossing-over between homoeologues heterozygous for the inversion. However, the arguments set forth by Lewis and John (1966) concerning the production of dicentric bridges with fragments by aberrant meiotic breakage and reunion phenomena unconnected with inversion heterozygosity (see p. 382 above) make it perfectly clear that a single observation of an anaphase bridge and fragment cannot be held to demonstrate the existence of a paracentric inversion in heterozygous condition.

Furthermore, neither in the text nor in the legends to the plates are the two alternative pathways (doubled = apomictic = "8-cell") and ("normal" = non-apomictic = "16-cell") clearly and consistently distinguished, although it appears from both description and illustrations that both are present. Bierhorst comments on the presence of some young sporangia with 16 or 24 spore mother cells, which he concludes (from the absence of mature sporangia with 64 or 96 spores) would be ultimately abortive, with abortion occurring after meiosis, since sporangia with 16 young tetrads are also present. However, he makes no mention of having observed meiosis in these 16-cell sporangia, though he states, "In certain other sporangia ... chromosome pairing is irregular in nearly every cell ... a variable number, usually 2 to 8, of unpaired univalents were seen. ... Such sporangia ... are interpreted as resulting from a breakdown in the pairing regulatory mechanism" (p. 452). It seems very likely that Bierhorst has failed to appreciate the probable true nature of these sporangia, since meiotic irregularity is usual in 16-cell sporangia, arising from the interspecific hybrid origin of most, if not all, fern apomicts. It is therefore not excluded, and is indeed much more probable, that the critical observation in one cell of anaphase bridge and fragment is in fact made on a cell derived from a "16-cell sporangium". In such a sporangium, in a plant of interspecific hybrid origin, whatever chromosome pairing occurs may indeed to a significant degree be between homoeologues, whenever truly homologous partners are not present, but these sporangia make no contribution to the apomictic life-cycle, and are therefore irrelevant to the establishment of Klekowski's hypothesis.

event, but in view of the number of mitoses involved in a complex organism, it will be statistically a regular phenomenon.

The real point of issue between Klekowski and myself is in effect not whether or not crossing-over between homoeologues can occur in Döpp-Manton apomicts, but when it occurs. Mitotic recombination in apomicts implies for the most part exchanges between homoeologues, simply because the great majority of apomicts are interspecific hybrids. Thus though the possibility of pairing of (and therefore also crossing-over between) homoeologues is excluded in "8-cell" sporangia following the Döpp-Manton scheme simply by the chromosomal mechanics of the process, pairing (whether partial or complete) and reciprocal exchange between homoeologues is possible as a rare mitotic event. However, the genetic effects resulting from these two processes (assuming that both were in fact possible) would not be distinguishable.

2. Vegetative apomixis

As interpreted by Gustaffson (1946, 1947a, b) apomixis also includes all processes of vegetative reproduction that permit a sexually sterile taxon to persist and propagate.

Vegetative apomixis in ferns has been reviewed by Walker (1966b) who distinguished four main forms of vegetative reproduction in ferns: (1) branching of rhizomes, (2) stolons, (3) root buds and (4) bulbils borne on the rachis or lamina of fronds.

Propagation by such means as bulbiferous and rooting frond apices (e.g. *Asplenium rhizophyllum, A. tripteropus*), by production of detachable bulbils all over the surface of the lamina (e.g. *A. bulbiferum*), is known as an accessory mode of reproduction in a great many fertile sexually reproducing species, and would permit any sexually sterile hybrids these species may produce (e.g. *A. bulbiferum* x *flaccidum*) to perpetuate themselves, assuming of course that the bulbiferous or proliferant character is expressed in the hybrid.* Walker (1966b) mentions an interesting case of a New Guinea species of *Ophioglossum* which builds up large populations in newly cleared garden lands by means of propagation by root-buds far more quickly than could be achieved through spores. Some species with regular apomictic spore production also possess the capacity to proliferate vegetatively, and also have the benefit of two modes of reproduction. Thus some strains of *Asplenium monanthes*, like the sexual *A. rhizophyllum*, have fronds which are proliferant at the apex.

Walker gives several examples from his field experience in the West Indies where sexually sterile hybrids are able to establish large populations by

---

* The proliferant character of *Asplenium rhizophyllum* is lost in its amphidiploid derivative, *A. pinnatifidum*. Equally, though the bulbiferous capacity of *Cystopteris bulbifera* is recognizable in its amphidiploid derivative, *C. tennesseensis*, it is non-functional in this species (Wagner, 1968).

vegetative reproduction (Walker, 1966b, pp. 158/160). No less than three different sterile *Blechnum* hybrids "have formed sizeable populations by means of stolons", and individual examples of a triploid *Gleichenia* hybrid have produced "quite extensive thickets locally by means of their profusely branching creeping rhizomes". In the case of *Asplenium* x *fawcettii*, the formation of bulbils has allowed this sterile triploid hybrid to produce a population covering quite a large area on the top of Blue Mountain Peak in Jamaica, where its putative parents, *A. harrisii* and *A. harpeodes*, come together. The example of *A. plenum* in Florida is similar, in that this species, which is extremely local in occurrence, sexually sterile, and believed to be of hybrid origin, is dependent on increase by formation of rhizome buds (Wagner, 1963; Morzenti, 1967).

Workers in the United States have described therefrom some other examples of peculiar interest. *Polypodium dispersum*, which increases by root proliferations as well as by apomictic spores, has particular interest in that most Florida populations (Florida is the northerly limit of this essentially tropical species) consist of dense mats of juvenile plants, with none of them soriferous, indicating that in these localities this species is dependent on vegetative apomixis (Evans, 1968).

This latitudinal effect is mirrored in a peculiar phenomenon which has been detected in the southern Appalachian mountains of the south-eastern USA, where several species of tropical ferns (including species of *Grammitis*, *Trichomanes* and *Vittaria*) have been discovered persisting as gametophyte populations, without producing sporophytes. These gametophyte populations maintain and increase themselves by means of vegetative gemmae (Farrar, 1967). (The regular production of gemmae by the gametophyte is well known in many species of *Trichomanes*, e.g. *T. holopterum* (Farrar and Wagner, 1968) and *T. pinnatum* (Bierhorst, 1975).) In the case of *Vittaria* cf. *lineata*, the gametophytes have apparently lost the capacity to produce sexual organs. Even in tropical latitudes this species produces extensive sterile mats of prothalli (Wagner and Sharp, 1963).

Walker (1966b) has concisely summarized the biological importance of vegetative reproduction in ferns under three counts. It may permit: (1) rapid colonization of newly available sites, (2) perpetuation and spread of favourable genotypes in an unaltered form, and (3) persistence of vigorous but sterile hybrid combinations until such time as either doubling of chromosome number creates a fertile amphidiploid derivative, or selection can operate to improve their fertility.

Can vegetative reproduction also sometimes permit a genotype to persist until it acquires by mutation a genetic factor or factors essential for effective sporangial apomixis, or does regular sporangial apomixis always depend on a glorious accident, the synthesis of a particular felicitous hybrid combination which happens to unite all the necessary genetic factors?

G. CONCLUSION

Our knowledge of the patterns of microevolution and processes of speciation operative in ferns is still very limited, but inasfar as generalizations are yet possible, it is evident that it would be difficult to overestimate the importance of polyploidy, at least for homosporous ferns. The processes of evolution by polyploidy involved in ferns are not essentially different from those operating in flowering plants. Moreover, the example of the *Adiantum caudatum* complex, in which all three fundamental categories (genomic allopolyploidy, segmental alloploidy and autopolyploidy) are found, shows that these processes are the same in tropical floras as in temperate floras.

It seems that the proportion of genomic allopolyploids may be higher in ferns than in flowering plants, though this might only be an accident of sampling, just as lack of information regarding autoalloploids amongst ferns is clearly in part attributable to a concentration of effort at the tetraploid level of ploidy, which is intrinsically easier to resolve. At least some of the hexaploid and octoploid taxa currently under investigation (G. Vida, research in progress) in *Ceterach* and *Cystopteris* may prove to belong to this category of polyploids, and it is difficult to believe that it is not present amongst the very high levels of ploidy found in the intractable *Asplenium aethiopicum* complex.

Present knowledge appears to indicate very distinct differences between genera with regard to the frequency of the different types of polyploids. Thus in *Asplenium* genomic alloploids and autoploids are very conspicuous in both Europe and North America, but no clear example of a segmental alloploid is yet known in the boreal zone. In contrast, segmental alloploidy has clearly been of importance in the evolution of *Cystopteris*, while it appears to be almost ubiquitous in *Polystichum*, to an extent which suggests some special idio-syncracy in this latter genus. Perhaps the segmental alloploidy is residual in character here. It may take the form of unusual conservatism in parts of the ancestral *Polystichum* genome; significant portions of some chromosomes, large enough to permit pairing of homoeologues, may have been maintained by selection essentially unaltered throughout the evolution of the genus, because of the presence of particularly felicitous complex linkages.

The interpretation of relationships within the Appalachian *Dryopteris spinulosa* complex and the origin of the supernumerary bivalents found in various hybrids therein has now become controversial (cf. Wagner, 1971; Hickok and Klekowski, 1975). However, if it is accepted that the established explanation is in principle correct, as is surely more probable, then the supernumerary pairing occurring between homoeologues in various hybrids can be regarded as a rough measure of the relative proportion of quasihomologous segments of functional (i.e. synaptically recognizable) size persisting with respect to different combinations of genomes. Just as the size and number of such segments present will vary from hybrid to hybrid, so can the amount of supernumerary homoeologous pairing be expected to: (1) vary between different hybrid combinations, (2) vary in any one hybrid, and (3) be more variable in some combinations than in others. This indeed is found to be the case. In other words, the Appalachian *Dryopteris* complex shows us what we might, from first principles, have expected to find to be generally the case. The level of differentiation between diploid genomes is high, but a range of levels of recognition between genomes is found, not the consistent non-recognition generally encountered in temperate *Asplenium*, or the surprisingly uniform higher level of homoeologous pairing found in *Polystichum* hybrids.

The apparent overall prevalence of genomic allopolyploidy in ferns, if a real phenomenon, must be correlated with particularly effective chromosomal isolation of species at the diploid level. It may well be significant that the phenomena which accompany the coincidence of interfertility and morphological differentiation between species, i.e. the formation of hybrid swarms and the occurrence of introgression, appear to be exceedingly rare in ferns, only one example, that in the *Pteris quadriaurita* complex in Ceylon, having yet been adequately investigated.

However, there can be little doubt that when more complexes have been studied in sufficient detail, known examples of incomplete speciation, such as is apparent in *Ceratopteris richardii* and *C. pteridoides*, and of incipient speciation, as observed in the overhang ecotype/ecospecies of *Asplenium trichomanes*, will become more numerous.

The extensive investigations carried out on European *Asplenium*, apart from elucidating interspecific relationships, have revealed, and to some extent elucidated, different levels of intraspecific differentiation in *A seelosii, sensu lato*, in the *A. foresiense/macedonicum* complex, the *A. haussknechtii/lepidum* complex, and in the *A. trichomanes* complex.

The recent investigations carried out in the isolated genus *Ceratopteris*, though as yet inevitably incomplete, suggest that this morphologically unique group may be distinct amongst fern genera in its biosystematic structure, in that all of its extant taxa may prove to be genetically very closely related, with little development of isolating barriers other than that provided by polyploidy.

The genetic study of reproductive mechanisms and breeding systems in ferns is still in an early stage of development. The concept that, in response to the potential problems of genetic fitness implicit in the propensity present in the fern life-cycle for the production of completely homozygous sporophytes, ferns have evolved a genetic system that permits a substantial level of homoeologous pairing within a diploidized meiosis is clearly highly controversial. The available evidence (which, since it is as yet all obtained from the unique genus *Ceratopteris*, it would be in any case hazardous to extrapolate to ferns in general) does not demonstrate levels of homoeologous exchange beyond that generally accepted as possible in flowering plants, which do not have to contend with this problem. It is much more probable that no special meiotic system exists in ferns, and that the problem of homozygosity is solved by the conventional means of elimination of unfit forms by natural selection.

The special features of the fern life-cycle, i.e. the independence of the gametophyte etc., are responsible for the fact that apomixis, though an important mode of reproduction amongst ferns, does not show the same diversity of mechanisms such as are known in the flowering plants, only two different systems of sporangial apomixis having been demonstrated. Moreover, apomixis is apparently always obligate, there being no development known in the ferns comparable to the facultatively apomictic systems so important in some angiosperm groups.

ACKNOWLEDGEMENTS

My grateful and appreciative thanks are owed to my colleagues Alan Wesley and Dr Chris Hill for their advice, criticism and much valuable assistance with regard to the palaeobotanical literature. I also wish to thank Jim Crabbe and Clive Jermy for help in correlating their list of fern genera with that of Copeland, Professor Irene Manton and Dr Arthur Sledge respectively for help in tracing cytogenetic and systematic literature, Dr Stanley Walker and Mrs Mary Gibby for assistance with *Dryopteris*, also Professor R. E. Holttum and Dr Trevor Walker for valuable comments and assistance with literature relating to aspects of Section IV. None of these persons should be held responsible for any of the opinions expressed in this review.

REFERENCES

Abraham, A., Ninan, C. A. and Matthew, P. M. (1962). *J. Indian bot. Soc.* **51**, 339-421.
Allen, R. F. (1914). *Trans. Wis. Acad. Sci. Arts Lett.* **17**, 1-56.
Alvin, K. L. (1968). *J. Linn. Soc. (Bot.)* **61**, 87-92.
Alvin, K. L. (1971). *Mém. Mus. r. Hist. nat. Belg.* **166**, 33pp., ix pls.
Alvin, K. L. (1975). *Palaeontology* **17**, 587-598.
Alston, A. H. G. (1956). *Taxon* **5**, 23-25.
Arnold, C. A. (1964). *Mem. Torrey bot. Club* **21**, 58-66.
Arnold, C. A. and Daugherty, L. H. (1963). *Contrib. Mus. Paleont. Univ. Mich.* **18**, 205-227.
Arnold, C. A. and Daugherty, L. H. (1964). *Contrib. Mus. Paleont. Univ. Mich.* **19**, 65-88.
Babcock, E. B. and Jenkins, J. A. (1943). *Univ. Calif. Publs Bot.* **18**, 241-292.
Bailey, I. W. and Swamy, B. G. L. (1951). *Am. J. Bot.* **38**, 373-379.
Ball, O. M. (1931). *Bull. Tex. agric. mech. Coll. ser.* **4** 2 (5).
Barber, H. N. (1954). *Victorian Nat.* **2**, 97-99.
Barber, H. N. (1957). *Proc. Linn. Soc. N.S.W.* **82**, 201-208.

Becherer, A. (1962). *Bauhinia* 2, 55-58.
Beddome, R. H. (1873). "The Ferns of Southern India", 2nd Edn. Higginbotham, Madras.
Bell, P. R. (1960). *New Phytol.* 59, 53-59.
Benoit, P. M. (1966). *Br. Fern Gaz.* 9, 277-282.
Berry, E. W. (1919), *Prof. pap. U.S. geol. Surv.* 112.
Bierhorst, D. W. (1968a). *Am. J. Bot.* 55, 87-108.
Bierhorst, D. W. (1968b). *Phytomorphology* 18, 232-268.
Bierhorst, D. W. (1969). *Am. J. Bot.* 56, 160-174.
Bierhorst, D. W. (1971). "Morphology of Vascular Plants". Macmillan, (New York).
Bierhorst, D. W. (1973). *In* "The Phylogeny and Classification of the Ferns" (Eds A. C. Jermy *et al.*). *Bot. J. Linn. Soc.* 67 *Suppl.* 1, 45-58.
Bierhorst, D. W. (1975). *Am. J. Bot.* 62, 448-456.
Bir, S. S. (1965). *Caryologia* 18, 107-115.
Bir, S. S. (1971). *Nucleus, Calcutta.* 14, 56-62.
Bir, S. S. (1973). *In* "Glimpses in Plant Research", Vol. 1, pp. 28-119. Vikas Publishing House, Delhi, and Indian Book Society; Lucknow.
Blasdell, R. F. (1963). *Mem. Torrey bot. Club* 21 (4), 102 pp.
Borrill, M. (1961). *J. Linn. Soc. (Bot.)* 56, 441-452.
Bose, M. N. and Sukh Dev (1961). *Palaeobotanist* 8, 57-64.
Bouharmont, J. (1966). *Bull. Jard. bot. État Brux.* 36, 383-391.
Bouharmont, J. (1972a). *Bull. Jard. bot. natn. Belg.* 42, 375-383.
Bouharmont, J. (1972b). *Br. Fern Gaz.* 10, 237-240.
Bouharmont, J. (1972c). *Chromosomes Today* 3, 253-258.
Boureau, E. (1970). "Traité de Paléobotanique", Vol. 4: I. Filicophyta. Masson, Paris.
Bower, F. O. (1923). "The Ferns", Vol. 1: Analytical examination of the criteria of comparison. Cambridge University Press.
Bower, F. O. (1926). "The Ferns", Vol. 2: The Eusporangiatae and other relatively primitive ferns. Cambridge University Press.
Bower, F. O. (1928). "The Ferns", Vol. 3: The leptosporangiate ferns. Cambridge University Press.
Bradshaw, M. E. (1963). *Watsonia* 5, 304-320.
Bradshaw, A. D. (1971). *In* "Ecological Genetics and Evolution" (Ed. R. Creed), pp. 20-50. Blackwell Scientific Publications, Oxford and Edinburgh.
Braithwaite, A. F. (1964a). *New Phytol.* 63, 293-305.
Braithwaite, A. F. (1964b). "A cytotaxonomic investigation on the Asplenium aethiopicum complex in Africa". Unpublished Ph.D. thesis, University of Leeds.
Braithwaite, A. F. (1969). *Br. Fern Gaz.* 10, 81-91.
Braithwaite, A. F. (1972a). *Jl S. Afr. Bot.* 38, 1-7.
Braithwaite, A. F. (1972b). *Jl S. Afr. Bot.* 38, 9-27.
Braithwaite, A. F. (1973). *Br. Fern Gaz.* 10, 293-303.
Braithwaite, A. F. (1975). *Bot J. Linn. Soc.* 71, 167-189.
Britton, D. M. (1953). *Am. J. Bot.* 40, 575-583.
Britton, D. M. (1965). *Mich. Bot.* 4, 3-9.
Britton, D. M. (1964). *Can. J. Bot.* 42, 1349-1354.
Britton, D. M. and Jermy, A. C. (1974). *Can. J. Bot.* 52, 1923-1926.
Britton, D. M. and Widén, C.-J. (1974). *Can. J. Bot.* 52, 627-638.
Brögger, A. (1960). *Blyttia* 18, 33-48.

Brown, D. F. M. (1964). "A monographic study of the fern genus *Woodsia*". Beihefte zur *Nova Hedwigia*, Heft 16, 154 pp., 40 pls. J. Cramer, Weinheim.

Brown, K. S. Jr., Sheppard, P. M. and Turner, J. R. G. (1974). *Proc. R. Soc. Lond. B* **187**, 369-378.

Brown, R. A. (1940). *J. Wash. Acad. Sci.* **30**, 344-356.

Brown, R. W. (1950). *Prof. pap. U.S. geol. Surv.* **221-D**, 45-66.

Brown, R. W. (1962). *Prof. pap. U.S. geol. Surv.* **375**, 119 pp.

Brownlie, G. (1957). *New Phytol.* **56**, 207-209.

Brownlie, G. (1958). *Trans. R. Soc. N.Z.* **85**, 212-216.

Brownlie, G. (1961). *Trans. R. Soc. N.Z.* **88**, 1-4.

Brownlie, G. (1965). *Pacif. Sci.* **19**, 493-497.

Brownsey, P. J. (1973). "An evolutionary study of the *Asplenium lepidum* complex". Unpublished Ph.D. thesis, University of Leeds.

Brownsey, P. J. (1975). *N.Z. Jl Bot.* **13**, 355-360.

Brownsey, P. J. (1976a). *New Phytol.* **76**, 523-542.

Brownsey, P. J. (1976b). *Bot. J. Linn. Soc.* **72**, 235-267.

Brownsey, P. J. (1977). *N.Z. Jl Bot.* **15**, in press.

Cadevall, J. and Font Quer, P. (1936). "Flora de Catalunya", Vol. 6, 445 pp., Institut d'Estudis Catalans. Barcelona.

Callé, J., Lovis, J. D. and Reichstein, T. (1975). *Candollea* **30**, 189-201.

Campbell, D. H. (1918). "The Structure and Development of Mosses and Ferns", 3rd Edn. London.

Chambers, T. C. (1955). *Nature, Lond.* **175**, 215.

Chandler, M. E. J. (1962). "The Lower Tertiary Floras of Southern England", Vol. 2: Flora of the pipe-clay series of Dorset (Lower Bagshot). British Museum (Nat. Hist.), London.

Chesters, K. I. M., Gnauck, F. R. and Hughes, N. F. (1967). *In* "The Fossil Record" (Eds W. B. Harland *et al.*), pp. 269-288.

Chiarugi, A. (1960). *Caryologia* **13**, 27-150.

Ching, R. C. (1940). *Sunyatsenia* **5**, 201-268.

Ching, R. C. (1966). *Acta Phytotax. Sin.* **11**, 25-29.

Chinnock, R. J. (1975). *N.Z. Jl Bot.* **13**, 743-768.

Chopra, N. (1960). *Castanea* **25**, 116-119.

Christensen, C. (1905/06). "Index Filicum". Hagerup, Copenhagen.

Christensen, C. (1913). "Index Filicum, Supplementum (1906-1912)". Hagerup, Copenhagen.

Christensen, C. (1917). "Index Filicum. Supplementum Preliminaire pour les années 1913, 1914, 1915, 1916". Bogtrykkeri, Copenhagen.

Christensen, C. (1934). "Index Filicum. Supplementum Tertium pro annis 1917-1933". Hagerup, Copenhagen.

Christensen, C. (1938). *In* "Manual of Pteridology" (Ed. F. Verdoorn), pp. 522-550. Nijhoff, The Hague.

Clapham, A. R., Tutin, T. G. and Warburg, E. F. (1962). "Flora of the British Isles", 2nd Edn. Cambridge University Press.

Clark, W. B., Bibbins, A. B. and Berry, E. W. (1911). "Maryland Geological Survey. Lower Cretaceous", *Md. geol. Surv. gen. Ser.* Johns Hopkins Press, Baltimore.

Clausen, J., Keck, D. D. and Hiesey, W. M. (1945). *Publs Carnegie Instn* **564**, 174 pp.

Copeland, E. B. (1938). *Philipp. J. Sci.* **67**, 1-110, 11 pls.

Copeland, E. B. (1947). "Genera Filicum". Chronica Botanica, Waltham, Mass.

Couper, R. A. (1953). *N.Z. geol. Surv. pal. Bull.* 22.
Couper, R. A. (1958). *Palaeontographica B* 103, 75-179.
Crabbe, J. A., Jermy, A. C. and Mickel, J. T. (1975). *Fern Gaz.* 11, 141-162.
Cronquist, A. (1974). *In* "Symposium on Origin and Phytogeography of Angiosperms", pp. 19-24. Birbal Sahni Institute of Palaeobotany Special Publication No. 1, Lucknow.
Daigobo, S. (1974). *J. Jap. Bot.* 49, 371-377.
Darlington, C. D. (1963). "Chromosome Botany and the Origins of Cultivated Plants", 2nd Edn. Allen and Unwin, London.
de la Sota, E. R. (1973). *In* "The Phylogeny and Classification of the Ferns" (Eds A. C. Jermy *et al.*). *Bot. J. Linn. Soc.* 67, *suppl.* 1, 229-244.
de Wet, J. M. J. and Harlan, J. R. (1970). *Evolution, Lancaster, Pa.* 23, 270-277.
de Wet, J. M. J. and Harlan, J. R. (1972). *Taxon* 21, 67-70.
Diels, L., *in* Engler, A. and Prantl, K. (1899/1902). "Die naturlichen Pflanzenfamilien", Vol. 1 (4): Pteridophyta. Engelmann, Leipzig.
Döpp, W. (1932). *Planta* 17, 86-152.
Döpp, W. (1938). *In* "Manual of Pteridology" (Ed. F. Verdoorn), pp. 233-283. Nijhoff, The Hague.
Döpp, W. (1941). *Ber. dt. bot. Ges.* 59, 423-426.
Döpp, W. (1949). *Ber. dt. bot. Ges.* 62, 61-68.
Döpp, W. (1955). *Planta* 46, 70-91.
Döpp, W. (1961). *Planta* 57, 8-12.
Doyle, J. A. (1969). *J. Arnold Arbor.* 50, 1-35.
Ehrendorfer, F. (1959). *Cold Spring Harb. Symp. quant. Biol.* 24, 141-152.
Emberger, L. (1944). "Les Plantes Fossiles dans leurs Rapports avec les Végétaux Vivants". Masson, Paris.
Emberger, L. (1968). "Les Plantes Fossiles dans leurs Rapports avec les Végétaux Vivants", 2nd Edn. Masson, Paris.
Emmott, J. I. (1964). *New Phytol.* 63, 306-318.
Engler, A. and Prantl, K. (1898/1902). "Die naturlichen Pflanzenfamilien", Vol, 1 (4): Pteridophyta. Engelmann, Leipzig.
Erdtman, G. and Sorsa, P. (1971). "Pollen and Spore Morphology/Plant Taxonomy", Vol. 4: Pteridophyta. Almqvist and Wiksell, Stockholm.
Evans, A. M. (1964). *Science, N.Y.* 143, 261-263.
Evans, A. M. (1968). *Am. Fern J.* 58, 169-175.
Evans, A. M. (1969a). *Ann. Mo. bot. Gdn* 55, 193-293.
Evans, A. M. (1969b). *Bioscience* 19, 708-711.
Fabbri, F. (1963). *Caryologia* 16, 237-335.
Fabbri, F. (1965). *Caryologia* 18, 675-731.
Fagerlind, F. (1937). *Acta Horti Bergiani* 11, 195-470.
Farrar, D. R. (1967). *Science, N.Y.* 155, 1266-1267.
Farrar, D. R. and Wagner, W. H. Jr. (1968). *Bot. Gaz.* 129, 210-219.
Favarger, C. and Contandriopoulos, J. (1961). *Ber. schweiz. bot. Ges.* 71, 384-408.
Fraser-Jenkins, C. R. (1976). *Fern Gaz.* 11, 263-267.
Fraser-Jenkins, C. R. and Corley, H. V. (1972). *Br. Fern Gaz.* 10, 221-231.
Fraser-Jenkins, C. R., Reichstein, T. and Vida, G. (1975). *Fern Gaz.* 11, 177-198.
Friebel, H. (1933). *Beitr. Biol. Pfl.* 21, 167-210.
Foster, A. S. and Gifford, E. M., Jr. (1974). "Comparative Morphology of Vascular Plants", 2nd Edn. W. H. Freeman, San Francisco.
Ganders, F. R. (1972). *Bot. J. Linn. Soc.* 65, 211-221.

Gardner, J. S. and Ettingshausen, Baron C. (1879/1882). "A Monograph of the British Eocene Flora", Vol. 1: Filices. *Palaeontogr. Soc.* (*Monogr.*). Palaeontographical Society, London.
Gastony, G. J. and Baroutsis, J. G. (1975). *Am. Fern J.* **65**, 71-75.
Geitler, L. (1937). *Z. indukt. Abstamm. u. VererbLehre* **73**, 182-197.
Ghatak, J. and Manton, I. (1971). *Br. Fern Gaz.* **10**, 183-192.
Goebel, K. (1915/1918). "Organographie der Pflanzen", Vol. 2 (2). Jena.
Gopal-Iyengar, A. R. (1957). *Proc. Indian Sci. Congr.* **44** (3), 249.
Graham, A. (1963). *Am. J. Bot.* **50**, 921-936.
Gustaffson, A. (1946). *Acta Univ. Lund.* **42**, 1-66.
Gustaffson, A. (1947a). *Acta Univ. Lund.* **43**, 71-178.
Gustaffson, A. (1947b). *Acta Univ. Lund.* **44**, 183-370.
Haffer, J. (1969). *Science, N.Y.* **165**, 131-137.
Haga, T. (1953). *Cytologia* **18**, 50-66.
Hagerup, O. (1932). *Hereditas* **16**, 19-40.
Harborne, J. B., Williams, C. A. and Smith, D. M. (1973). *Biochem. Systematics* **1**, 51-54.
Harland, W. B. *et al.* (Eds) (1967). "The Fossil Record". Geological Society, London.
Harris, T. M. (1961). "The Yorkshire Jurassic Flora", Vol. 1: Thallophyta-Pteridophyta. British Museum (Nat. Hist.), London.
Harris, T. M. (1973). *In* "The Phylogeny and Classification of the Ferns" (Eds A. C. Jermy *et al.*). *Bot. J. Linn. Soc.* **67**, *Suppl.* 1, pp. 41-44.
Harris, W. F. (1955). *Bull. N.Z. Dep. Scient. ind. Res.* **116**, 186 pp.
Hennipman, E. (1968). *Blumea* **16**, 105-108.
Hickok, L. G. and Klekowski, E. J. Jr. (1973). *Am. J. Bot.* **60**, 1010-1022.
Hickok, L. G. and Klekowski, E. J. Jr. (1974). *Evolution, Lancaster, Pa.* **28**, 439-446.
Hickok, L. G. and Klekowski, E. J. Jr. (1975). *Am. J. Bot.* **62**, 560-569.
Hill, C. and van Konijnenburg van Cittert, J. H. A. (1973). *Naturalist, Hull* **1973** (925), 59-63.
Hirabayashi, H. (1968). *J. Jap. Bot.* **43**, 157-159.
Hirabayashi, H. (1969). *J. Jap. Bot.* **44**, 113-119.
Hirabayashi, H. (1970). *J. Jap. Bot.* **45**, 11-19.
Hirabayashi, H. (1974). "Cytogeographic Studies on Dryopteris of Japan", Harashobo, Tokyo.
Hollick, A. (1906). *Monogr. U.S. geol. Surv.* **50**, 219 pp.
Hollick, A. (1936). *Prof. pap. U.S. geol. Surv.* **182**, 185 pp.
Holttum, R. E. (1947). *J. Linn. Soc.* (*Bot.*) **53**, 123-158.
Holttum, R. E. (1949). *Biol. Rev.* **24**, 267-296.
Holttum, R. E. (1956). *Kew Bull.* **3**, 551-553.
Holttum, R. E. (1957). *Reinwardtia* **4**, 257-280.
Holttum, R. E. (1959a). "Flora Malesiana", Ser. 2. "Pteridophyta", Vol. 1 (1): Gleicheniaceae, pp. 1-36. Noordhoff, Groningen.
Holttum, R. E. (1959b). "Flora Malesiana", Ser. 2. "Pteridophyta", Vol. 1 (1): Schizaeaceae, pp. 37-61. Noordhoff, Groningen.
Holttum, R. E. (1963). "Flora Malesiana", Ser. 2. "Pteridophyta", Vol. 1 (2): Cyatheaceae, pp. 65-176. Noordhoff, Groningen.
Holttum, R. E. (1968). *Blumea* **16**, 87-95.
Holttum, R. E. (1969). *Blumea* **17**, 5-32.
Holttum, R. E. (1971). *Blumea* **19**, 2-52.
Holttum, R. E. (1973a). *In* "The Phylogeny and Classification of the Ferns" (Eds A. C. Jermy *et al.*). *Bot. J. Linn. Soc.* **67**, *Suppl.* 1, 1-10.

Holttum, R. E. (1973b). *In* "The Phylogeny and Classification of the Ferns" (Eds A. C. Jermy *et al.*). *Bot. J. Linn. Soc.* **67**, *Suppl.* 1, 173-189.
Holttum, R. E. (1974). *Jl S. Afr. Bot.* **40**, 123-168.
Holttum, R. E. (1975). *Kew Bull.* **30**, 327-343.
Holttum, R. E. and Sen, U. (1961). *Phytomorphology* **11**, 406-420.
Hooker, W. J. and Baker, J. G. (1865/1868). "Synopsis Filicum", Hardwicke, London.
Jain, S. K. and Bradshaw, A. D. (1966). *Heredity, Lond.* **21**, 407-441.
Jalas, J. and Suominen, J. (Eds) (1972). "Atlas Florae Europae", Vol. 1: Pteridophyta (Psilotaceae to Azollaceae). Societas Biologica Fennica Vanamo, Helsinki.
Jarrett, F. M., Manton, I. and Roy, S. K. (1968). *Kew Bull.* **22**, 475-480.
Jermy, A. C., Crabbe, J. A. and Thomas, B. A. (Eds) (1973). *Bot. J. Linn. Soc.* **67**, *Suppl.* 1, 284 pp. Academic Press, London and New York.
Jones, K. and Jopling, C. (1972). *Bot. J. Linn. Soc.* **65**, 129-162. .
Kawakami, S. (1970). *Bot. Mag. Tokyo* **83**, 74-81.
Kawakami, S. (1971). *Bot. Mag. Tokyo* **84**, 180-186.
Klekowski, E. J. Jr. (1969). *Bot. J. Linn. Soc.* **62**, 361-377.
Klekowski, E. J. Jr. (1970a). *Bot. J. Linn. Soc.* **63**, 153-169.
Klekowski, E. J. Jr. (1970b). *Bot. J. Linn. Soc.* **63**, 171-176.
Klekowski, E. J. Jr. (1970c). *Am. J. Bot.* **57**, 1122-1138.
Klekowski, E. J. Jr. (1972). *Evolution, Lancaster, Pa.* **26**, 66-73.
Klekowski, E. J. Jr. (1973a). *Am. J. Bot.* **60**, 146-154.
Klekowski, E. J. Jr. (1973b). *Am. J. Bot.* **60**, 535-544.
Klekowski, E. J. Jr. and Baker, H. G. (1966). *Science, N.Y.* **153**, 305-307.
Klekowski, E. J. Jr. and Hickok, L. G. (1974). *Am. J. Bot.* **61**, 422-432.
Klekowski, E. J. Jr. and Lloyd, R. M. (1968). *J. Linn. Soc. (Bot.)* **60**, 315-324.
Knobloch, I. W. (1966). *Am. Fern J.* **56**, 163-167.
Knobloch, I. W. and Correll, D. S. (1962). "Ferns and Fern Allies of Chihuahua, Mexico" *Publ. Texas Res. Foundation* **3**, 198 pp. Renner, Texas.
Knowlton, F. H. (1910). *Smithson. misc. Collns* **52** (1884), 489-496.
Knowlton, F. H. (1930). *Prof. pap. U.S. geol. Surv.* **155**, 135 pp.
Kobayashi, T. and Yosida, T. (1944). *Japan J. Geol. and Geogr.* **19**, 255-271.
Kramer, K. U. (1957). *Acta bot. neerl.* **6**, 97-290.
Kramer, K. U. (1971). "Flora Malesiana", Ser. 2. "Pteridophyta", Vol. 1 (3): Lindsaea group, pp. 177-254. Noordhoff, Groningen.
Kruckeberg, A. R. (1954). *Ecology* **35**, 267-274.
Kryshtofovich, A. N. (1958). *Paleobotanika* **3**, 7-70, pl. 1-13.
Kurita, S. (1963). *J. Coll. Arts Sci. Chiba Univ. Nat. Sci. Ser.* **4**, 43-52, 3 pls.
Kurita, S. (1965). *J. Jap. Bot.* **40**, 358-362, 14 pls.
Kurita, S. (1967). *Annual Report, Foreign Students' College, Chiba University* **2**, 41-56.
Kurita, S. (1971). *Contr. Lab. phylogenetic Bot., Foreign Students' Coll., Chiba Univ.* **50**, 41-43.
Kurita, S. (1976). *La Kromosomo* **2**, in press.
Kurita, S. and Ikebe, C. (1976). *J. Jap. Bot.* **51** (12), in press.
Kurita, S. and Nishida, M. (1965). *Bot. Mag. Tokyo* **78**, 461-473.
Lang, F. A. (1969). *Madroño* **20**, 53-60.
Lang, F. A. (1971). *Madroño* **21**, 235-254.
Larsen, K. (1960). *Danske Vidensk. Selsk. Biol. Skrifter* **11**, 1-60.
Lewis, D. (1943). *J. Genet.* **45**, 171-185.
Lewis, D. (1949). *Biol. Rev.* **24**, 472-496.
Lewis, K. R. and John, B. (1966). *Chromosoma (Berl.)* **18**, 287-304.

Litardière, R. dc (1921). *Cellule* **31**, 255-473.
Ljungdahl, H. (1924). *Svensk. Bot. Tidskr.* **18**, 279-291.
Lloyd, R. M. (1971). *Univ. Calif. Publs. Bot.* **61**, 86 pp.
Lloyd, R. M. (1973). *Am. Fern J.* **63**, 43-48.
Lloyd, R. M. (1974a). *Ann. Mo. bot. Gdn* **61**, 318-331.
Lloyd, R. M. (1974b). *Brittonia* **26**, 139-160.
Lloyd, R. M. and Lang, F. A. (1964). *Br. Fern Gaz.* **9**, 168-177.
Lorch, J. (1967). *Israel J. Bot.* **16**, 131-155 & pls. 163-180.
Löve, A. (1970). "Islenzk Ferdaflóra". Almenna Bókafélagid, Reykjavik.
Löve, A. and Kjellqvist, E. (1972). *Lagascalia* **2**, 23-35.
Löve, A. and Löve, D. (1943). *Hereditas* **29**, 145-163.
Löve, A. and Löve, D. (1971). *Naturaliste can.* **98**, 469-494.
Löve, A. and Solbrig, O. T. (1964). *Taxon* **13**, 99-110.
Lovis, J. D. (1955a). *In* "Species Studies in the British Flora" (Ed. J. E. Lousley). *B.S.B.I. Conf. Rep.* **4**, 99-103.
Lovis, J. D. (1955b). *Proc. bot. Soc. Br. Isl.* **1**, 389-390.
Lovis, J. D. (1958). *Nature, Lond.* **181**, 1085.
Lovis, J. D. (1963). *Br. Fern Gaz.* **9**, 110-113.
Lovis, J. D. (1964a). *Br. Fern Gaz.* **9**, 147-160.
Lovis, J. D. (1964b). *Nature, Lond.* **203**, 324-325.
Lovis, J. D. (1968a). *Br. Fern Gaz.* **10**, 13-20.
Lovis, J. D. (1968b). *Nature, Lond.* **217**, 1163-1165.
Lovis, J. D. (1970). *Br. Fern Gaz.* **10**, 153-157.
Lovis, J. D. (1973). *In* "The Phylogeny and Classification of the Ferns" (Eds A. C. Jermy *et al.*). *Bot. J. Linn. Soc.* **67**, Suppl. 1, 211-277, pls 2.
Lovis, J. D. (1975). *Fern Gaz.* **11**, 137-140.
Lovis, J. D. and Reichstein, T. (1964). *Br. Fern Gaz.* **9**, 141-146.
Lovis, J. D. and Reichstein, T. (1968a). *Naturwissenschaften* **55**, 117-120.
Lovis, J. D. and Reichstein, T. (1968b). *Bauhinia* **4**, 53-63.
Lovis, J. D. and Reichstein, T. (1969). *Ber. schweiz. bot. Ges.* **79**, 335-345.
Lovis, J. D. and Roy, S. K. (1964). *Nature, Lond.* **201**, 1348.
Lovis, J. D. and Shivas, M. G. (1954). *Proc. bot. Soc. Br. Isl.* **1**, 97.
Lovis, J. D. and Vida, G. (1969). *Br. Fern Gaz.* **10**, 53-67.
Lovis, J. D., Melzer, H. and Reichstein, T. (1966). *Bauhinia* **3**, 87-101.
Lovis, J. D., Sleep, A. and Reichstein, T. (1969). *Ber. schweiz. bot. Ges.* **79**, 369-376.
Lovis, J. D., Brownsey, P. J., Sleep, A. and Shivas, M. G. (1972). *Br. Fern Gaz.* **10**, 263-268.
Loyal, D. S. (1962). *Res. Bull. Panjab Univ. Sci.* **13**, 25-30.
Loyal, D. S. (1963). *Mem. Indian bot. Soc.* **4**, 22-29.
Mangenot, S. and Mangenot, G. (1962). *Revue Cytol. Biol. vég.* **25**, 411-447.
Manton, I. (1950). "Problems of Cytology and Evolution in the Pteridophyta". Cambridge University Press.
Manton, I. (1951). *Nature, Lond.* **167**, 37.
Manton, I. (1953). *Symp. Soc. exp. Biol.* **7**, 174-185.
Manton, I. (1954a). *Nature, Lond.* **173**, 453.
Manton, I. (1954b). *In* "Flora of Malaya" (Ed. R. E. Holttum), Vol. 2: Ferns of Malaya, pp. 623-627 (Appendix). Government Printing Office, Singapore.
Manton, I. (1958a). *J. Linn. Soc. (Bot.)* **56**, 73-92.
Manton, I. (1958b). *In* "Systematics of Today" (Ed. O. Hedberg). *Uppsala Univ. Arsskr.* **1958** (6), 104-112.

Manton, I. (1959). *In* "The Ferns and Fern Allies of West Tropical Africa" (Ed. A. H. G. Alston), pp. 75-81. Crown Agents, London.

Manton, I. (1961). *In* "A Darwin Centenary" (Ed. P. J. Wanstall). *B.S.B.I. Conf. Rep.* **6**, 105-119.

Manton, I. (1969). *Biol. J. Linn. Soc.* **1**, 219-222.

Manton, I. (1973). *In* "The Phylogeny and Classification of the Ferns" (Eds A. C. Jermy *et al.*). *Bot. J. Linn. Soc.* **67**, *Suppl.* 1, 257-263.

Manton, I. and Reichstein, T. (1961). *Ber. schweiz. bot. Ges.* **71**, 370-383.

Manton, I. and Reichstein, T. (1965). *Bauhinia* **2**, 307-312.

Manton, I. and Shivas, M. G. (1953). *Nature, Lond.* **172**, 410.

Manton, I. and Sledge, W. A. (1954). *Phil. Trans. R. Soc. B* **238**, 127-185.

Manton, I. and Vida, G. (1968). *Proc. R. Soc. B* **170**, 361-379.

Manton, I. and Walker, S. (1954). *Ann. Bot.* **18**, 377-383.

Manton, I., Roy, S. K. and Jarrett, F. M. (1966). *Kew Bull.* **18**, 553-565.

Manton, I., Ghatak, J. and Sinha, B. M. B. (1967). *J. Linn. Soc. (Bot.)* **60**, 223-235.

Manton, I., Sinha, B. M. B. and Vida, G. (1970). *Bot. J. Linn. Soc.* **63**, 1-21.

Matsuura, H. (1950). *Cytologia* **16**, 48-57.

Masuyama, S. (1975). *J. Jap. Bot.* **50**, 105-114.

Mehra, P. N. (1961). *Proc. Indian Sci. Congr.* **48** (2), 1-24.

Mehra, P. N. and Bir, S. S. (1960). *Am. Fern J.* **50**, 276-295.

Mehra, P. N. and Singh, G. (1957). *J. Genet.* **55**, 379-393.

Mehra, P. N. and Verma, S. C. (1960). *Caryologia* **13**, 613-650.

Mettenius, G. (1856). "Filices Horti Botanici Lipsiensis". Voss, Leipzig.

Meyer, D. E. (1952). *Biblthca. bot.* **123**, 34 pp.

Meyer, D. E. (1959). *Willdenowia* **2**, 214-217.

Meyer, D. E. (1967). *Ber. dt. bot. Ges.* **80**, 28-39.

Meyer, D. E. (1968). *Ber. dt. bot. Ges.* **81**, 92-106.

Mickel, J. T. (1962). *Iowa St. J. Sci.* **36**, 349-482.

Mickel, J. T. (1973). *In* "The Phylogeny and Classification of the Ferns" (Eds A. C. Jermy *et al.*). *Bot. J. Linn. Soc.* **67**, *Suppl.* 1, 135-144.

Mickel, J. T., Wagner, W. H. Jr. and Chen, K. L. (1966). *Caryologia* **19**, 95-102.

Miner, E L. (1934). *Am. J. Bot.* **21**, 261-264.

Mitui, K. (1965). *J. Jap. Bot.* **40**, 117-124.

Mitui, K. (1968). *Sci. Rep. Tokyo Kyoiko Daig., Sect. B* **13**, 285-333.

Mitui, K. (1970). *J. Jap. Bot.* **45**, 84-90.

Mitui, K. (1971). *J. Jap. Bot.* **46**, 83-96.

Mitui, K. (1973). *J. Jap. Bot.* **48**, 247-253.

Mitui, K. (1975). *Bull. Nippon Dental Coll., Gen. Education* **4**, 221-271.

Mitui, K. (1976). *Bull. Nippon Dental Univ., Gen. Education* **5**, 133-140.

Moore, R. J. (1973). *Regnum veg.* **90**, 539 pp.

Moore, R. J. (1974). *Regnum veg.* **91**, 108 pp.

Morton, C. V. (1968). *Contr. U.S. natn. Herb.* **38**, 153-214.

Morton, J. K. (1961). *In* "Recent Advances in Botany", Vol. 1, pp. 900-903. University of Toronto Press, Toronto.

Morton, J. K. (1962). *J. Linn. Soc. (Bot.)* **58**, 231-283.

Morton, J. K. (1966). *Chromosomes Today* **1**, 73-76.

Morton, J. K. (1967). *J. Linn. Soc. (Bot.)* **60**, 167-221.

Morzenti, V. M. (1962). *Am. Fern J.* **52**, 69-78.

Morzenti, V. M. (1966). *Am. Fern J.* **56**, 167-177.

Morzenti, V. M. (1967). *Am. J. Bot.* **54**, 1061-1068.

Muller, J. (1970). *Biol. Rev.* **45**, 417-450.

Mulligan, G. A., Cinq-Mars, L. and Cody, W. J. (1972). *Can. J. Bot.* **50**, 1295-1300.
McNeilly, T. (1968). *Heredity, Lond.* **23**, 99-108.
Nakai, T. (1950). *Bull. Nat. Sci. Mus. Tokyo* **29**, 1-71.
Navashin, M. S. (1932). *Z. indukt. Abstamm. u. VererbLehre* **63**, 224-231.
Nayar, B. K. (1970). *Taxon* **19**, 229-236.
Newton, W. C. F. and Pellew, C. (1929). *J. Genet.* **20**, 405-467.
Niizeki, S., Nishida, M. and Kurita, S. (1963). *J. Jap. Bot.* **38**, 144-148.
Ninan, C. A. (1956a). *Cellule* **57**, 307-318.
Ninan, C. A. (1956b). *Curr. Sci.* **25**, 161-162.
Ninan, C. A. (1956c). *J. Indian bot. Soc.* **35**, 233-239.
Okabe, S. (1929). *Sci. Rep. Tohoku Univ. (Ser. 4)* **4**, 373-380.
Oishi, S. (1940). *J. Fac. Sci. Hokkaido Univ. Series IV, Geol. & Min.* **5**, 123-480, 48 pls.
Ornduff, R. (1967). *Regnum veg.* **50**, 128 pp.
Ornduff, R. (1968). *Regnum veg.* **55**, 126 pp.
Page, C. N. (1973). *Monographiae biol. Canarienses* **4**, 83-88.
Panigrahi, G. (1962). *Nucleus, Calcutta* **5**, 53-64.
Panigrahi, G. (1963). *Proc. natn. Inst. Sci. India B* **29**, 383-395.
Panigrahi, G. (1965). *Am. Fern J.* **48**, 136-142.
Panigrahi, G. and Manton, I. (1958). *J. Linn. Soc. (Bot.)* **55**, 729-743.
Panigrahi, G. and Patnaik, S. N. (1963). *Am. Fern J.* **53**, 145-148.
Panigrahi, G. and Patnaik, S. N. (1964a). *Mem. Indian Bot. Soc.* **4**, 6-19, 2 pls.
Panigrahi, G. and Patnaik, S. N. (1964b). *J. Indian bot. Soc.* **43**, 311-321.
Parkin, J. (1923). *Proc. Linn. Soc. Lond.* **153**, 51-64.
Patnaik, S. N. and Panigrahi, G. (1963). *Am. Fern J.* **53**, 40-46.
Perring, F. H. (Ed.) (1968). "Critical Supplement to the Atlas of the British Flora". Thomas Nelson, London.
Phillips, J. (1875). "Illustrations of the Geology of Yorkshire: or, a description of the strata and organic remains" (Ed. R. Etheridge), 3rd Edn. Part 1: The Yorkshire Coast. Murray, London.
Pichi-Sermolli, R. E. G. (1958). *In* "Systematics of Today" (Ed. O. Hedberg). *Uppsala Univ. Arsskr.* **1958** (6), 70-90.
Pichi-Sermolli, R. E. G. (1970). *Webbia* **25**, 219-297.
Pichi-Sermolli, R. E. G. (1973). *In* "The Phylogeny and Classification of the Ferns" (Eds A. C. Jermy *et al.*). *Bot. J. Linn. Soc.* **67**, *Suppl.* **1**, 11-40 & 19 pls.
Pichi-Sermolli, R. E. G. *et al.* (1965). *Regnum veg.* **37**, 370 pp.
Pignataro, L. D. G. (1971). *Atti Ist. bot. Univ. Lab. crittogam. Pavia (Ser. 6)* **7**, 29-31.
Platt, J. R. (1964). *Science, N.Y.* **146**, 347-353.
Pontecorvo, G. (1959). "Trends in Genetic Analysis". Columbia University Press, New York.
Pontecorvo, G. and Kafer, E. (1958). *Adv. Genet.* **9**, 71-104.
Prance, G. T. (1973). *Acta Amazonica* **3**, 5-28.
Presl, K. B. (1843). *Abh. K. bohm. Ges. Wiss.* **3**, 93-162.
Rao, R. S., Kammathy, R. V. and Raghavan, R. S. (1968). *J. Linn. Soc. (Bot.)* **60**, 357-380.
Reichstein, T., Lovis, J. D., Greuter, W. and Zaffran, J. (1973). *Ann. Mus. Goulandris* **1**, 133-163.
Riley, R. (1960). *Heredity, Lond.* **15**, 407-429.
Riley, R. and Chapman, V. (1958). *Nature, Lond.* **182**, 713-715.

Riley, R. and Chapman, V. (1964). *Wheat Inform. Serv.* **17**, 12-15.
Riley, R. and Law, C. N. (1965). *Adv. Genet.* **13**, 57-114.
Riley, R., Chapman, V. and Kimber, G. (1959). *Nature Lond.* **183**, 1244-1246.
Riley, R., Chapman, V. and Kimber, G. (1960). *Nature, Lond.* **186**, 259-260.
Rivas-Martinez, S. (1967). *Bull. Jard. bot. natn. Belg.* **37**, 329-334.
Rothmaler, W. (1963). "Exkursionflora von Deutschland", Vol. 4, Kritischer Erganzungsband.
Roberts, R. H. (1966). *Br. Fern Gaz.* **9**, 283-287.
Roberts, R. H. (1970). *Watsonia* **8**, 121-134.
Roy, S. K. and Holttum, R. E. (1965a). *Am. Fern J.* **55**, 35-37.
Roy, S. K. and Holttum, R. E. (1965b). *Am. Fern J.* **55**, 154-158.
Roy, S. K. and Holttum, R. E. (1965c). *Am. Fern J.* **55**, 158-164.
Roy, S. K. and Manton, I. (1965). *New Phytol.* **64**, 286-292.
Roy, S. K. and Manton, I. (1966). *J. Linn. Soc. (Bot.)* **59**, 343-347.
Sears, E. R. and Okamoto, M. (1959). *Int. Congr. Genet. 10th, Montreal, 1958* **2**, 258-259.
Seward, A. C. (1900). "Catalogue of the Mesozoic Plants in the Department of Geology, British Museum (Natural History)", Vol. 3: The Jurassic Flora. I: The Yorkshire Coast. British Museum (Nat. Hist.), London.
Seward, A. C. (1904). *Rec. geol. Surv. Victoria* **1**, 155-210.
Seward, A. C. (1910). "Fossil Plants", Vol. 2. Cambridge University Press.
Seward, A. C. (1926). *Phil. Trans. R. Soc. B* **215**, 57-176.
Seward, A. C. (1933). "Plant Life Through the Ages", 2nd Edn. Cambridge University Press.
Seward, A. C. and Holttum, R. E. (1921). *Bull. geol. Surv. Sth. Rhod.* **8**, 39-45.
Seward, A. C. and Holttum, R. E. (1922). *Q. Jl geol. Soc. Lond.* **78**, 271-277.
Seward, A. C. and Holttum, R. E. (1924). *In* "Tertiary and Post-tertiary Geology of Mull, Loch Aline and Oban" (Eds E. B. Bailey *et al.*), pp. 67-90. *Mem. Geol. Surv. U.K.* (Scotland). HMSO, Edinburgh.
Sharma, B. D. (1971). *Palaeontographica B* **133**, 61-71, pls. 19-20.
Shivas, M. G. (1961a). *J. Linn. Soc. (Bot.)* **58**, 13-25.
Shivas, M. G. (1961b). *J. Linn. Soc. (Bot.)* **58**, 27-38.
Shivas, M. G. (1969). *Br. Fern Gaz.* **10**, 68-80.
Sinha, B. M. B. and Manton, I. (1970). *Bot. J. Linn. Soc.* **63**, 247-264.
Sleep, A. (1966). "Some cytotaxonomic problems in the fern genera Asplenium and Polystichum". Unpublished Ph.D. thesis, University of Leeds.
Sleep, A. (1967). *Br. Fern Gaz.* **9**, 321-329.
Sleep, A. and Reichstein, T. (1967). *Bauhinia* **3**, 299-314.
Sleep, A., Scannell, M. J. P., Synnott, D., McClintock, D. and Reichstein, T. (1977). *Proc. R. Ir. Acad.* In press.
Smith, A. R. (1971). *Univ. Calif. Publs Bot.* **59**, 136 pp.
Smith, A. R. (1974). *Am. Fern J.* **64**, 83-95.
Smith, D. M. and Levin, D. A. (1963). *Am. J. Bot.* **50**, 952-958.
Smith, D. M., Bryant, T. R. and Tate, D. E. (1961). *Brittonia* **13**, 289-292.
Solbrig, O. T. (1968). *In* "Modern Methods in Plant Taxonomy" (Ed. V. H. Heywood). *B.S.B.I. Conf. Rep.* **10**, pp. 77-96.
Sorsa, V. (1963). *Hereditas* **49**, 337-344.
Sorsa, V. (1968). *Caryologia* **21**, 97-103.
Sorsa, V. and Widén, C.-J. (1968). *Hereditas* **60**, 273-293.
Sporne, K. R. (1970). "The Morphology of Pteridophytes", 3rd Edn. Hutchinson, London.
Stebbins, G. L. (1947). *Adv. Genet.* **1**, 403-429.

Stebbins, G. L. (1950). "Variation and Evolution in Plants". Columbia University Press, New York.
Stebbins, G. L. (1971). "Chromosomal Evolution in Higher Plants". Edward Arnold, London.
Stebbins, G. L. and Zohary, D. (1959). *Univ. Calif. Publs Bot.* **31**, 40 pp.
Steil, W. N. (1919). *Ann. Bot.* **33**, 109-132, 6 pls.
Stern, C. (1936). *Genetics* **21**, 625-730.
Stokey, A. G. (1945). *Bot. Gaz.* **106**, 402-411.
Stokey, A. G. and Atkinson, L. R. (1956). *Phytomorphology* **6**, 249-261.
Stone, W. S. (1962). *Univ. Texas Publ.* **6205**, 507-538.
Surange, K. R. (1966). "Indian Fossil Pteridophytes". Botanical Monograph 4. Council of Scientific and Industrial Research, New Delhi.
Swanson, C. R. (1940). *Genetics* **25**, 438-465.
Tagawa, M. (1959). "Coloured Illustrations of the Japanese Pteridophyta". Hoikusha, Osaka, Japan.
Takei, M. (1969). *Bot. Mag. Tokyo* **82**, 482-487.
Takhtajan, A. (1969). "Flowering Plants. Origin and Dispersal" (Transl. C. Jeffrey). Oliver and Boyd, Edinburgh.
Tarr, R. S. and Butler, B. S. (1909). *Prof. pap. U.S. geol. Surv.* **64**, 183 pp.
Tattersall, J. A. (1961). *Ann. Mag. nat. Hist. (Ser. 13)* **4**, 349-352, 12 pls.
Tatuno, S. and Kawakami, S. (1969). *Bot. Mag. Tokyo* **82**, 436-444.
Tatuno, S. and Okada, H. (1970). *Bot. Mag. Tokyo* **83**, 202-210.
Tatuno, S. and Takei, M. (1969a). *Bot. Mag. Tokyo* **82**, 121-129.
Tatuno, S. and Takei, M. (1969b). *Bot. Mag. Tokyo* **82**, 403-408.
Tatuno, S. and Takei, M. (1970). *Bot. Mag. Tokyo* **83**, 67-73.
Tatuno, S. and Yoshida, H. (1966). *Bot. Mag. Tokyo* **79**, 244-252.
Tatuno, S. and Yoshida, H. (1967). *Bot. Mag. Tokyo* **80**, 130-138.
Tavel, F. von (1937). *Verh. schweiz. naturf. Ges.* pp. 153-154.
Taylor, R. L. and Brockman, R. P. (1966). *Can. J. Bot.* **44**, 1069-1103.
Taylor, T. N. and Mickel, J. T. (Eds) (1974). *Ann. Mo. bot. Gdn.* **61**, 307-482.
Teixeira, C. (1945). *Bolm. Soc. geol. Port.* **4** (3), 1-4, pls. 1-3.
Teixeira, C. (1948). "Flora Mesozoica Portuguesa", Vol. 1. Servico Geologico Portugal, Lisbon.
Thompson, J. McLean (1917). *Trans. R. Soc. Edin.* **52**, 157-165.
Tidwell, W. D., Rushforth, S. R. and Reveal, J. L. (1967). *Brigham Young Univ. geol. Stud.* **14**, 237-240.
Tischler, G. (1935). *Bot. Jahrb.* **67**, 1-36.
Tischler, G. (1955). *Cytologia* **20**, 101-118.
Tobgy, H. A. (1943). *J. Genet.* **45**, 67-111.
Tryon, A. F. (1961). *Rhodora* **63**, 91-102.
Tryon, A. F. (1964). *Am. J. Bot.* **51**, 939-942.
Tryon, A. F. (1970). *Contr. Gray Herb. Harv.* **200**, 54-189.
Tryon, A. F. and Feldman, L. J. (1975). *Can. J. Bot.* **53**, 2260-2273.
Tryon, A. F. and Vida, G. (1967). *Science, N.Y.* **156**, 1109-1110.
Tryon, A. F., Bautista, H. P. and Silva Araujo, I. da (1975). *Acta Amazonica* **5**, 35-43.
Tryon, R. (1962). *Contr. Gray Herb. Harv.* **191**, 91-107.
Tryon, R. (1970). *Contr. Gray Herb. Harv.* **200**, 1-53.
Tsai, J.-L. (1973). *J. Sci. & Eng.* [Taichung] **10**, 261-275.
Turesson, G. (1943). *Bot. Notiser* **1943**, 413-427.
Tutin, T. G. (1932). *Ann. Bot.* **46**, 503-508, pl. xvi.
Tutin, T. G. *et al.* (Eds) (1964). "Flora Europaea", Vol. 1: Lycopodiaceae to Platanaceae. Cambridge University Press.

Tutin, T. G. *et al.* (Eds) (1972). "Flora Europaea", Vol. 3: Diapensaceae to Myoporaceae. Cambridge University Press.

Upcott, M. (1939). *J. Genet.* **39**, 79-100.

van Cittert, J. H. A. (1966). *Acta bot. neerl.* **15**, 284-289.

van Cotthem, W. R. J. (1970). *Bull. Jard. bot. natn. Belg.* **40**, 81-239.

van Cotthem, W. R. J. (1973). *In* "The Phylogeny and Classification of the Ferns" (Eds A. C. Jermy *et al.*). *Bot. J. Linn. Soc.* **67**, *Suppl.* 1, pp. 59-71.

Vanzolini, P. E. and Williams, E. E. (1970). *Arq. Zool., Sao Paulo* **19**, 1-298.

Verdoorn, F. (Ed.) (1938). "Manual of Pteridology". Nijhoff, The Hague.

Verma, S. C. (1956). *Curr. Sci.* **25**, 398-399.

Verma, S. C. (1957). *Cytologia* **22**, 393-403.

Vessey, J. and Barlow, B. A. (1963). *Proc. Linn. Soc. N.S.W.* **88**, 301-306.

Vida, G. (1963). *Acta bot. hung.* **9**, 197-215.

Vida, G. (1964). *Acta bot. hung.* **11**, 281-285.

Vida, G. (1970). *Caryologia* **23**, 525-547.

Vida, G. (1972). *In* "Evolution in Plants" (Ed. G. Vida). *Symposia Biologica Hungarica* **12**, pp. 51-60. Akademiai Kiado, Budapest.

Vida, G. (1974). *Acta bot. hung.* **20**, 181-192.

Vida, G., Page, C. N., Walker, T. G. and Reichstein, T. (1970). *Bauhinia* **4**, 223-253.

Wace, N. M. and Dickson, J. H. (1965). *Phil. Trans. R. Soc. B* **249**, 273-360.

Wagner, W. H. Jr. (1954). *Evolution, Lancaster, Pa.* **8**, 103-118.

Wagner, W. H. Jr. (1955). *Rhodora* **57**, 219-240.

Wagner, W. H. Jr. (1956). *Am. Fern J.* **46**, 75-82.

Wagner, W. H. Jr. (1963a). *Am. Fern J.* **53**, 1-16.

Wagner, W. H. Jr. (1963b). *Castanea* **28**, 113-150.

Wagner, W. H. Jr. (1966). *Am. Fern J.* **56**, 3-17.

Wagner, W. H. Jr. (1968). *In* "Modern Methods in Taxonomy" (Ed. V. H. Heywood). *B.S.B.I. Conf. Rep.* **10**, pp. 113-138.

Wagner, W. H. Jr. (1971). *Virginia Polytechnic Institute and State University Research Division Monograph* **2**, 147-192.

Wagner, W. H. Jr. (1973a). *Am. Fern J.* **63**, 99-115.

Wagner, W. H. Jr. (1973b). *In* "The Phylogeny and Classification of the Ferns" (Eds A. C. Jermy *et al.*). *Bot. J. Linn. Soc.* **67**, *Suppl.* 1, 243-256.

Wagner, W. H. Jr. and Boydston, K. (1958). *Am. Fern J.* **48**, 146-159.

Wagner, W. H. Jr. and Boydston, K. (1961). *Brittonia* **13**, 286-289.

Wagner, W. H. Jr. and Darling, T. Jr. (1957). *Brittonia* **9**, 57-63.

Wagner, W. H. Jr. and Hagenah, D. J. (1956). *Rhodora* **58**, 79-87.

Wagner, W. H. Jr. and Hagenah, D. J. (1962). *Brittonia* **14**, 90-100.

Wagner, W. H. Jr. and Sharp, A. J. (1963). *Science, N.Y.* **142**, 1483-1484.

Wagner, W. H. Jr. and Wagner, F. S. (1965). *Proc. Rochester Acad. Sci.* **11**, 57-104.

Wagner, W. H. Jr. and Wagner, F. S. (1966). *Castanea* **31**, 121-140.

Wagner, W. H. Jr. and Wagner, F. S. (1969). *Brittonia* **21**, 178-186.

Wagner, W. H. Jr., Wagner, F. S. and Hagenah, D. J. (1969). *Mich. Bot.* **8**, 137-145.

Wakefield, N. A. (1943). *Vict. Nat.* **60**, 142.

Walker, S. (1955). *Watsonia* **3**, 193-209.

Walker, S. (1961). *Am. J. Bot.* **48**, 607-614.

Walker, S. (1962a). *Am. J. Bot.* **49**, 497-503.

Walker, S. (1962b). *Am. J. Bot.* **49**, 971-974.

Walker, S. (1969). *Br. Fern Gaz.* **10**, 97-99.

Walker, T. G. (1957). *Kew Bull.* **3**, 429-432.
Walker, T. G. (1958). *Evolution, Lancaster, Pa.* **12**, 82-92.
Walker, T. G. (1962). *Evolution, Lancaster, Pa.* **16**, 27-43.
Walker, T. G. (1966a). *Trans. R. Soc. Edinb.* **66**, 169-237.
Walker, T. G. (1966b). *In* "Reproductive Biology and Taxonomy of Vascular Plants" (Ed. J. G. Hawkes). *B.S.B.I. Conf. Rep.* **9**, 152-161.
Walker, T. G. (1968). *Proc. Linn. Soc. Lond.* **179**, 279-286.
Walker, T. G. (1973a). *In* "The Phylogeny and Classification of the Ferns" (Eds A. C. Jermy *et al.*). *Bot. J. Linn. Soc.* **67**, *Suppl.* 1, 91-110.
Walker, T. G. (1973b). *Trans. R. Soc. Edinb.* **69**, 109-135.
Ward, L. F., Fontaine, W. M., Bibbins, A. and Wieland, G. R. (1905). *Monogr. U.S. geol. Surv.* **48**, 616 pp., 119 pls.
Watson, J. (1969). *Bull. Br. Mus. nat. Hist. (Geol.)* **17**, 207-256, 6 pls.
White, M. J. D. (1945). "Animal Cytology and Evolution", 1st Edn. Cambridge University Press.
White, M. J. D. (1973). "Animal Cytology and Evolution", 3rd Edn. Cambridge University Press.
Widén, C.-J. (1969). *Ann. Acad. Sci. fenn. Ser. A, IV Biol.* **143**, 1-19.
Widén, C.-J. and Britton, D. M. (1969). *Can. J. Bot.* **47**, 1337-1344.
Widén, C.-J. and Britton, D. M. (1971a). *Can. J. Bot.* **49**, 247-258.
Widén, C.-J. and Britton, D. M. (1971b). *Can. J. Bot.* **49**, 1141-1154.
Widén, C.-J. and Britton, D. M. (1971c). *Can. J. Bot.* **49**, 1589-1600.
Widén, C.-J. and Sorsa, V. (1966). *Hereditas* **56**, 377-381.
Widén, C.-J. and Sorsa, V. (1969). *Hereditas* **62**, 1-13.
Widén, C.-J., Sorsa, V. and Sarvela, J. (1970). *Acta bot. fenn.* **91**, 1-30.
Widén, C.-J., Vida, G., Euw, J. von and Reichstein, T. (1971). *Helv. chim. Acta* **54**, 2824-2850.
Widén, C.-J., Fraser-Jenkins, C. R., Lounasmaa, M., Euw, J. von and Reichstein, T. (1973). *Helv. chim. Acta* **56**, 831-838.
Widén, C.-J., Britton, D. M., Wagner, W. H. Jr. and Wagner, F. S. (1975). *Can. J. Bot.* **53**, 1554-1567.
Widén, C.-J., Lounasmaa, M., Vida, G. and Reichstein, T. (1975). *Helv. chim. Acta* **58**, 880-904.
Wieffering, J. H., Fikenscher, L. H. and Hegnauer, R. (1965). *Pharmac. Weekblad* **100**, 737-754.
Wilkie, D. (1956). *Heredity, Lond.* **10**, 247-256.
Willis, J. C. (1966). "A Dictionary of the Flowering Plants and Ferns" (Revised by H. K. Airy Shaw), 7th Edn. Cambridge University Press.
Willis, J. C. (1973). "A Dictionary of the Flowering Plants and Ferns" (Revised by H. K. Airy Shaw), 8th Edn. Cambridge University Press.
Wood, C. C. (1973). *In* "The Phylogeny and Classification of the Ferns" (Eds A. C. Jermy *et al.*). *Bot. J. Linn. Soc.* **67**, *Suppl.* 1, 191-202.

NOTES ADDED IN PROOF

1. (p. 259). Holttum has examined living plants of *Hemidictyum margi-natum* now in cultivation at Kew. He writes (pers. comm., 3.9.1976) that "I am quite sure that it is not a Thelypteroid fern. In my judgement it comes nearest to *Diplazium*, but is certainly a distinct genus." It is thus evident that the transfer in this review of *Hemidictyum* to Thelypteridaceae was unfortunate. It seems that *Hemidictyum* must be yet another example of drastic aneuploid reduction in ferns (cf. pp. 295 et seq., & 306, footnote, below).

2. (p. 281). Information given in Löve, A., Löve, D. and Pichi-Sermolli, R. E. G., (1977), *"Cytotaxonomical Atlas of the Pteridophyta"*, (Cramer, Vaduz), indicates that the following base numbers can be added to the list of genera and chromosome numbers given on pp. 265-281: *Selliguea*, 37; *Neolepisorus*, 36 (ex *Neocheiropteris*, no count is known for *Neocheiropteris* sensu stricto); *Paragramma*, 36; *Hyalotricha*, 37; *Rhachidosorus*, 40; *Kuni-watsukia*, 40; *Arthrobotrya*, c.40.

3. (p. 292). Recent counts reported are $n = 66$ in *Plagiogyria mat-sumureana* (Mitui, 1975), $n = 75$ in *P. formosana* & *P. stenoptera*, $n = 125$ in *P. euphlebia* (Tsai, 1973).

4. (p. 294). Dr. Trevor Walker has found *Dicksonia blumei* from Java to be tetraploid with $n = 130$ (*pers. comm.*).

5. (p. 310). The statement that $n = 33$ in Cheiropleuriaceae (de la Sota, 1973, p. 233) was based on a misunderstanding (de la Sota, pers. comm., 1.11.1976), (cf. *Contr. Gray Herb. Harv.* (1959) **185**, p. 112). The chromosome number of *Cheiropleuria* is still unknown.

6. (p. 322). See Note 4, above.

7. (p. 355). Analyses of meiotic pairing in 11 triploid and 8 tetraploid wild interspecific hybrids reported by Daigobo (1974) suggest the phenomenon is widespread in Japanese *Polystichum*.

Author Index

Page numbers in ordinary figures are text references; page numbers in italic are bibliographical references

Subject Index

A

Abscisic acid (ABA)
    effect on stomata, 138-139, 141, 143, 146, 184
    synthesis, 139
Acacia harpophylla, leaf conductance and water potentials, 147
Acetobacter suboxydans, cytochrome, 76
A. T71 H$^+$/O ratio, 89
Acrostichum preaureum, fossil structure, 251
Actiniopteris, apomixis, 390
Actinostachys, cytology, 290
Adenosine diphosphate (ADP), control of respiration, 91-97
Adenosine triphosphatase (ATPase)
    inhibitor, 98-99
    location on membranes, 84-85
    stoichiometry, 87
Adiantaceae
    base numbers and classification, 283
    polyploidy, 322
Adiantum
    apomixis, 390
    fossil record, 234, 235, 252
    polyploidy, 256, 322
    A. caudatum, genome analysis, 333, 334, 337
    A. caudatum complex, cytogenetics of
        A. incisum, 371-376
        A. indicum, 371, 376
        A. malesianum, 371-376
        A. zollingerii, 371-376
        "*B396*", 371-376
    A. poirettii, polyploidy, 329
    A. reniforme, polyploidy, 324
Aerobacter aerogenes
    P/O ratio, 85
    respiratory control, 96

Alcaligenes faecalis, respiratory control, 94
Allantodiopsis erosa, fossil record, 251
Alloploidy in ferns, 320, 321
Allosyndesis, 355
Alocasia macrorrhiza, stomatal behaviour, 186, 187
Amaryllidaceae, polyploidy, 326
Amoeba, and bacterial symbiosis, 52-55, 106, 107
Anemia, polyploidy, 255, 322
Aneuploidy
    in *Drosophila*, 296
    mechanisms, 296
    significance in modern ferns, 295-307
Angiopteris neglecta, fossil record, 236
Angiosperm origins, 253
Antimycin A, inhibition of bacterial electron transport, 77
Apomixis in ferns
    reproductive apomixis, 387-398
    vegitative apomixis, 398-399
Appalachian spleenwort complex, chromosome pairing, 330, 331
Aspidistes beckerii, fossil structure, 242-243
A. sewardii, fossil structure, 242-243
A. thomasii
    fossil record, 237-242
    fossil structure, 237-239, 247
Aspleniaceae, cytogenetics, 360-371
Asplenium
    fossil record, 234, 235, 249, 252
    polyploidy, 322, 324, 328-329
    A. adiantum-nigrum, parentage, 361
    A. adulterinum
        cytogenetic analysis, 362
        distribution, 363
        parentage, 361
    A. aegaeum, intraspecific differentiation, 367

Fungi—*cont.*

    Ceratocystis fimbriata, specificity, 10

    Cladosporium cucumerinum, mechanical resistance, 5

    Colletotrichum lindemuthianum, secretion of glucanase inhibitor, 5

    Erysiphe graminis, specificity, 9

    Helminthosporium sacchari, phytotoxin production, 4

    Ustilago sphaerogena, cytochrome, 79

Fungitoxic and fungistatic compounds in disease resistance, 4

G

Gene-for-gene interaction, 30, 33-41

    experimental approach, 42-44

    model of gene control of resistance, 35-41

Genetics of host-parasite interactions

    gene-for-gene relationship, 3, 7-8, 30

    quadratic check, 7-8

Gleichenia, vegitative apomixis, 399

Gleicheniaceae

    cytology and phylogeny, 311, 314

    polyploidy, 322

Gleicheniales, cytology, 288-289

Gleicheniopsis

    fossil record, 243-245

    G. fecunda, sporangia, 243

    G. sewardii, sporangia, 243

Gleichenites, fossil record, 244

Glycine max, leaf water potential and evaporation rate, 149

Glycoprotein, carbohydrate binding (see lectins)

Glycosyl transferases, products of parasite avirulence genes, 34, 39

Gossypium hirsutum

    assimilation rate, 185

    water relations, 152, 153, 158, 159

Guard cell metabolism

    malate synthesis, 140-141

    malic dehydrogenase, 141

    phospho-enol pyruvate (PEP) carboxylase, 140-141

H

Haemophilus parainfluenzae, cytochromes, 75

Hammada scoparia, stomatal response to humidity, 169-173

Hapten inhibition technique, for lectin specificity, 26

Hatch-Slack pathway plants

    Amaranthus viridis, assimilation rate, 178

    Atriplex hymenelytra, assimilation (at high temperature and irradiance) 208

    Atriplex rosea, assimilation rate, 178

    Atriplex spongiosa, assimilation rate, 178

    Pennisetum purpureum, leaf conductance, 183

    Pennisetum typhoides, leaf conductance, 183

    Sorghum sudanense, CO_2 feedback loop, 180-181 "superconductivity", 205-209

    Tidestromia oblongifolia, assimilation rate, 178, 206-208

    Water use efficiency in C_3 and C_4 plants, 195-198

    Zea mays, 132, 133, 141, 142, 149, 180, 184, 208

Haussmannia, fossil record, 237

Hedera helix, vascular tissue, 122

Helianthus annuus, leaf water potential and evaporation rate, 148-150

Helminthosporoside, 32-33

Helminthosporium sacchari, 32

H. victoriae

    toxin-host specificity, 32

    toxin production 32

Hemionites, apomixis, 390

Hydrogenomonas eutropha

    H^+/O ratio, 88

    respiratory control, 96

Hymenophyllales, cytology, 285

Hymenophyllum

    aneuploidy, 295

    fossil record, 286-288

H. javaricum, apomixis, 391

Hypersensitive response, 30, 33, 34, 35, 39

Hypolepis, polyploidy, 322